MW01377350

Dynamics:
Models and Kinetic Methods for Non-equilibrium Many Body Systems

NATO Science Series

A Series presenting the results of activities sponsored by the NATO Science Committee. The Series is published by IOS Press and Kluwer Academic Publishers, in conjunction with the NATO Scientific Affairs Division.

A. Life Sciences	IOS Press
B. Physics	Kluwer Academic Publishers
C. Mathematical and Physical Sciences	Kluwer Academic Publishers
D. Behavioural and Social Sciences	Kluwer Academic Publishers
E. Applied Sciences	Kluwer Academic Publishers
F. Computer and Systems Sciences	IOS Press

1. Disarmament Technologies	Kluwer Academic Publishers
2. Environmental Security	Kluwer Academic Publishers
3. High Technology	Kluwer Academic Publishers
4. Science and Technology Policy	IOS Press
5. Computer Networking	IOS Press

NATO-PCO-DATA BASE

The NATO Science Series continues the series of books published formerly in the NATO ASI Series. An electronic index to the NATO ASI Series provides full bibliographical references (with keywords and/or abstracts) to more than 50000 contributions from international scientists published in all sections of the NATO ASI Series.
Access to the NATO-PCO-DATA BASE is possible via CD-ROM "NATO-PCO-DATA BASE" with user-friendly retrieval software in English, French and German (WTV GmbH and DATAWARE Technologies Inc. 1989).

The CD-ROM of the NATO ASI Series can be ordered from: PCO, Overijse, Belgium

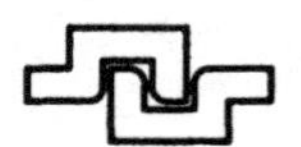

Series E: Applied Sciences – Vol. 371

Dynamics: Models and Kinetic Methods for Non-equilibrium Many Body Systems

edited by

John Karkheck
Department of Physics,
Marquette University,
Milwaukee, WI, U.S.A.

Springer-Science+Business Media, B.V.

Proceedings of the NATO Advanced Study Institute on
Dynamics: Models and Kinetic Methods for Non-equilibrium Many Body Systems
Leiden, The Netherlands
27 July–8 August 1998

A C.I.P. Catalogue record for this book is available from the Library of Congress.

ISBN 978-0-7923-6553-2 ISBN 978-94-011-4365-3 (eBook)
DOI 10.1007/978-94-011-4365-3

Printed on acid-free paper

DEDICATED TO
GEORGE STELL

Table of Contents

4. Quantum Kinetic Theory

Preface

The idea for a NATO ASI on kinetic theory was born in the summer of 1983 when Henk van Beijeren, Ignatz de Schepper and I spent many pleasant hours discussing the kinetic theory of liquids at my parents-in-law summer beach house in Baiting Hollow, Long Island, New York. Over the years our situations changed, but not the dream. That it took so long for that idea to come to fruition, finally in the summer of 1998, is, perhaps, appropriate, for progress in the subject itself is characterized by long time scales.

The conference, held in the Lorentz Institute at Leiden University from July 27 to August 7, 1998, extended far beyond that original glimmer. The past years have seen a resurgence of the kinetic theory approach to dynamical many-body problems that was developed by Boltzmann, Maxwell, Smoluchowski, Enskog, Chapman, and others in the last few decades of the nineteenth and first few decades of the twentieth centuries. Modern kinetic theory offers a unifying theoretical framework within which a great variety of seemingly unrelated physical systems that exhibit complex dynamical behavior can be explored in a coherent manner. For example, these methods are being applied in such diverse areas as the dynamics of colloidal suspensions, flows of granular material, transport of electrons in mesoscopic systems, and the calculation of Lyapunov exponents and other characteristic properties of classical many-body systems that are characterized by chaotic behavior. Many of these exciting developments were captured in our program which focused on four subject areas: Brownian Motion, Dynamical Systems, Granular Flows, and Quantum Kinetic Theory.

The ASI was designed to bring together experts in these four areas and young scientists who had interest in learning kinetic theory methods in a variety of contexts, as well as about the results that can be achieved by them. Ten lectures devoted to each subject area included a series of opening pedagogical lectures, that served to introduce the basic techniques and fundamental issues, followed by in-depth lectures on specific problems and applications. In this volume, the lecture papers are presented in the sequence in which the lectures occurred within each subject area. At the ASI, the lectures among subject areas were intermingled. On each of four nights, a poster session was devoted to each subject area. This greatly broadened the scope of the program and engendered lively discussions. Many breakout sessions were held, often to expand upon controversial issues or to penetrate more deeply into specialized topics. The poster and breakout topics are not included here. Three evening sessions were devoted to subjects of broad appeal, and the papers are included here.

This volume is arranged into four subject-area sections separated by three interludes, the first a discussion of Dutch art, the second a reflection on Ludwig Boltzmann and his ideas in the context of modern dynamical systems theory, and the third a historic note about Paul Ehrenfest.

Steeped so deeply in the history of statistical physics, and in Dutch history, Leiden was an ideal locale for this ASI. The new Lorentz Center provided a congenial setting that promoted strong scientific interaction among the participants, and afforded a relaxing social atmosphere with warm Dutch hospitality, while being convenient to the cultural and historical sites of Leiden, many of which were included in the social program. The conference spanned a twelve day period, with one day off for a lengthy excursion to the seashore and the Delta Works, and the old city of Zierikzee.

Special thanks go to Jean-Pierre Hansen, Bob Dorfman, Matthieu Ernst, Ted Kirkpatrick, Ubbo Felderhof, and Eivind Hauge for serving as session leaders, to Bob Dorfman and Eddie Cohen for their splendid Interlude presentations, to all the fine speakers who also devoted many hours interacting with the student participants. The Lorentz Center staff, Wim van Saarloos, Annette Vermond, and Hans van Bemmel, made the program flow with nary a turbulent eddy. Deep gratitude goes to Hans van Leeuwen for hosting a reception at the Academiegebouw, to Gerard 't Hooft for his generosity, and to our benefactors, Lorentz Fonds, KNAW, and especially the NATO Science Committee. Appreciation goes to the organizers of STATPHYS 20 who granted our NATO ASI the status of satellite conference. I am deeply indebted to Henk van Beijeren for keeping the faith all these years. My student helpers Miriam Jones, who created the art work for our poster and assisted with its distribution, and Christopher Sloane, who worked on the early stages of editing, made my work more enjoyable and my tasks much less burdensome. Without the extraordinary dedication and talented assistance of Stephanie Witczak, this volume would not be.

John Karkheck
Milwaukee, Wisconsin

Organizing Committee
J.R. Dorfman
M.H. Ernst
J.P. Hansen
T.R. Kirkpatrick

Co-Directors
H. van Beijeren
J. Karkheck

DYNAMICS OF COLLOIDAL SYSTEMS: BEYOND THE STOCHASTIC APPROACH

LYDÉRIC BOCQUET
Laboratoire de Physique, Ecole Normale Supérieure de Lyon
69364 LYON Cedex 07, France

AND

JEAN-PIERRE HANSEN
Department of Chemistry, University of Cambridge
Cambridge CB2 1EW, United Kingdom

1. Colloidal Dispersions: Mesoscale Dynamics

Colloidal dispersions are ubiquitous in everyday life. Examples include ink, paints, milk, lubricants or fog, and technological applications are common in the cosmetic, food, oil and pharmaceutical industries. The dispersions are essentially two-phase systems, involving mesoscopic solid or liquid particles, with typical sizes in the range 10-10^3 nm, suspended in a liquid or a gas (aerosols) of much smaller molecules, often improperly referred to as the "solvent".

Restriction will be made here to solid, spherical colloidal particles, like polymethylmethacrylate (PMMA) particles, sterically stabilized by adsorbed polymer "brushes", dispersed in an organic solvent. The large asymmetry of such binary "mixtures" of mesoscopic and microscopic particles poses a severe challenge to a statistical description of static and dynamic properties. Clearly, some coarse-graining is warranted. As regards the static properties (structure, phase behavior, etc.), this may be achieved by integrating out the microscopic degrees of freedom associated with the solvent molecules and any additional component, like ions or polymer coils; this procedure leads to effective (solvent-averaged) interactions between colloidal particles, which have an entropic contribution, and are state-dependent [1].

Such coarse-graining is more delicate when it comes to dynamical properties, in particular the "Brownian" motion of individual or interacting colloidal particles. During the first two decades of this century Einstein,

J. Karkheck (ed.),
Dynamics: Models and Kinetic Methods for Non-equilibrium Many Body Systems, 1–16.

Langevin, Lorentz, Planck, Smoluchovski and others developed the stochastic approach to Brownian motion, based on the intuitively natural assumption of a complete separation of time scales. To be specific we henceforth restrict the discussion to the highly simplified model of a binary "mixture" of large spheres (diameter Σ, mass M), representing the colloidal particles, and small spheres (σ, m), for the molecules of the (suspending) fluid. The size ratio Σ/σ is typically in the range 30 - 3000, and the mass ratio is accordingly of the order of 10^4 - 10^{10}. The microscopic time scale of the fluid is essentially governed by the Enskog mean collision time $\tau_E \sim 1/(n\sigma^2 v_T)$ (where n is the number density of the fluid, and v_T is the thermal velocity), which for a dense fluid (e.g. water) at room temperature is of the order of 10^{-13} sec. This should be compared to the characteristic relaxation time of the velocity of a large sphere, or Brownian (B) particle in a fluid of shear viscosity η, $\tau_B = 1/\zeta$, where the friction coefficient ζ is given by Stokes' law,

$$\zeta = \frac{3\pi\eta\Sigma}{M}. \tag{1}$$

For large B-particles ($\Sigma \approx 10^3 nm$), τ_B is typically of the order of 10^{-8} sec which is 5 orders of magnitude longer than τ_E, thus strongly suggesting complete separation of time scales. The configurational relaxation time of a suspension of B-particles may be roughly estimated from the time it takes a B-particle to diffuse over a distance of the order of its diameter, i.e. $\tau_C \approx \Sigma^2/D$, where the diffusion constant D may be estimated from Einstein's relation,

$$D = \frac{k_B T}{M\zeta}. \tag{2}$$

This leads to characteristic times of order 10^{-3} sec or longer, yet another 5 orders of magnitude larger than τ_B (Smoluchovski regime); this separation is the basis of the classic derivation of Smoluchovski's equation from the Fokker-Planck equation [2,3].

The situation is not, however, as clear-cut, as already pointed out by Lorentz [4]. Indeed collective dynamics of the fluid may be characterized by the viscous relaxation time, i.e. the time it takes a shear disturbance in the fluid, triggered, e.g. by the motion of the B-particle, to diffuse over a distance of the order of the diameter of the latter, i.e. $\tau_\eta = \Sigma^2/\nu$, where $\nu = \eta/\rho$ is the kinematic viscosity, and $\rho = n \times m$ (fluid mass density). Using Stokes' law (1) and the mass density of a B-particle, $\rho_B = M/(\pi\Sigma^3/6)$, it is easily established that

$$\tau_B = \frac{\rho_B}{18\rho}\tau_\eta. \tag{3}$$

The two relaxation times are hence of the same order of magnitude under the physically relevant condition where $\rho_B \approx \rho$. The **collective** dynamics of the fluid, and the individual motion of B-particles are hence entangled, thus questioning the validity of the Fokker-Planck (FP) level description of Brownian motion.

These lecture notes summarize our recent work with J. Piasecki on Brownian motion, based on rigorous multiple time scale analysis [5]. The prime object is to analyse the range of validity of the traditional stochastic description, and to point out its limitations, and possible generalization. For all calculational details, the reader is referred to the original papers [6,7,8,9,10,11]. The emphasis in these lecture notes is on the main conceptual and physical ideas, and on a pedagogical statement of the key results. The long term objective would be to derive hydrodynamic interactions between Brownian particles from first principles, i.e. from Liouville's equation for the full two-component system.

2. Diffusion of Tagged and Brownian Particles

Consider a fluid "bath" of N identical atoms of coordinates and velocities $i \equiv (\mathbf{r}_i, \mathbf{v}_i)$, and one B-particle with $B \equiv (\mathbf{R}, \mathbf{V})$. The total Hamiltonian is of the form

$$\begin{aligned} H &= H_b + H_B \\ &= H_b(1 \ldots N) + \frac{M}{2}V^2 + \sum_i v(|\mathbf{r}_i - \mathbf{R}|). \end{aligned} \tag{4}$$

The Liouville operator accordingly splits into two parts,

$$\begin{aligned} L &= L_b + L_B \\ &= \{H_b, \ldots\} + \{H_B, \ldots\}, \end{aligned} \tag{5}$$

where $\{,\}$ denotes the usual Poisson bracket. The time evolution of the distribution function $f_{N+1}(B, 1 \ldots N; t)$ is determined by Liouville's equation,

$$\left(\frac{\partial}{\partial t} + L\right) f_{N+1} = 0, \tag{6}$$

while any dynamical variable $A(t)$, which is a function of the instantaneous values of the particle positions $\{\mathbf{r}_i(t)\}$, $\mathbf{R}(t)$ and velocities $\{\mathbf{v}_i(t)\}$, $\mathbf{V}(t)$ satisfies

$$\left(\frac{\partial}{\partial t} + L\right) A(t) = 0. \tag{7}$$

Using the Mori-Zwanzig projection operator formalism [12], eq. (7) may be transformed into an exact "generalized" Langevin equation. Consider first the case where the B-particle is a "tagged" particle, identical to the N fluid (or "bath") atoms. The generalized Langevin equation for the tagged-particle velocity ($A(t) = \mathbf{v}(t)$) reads [12] :

$$m\frac{d\mathbf{v}(t)}{dt} + m\int_0^t ds\ \zeta(t-s)\ \mathbf{v}(s) = \mathbf{R}(t) \tag{8}$$

where the memory function, or non-local friction coefficient $\zeta(t)$, and the "random" force $\mathbf{R}(t)$ have well defined statistical definitions involving projected time evolution operators [12]. The Laplace transforms of the normalized velocity autocorrelation function (ACF),

$$Z(t) = \frac{\langle \mathbf{v}(t)\cdot\mathbf{v}(0)\rangle}{\langle v^2\rangle} = \frac{m}{3k_BT}\langle \mathbf{v}(t)\cdot\mathbf{v}(0)\rangle, \tag{9}$$

and of $\zeta(t)$ (which is itself the ACF of $\mathbf{R}(t)$) are simply related by

$$\begin{aligned}\tilde{Z}(z) &= \int_0^\infty dt\ e^{izt} Z(t) \\ &= \frac{1}{-iz + \tilde{\zeta}(z)}.\end{aligned} \tag{10}$$

The diffusion constant of the tagged particle (or self-diffusion constant) is given in terms of the velocity ACF by

$$D = \frac{k_BT}{m} \lim_{z\to 0} \tilde{Z}(z), \tag{11}$$

where the $z \to 0$ limit must be taken after the thermodynamic limit.

In the case under consideration, the dynamics of the tagged particle and of the fluid atoms are identical, so that $Z(t)$ and $\zeta(t)$ relax on similar time scales, of the order of 10^{-13} sec in a dense fluid. In the case of a colloidal Brownian particle, $\Sigma \gg \sigma$ and $M \gg m$, and the traditional approach is to **assume** a complete separation of time scales, i.e. to assume that the "random" force fluctuates on the microscopic time scale of the fluid atoms, which is negligibly short compared to the much longer relaxation time of the B-particle. This is embodied in the familiar ansatz

$$\zeta(t) = \langle \mathbf{R}(t)\cdot\mathbf{R}(0)\rangle = \zeta\delta(t) \tag{12}$$

leading directly to Langevin's local equation

$$M\frac{d\mathbf{V}}{dt} = -M\zeta\mathbf{V}(t) + \mathbf{R}(t). \tag{13}$$

$\mathbf{R}(t)$ is assumed to be a Markovian stochastic process with infinitely short correlation time. The resulting velocity ACF decays exponentially, as expected from Doob's theorem,

$$Z(t) \sim \exp(-\zeta t), \tag{14}$$

and eq. (11) leads then directly back to Einstein's relation (2).

The Markovian nature of the stochastic process may also be exploited to derive the corresponding evolution equation for the distribution function of the B - particle, namely the Fokker-Planck (FP) equation [2],

$$\left(\frac{\partial}{\partial t}+\mathbf{V}\cdot\frac{\partial}{\partial \mathbf{R}}+\frac{\mathbf{F}}{M}\cdot\frac{\partial}{\partial \mathbf{V}}\right)f(\mathbf{R},\mathbf{V};t)=\zeta\,\frac{\partial}{\partial \mathbf{V}}\cdot\left(\mathbf{V}+\frac{k_BT}{M}\frac{\partial}{\partial \mathbf{V}}\right)f(\mathbf{R},\mathbf{V};t), \tag{15}$$

where $\mathbf{F}$ denotes any external force field. From eq. (14) the velocity of the B-particle relaxes on the time scale $\tau_B = 1/\zeta$ which is typically of the order of 10^{-8} sec. The B-particle position $\mathbf{R}$ changes on the even longer time scale τ_C. The relaxation of any spatial inhomogeneity of B-particles is governed by Smoluchovski's evolution equation for the local density,

$$\rho(\mathbf{R};t)=\int d\mathbf{V} f(\mathbf{R},\mathbf{V};t), \tag{16}$$

namely [2],

$$\frac{\partial}{\partial t}\rho(\mathbf{R},t)=\frac{\partial}{\partial \mathbf{R}}\cdot\left\{D\frac{\partial}{\partial \mathbf{R}}-\frac{1}{M\zeta}\mathbf{F}\right\}\rho(\mathbf{R},t). \tag{17}$$

Equation (17) is traditionally derived from the FP equation (15) by Kramer's expansion [2,13].

The key objective of these lectures is to derive the FP equation for a single, or for interacting, Brownian particles from first principles, i.e. without making any a priori assumption on the separation of time scales, and to specify the precise domain of validity of eqs. (15) and (17).

3. Multiple Time-Scale Analysis

The basic tool throughout these lecture notes is the method of multiple time-scales [5], which is routinely used in Celestial or Fluid Mechanics, in order to avoid secular divergence in dynamical problems involving widely separated "natural" time scales. The method seems to be less familiar in Statistical Mechanics, and appears to have been first applied to the Brownian motion problem by Cukier and Deutch [14].

Consider a function $f(t)$ (e.g. the distribution functions introduced in the previous section), which satisfies an evolution equation

$$\frac{\partial}{\partial t}f(t)=\hat{O}f(t), \tag{18}$$

where the evolution operator (e.g. the Liouville operator) acts on the other unspecified variables (e.g. positions and momenta) of the function f. The formal solution of eq. (18) is

$$f(t) = \exp(\hat{O}t)f(0), \tag{19}$$

where $\exp(\hat{O}t)$ is the propagator acting on the function at the initial time $t = 0$.

If, for physical reasons, it is expected that the system involves well-separated time scales, $\tau_0 \ll \tau_1 \ll \ldots \ll \tau_n$, then it proves useful to introduce the auxiliary function

$$f^{(\epsilon)}(t_0, t_1, \ldots, t_n), \tag{20}$$

where $t_0 = t/\tau_0 \gg t_1 = t/\tau_1 \gg \ldots \gg t_n = t/\tau_n$. The variable t_0 characterizes the fastest process, while t_n is associated with the slowest relaxation. It is assumed that the ordering of the time scales is controlled by a small dimensionless parameter ϵ, i.e. $\tau_j = \tau_0/\epsilon^j$. The auxiliary function (20) satisfies the generalized evolution equation

$$\left(\frac{\partial}{\partial t_0} + \epsilon\frac{\partial}{\partial t_1} + \ldots + \epsilon^n\frac{\partial}{\partial t_n}\right) f^{(\epsilon)}(t_0, t_1, \ldots, t_n) = \hat{O}f^{(\epsilon)}(t_0, t_1, \ldots, t_n). \tag{21}$$

On the "physical axis", the $(n+1)$ time variables are simply related by

$$t_0 = t; t_1 = \epsilon t; \ldots t_n = \epsilon^n t, \tag{22}$$

and the solution of the evolution equation (18) is obtained by taking the function $f^{(\epsilon)}$ of $(n+1)$ initially independent variables to the physical axis (22).

One may wonder what has been gained by this apparent detour. In fact the auxiliary function $f^{(\epsilon)}$ allows the flexibility of choosing convenient boundary conditions away from the physical axis. The liberty can be put to good use to eliminate secular divergences that arise when attempts are made to expand $f(t)$ in powers of the small parameter ϵ. If no special care is taken, such expansions are not, generally, uniformly convergent, i.e. for times $t > 1/\epsilon^n$, the nth order term in the expansion becomes larger than the preceding, lower order terms. This lack of uniform convergence is precisely dealt with by the multiple time scale analysis, as may be illustrated by the trivial example of a weakly damped harmonic oscillator [15], or by the example of Kramer's $1/\zeta$ expansion for deriving [3] Smoluchovski's equation (17) from the FP equation (15).

4. From Liouville to Fokker-Planck

The task of deriving the irreversible time evolution and dissipative behavior of many-particle systems on mesoscopic scales, as described, e.g., by the FP equation, from the fully time-reversible dynamics on the microscopic scale, as embodied in Liouville's equation, is one of the fundamental, largely unsolved problems in Statistical Mechanics. Early attempts to bypass the stochastic assumptions involved in the phenomenological description of Brownian motion go back to Lebowitz, Résibois and collaborators [16], but their derivation of the FP equation from Liouville's equation still involves an implicit assumption of a separation of time scales. The first fully satisfactory derivation, based on a multiple time-scale analysis, avoiding any ad hoc assumption, is due to Cukier and Deutch [14]. The microscopic time evolution of a system of N bath particles and one B-particle is governed by the Liouville equation (6). Switching to momenta $\mathbf{p}_i = m\mathbf{v}_i$, $\mathbf{P} = M\mathbf{V}$, the two terms L_b and L_B in eq. (5) become

$$\begin{aligned} L_b &= \sum_i \left(\frac{\mathbf{p}_i}{m} \cdot \frac{\partial}{\partial \mathbf{r}_i} + \mathbf{F}_i \cdot \frac{\partial}{\partial \mathbf{p}_i} \right), \\ L_B &= \frac{\mathbf{P}}{M} \cdot \frac{\partial}{\partial \mathbf{R}} + \mathbf{F} \cdot \frac{\partial}{\partial \mathbf{P}}, \end{aligned} \tag{23}$$

where $\mathbf{F}_i$ and $\mathbf{F}$ denote the total force acting on bath atom i and of the bath on the B-particle, respectively. The natural "smallness" parameter in the problem is $\epsilon = (m/M)^{1/2}$ [16], and the Brownian limit is defined as $\epsilon \to 0$ for a **fixed** size ratio σ/Σ. Introducing a scaled momentum of the B-particle, $\mathbf{p} = \epsilon \mathbf{P}$, the kinetic energy of the latter becomes $P^2/2M = p^2/2m$ and L_B scales as

$$L_B = \epsilon \left(\frac{\mathbf{p}}{m} \cdot \frac{\partial}{\partial \mathbf{R}} + \mathbf{F} \cdot \frac{\partial}{\partial \mathbf{p}} \right) \equiv \epsilon L'_B. \tag{24}$$

The objective is to obtain a closed kinetic equation for the B-particle distribution function $f(B;t)$ which results from integration of the full phase-space density f_{N+1} over all bath variables; carrying out this integration on both sides of the Liouville equation (6), one arrives at the **exact** evolution equation for f,

$$\frac{\partial}{\partial t} f(\mathbf{R};\mathbf{p};t) = \epsilon \frac{\mathbf{p}}{m} \cdot \frac{\partial}{\partial \mathbf{R}} f - \epsilon \int d\mathbf{r}^N \, d\mathbf{p}^N \, \mathbf{F} \cdot \frac{\partial}{\partial \mathbf{p}} f_{N+1}. \tag{25}$$

We introduce auxiliary functions $f_{N+1}^{(\epsilon)}(t_0, t_1, t_2, \ldots)$ and $f^{(\epsilon)}(t_0, t_1, t_2, \ldots)$, where on the physical axis, the time-scales $t_0, t_1, t_2, \ldots$ are ordered in powers of $\epsilon = (m/M)^{1/2}$ according to eq. (22). The multiple time-scale analysis sketched in the previous section is now applied to the evolution equations (6) and (25), e.g.,

$$\left(\frac{\partial}{\partial t_0} + \epsilon \frac{\partial}{\partial t_1} + \epsilon^2 \frac{\partial}{\partial t_2} + \ldots \right) f_{N+1}^{(\epsilon)} = -(L_b + \epsilon L'_B) f_{N+1}^{(\epsilon)}. \tag{26}$$

In practice, $f_{N+1}^{(\epsilon)}$ and $f^{(\epsilon)}$ are expanded in powers of ϵ, e.g.,

$$f^{(\epsilon)}(B;t_0,t_1,\ldots) = f^{(0)}(B;t_0,t_1,\ldots) + \epsilon\, f^{(1)}(B;t_0,t_1,\ldots) + \epsilon^2\, f^{(2)}(B;t_0,t_1,\ldots) + \ldots \qquad (27)$$

These expansions are substituted into the generalized evolution equation (26) and the corresponding form of (25), and equal powers of ϵ are identified. Boundary conditions on the auxiliary functions at $t_0 = 0$ and for $t_0 \to \infty$ are chosen such as to eliminate secular divergences at each order in ϵ, thus ensuring a uniformly convergent expansion. Details are given in refs. [14] and [17]. The FP equation (15) is recovered when the expansion is truncated after order ϵ^2 and upon returning to the physical axis (22). The analysis also yields a microscopic expression for the friction coefficient appearing in the FP and Langevin equations, in terms of the ACF of the force exerted by the bath on the B-particle,

$$\zeta = \frac{1}{3Mk_BT}\int_0^\infty dt\ \langle \mathbf{F}(t)\cdot\mathbf{F}(0)\rangle_{bath|\mathbf{R}}, \qquad (28)$$

where the statistical average is over bath degrees of freedom, in equilibrium around a B-particle fixed at $\mathbf{R}$. This result was given by Kirkwood already in 1946 [18], but care must be exerted in taking the thermodynamic limit of the ACF before taking the upper limit in the time integration to infinity. This poses some practical problems in attempts to extract ζ from Molecular Dynamics (MD) simulations [17, 7], which will be discussed in Section 6.

In collaboration with J. Piasecki, we have recently extended the above multiple time-scale derivation of the FP equation to the case where the bath particles and the B-particle are elastic hard spheres [6, 7, 8, 9, 10]. The dynamics of such a system reduce to a succession of instantaneous elastic collisions, as already considered by L. Boltzmann in 1872! The HS model has many advantages:

· simplicity of the interactions: collisions are strictly binary

· MD simulations generate exact, reversible trajectories in phase space, except for computer round-off errors; moreover, such simulations are very fast!

· absence of an energy scale: the only relevant thermodynamic variable of such an "athermal" system is the packing fraction

· kinetic theory of HS is well advanced, and much progress has been made since the early work of Boltzmann and Enskog; in particular, processes involving correlated collisions are now well understood [20]

· HS are, since the pioneering work of van der Waals, the fundamental model of Liquid State theory [12].

The HS model has, however, one drawback: the simple differential Liouville operator (23) for continuous interactions between particles is replaced by a singular pseudo-Liouville operator [21], which is not self-adjoint and renders formal manipulations technically more complicated. Allowing for these complications, the multiple time-scale analysis sketched above carries through, and leads once more back to the FP equation (15) if the expansion in powers of $\epsilon = (m/M)^{1/2}$ is truncated after second order. With the usual caveat concerning the order of limits (thermodynamic limit of the statistical averages over a bath ensemble, for a fixed position $\mathbf{R}$ of the B-particle, to be taken before the upper time integration limit is taken to infinity), the friction coefficient is still given by Kirkwood's formula (28). However, due to the instantaneous nature of the HS collisions, the "force" $\mathbf{F}$ is given by the rate of momentum transfer from the bath to the immobile B-particle [22, 7] :

$$\mathbf{F}(t) = \sum_{(c)} (-2m)(\mathbf{v}_c \cdot \mathbf{r}_c)\hat{\mathbf{r}}_c \delta(t - t_c), \tag{29}$$

where $\mathbf{r}_c$ is the position of a bath particle relative to the B-particle at the instant of collision ($\hat{\mathbf{r}}_c = \mathbf{r}_c/d$ with $d = (\Sigma + \sigma)/2$ the distance of closest approach), $\mathbf{v}_c$ is the bath particle velocity just prior to collision, and the sum is over the succession of bath - B-particle collisions, taking place at discrete times t_c.

For HS, ζ naturally splits into a "static", Enskog contribution ζ_1, which follows if successive bath - B-particle collisions are assumed to be uncorrelated, and into a "dynamic" contribution ζ_2 arising from correlated collisions [6]. If ν_c denotes the bath - B-particle collision rate (which may be expressed in terms of the equilibrium density of bath particles at contact with the B-particle),

$$\zeta_1 = \frac{4}{3}\frac{m}{M}\nu_c. \tag{30}$$

It is important to stress that the dynamical contribution is negative, and is non-zero even if the bath reduces to a Boltzmann gas, i.e. if collisions between bath molecules are uncorrelated ("molecular chaos") [8].

5. Interacting Brownian Particles

The multiple time-scale analysis may be generalized to the case where n Brownian particles are suspended in a bath of much lighter particles [23, 9]. This leads to the following FP equation for the distribution function of n interacting B-particles:

$$\left(\frac{\partial}{\partial t} + \sum_{a=1}^{n}\left[\mathbf{V}_a \cdot \frac{\partial}{\partial \mathbf{R}_a} + \frac{1}{M}\left(\langle \mathbf{F}(\mathbf{R}_a;t)\rangle_{bath} + \right.\right.\right.$$

$$\left.\left.\left.\sum_{b=1}^{n} \mathbf{F}_{ab}\right) \cdot \frac{\partial}{\partial \mathbf{V}_a}\right]\right) f_n(\mathbf{R}_1, \mathbf{V}_1, \ldots, \mathbf{R}_n, \mathbf{V}_n; t) =$$

$$= \sum_{a,b=1}^{n} \bar{\bar{\zeta}}(a,b) : \frac{\partial}{\partial \mathbf{V}_a} \left(\frac{\partial}{\partial \mathbf{V}_b} + \frac{k_B T}{M} \mathbf{V}_b \right) f_n(\mathbf{R}_1, \mathbf{V}_1, \ldots, \mathbf{R}_n, \mathbf{V}_n; t). \quad (31)$$

In the HS case, the force $\mathbf{F}_{ab}$ acting between two B-particles must be replaced by the appropriate collision operator. The n-particle FP equation (31) involves an $n \times n$ matrix of 3×3 friction tensors, $\bar{\bar{\zeta}}(a,b)$, the expression of which involves the time-integral of the correlation function of the fluctuating forces exerted by the bath on B-particles. In the HS case, $\bar{\bar{\zeta}}$ again naturally splits into a "static" Enskog part, and a "dynamic" contribution due to correlated collisions [9].

The l.h.s. of eq. (31) involves, apart from the direct forces between B - particles, a bath-mediated force $\langle \mathbf{F}(\mathbf{R}_a; t) \rangle_{bath}$, which does not vanish due to the anisotropy of the local fluid density around any given B-particle, due to the presence of other B-particles. This is precisely the depletion force, which is traditionally introduced on the basis of entropic arguments [24]. Ref [9] provides the exact expression for the depletion force, in terms of the local equilibrium density of the bath, valid for any packing fraction of the latter.

6. Molecular Dynamics and Friction Tensor

For a single B-particle, a hydrodynamic calculation gives the well-known Stokes' law for the friction coefficient, eq. (1). On the other hand, in the case of many B-particles, the friction tensor $\bar{\bar{\zeta}}(a,b)$ is a rather complicated quantity since it depends on the configuration of the whole set of B-particles. Due to the presence of the solvent, the suspended particles interact through long-range, many-body, hydrodynamic forces. The "standard" approach consists in treating the suspending fluid at the level of Stokes equations [25], which can be formally solved to give an explicit solution using the method of induced forces [26]. We shall not develop further this aspect here, which has been widely discussed elsewhere in the literature (see e.g. ref [1] and references therein).

These approaches are valid when the fluid can be considered as a continuum. More precisely, the mean free-path in the fluid is assumed to be much less than the diameter of B-particles. If this assumption is relaxed (for example in a dilute gas, or when the diameter of the B-particle is of the same order as that of the fluid particles), the calculation of the friction coefficient requires a full microscopic description of the dynamics. Such a task can be carried out using Molecular Dynamics (MD) simulations, which allow for a step-by-step determination of the trajectories of all particles in the system. The Kirkwood formula which expresses the friction coefficient in terms of the force ACF (such as eq. (28) for a single particle) then provides a natural starting point for a numerical estimate of the friction coefficient (or tensor). This straightforward "recipe" is ineffective, however, due to delicate problems associated with the order of thermodynamic and infinite time limit. We shall here emphasize this aspect which can lead to spurious estimates of the friction tensor.

Let us consider first the case of the friction coefficient ζ on a single B-particle. We introduce $\gamma = M\zeta$ which remains finite when $M \to \infty$. The Kirkwood formula expresses γ in terms of the force ACF acting on the **fixed** B-particle. Thus, in typical MD simulations one is interested in the dynamics of N fluid particles ($N \approx 10^2 - 10^4$) evolving in the presence of one fixed B-particle. It can be then easily understood that for such a finite system, the direct application of the Kirkwood formula yields invariably $\gamma_N = 0$. The main reason is that in a finite system the force due to the B-particle can be identified with the time-derivative of the total fluid momentum $\mathbf{P}(t) : \mathbf{F}(t) = -\dot{\mathbf{P}}(t)$. This is not the case in a strictly infinite system, where momentum is carried off to infinity. Since the force is the time derivative of momentum, the force ACF in the Kirkwood expression can be explicitly integrated, yielding

$$\gamma_N = \frac{1}{3k_BT} \lim_{t\to\infty} \left(\langle \mathbf{P}(t) \cdot \mathbf{F}(0) \rangle_N - \langle \mathbf{P}(0) \cdot \mathbf{F}(0) \rangle_N \right) = 0, \tag{32}$$

which, of course, only holds if the infinite time limit is taken **before** the thermodynamic limit $N \to \infty$. A simple argument allows a better understanding of this point. In equilibrium, the total momentum of the fluid, $\mathbf{P}(t)$, fluctuates because of collisions of the bath with the B-particle. According to Onsager's principle, the regression of the fluctuations of the momentum must be governed by the (non-equilibrium) phenomenological equations. The latter state that the force exerted by the flowing fluid on the test particle is proportional to the fluid velocity $\mathbf{v}(t)$,

$$\mathbf{F}(t) = \gamma \mathbf{v}(t). \tag{33}$$

In our case, the velocity $\mathbf{v}(t)$ can be identified with the center-of-mass velocity of the fluid $\mathbf{v}(t) = \mathbf{P}(t)/Nm$, so that one gets

$$\dot{\mathbf{P}}(t) = -\frac{\gamma}{Nm}\mathbf{P}(t). \tag{34}$$

This equation shows that the total momentum correlation function relaxes *exponentially*, with a relaxation time given in terms of the friction coefficient : $\tau_N = Nm/\gamma$. Within this model, the Kirkwood expression can be explicitly computed to yield

$$\frac{1}{3k_BT} \int_0^t ds \, \langle \mathbf{F}(s) \cdot \mathbf{F}(0) \rangle = \frac{1}{3k_BT} \langle \dot{\mathbf{P}}(t) \cdot \mathbf{P}(0) \rangle_N = \gamma \, \exp(-t/\tau_N). \tag{35}$$

Thus, in a finite system, τ_N is finite and the Kirkwood formula vanishes when the upper time limit goes to infinity. On the other hand, if the thermodynamic limit is taken before time going to infinity, τ_N goes to infinity and eq. (35) gives a finite result.

This simple argument shows that no sensible result can be obtained using the Kirkwood formula in a finite system. However, as indicated above, the existence of a friction on the B-particle is of course still present in the dynamics of the fluid, and the friction coefficient $\gamma = M\zeta$ can be measured by computing the ACF of the total momentum in the fluid $\langle \mathbf{P}(t) \cdot \mathbf{P}(0) \rangle$. According to the previous discussion, the latter is predicted to decay exponentially with a characteristic time $\tau_N = Nm/\gamma$, providing an indirect measure of γ. This can be indeed verified in the simulations [19, 7].

When extended to the computation of friction tensors for many B-particles, the situation is in fact even worse. Consider a system of N fluid particles evolving in the presence of *two* B-particles fixed at their positions. The problem now involves a self friction tensor, $\bar{\bar{\zeta}}_{11}$, defined in terms of the ACF of the force acting on the same particle $\langle \mathbf{F}_1(t)\mathbf{F}_1(0) \rangle$, and a mutual friction tensor, $\bar{\bar{\zeta}}_{12}$, defined in terms of the correlation function of the force acting on two different particles $\langle \mathbf{F}_1(t)\mathbf{F}_2(0) \rangle$. The derivative of the total momentum of the fluid is now equal to the sum of the forces due to both fixed particles, $\mathbf{F}_1(t)$ and $\mathbf{F}_2(t)$: $\mathbf{F}_1(t) + \mathbf{F}_2(t) = -\dot{\mathbf{P}}(t)$. Thus the sum of the self and mutual friction tensor $\bar{\bar{\zeta}}_{11} + \bar{\bar{\zeta}}_{12}$ involves the integral over time of the time derivative of $\mathbf{P}(t)$. As a consequence, in the same spirit as for the single B-particle case, one obtains in a finite system $\bar{\bar{\zeta}}_{11}^{N} + \bar{\bar{\zeta}}_{12}^{N} = 0$, when Kirkwood formulae are used! This result is of course not correct and only follows from the conservation of total momentum in the finite system. By analysing more carefully the dynamics in a finite system, the following surprising result can be derived:

$$\bar{\bar{\zeta}}_{11}^{N} = -\bar{\bar{\zeta}}_{12}^{N} = \frac{\bar{\bar{\zeta}}_{11}^{N=\infty} - \bar{\bar{\zeta}}_{12}^{N=\infty}}{2} + \mathcal{O}\left(\frac{1}{N}\right), \tag{36}$$

where the index N stands for a calculation in a system of finite size N, while $N = \infty$ refers to the thermodynamic limit. In other words, the friction tensors computed in any finite system, $\bar{\bar{\zeta}}_{ab}^{N}$, are not good approximations for the friction tensors in an infinite system, $\bar{\bar{\zeta}}_{ab}^{N=\infty}$. As follows from eq. (36), the difference $\bar{\bar{\zeta}}_{ab}^{N} - \bar{\bar{\zeta}}_{ab}^{N=\infty}$ is indeed of **order one**. This result summarizes the spirit of these numerical studies: in computing friction tensors with MD simulations, one should remember that finite size effects are not of order $\mathcal{O}\left(\frac{1}{N}\right)$ as usually expected, but $\mathcal{O}(1)$. A specific analysis has to be done to extract the physical information from the simulations.

We refer to ref. [10] for more details and a full discussion of the physical results for the two B-particles case.

7. Break-down of the Fokker-Planck Description: Non-Markovian Effects

In the previous paragraphs, we have shown that the Fokker-Planck description can be derived from first principles in the limit where the Brownian particle is much heavier than the fluid particles. In this limit, the mass density of the B-particle ρ_B is much larger than the mass density of the fluid ρ (since both diameters are kept fixed). Thus, as shown using eq. (3), there exists a wide time-scale separation between the fluid and the B-particle dynamics : $\tau_B \gg \tau_\eta$. However, in the experimental case, these two mass densities are usually taken of the same order of magnitude in order to avoid sedimentation effects, so that there is no wide time-scale separation. The foundations for the stochastic approach should be questioned. In fact, only part of it is affected.

First, it is easy to check that the time scale for configurational relaxation, τ_C, is still much larger than $\tau_\eta \sim \tau_B$. Therefore, the configurational distribution function is still expected [27] to obey the Smoluchovski equation, eq. (17). In other words, the process of spatial relaxation is markovian. Only the Fokker-Planck (or Kramers) equation involving the full $(\mathbf{R},\mathbf{V})$ process is expected to break down. One amazing point is that the Smoluchovski equation is traditionally derived from the Fokker-Planck equation, although the latter has less validity than the former.

Secondly, the time-scale τ_η characterizes the *collective* motion of the fluid, so that a "local" (in space) separation of time scales does still exist between the fluid and the B-particle. One important consequence is that one should be able to still eliminate the fluid variables from the description. This has been shown at the level of fluctuating hydrodynamics by a number of authors (see e.g. [28, 29]) In this approach, the fluid is assumed to obey Navier-Stokes equations of hydrodynamics, to which a fluctuating term has been added. The fluid is coupled to the B-particles through the boundary conditions at their surface. In the case of one B-particle, this leads to a non-Markovian Langevin equation,

$$M\frac{d\mathbf{V}}{dt} = -M\int_0^t d\tau\ \zeta(t-\tau)\ \mathbf{V}(\tau) + \tilde{\mathbf{F}}(t). \tag{37}$$

In eq. (37), $\zeta(t)$ is the time-dependent friction coefficient, defined as:

$$\zeta(t) = \frac{1}{3Mk_BT}\langle\mathbf{F}\cdot\mathbf{F}(-s)\rangle_{bath}. \tag{38}$$

However a full microscopic approach was still lacking up to now. We have thus reconsidered the problem, by starting from a microscopic description of the system in the small mass ratio, $\frac{m}{M} \ll 1$, but now **supplemented by the condition of equivalent mass densities**, $\rho \sim \rho_B$. The main objective of this work was to find out what equation would replace the Fokker-Planck equation. As in our previous work, we made use of the multiple time scale analysis. We only point out the main results.

Two important time-scales emerge from the dynamics:

(1) on the first time scale, $t \sim 1/\zeta$, the Brownian particle "does not move", whereas its velocity distribution relaxes in a thermalization process. A closed equation (*i.e.* free of any fluid variable) controlling the relaxation of the B particle distribution function is found, which does not reduce to the Fokker-Planck form. In contradistinction to the latter, the reduced equation controlling thermalization is found to be non-local both in time and velocity space, owing to correlated recollision events between the fluid and particle B. The latter reads

$$\frac{\partial}{\partial t} f(B;t) = \tag{39}$$

$$\int_0^t ds\ \zeta(t-s) \frac{\partial}{\partial \mathbf{V}} \cdot \exp\left\{ - \int_s^t ds'\ \mathcal{L}_B(s') \right\} \left(\mathbf{V} + \frac{k_B T}{M} \frac{\partial}{\partial \mathbf{V}} \right) f(B;s),$$

with $\mathcal{L}_B = \frac{\partial}{\partial \mathbf{V}} \cdot \mathcal{F}(B;t)$, $\mathcal{F}(B;t)$ being the dynamical friction force due to the fluid, acting on the B particle during its relaxation (see eqs. (65)-(66) of ref. [11] for a complete definition). An important point is that, in spite of this complex dynamical behavior, the diffusion constant of B is still given by the Stokes-Einstein relation (2).

The presence of memory terms in (39) results from the building up of the friction force by the reaction of the suspending fluid to the motion of B. Indeed, this reaction takes a finite time to occur (compared to the relaxation time of the velocity of particle B), and the friction force due to the fluid is accordingly displaced in time and velocity space. Moreover, one can show that this non-markovian effect leads to a "slow" thermalization, algebraic in time, in contradistinction to the exponential decay predicted by the Langevin equation. This non-exponential behavior is in complete agreement with the predictions of the fluctuating hydrodynamics approaches [28, 30]. Numerical simulations of colloidal suspensions, based on fluctuating Lattice Bolztmann techniques [31], do confirm the presence of the so-called "long-time tails" in the velocity autocorrelation function of the Brownian particles. Moreover this algebraic decay has been observed experimentally in the "short-time" dynamics (*i.e.* on the scale of the relaxation of the velocity of the Brownian particles) of colloidal suspensions, using Diffusing Wave Spectroscopy (DWS) techniques [32].

(2) on the second time scale, $t \sim \Sigma^2/D$, spatial diffusion takes place and is still described by the Smoluchovski equation, *i.e.* no memory effect appears and the corresponding process is Markovian.

In other words, the spatial relaxation of the Brownian particle is a Markov process, while the thermalization of the velocity of the Brownian particle, which occurs on a shorter time scale, is not. We refer to ref. [11] for further details and comments.

Acknowledgement

We are very grateful to Jarek Piasecki for his constant collaboration throughout the work reported in this review.

References

1. For recent reviews, see e.g. the papers by W. K. Poon and P. N. Pusey, and by J. P. Hansen, in *Observation, Prediction and Simulation of Phase Transitions in Complex Fluids*, M. Baus, L. F. Rull, and J. P. Ryckaert, eds., (Kluwer, Dordrecht, 1995).
2. See e.g. N. G. van Kampen, *Stochastic Processes in Physics and Chemistry*, (North Holland, Amsterdam, 1990).
3. For a recent, rigorous derivation, see L. Bocquet, Am. J. Phys. **65**, 140 (1997)
4. H. A. Lorentz, *Lessen over Theoretische Natuurkunde. Vol. V. Kinetische Problemen*, (E. J. Brill, Leiden, 1921).
5. J. Piasecki, *Echelles de Temps en Théorie Cinétique*, (Presses polytechniques et Universitaires romandes Lausanne, 1997).
6. L. Bocquet, J. Piasecki, and J. P. Hansen, J. Stat. Phys. **76**, 505 (1994).
7. L. Bocquet, J. P. Hansen, and J. Piasecki, J. Stat. Phys. **76**, 527 (1994).
8. L. Bocquet, J. P. Hansen, and J. Piasecki, Nuovo Cim. **16D**, 981 (1994).
9. J. Piasecki, L. Bocquet, and J. P. Hansen, Physica A **218**, 125 (1995).
10. L. Bocquet, J. P. Hansen, and J. Piasecki, J. Stat. Phys. **89**, 322 (1997).
11. L. Bocquet and J. Piasecki, J. Stat. Phys. **87**, 1005 (1997).
12. See e.g. J. P. Hansen and I. R. McDonald, *Theory of Simple Liquids*, 2nd ed. (Academic Press, London, 1986).
13. H. A. Kramers, Physica **7**, 284 (1940).
14. R. I. Cukier and J. M. Deutch, Phys. Rev. **177**, 240 (1969).
15. J. L. Anderson, Am. J. Phys. **60**, 923 (1992).
16. J. L. Lebowitz and E. Rubin, Phys. Rev. **131**, 2381 (1963); P. Résibois and H. T. Davis, Physica **30**, 1077 (1964); J. L. Lebowitz and P. Résibois, Phys. Rev. **139**, 1101 (1963).
17. L. Bocquet and J. P. Hansen, in *The Physics of Complex Systems*, F. Mallamace and H. E. Stanley, eds., (IOS Press, Amsterdam, 1997).
18. J. Kirkwood, J. Chem. Phys. **14**, 180 (1946).
19. P. Español and I. Zuniga, J. Chem. Phys. **98**, 574 (1993).
20. E.G.D. Cohen, Physica A **194**, 229 (1993).
21. M. H. Ernst, J. R. Dorfman, W. Hoegy, and J. M. J. van Leeuwen, Physica **45**, 127 (1965); see also P. Résibois and M. de Leener, *Classical Kinetic Theory of Fluids*, (Wiley, New York, 1977).
22. B. J. Alder and W. E. Alley, in *Molecular Structure and Dynamics*, M. Balaban, ed., (International Science Services, 1980).

23. The case of continuous interactions between all particles was considered by R. M. Mazo, J. Stat. Phys. **1**, 559 (1969).
24. S. Asakura and F. Oosawa, J. Polymer Sci. **33**, 183 (1958); for a more modern presentation, see M. Dijkstra, R. van Roij, and R. Evans, Phys. Rev. Lett. **81**, 2268 (1998).
25. J. Happel and H. Brenner, *Low Reynolds Number Hydrodynamics* (Martinus Nijhoff, Dordrecht, 1986).
26. P. Mazur and W. van Saarloos, Physica A **115**, 21 (1982).
27. J.N. Roux, Physica A **188**, 526 (1992).
28. E.H. Hauge and A. Martin-Löf, J. Stat. Phys. **7**, 259 (1973).
29. E.J. Hinch, J. Fluid. Mech. **72**, 499 (1975).
30. D. Bedeaux and P. Mazur, Physica A **76**, 247 (1974).
31. A.J.C. Ladd, Phys. Rev. Lett. **70**, 1339 (1993); J. Fluid Mech. **271**, 311 (1994).
32. P.D. Fedele and Y.W. Kim, Phys. Rev. Lett. **44**, 691 (1980); J.X. Zhu, D.J. Durian, J. Mueller, D.A. Weitz, and D.J. Pine, Phys. Rev. Lett. **68**, 2559 (1992).

LATTICE-BOLTZMANN SIMULATIONS OF HYDRODYNAMICALLY INTERACTING PARTICLES

A. J. C. LADD
Department of Chemical Engineering, University of Florida
Gainesville, FL 32611-6005, USA

Abstract. Lattice-Boltzmann methods are being increasingly used to solve problems in computational fluid dynamics. The combination of robustness and simplicity has made it the method of choice for problems involving fluid flow through geometrically complex structures. Progress in several areas is summarized in a recent review article [1]. In this paper I will focus on applications of the lattice-Boltzmann method to simulations of particle-fluid suspensions. Since the basic principles of the method have already been described [2, 3], the focus of this article will be a review of recent developments and a discussion of some of the fine points that arise in practical applications of the method. In particular I will contrast the accuracy and complexity of different solid-fluid boundary conditions, discuss the application of an external pressure gradient, and describe additional complications arising from particle motion.

1. Lattice-Boltzmann Equation

In the lattice-Boltzmann model, the fundamental quantity is the discretized one-particle velocity distribution function $n_i(\mathbf{r}, t)$, which describes the number of particles at a particular node of the lattice $\mathbf{r}$, at a time t, with a velocity $\mathbf{c}_i$; $\mathbf{r}$, t, and $\mathbf{c}_i$ are discrete, whereas n_i is continuous. The hydrodynamic fields, mass density ρ, momentum density $\mathbf{j}$, and momentum flux $\mathbf{\Pi}$, are moments of this discrete velocity distribution:

$$\rho = \sum_i n_i, \quad \mathbf{j} = \sum_i n_i \mathbf{c}_i, \quad \mathbf{\Pi} = \sum_i n_i \mathbf{c}_i \mathbf{c}_i. \tag{1}$$

The computational utility of the lattice-Boltzmann equation is related to the realization that only a small set of velocities is necessary to simulate

J. Karkheck (ed.),
Dynamics: Models and Kinetic Methods for Non-equilibrium Many Body Systems, 17–30.

the Navier-Stokes equations [4]. Note that the velocities are such that all particles move from node to node simultaneously.

In order for the viscous stresses to be isotropic the set of velocities chosen for the lattice-Boltzmann model must satisfy the condition

$$\sum_i c_{i\alpha} c_{i\beta} c_{i\gamma} c_{i\delta} = \{\delta_{\alpha\beta}\delta_{\gamma\delta} + \delta_{\alpha\gamma}\delta_{\beta\delta} + \delta_{\alpha\delta}\delta_{\beta\gamma}\}. \tag{2}$$

This isotropy condition is satisfied by several different lattice-Boltzmann models, among them the 18-velocity model described in reference [2]. This model uses the [100] and [110] directions of a simple cubic lattice with twice the density of particles moving in [100] directions as in [110] directions; alternatively a 14-velocity model can be constructed from the [100] and [111] directions with a density ratio of 7:1. Although the 14-velocity model requires less computation and less memory than the 18-velocity model, it suffers from additional "checkerboard" invariants [5]. If the lattice nodes are split into two groups, one where the sum of the nodal coordinates is even and one where it is odd, then, in the 14-velocity model, all the population density in one group moves to the other group at the next time step. This spurious invariant is not present in the 18-velocity model, which makes it preferable in practice to the 14-velocity model [5]. The 18-velocity model can be augmented by including a density of stationary particles, which enables it to correctly model small deviations from the incompressible limit. This 19-velocity model is preferable to the 18-velocity model described in Ref. [2] although in actual simulations the differences are small.

The time evolution of the distribution function is described by a discrete analogue of the Boltzmann equation [6],

$$n_i(\mathbf{r} + \mathbf{c}_i, t + 1) = n_i(\mathbf{r}, t) + \Delta_i(\mathbf{r}, t), \tag{3}$$

where Δ_i is the change in n_i due to instantaneous molecular collisions at the lattice nodes. The post-collision distribution $n_i + \Delta_i$ is propagated for one time step, in the direction $\mathbf{c}_i$. The collision operator $\Delta_i(n)$ depends on all the population densities at the node, collectively denoted by $n(\mathbf{r}; t)$; it can take any form, subject to the constraints of mass conservation, momentum conservation, and isotropy. In the kinetic theory of gases, the collision operator is constructed by averaging individual particle-particle collisions, under the assumption that the distribution functions are uncorrelated from those at previous times. One then finds the distribution function that is invariant under this collision operator; for classical gases this is the Maxwell-Boltzmann distribution. However, the equilibrium distribution for lattice gases contains artifacts caused by the lack of Galilean invariance in the model, which show up in the macroscopic hydrodynamic equations [6]. To avoid this, we will first seek an equilibrium distribution function, n_i^{eq}, that is Galilean invariant and then construct a suitable collision operator to describe the relaxation of the non-equilibrium distribution, $n_i^{neq} = n_i - n_i^{eq}$.

The form for the equilibrium distribution is constrained by the moment conditions that are required to reproduce the Navier-Stokes equations on large space and time scales (*cf.* Eq. (1)):

$$\rho = \sum_i n_i^{eq} \tag{4}$$

$$\rho \mathbf{u} = \sum_i n_i^{eq} \mathbf{c}_i \tag{5}$$

$$\rho c_s^2 \mathbf{1} + \rho \mathbf{u}\mathbf{u} = \sum_i n_i^{eq} \mathbf{c}_i \mathbf{c}_i. \tag{6}$$

Equations (4) and (5) result from the requirements of mass and momentum conservation during the collision process, or in other words

$$\sum_i n_i^{neq} = \sum_i n_i^{neq} \mathbf{c}_i = 0. \tag{7}$$

The pressure in Eq. (6), $p = \rho c_s^2$, takes the form of an ideal gas equation of state with adiabatic sound speed c_s. It is also valid in the dense liquid phase if the density fluctuations are small (*i.e.* the Mach number $M = u/c_s << 1$). In contrast to the equilibrium distribution for the lattice-gas model [4], Eq. (6) ensures that the inviscid hydrodynamic equations are correctly reproduced; the viscous terms come from the non-equilibrium distribution as in the Chapman-Enskog approach.

A suitable form for the equilibrium distribution of the 19-velocity model that satisfies Eqs. (4) and (5), as well as the isotropy condition (Eq. (2)), is [7]

$$n_i^{eq} = a^{c_i} \rho \left[1 + \frac{\mathbf{u} \cdot \mathbf{c}_i}{c_s^2} + \frac{(\mathbf{u} \cdot \mathbf{c}_i)^2}{2c_s^4} - \frac{u^2}{2c_s^2} \right], \tag{8}$$

where $c_s = \sqrt{1/3}$ and the densities of the three speeds are

$$a^0 = \frac{1}{3}, \quad a^1 = \frac{1}{18}, \quad a^{\sqrt{2}} = \frac{1}{36}. \tag{9}$$

Next we construct a collision operator, $\Delta_i(n)$, such that $\Delta_i(n^{eq}) = 0$. A computationally useful form is obtained by linearizing the collision operator about the local equilibrium n^{eq} [8], *i.e.*

$$\Delta_i(n) = \sum_j \mathcal{L}_{ij}(n_j - n_j^{eq}), \tag{10}$$

where $\mathcal{L}$ is the linearized collision operator. We require that the linearized collision operator satisfies the following eigenvalue equations;

$$\sum_i \mathcal{L}_{ij} = 0, \quad \sum_i \mathbf{c}_i \mathcal{L}_{ij} = 0, \quad \sum_i \overline{\mathbf{c}_i \mathbf{c}_i} \mathcal{L}_{ij} = \lambda \overline{\mathbf{c}_j \mathbf{c}_j}, \quad \sum_i c_i^2 \mathcal{L}_{ij} = \lambda_B c_j^2, \tag{11}$$

where $\overline{\mathbf{c}_j \mathbf{c}_j}$, indicates the traceless part of $\mathbf{c}_j \mathbf{c}_j$. The first two equations follow from conservation of mass and momentum and the last two equations describe the isotropic relaxation of the stress tensor; the eigenvalues λ and λ_B are related to the shear and bulk viscosities. Equation (11) accounts for 10 of the eigenvectors of $\mathcal{L}$. The remaining 8 modes (or 9 modes if stationary particles are included) result from higher-order moments of $\mathcal{L}$ that are not relevant to simulations of the Navier-Stokes equations, but which do affect the boundary conditions at the solid-fluid interfaces. In Ref. [2] the eigenvalues of these kinetic modes were set to -1, which both simplifies the simulation and ensures a rapid relaxation of the non-hydrodynamic modes. In that work the eigenvalue of the bulk viscosity mode was also set to -1.

The collision operator can be further simplified by taking a single eigenvalue for both the viscous and the non-hydrodynamic modes [7]. This exponential relaxation time (ERT) approximation, $\Delta_i = -n_i^{neq}/\tau$, has become the most popular form for the collision operator because of its simplicity and computational efficiency. However significant errors have been observed in two-dimensional channel flow when $\tau > 1$ [9, 10], and results shown below indicate that these errors are caused by the collision operator rather than the boundary conditions. More general collision operators, such as the one given in Ref. [2], have significantly smaller errors for large values of the viscosity. In addition it is possible to tune the eigenvalues of the non-hydrodynamic modes so as to minimize the errors at the solid-fluid interfaces [11].

To find the long-time, long-wavelength dynamics, a scaling parameter ϵ is introduced, defined as the ratio of the lattice spacing to a characteristic macroscopic length; the hydrodynamic limit corresponds to $\epsilon \ll 1$. In a molecular gas the appropriate scaling parameter is the Knudsen number, the ratio of the mean-free path between collisions to the macroscopic length scale. The hydrodynamic behavior of a particular lattice-Boltzmann model can be determined via a two-time-scale expansion in powers of ϵ, similar to that described by Professor Hansen [12]. The details are given in innumerable sources and will not be repeated here. Thermal fluctuations in the fluid can be included by a straightforward extension of the lattice-Boltzmann model [13]. This makes it possible to simulate Brownian motion in colloidal suspensions with hydrodynamic memory effects included [13, 14, 15].

2. Boundary Conditions

To simulate the hydrodynamic interactions between solid particles in suspension, the lattice-Boltzmann model must be modified to incorporate the boundary conditions imposed on the fluid by the solid particles. It is also necessary to calculate the stresses exerted by the fluid on the particle surfaces. Fixed solid objects were first introduced into lattice-gas models by replacing the normal collision rules at a specified set of "boundary nodes" by the "bounce-back" collision rule [6] in which incoming particles are reflected back towards the nodes they came from. This rule replaces the normal collision rules at the boundary nodes, and sets up an approximate zero-velocity boundary condition around the surface described by these nodes. Forces are calculated from the momentum transfer at each boundary node and summed to give the force (and torque) on each object. In the finite-difference and finite-element methods, interpolation is necessary to calculate surface stresses from velocity gradients in the fluid, which in turn requires calculation of local surface normals. The bounce-back rule eliminates these complications, which are severe for irregularly shaped objects.

Detailed theoretical analysis of the bounce-back rule for two-dimensional Poiseuille flow has shown that the location of the zero-velocity plane is shifted from the location of the boundary nodes into the fluid, by an amount $0.5 + \alpha$ [11, 16]. The quantity α depends on the eigenvalues of the collision operator, primarily the shear viscosity: α is close to zero when the kinematic shear viscosity is around one-sixth. Although it is possible to obtain second-order convergence in a particular geometry by constructing a collision operator such that $\alpha = -1/2$, since α depends on the orientation of the wall, it is not possible to obtain second-order convergence by this method for more complex objects [11]. Alternatively, one can interpret the solution in terms of a hydrodynamic boundary that is displaced by half a lattice spacing from the physical one, and then choose a viscosity close to $1/6$ so that α is approximately zero [17]. Although the convergence of this method is in principle still only first-order, the scheme works quite well in practice [18, 19].

A fundamental improvement to the "nodal bounce-back" rule is to place the boundary nodes midway between interior (solid) and exterior (fluid) nodes [20, 21]. The normal collision rules are carried out at all fluid nodes and augmented by bounce-back rules at the center of links connecting lattice nodes on either side of the particle surface. The key difference is that a particle at a node adjacent to the solid surface hits the surface and returns (with opposite velocity) in one time step, whereas it takes two time steps to return when the boundary nodes are located at lattice nodes. In this case it can be shown that the hydrodynamic boundary is now located at the boundary nodes (*i.e.* midway between lattice nodes) with deviations that are second-order in the mesh resolution. In Poiseuille flow, the "link bounce-back" rule gives velocity fields that deviate from the exact solution by a constant slip velocity $u_s = u_{LBE} - u_{Exact}$ [10],

$$u_s/u_c = \beta/L^2, \tag{12}$$

where L is the channel width and $u_c = L^2\nabla p/8\nu$ is the exact velocity at the center of the channel. The value of β is dependent on the collision operator. For the ERT model [10],

$$\beta = 48\nu^2 - 4\nu - 1, \tag{13}$$

whereas for the linear collision operator in Ref. [2]

$$\beta = 4\nu - 1. \tag{14}$$

For small viscosities, ν, the slip velocity tends to a constant value ($-L^{-2}$) independent of collision model. However the slip velocity obtained with the ERT collision operator is sensitive to viscosity and diverges as ν^2 for large viscosities. Thus, the large slip velocity found for $\tau > 1$ [9, 10] is a consequence of the ERT model rather than the boundary conditions; the dependence is much weaker for other collision operators.

Although the link bounce-back rule was proposed at an early stage [20, 21], it was inconvenient to implement in a lattice-gas simulation [18, 22] because it imposed a large computational overhead on the bit-wise update. Furthermore, the improved convergence obtained with the link bounce-back method was not fully appreciated until recently [10]. Various interpolation schemes have been proposed to obtain second-order boundary conditions [23, 24, 25], acting under the assumption that the bounce-back method is inherently first order convergent. A variant of these schemes uses the full velocity distribution function at a node to deduce the local velocity gradient [26]. These methods share the drawback that they require information about the shape of the particle surface. For general three-dimensional objects, the resulting algorithms are complex and not necessarily well-defined without additional constraints [24]. As a result these methods have been applied exclusively to planar surfaces and two-dimensional systems. By contrast, the bounce-back rule can be applied to surfaces of arbitrary shape, without additional complications.

3. External Pressure Gradients

External pressure gradients are maintained by applying a force density to each fluid node. Typically a uniform increment of momentum is added to each node at each time step. It has been pointed out that an additional term is necessary to correctly represent a spatially varying body force [27]. However, for a homogeneous force density this additional term is of the same order as the error terms and is therefore unnecessary in this case.

A further complication arises from the discrete time step of the lattice-Boltzmann model. In the presence of a constant force density, $\mathbf{f}$, the lattice-Boltzmann update is given by (*cf.* Eq. (3))

$$n_i(\mathbf{r} + \mathbf{c}_i, t+1) = n_i(\mathbf{r}, t) + \Delta_i(\mathbf{r}, t) + \frac{a^{c_i}}{c_s^2}\mathbf{f} \cdot \mathbf{c}_i. \tag{15}$$

Although in most published work the velocity field is measured before the application of the force density, it could equally well be measured afterwards. The velocity fields before and after applying the force density are related,

$$\rho \mathbf{u}_+ = \rho \mathbf{u}_- + \mathbf{f}. \tag{16}$$

Thus, the slip velocities defined in Eqs. (13) and (14) depend on when the velocity field is measured. To decide on the correct choice for the definition of the velocity field, we can compare results obtained with an external pressure gradient with those obtained by another method of driving the fluid flow. In Ref. [3], flows over periodic arrays of spheres were driven by an external pressure gradient (using Eq.(15)) and also by a constant velocity boundary condition. A quasi-periodic system with several unit cells in the flow direction was used to obtain the proper inlet and outlet boundary conditions at a central cell. The mean flow velocity and drag force were measured for this central cell only; different numbers of cells were taken to ensure that there were no artifacts introduced by the boundary conditions at the ends of the system. The results obtained for the quasi-periodic systems agreed exactly with results obtained with systems driven by a pressure gradient *if* the mean of the velocity field before and after forcing, $(\mathbf{u}_+ + \mathbf{u}_-)/2$, was used [3]. The corrected slip velocities from Eqs. (13) and (14) are therefore

$$\beta = 48\nu^2 - 1 \tag{17}$$

for the ERT model, and

$$\beta = 8\nu - 1 \tag{18}$$

for the linear collision operator in Ref. [2]. We note that boundary conditions that have been tuned to give $\mathbf{u}_- = 0$ at the solid-fluid interface [10] are not exact if we take the more accurate velocity field

$$\rho \mathbf{u} = \rho \mathbf{u}_- + \frac{1}{2}\mathbf{f}. \tag{19}$$

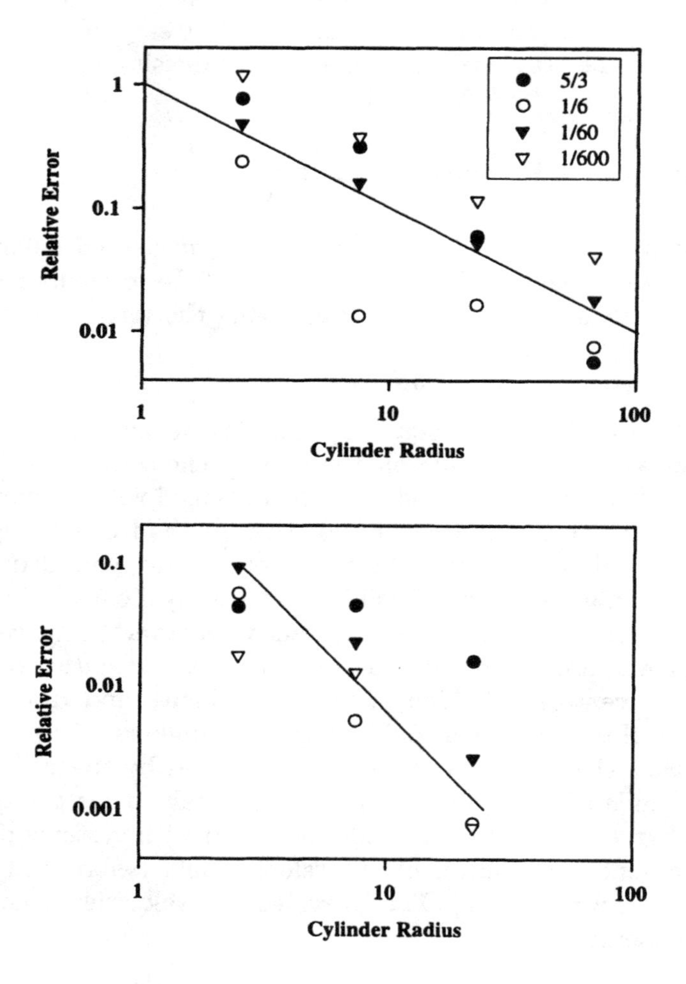

Figure 1. Relative errors in the drag coefficient of a periodic array of cylinders at zero Reynolds number. The errors are computed for different cylinder radii and for different fluid viscosities using the numerical solutions of Ref. [28] as a reference. The convergence obtained using the nominal radius of the cylinder (upper figure) can be compared with that obtained using the hydrodynamic radius of the cylinder (lower figure). The different symbols correspond to the values of the kinematic viscosity indicated in the legend. The straight lines in the upper and lower figures have slopes of -1 and -2, respectively.

4. Flow in a Periodic Array of Cylinders

The flow of fluid through a periodic array of cylinders has been used as a test of the effectiveness of the bounce-back rule for a typical suspension geometry. The present calculations are an extension of previously published work [3] to include simulations with a 1000-fold variation in fluid viscosity. Cylinders of nominal radius 2.5, 7.5, 22.5, and 67.5 lattice units have been used to simulate flow at an area fraction of approximately 40%; the ratio of the drag force to the mean flow velocity was compared with essentially exact results [28] as described in Ref. [3]. The relative errors shown in Fig. 1 are based on the nominal radius of the disk (upper plot) and on the hydrodynamic radius (lower plot), obtained by fitting the measured drag coefficient at an area fraction of 10%.

If the comparisons with Ref. [28] are based on the nominal radius of the disk, approximately linear convergence with increasing resolution is observed. The errors are large except in the case $\nu = 1/6$ where the nominal radius and hydrodynamic radius are similar. It is clear that the shift in the location of the hydrodynamic boundary must be accounted for if quantitative results are to be obtained with computationally useful particle sizes (typically less than 10 lattice units) and viscosities significantly different from 1/6. Once the hydrodynamic radius is taken into account, the errors are reduced by an order of magnitude. Also the rate of convergence is increased to approximately quadratic dependence on resolution, except for the largest fluid viscosity, $\nu = 5/3$.

It was shown in Ref. [3] that the hydrodynamic radius depends only on the fluid viscosity and not on the particle configuration, flow geometry or Reynolds number. The location of the hydrodynamic boundary, as determined from simulations at an area fraction of 10%, roughly follows the movement of the zero-velocity plane in a two-dimensional channel flow [3]. This suggests that if we can eliminate the channel-flow slip velocity for a range of viscosities, it may be possible to eliminate the viscosity dependence of the hydrodynamic radius altogether.

5. Moving Boundary Conditions

The generalization of the bounce-back rule to describe moving surfaces [21] made quantitative lattice-gas and lattice-Boltzmann simulations of particle suspensions possible. Prior to that, the moving boundary condition was implemented in an *ad hoc* fashion by adjusting the local equilibrium distribution function [29]; unfortunately, closely related methods persist to this day. To understand the physics of the moving boundary condition, one can imagine an ensemble of particles, moving at constant velocity u, impinging on a massive wall oriented perpendicular to the particle motion. The wall is itself moving with velocity $u_w \ll u$. The velocity of the particle after collision with the wall is $-u+2u_w$ and the force exerted on the wall is proportional to $u-u_w$. Since the velocities in the lattice-Boltzmann model are discrete, the desired boundary condition cannot be implemented directly, but we can instead modify the density of returning particles so that the momentum transferred to the wall is the same as in the continuous velocity case. This is the essence of the physics, described somewhat differently, in Refs. [2] and [21]. The stochastic version of the rule was derived first in [21], allowing for the discrete population density that characterizes lattice-gas models.

In accounting for the momentum transfer at a moving wall, local mass conservation is violated. It can be restored by including fluid in the interior region of the particle and then allowing an appropriate flux of fluid particles to cross the solid surface. Thus, the particles, rather than being solid, comprise a solid shell of given mass and inertia and contain fluid of the same mass density as the bulk fluid. It has been suggested that this model is unphysical [30], but the more pertinent question is to what extent do the dynamics of these fluid-filled particles differ from the dynamics of solid particles. This question was addressed in section 5 of Ref. [3]; here we summarize the important findings. The parameter characterizing the motion of the interior fluid is the dimensionless frequency $\omega^* = \omega a^2/\nu_{int}$ where a is the sphere radius and ν_{int} is the kinematic viscosity of the interior fluid, which can be different from that in the exterior flow region. Numerical results show that the drag force on an oscillating sphere is independent of the interior fluid as long as $\omega^* < 1$. The reduced frequency can be made arbitrarily small by increasing the viscosity of the interior fluid. Furthermore, it was shown that to lowest order in the frequency, flow in the interior fluid contributes a viscosity independent term that is exactly equal to the inertia of the interior fluid. Deviations from the inertial drag are generally small, proportional to the square of the frequency in leading order. Thus, the particle behaves dynamically as if its mass is the sum of the shell mass, used to compute the change in shell velocity arising from hydrodynamic forces, and the mass of the interior fluid. Effects of the interior fluid on the particle dynamics are small so long as the contribution of the interior fluid to the inertia of the particle is taken into account. In most practical calculations, the interior fluid takes the same viscosity as the exterior fluid, but the interior fluid can be made more viscous if necessary.

Alternative methods have been proposed in which fluid is excluded from the interior of the particle [30, 31], which then behaves as a solid object described by its shell mass and inertia only. Although an acceptable velocity boundary condition can be implemented in the lattice-Boltzmann method without using interior fluid, this "one-sided" scheme does not maintain local conservation of mass. The change in local mass density is proportional to the local velocity of the boundary node and averages to zero when summed over the whole particle surface. An attempt to maintain local mass conservation with a one-sided boundary condition [30] is fundamentally flawed and leads to incorrect pressure distributions around the solid particle. More recent simulations [31] show excellent agreement between lattice-Boltzmann simulations and finite-element calculations [32], independent of the presence or absence of interior fluid, so long as the shell mass and inertia are adjusted accordingly. Thus, we conclude that accurate simulations can be carried out with or without interior fluid.

There are drawbacks to both of these schemes. The inclusion of interior fluid limits the simulations to suspensions where the particles are at least neutrally buoyant with respect to the fluid. In order to simulate low-mass particles such as bubbles, the interior fluid must be excluded. The drawbacks of excluding interior fluid are that local mass conservation is lost, and that it significantly adds to the complexity of the calculations. The additional complication comes from the necessity to add fluid with the correct mass and momentum density to solid sites being vacated as the particle moves and to similarly remove fluid from sites being covered by the particle. Moreover, a substantial perturbation is thereby added to the fluid flow whenever a particle changes position. Nevertheless it should be emphasized that both methods give results in good agreement with one another and with independent finite-element simulations [31].

6. Particle Motion

In Ref. [3] particle velocities were determined from an explicit update,

$$\mathbf{U}(t+1) = \mathbf{U}(t-1) + 2\mathbf{F}(t+1)/M; \tag{20}$$

velocities are updated every other time step to minimize the effects of the staggered momentum invariants [33]. There is a stability criterion for Eq. (20) and its rotational equivalent [3], which can be expressed as a condition on the effective mass density of the solid particle (including the mass of the interior fluid),

$$\frac{\rho_s}{\rho_f} > 1 + 10/a. \tag{21}$$

This stability criterion imposes a serious constraint on the size of particles which can be used to simulate realistic particle-fluid systems for which $\rho_s/\rho_f \approx 2$. However, an implicit velocity update has been proposed [34] that is unconditionally stable. Schematically,

$$U(t+1) - U(t) = \frac{F_0(t) + \lambda U(t+1)}{M}, \tag{22}$$

where $F_0(t)$ is the part of the particle force arising from the zero-velocity bounce-back rule and $\lambda U(t+1)$ comes from the velocity-dependent part of the boundary update [3]. An explicit update replaces the unknown $U(t+1)$ on the right-hand side of Eq. (22) by $U(t)$ (as in Eq. (20)), but an unconditionally stable update can be achieved by solving Eq. (22) for $U(t+1)$ directly. The drawback of this new scheme is that it takes two sweeps through the boundary nodes. The first sweep is used to determine F_0 so that $U(t+1)$ can be found; this involves solving six coupled simultaneous equations for the translational and rotational velocities of each particle. The second sweep uses the new velocities to finish the boundary node updates.

Further complications arise when we consider the effects of particle displacements through the lattice. One possibility is to keep track of the actual particle coordinates but only update the boundary node positions when the particle center moves nearer to an adjacent lattice site than to its current one. Thus, the boundary-node map is always centered on a lattice node and the particle shape is constant in time (for spherical particles). This approach has two drawbacks. The first problem is that the translation of the boundary node map by a distance of the order of one lattice spacing introduces a significant discontinuity into the fluid flow. The second problem is that an additional check is required before the boundary node map is moved, to ensure that there are no overlapping maps. Even in moderately dense suspensions this leads to "traffic jams" with boundary node maps that are unable to move to the correct locations. A better solution is to move the boundary-node maps more or less continuously in time so that discontinuities in fluid motion are small and traffic jams do not arise. In this case the particle shape fluctuates with changes in particle position, but the effects of these shape fluctuations are small if the particle is larger than about 5 lattice units.

7. Conclusions

The usefulness of a numerical algorithm depends on 5 factors: accuracy, reliability, efficiency, simplicity, and flexibility. Finite-element methods are potentially more accurate than lattice-Boltzmann schemes because an adjustable mesh is better able to resolve fluid flow in the small gaps between the particles. Although lattice-Boltzmann simulations can incorporate more complicated meshes [1], in my opinion these additional complications contradict the essential simplicity of the method. In practical finite-element calculations, high-aspect-ratio meshes, necessary to resolve the flow in small gaps, can lead to large and unpredictable errors. Moreover, the small elements tend to impose strict limits on the size of the time step that can be used. Lattice-Boltzmann simulations have been shown by comparison both with finite-element simulations of particle suspensions [3, 31] and spectral simulations of homogeneous turbulence to be capable of a high degree of accuracy [35]. The simplicity of the code and absence of tuning parameters present in finite-element codes make the lattice-Boltzmann scheme more robust and reliable in practice. This is attested to by the large number of successful numerical tests of the lattice-Boltzmann method in a variety of flow situations [1]. Tests of the computational efficiency of the lattice-Boltzmann method have shown that it is competitive with spectral methods for homogeneous flows [36] and much faster than finite-element methods [32] for particle suspensions. The simplicity of the lattice-Boltzmann algorithm makes it accessible to a wide variety of potential users, and suggests that the success of molecular dynamics as a widely used numerical tool may be at least partially repeated for the lattice-Boltzmann method. The most serious drawback of the lattice-Boltzmann scheme is its lack of flexibility with respect to variations in the underlying continuum equations. While a finite-element code can be used for a variety of fluid flows, both Newtonian and non-Newtonian, perhaps with the addition of thermal diffusion or radiation transport, lattice-Boltzmann simulations are not readily extended beyond incompressible viscous flows, although some extensions have been proposed[1].

References

1. S. Chen and G.D. Doolen, Ann. Rev. Fl. Mech. **30**, 329 (1998).
2. A. J. C. Ladd, J. Fluid Mech. **271**, 285 (1994).
3. A. J. C. Ladd, J. Fluid Mech. **27**, 311 (1994).
4. U. Frisch, B. Hasslacher, and Y. Pomeau, Phys. Rev. Lett. **56**, 1505 (1986).
5. A. Koponen, *Simulations of Fluid Flow in Porous Media by Lattice-Gas and Lattice-Boltzmann Methods.* PhD thesis, University of Jyväkylä, Finland, 1998.
6. U. Frisch, D. d'Humières, B. Hasslacher, P. Lallemand, Y. Pomeau, and J-P. Rivet, Complex Systems **1**, 649 (1987).
7. Y. H. Qian, D. d'Humières, and P. Lallemand, Europhys. Lett. **17**, 479 (1992).
8. F. Higuera, S. Succi, and R. Benzi, Europhys. Lett. **9**, 345 (1989).
9. M. A. Gallivan, D. R. Noble, J. G. Georgiadis, and R. O. Buckius, Int J. Numer. Meth. Fluids **25**, 249 (1997).
10. X. He, Q. Zou, L-S. Luo, and M. Dembo, J. Stat. Phys. **87**, 115 (1997).
11. I. Ginzbourg and P. M. Adler, J. Phys. II (France) **4**, 191 (1994).
12. J.P.Hansen, this volume, p1.
13. A. J. C. Ladd, Phys. Rev. Lett. **70**, 1339 (1993).
14. A. J. C. Ladd, H. Gang, J. X. Zhu, and D. A. Weitz, Phys. Rev. Lett. **74**, 318 (1995).
15. A. J. C. Ladd, H. Gang, J. X. Zhu, and D. A. Weitz, Phys. Rev. E **52**, 6550 (1995).
16. R. Cornubert, D. d'Humières, and C. D. Levermore, Physica D **47**, 241 (1991).
17. D. P. Ziegler, J. Stat. Phys. **71**, 1171 (1995).
18. A. J. C. Ladd and D. Frenkel, Phys. Fluids A **2**, 1921 (1990).
19. O. P. Behrend, Phys. Rev. E **52**, 1164 (1995).
20. J. A. Somers and P. C. Rem, in *Shell Conference on Parallel Computing*, G. A. van der Zee, ed., (Lecture Notes on Computer Science, 1988).

21. A. J. C. Ladd and D. Frenkel, in *Cellular Automata and Modeling of Complex Physical Systems*, P. Manneville, N. Boccara, G. Y. Vichniac, and R. Bidaux, eds., (Springer-Verlag, Berlin, 1989).
22. M. A. van der Hoef, D. Frenkel, and A. J. C. Ladd, Phys. Rev. Lett. **67**, 3459 (1991).
23. P. A. Skordos, Phys. Rev. E **48**, 4823 (1993).
24. D. R. Noble, S. Y. Chen, J. G. Georgiadis, and R. O. Buckius, Phys. Fluids **7**, 203 (1995).
25. S. Chen, D. Martinez, and R. Mei, Phys. Fluids **8**, 2527 (1996).
26. I. Ginzbourg and D. d'Humières, J. Stat. Phys. **84**, 927 (1996).
27. L-S. Luo, Phys. Rev. Lett. **81**, 1618 (1998).
28. A. S. Sangani and A. Acrivos, Int. J. Multiphase Flow **8**, 193 (1982).
29. A. J. C. Ladd, M. E. Colvin, and D. Frenkel, Phys. Rev. Lett. **60**, 975 (1988).
30. C. K. Aidun and Y. N. Lu, J. Stat. Phys. **81**, 49 (1995).
31. C. K. Aidun, Y. N. Lu, and E. Ding, J. Fluid Mech. **373**, 287 (1998).
32. J. Feng, H. H. Hu, and D. D. Joseph, J. Fluid Mech. **261**, 95 (1994).
33. G. R. McNamara and G. Zanetti. Phys. Rev. Lett. **61**, 2332 (1988).
34. C. P. Lowe, D. Frenkel, and A. J. Masters, J. Chem. Phys. **76**, 1582 (1995).
35. D. O. Martinez, W. H. Matthaes, S. Chen, and D. C. Montgomery, Phys. Fluids **6**, 1285 (1994).
36. S. Chen, Z. Wang, X. Shan, and G. D. Doolen, J. Stat. Phys **68**, 379 (1992).

ORIENTATIONAL RELAXATION AND BROWNIAN MOTION

B.U. FELDERHOF
Institut für Theoretische Physik A, R.W.T.H. Aachen
Templergraben 55, 52056 Aachen, Germany

AND

R.B. JONES
Queen Mary and Westfield College, Department of Physics
Mile End Road, London E1 4NS, United Kingdom

1. Introduction

The theory of Brownian motion was developed first by Einstein [1] for a single spherical particle immersed in a fluid and subjected to an external potential. Einstein realized that on a sufficiently slow time scale the observed random displacements of the particle can be described by a generalized diffusion equation. The particle momentum can be ignored, since it changes on a much faster time scale. Its probability distribution rapidly becomes nearly maxwellian, corresponding to the temperature of the fluid [2, 3]. Einstein's simple equation for a single particle was generalized by Smoluchowski to the mutual diffusion of a pair of particles, interacting with a central potential [4]. He also considered the possibility of reaction between the two partners [5]. The modern theory of interacting Brownian particles is based on a generalized Smoluchowski equation for the probability distribution of the entire configuration of particles [6]. Thus, if we consider N Brownian particles in volume V, which at time t have the configuration

$$\mathbf{X} \equiv (\mathbf{R}_1, \cdots, \mathbf{R}_N), \tag{1}$$

where $\mathbf{R}_j$ denotes the position of the center of particle j, then the Brownian motion of the whole system is described by a generalized diffusion equation in the $3N$-dimensional configuration space. The generalized Smoluchowski

J. Karkheck (ed.),
Dynamics: Models and Kinetic Methods for Non-equilibrium Many Body Systems, 31–38.

equation for the probability distribution $P(\mathbf{X},t)$ reads

$$\frac{\partial P}{\partial t} = \frac{\partial}{\partial \mathbf{X}} \cdot \mathsf{D} e^{-\beta\Phi} \cdot \frac{\partial}{\partial \mathbf{X}} (e^{\beta\Phi} P), \tag{2}$$

with $3N \times 3N$ diffusion matrix $\mathsf{D}(\mathbf{X})$ and potential $\Phi(\mathbf{X})$. The diffusion matrix is symmetric and given by the generalized Einstein relation

$$\mathsf{D}(\mathbf{X}) = k_B T \boldsymbol{\mu}(\mathbf{X}), \tag{3}$$

where the mobility matrix $\boldsymbol{\mu}(\mathbf{X})$ follows from the solution of Stokes' linear steady state hydrodynamic equations. The explicit calculation of the diffusion matrix is difficult and requires an elaborate numerical algorithm [7]. In contrast, the potential $\Phi(\mathbf{X})$ may be assumed known. It is convenient to impose periodic boundary conditions. Then the wall potential can be omitted. The potential $\Phi(\mathbf{X})$ is usually assumed to be given by a sum of pair interactions. In the course of time any probability distribution $P(\mathbf{X}, t)$ tends to the equilibrium distribution

$$P_{eq}(\mathbf{X}) = \exp[-\beta\Phi(\mathbf{X})]/Z(\beta), \tag{4}$$

where the factor $1/Z(\beta)$ normalizes the distribution to unity. The equilibrium distribution is independent of the hydrodynamic interactions.

The generalized Smoluchowski equation (2) provides a quite detailed description on a microscopic level, even though the fluid degrees of freedom and the momenta of the Brownian particles have been eliminated. Much work remains to extract information on observable quantities such as the dynamic scattering function observed in light scattering or the frequency-dependent shear viscosity of the suspension. To that purpose theoretical tools have been developed which allow definite predictions, at least for semidilute suspensions. For such suspensions the problem can be reduced to the solution of a Smoluchowski equation for a pair of interacting Brownian particles.

2. Internal Degrees of Freedom

The theory can be extended to apply to suspensions of Brownian particles with internal degrees of freedom. For example, in ferrofluids the particles are spherical, but carry a permanent magnetic moment. Its orientation provides two additional degrees of freedom per particle. We can also consider mixtures and regard the species label as an internal degree of freedom. In that case number-conserving chemical reactions between species can be described as transitions between states [8]. Thus, in generalization of Eq.(1), the microscopic situation at time t is taken to be fully specified by the configuration

$$\mathbf{X}, \mathbf{s} \equiv (\mathbf{R}_1, s_1, \mathbf{R}_2, s_2, \cdots, \mathbf{R}_N, s_N), \tag{5}$$

where s_j denotes the state of particle j. If we consider ν possible states, then s_j may be represented as a vector with ν components $\{s_{j\alpha}\}$ with component $s_{j\alpha}$ equal to one if molecule j is in the state denoted by s_j and zero otherwise. By discretization of the directions of the moment we can map the case of particles with orientation onto the same mathematical description. The potential may be taken to have the form

$$\Phi(\mathbf{X}, \mathbf{s}) = \sum_{j=1}^{N} \varepsilon(s_j) + \frac{1}{2} \sum_{i \neq j}^{N} v(s_i, s_j, |\mathbf{R}_i - \mathbf{R}_j|), \tag{6}$$

where the first term represents the binding energy and the second one the sum of pair interactions. The diffusion of particles is governed by a symmetric $3N\nu \times 3N\nu$ diffusion matrix $\mathsf{D}(\mathbf{X}, \mathbf{s})$.

Statistically the system is described by a time-dependent probability distribution $P(\mathbf{X}, \mathbf{s}, t)$ on the configuration space. The distribution satisfies the master equation

$$\frac{\partial P}{\partial t} = \mathcal{W} P \tag{7}$$

with a master operator

$$\mathcal{W} = \mathcal{D}_T + \mathcal{W}_R, \tag{8}$$

where the diffusion operator $\mathcal{D}_T$ has the Smoluchowski form (2) and the operator $\mathcal{W}_R$ is defined by a kernel $W_R(\mathbf{X}, \mathbf{s}|\mathbf{X}', \mathbf{s}')$ describing local transitions between states. The transition probabilities $W_R(\mathbf{X}, \mathbf{s}|\mathbf{X}', \mathbf{s}')$ are assumed to satisfy the detailed balance condition

$$W_R(\mathbf{X}, \mathbf{s}|\mathbf{X}', \mathbf{s}') P_{eq}(\mathbf{X}', \mathbf{s}') = W_R(\mathbf{X}', \mathbf{s}'|\mathbf{X}, \mathbf{s}) P_{eq}(\mathbf{X}, \mathbf{s}). \tag{9}$$

In practice one considers only transitions in which one particle changes its state locally, or a pair of particles changes states locally.

The detailed balance condition also holds [9] for the Smoluchowski diffusion operator $\mathcal{D}_T$. The implied symmetry permits the application of the Mori projection operator formalism [10]. The average of a variable $A(\mathbf{X}, \mathbf{s})$, defined on the set of states, is given by

$$\overline{A}(t) = \sum_{\mathbf{s}} \int d\mathbf{X}\, \mathrm{A}(\mathbf{X}, \mathbf{s}) P(\mathbf{X}, \mathbf{s}, t). \tag{10}$$

Its time-evolution may be expressed with the transpose of the master operator $\mathcal{W}$. We define the function $f(\mathbf{X}, \mathbf{s}, t)$ by

$$P(\mathbf{X}, \mathbf{s}, t) = f(\mathbf{X}, \mathbf{s}, t) P_{eq}(\mathbf{X}, \mathbf{s}). \tag{11}$$

It satisfies the equation

$$\frac{\partial f}{\partial t} = \mathcal{L}f, \tag{12}$$

where the evolution operator $\mathcal{L}$ is the transpose of $\mathcal{W}$. In our case the operator $\mathcal{L}$ is a sum of two terms

$$\mathcal{L} = \mathcal{L}_T + \mathcal{L}_R. \tag{13}$$

The operator $\mathcal{L}_T$, corresponding to diffusion, takes the form

$$\mathcal{L}_T = e^{\beta\Phi}\frac{\partial}{\partial \mathbf{X}} \cdot \mathsf{D}\, e^{-\beta\Phi} \cdot \frac{\partial}{\partial \mathbf{X}}. \tag{14}$$

By use of the detailed balance condition (9) one finds for the kernel $L_R(\mathbf{X}, \mathbf{s}|\mathbf{X}', \mathbf{s}')$ corresponding to the reaction operator $\mathcal{L}_R$

$$L_R(\mathbf{X}, \mathbf{s}|\mathbf{X}', \mathbf{s}') = W_R(\mathbf{X}', \mathbf{s}'|\mathbf{X}, \mathbf{s}). \tag{15}$$

We define the scalar product between two variables $A(\mathbf{X}, \mathbf{s})$ and $B(\mathbf{X}, \mathbf{s})$ as

$$(A, B) = \sum_{\mathbf{s}} \int d\mathbf{X} \quad P_{eq}(\mathbf{X}, \mathbf{s}) A(\mathbf{X}, \mathbf{s}) B(\mathbf{X}, \mathbf{s}). \tag{16}$$

Clearly the product is symmetric, $(A, B) = (B, A)$. It is easily checked, by use of detailed balance, that the operator $\mathcal{L}$ is self-adjoint in this scalar product,

$$(A, \mathcal{L}B) = (\mathcal{L}A, B). \tag{17}$$

This formal property is very useful. We consider a set of variables A_α,where the label α takes the values $1, \cdots, \nu$. The corresponding time-dependent observable is defined by

$$A_\alpha(t) = e^{\mathcal{L}t} A_\alpha(0), \qquad A_\alpha(0) = A_\alpha. \tag{18}$$

It is convenient to introduce vector notation $\mathbf{A} = (A_1, \cdots, A_\nu)$ for the set of observables. By use of Mori's projection operator formalism one can now show [8] that the relaxation matrix

$$\mathsf{F}(t) = (\mathbf{A}(t), \mathbf{A}^*) \tag{19}$$

satisfies the equation

$$\frac{d\mathsf{F}}{dt} = \Omega \cdot \mathsf{F}(t) - \int_0^t \mathsf{M}(t - t') \cdot \mathsf{F}(t')dt', \tag{20}$$

with instantaneous rate matrix Ω and memory matrix $\mathsf{M}(t)$. We specify these matrices in the next section after a suitable choice of variables.

3. Relaxation Matrix

In the present context it is natural to consider the time-correlation function of plane wave density variations of the various species. Thus, we consider the set of variables

$$n_\alpha(\mathbf{q}) = \sum_{j=1}^{N} \theta_\alpha(s_j) e^{i\mathbf{q}\cdot\mathbf{R}_j}, \qquad \alpha = 1, \cdots, \nu, \tag{21}$$

where $\theta_\alpha(s_j)$ takes the value 1 if s_j corresponds to the state α, and zero otherwise. It is of interest to study the $\nu \times \nu$ relaxation matrix $\mathsf{F}(\mathbf{q}, t)$ with elements

$$F_{\alpha\beta}(\mathbf{q}, t) = \lim_{\substack{N\to\infty \\ V\to\infty}} \frac{1}{N}(n_\alpha(\mathbf{q}, t), n_\beta(-\mathbf{q})) \tag{22}$$

in the thermodynamic limit $N \to \infty$, $V \to \infty$ at constant $n_0 = N/V$. For the value of the relaxation matrix at time $t = 0$ we introduce the separate notation

$$\mathsf{S}(\mathbf{q}) = \mathsf{F}(\mathbf{q}, 0). \tag{23}$$

Then the $\nu \times \nu$ matrix $\Omega(\mathbf{q})$ in Eq. (20) has elements

$$\Omega_{\alpha\beta}(\mathbf{q}) = \sum_{\gamma} \frac{1}{N}(\mathcal{L} n_\alpha(\mathbf{q}), n_\gamma(-\mathbf{q}))[\mathsf{S}(\mathbf{q})^{-1}]_{\gamma\beta}, \tag{24}$$

and the memory function $\mathsf{M}(\mathbf{q}, t)$ is given by the expression

$$\mathsf{M}(\mathbf{q}, t) = -\frac{1}{N}(\mathbf{f}(\mathbf{q}, t), \mathbf{f}(-\mathbf{q})) \cdot \mathsf{S}(\mathbf{q})^{-1} \tag{25}$$

with the so-called random force $\mathbf{f}(\mathbf{q}, t)$ and $\mathbf{f}(-\mathbf{q}) = \mathbf{f}(-\mathbf{q}, 0)$. In the Mori-formalism one introduces a projector $\mathcal{P}_\mathbf{q}$ on the chosen set of variables $\{n_\alpha(\mathbf{q})\}$ by the definition

$$\mathcal{P}_\mathbf{q} A = \sum_{\beta\gamma} \frac{1}{N} < A\, n_\beta(-\mathbf{q}) >_N [\mathsf{S}(\mathbf{q})^{-1}]_{\beta\gamma} n_\gamma(\mathbf{q}) \tag{26}$$

where the angle brackets indicate an average over the equilibrium distribution $P_{eq}(\mathbf{X}, \mathbf{s})$. The orthogonal projector $\mathcal{Q}_\mathbf{q}$ is defined by

$$\mathcal{Q}_\mathbf{q} = 1 - \mathcal{P}_\mathbf{q}. \tag{27}$$

The random force $\mathbf{f}(\mathbf{q})$ has components

$$f_\alpha(\mathbf{q}) = \mathcal{Q}_\mathbf{q} \mathcal{L} n_\alpha(\mathbf{q}). \tag{28}$$

Its time-dependence in Eq.(25) is governed by projected dynamics

$$f_\alpha(\mathbf{q},t) = \exp(\mathcal{Q}_\mathbf{q}\mathcal{L}t)\mathcal{Q}_\mathbf{q}\mathcal{L}\, n_\alpha(\mathbf{q}). \tag{29}$$

This can be re-expressed in terms of dynamics governed by the actual time-evolution operator $\mathcal{L}$. The relation is expressed conveniently in terms of Laplace transforms.

The Laplace transform of the relaxation matrix is defined by

$$\hat{\mathsf{F}}(\mathbf{q},p) = \int_0^\infty e^{-pt}\mathsf{F}(\mathbf{q},t)\,dt. \tag{30}$$

It can be expressed as

$$\hat{F}_{\alpha\beta}(\mathbf{q},p) = \lim_{\substack{N\to\infty \\ V\to\infty}} \frac{1}{N} < n_\alpha(\mathbf{q})(p-\mathcal{L})^{-1}n_\beta(-\mathbf{q}) >_N . \tag{31}$$

From Eq. (20) it follows that we have also

$$\hat{\mathsf{F}}(\mathbf{q},p) = [p\mathsf{I} - \boldsymbol{\Omega}(\mathbf{q}) + \hat{\mathsf{M}}(\mathbf{q},p)]^{-1} \cdot \mathsf{S}(\mathbf{q}). \tag{32}$$

Following the procedure developed by Mori [10] we may cast the memory function in the form

$$\hat{\mathsf{M}}(\mathbf{q},p) = -\hat{\boldsymbol{\Phi}}(\mathbf{q},p)\{\mathsf{I} + [p\mathsf{I} - \boldsymbol{\Omega}(\mathbf{q})]^{-1}\hat{\boldsymbol{\Phi}}(\mathbf{q},p)\}^{-1}, \tag{33}$$

where the matrix $\hat{\boldsymbol{\Phi}}(\mathbf{q},p)$ has elements

$$\hat{\Phi}_{\alpha\beta}(\mathbf{q},p) = \sum_\gamma \frac{1}{N} < J_\gamma(-\mathbf{q})(p-\mathcal{L})^{-1}J_\alpha(\mathbf{q}) >_N [\mathsf{S}(\mathbf{q})^{-1}]_{\gamma\beta}, \tag{34}$$

with the current variable $J_\alpha(\mathbf{q})$ defined by

$$J_\alpha(\mathbf{q}) = \mathcal{Q}_\mathbf{q}\mathcal{L}n_\alpha(\mathbf{q}). \tag{35}$$

This is identical with the random force defined in Eq.(28). We use a different notation to indicate that the time-dependent variable $J_\alpha(\mathbf{q},t)$ is defined with the actual evolution operator $\mathcal{L}$, rather than with the projected one as in Eq.(29). It follows from Eqs.(24) and (26) that

$$J_\alpha(\mathbf{q}) = \mathcal{L}n_\alpha(\mathbf{q}) - \sum_\beta \Omega_{\alpha\beta}(\mathbf{q})n_\beta(\mathbf{q}), \tag{36}$$

which shows that the instantaneous motion corresponding to the matrix $\Omega(\mathbf{q})$ is subtracted.

4. Approximate Dynamics

So far our equations are exact. However, in order to arrive at explicit results we are forced to make approximations at this stage. It will be assumed that the equilibrium pair correlation functions of the suspension are known. This allows one to calculate the structure matrix $\mathsf{S}(\mathbf{q})$, defined by Eq.(23). With some more effort the matrix $\Omega(\mathbf{q})$ can also be calculated, since according to its definition Eq. (24) it is determined by an equilibrium average. The difficult part is the calculation of the memory matrix given by Eqs. (33) and (34). The average in Eq. (34) is the Laplace transform of a time-dependent current correlation function. We assume that it can be calculated with sufficient accuracy from the solution of the pair diffusion problem. Formally, the approximation is defined by a cluster expansion of the average and a truncation of the infinite sum at the pair level. Physically this implies that the exact time-evolution of the correlation function is decomposed into dynamical events involving only pairs of particles.

Thus, rewriting Eq.(34) in matrix notation as

$$\hat{\mathbf{\Phi}}(\mathbf{q},p) = \hat{\mathbf{\Lambda}}(\mathbf{q},p) \cdot \mathsf{S}(\mathbf{q})^{-1}, \tag{37}$$

we approximate

$$\hat{\mathbf{\Lambda}}(\mathbf{q},p) \approx \hat{\mathbf{\Lambda}}_2(\mathbf{q},p), \tag{38}$$

where $\hat{\mathbf{\Lambda}}_2(\mathbf{q},p)$ can be calculated from the pair diffusion-reaction problem. It can be shown [8] that the current variable for two particles can be expressed as

$$\mathbf{J}(1,2;\mathbf{q}) = \exp[i\mathbf{q} \cdot \frac{\mathbf{R}_1 + \mathbf{R}_2}{2}]\mathbf{J}_r(1,2;\mathbf{q}), \tag{39}$$

where the second factor depends only on the relative distance vector $\mathbf{r} = \mathbf{R}_2 - \mathbf{R}_1$. The two-particle cluster integral $\hat{\mathbf{\Lambda}}_2(\mathbf{q},p)$ can be expressed as

$$\hat{\mathbf{\Lambda}}_2(\mathbf{q},p) = \frac{1}{2} n_0 \sum_{s_1,s_2} x_{s_1} x_{s_2} \int g_{s_1 s_2}(r) \mathbf{J}_r(1,2;\mathbf{q}) \boldsymbol{\psi}_r(1,2;-\mathbf{q}) d\mathbf{r}, \tag{40}$$

where x_α is the equilibrium mole fraction of species α, $g_{\alpha\beta}(r)$ is the equilibrium radial distribution function of species α and β, and $\boldsymbol{\psi}_r(1,2;\mathbf{q})$ is the solution of the equation

$$[p - \mathcal{L}_r(1,2;\mathbf{q})]\boldsymbol{\psi}_r(1,2;\mathbf{q}) = \mathbf{J}_r(1,2;\mathbf{q}). \tag{41}$$

Here the matrix $\mathcal{L}_r(1,2;\mathbf{q})$ is a $\nu^2 \times \nu^2$-dimensional matrix operator. In specific applications the dimensionality can be reduced by symmetry arguments.

Thus, by the approximation (38) we have reduced the problem of calculation of the density relaxation matrix defined in Eq.(22) to the solution of the pair reaction-diffusion problem (41). The equilibrium structure of the suspension is assumed known. The detailed solution of Eq.(41) is not trivial, but can be achieved. Even for a system with only two species the solution contains a wealth of detail, which has only partly been analyzed [11, 12]. Such a system might be called a colloidal kinetic Ising model. For spheres with continuously varying orientation [13] one may consider a range-dependent Heisenberg pair interaction. For this system the pair problem (41) can also be solved exactly [14]. For both the colloidal kinetic Ising model and the kinetic Heisenberg model it would be desirable to compare the theory with computer simulations. For the kinetic Ising model such simulations should be relatively easy. If desired, the continuous space diffusion could be replaced by random walk on a lattice. Further work on the model would provide useful insight into systems with reversible diffusion-controlled reactions.

References

1. A. Einstein, Ann. Physik **17**, 549 (1905), translated in *Investigations on the Theory of the Brownian Movement*, R. Fürth, ed., (Dover, New York, 1956).
2. U.M. Titulaer, Physica A **91**, 321 (1978); **100**, 234, 251 (1980).
3. N.G. van Kampen, Physics Reports **124**, 69 (1985).
4. M. v. Smoluchowski, Phys. Z. **17**, 557 (1916).
5. M. v. Smoluchowski, Z. Phys. Chem. **92**, 129 (1917).
6. P.N. Pusey, in *Liquids, Freezing and Glass Transition*, J.P. Hansen, D. Levesque, and J. Zinn-Justin, eds., (North-Holland, Amsterdam, 1991), p.763.
7. B. Cichocki, B.U. Felderhof, K. Hinsen, E. Wajnryb, and J. Blawzdziewicz, J. Chem.Phys. **100**, 3780 (1994).
8. B.U. Felderhof and R.B. Jones, J. Chem. Phys. **103**, 10201 (1995).
9. N.G. van Kampen, *Stochastic Processes in Physics and Chemistry*, (North-Holland, Amsterdam, 1992), p.197.
10. H. Mori, Prog. Theor. Phys. **33**, 423 (1965).
11. B.U. Felderhof and R.B. Jones, J. Chem. Phys. **106**, 967 (1997).
12. B.U. Felderhof and R.B. Jones, J. Chem. Phys. **106**, 5006 (1997).
13. B.U. Felderhof and R.B. Jones, Phys. Rev. E **48**, 1084 (1993).
14. B.U. Felderhof and R.B. Jones, Phys. Rev. E **48**, 1142 (1993).

VISCOSITY AND DIFFUSION OF CONCENTRATED HARD-SPHERE-LIKE COLLOIDAL SUSPENSIONS

R. VERBERG
Chemical Engineering Department, University of Florida
Gainesville, FL 32611, USA

I. M. DE SCHEPPER
I. R. I. Delft University of Technology
2629 JB Delft, The Netherlands

AND

E. G. D. COHEN
Center for the Studies in Physics and Biology
The Rockefeller University
New York, NY 10021, USA

Abstract. We present explicit theoretical results for the viscosity and diffusion coefficient of concentrated hard-sphere-like colloidal suspensions. Our results are based on two relevant physical processes that take place on two widely separated time scales. At short times, $\tau_B \sim t \ll \tau_P$, with the Brownian time $\tau_B \sim 1$ ns and the Péclet time $\tau_P \sim 1$ ms, the dominant process is the so-called cage-diffusion. The colloidal particles are locked up in cages and the difficulty to escape out of one cage and into the next is related to the deformability of the cage. This process has a collective character reflected in the fact that each particle inside a cage is at the same time a wall particle of a neighboring cage and the escape rate is determined by the short-time collective diffusion coefficient for which we present an explicit expression. At long times, $t \gg \tau_P$, the dominant process is a coupled relaxation mechanism as described by the mode-coupling theory, via two slowly decaying modes associated with conserved single-particle or collective dynamical variables. We present closed expressions for the long-time wavenumber dependent self and collective diffusion coefficients and for the Newtonian and frequency dependent viscosity and compare them with a variety of experimental and computational results.

J. Karkheck (ed.),
Dynamics: Models and Kinetic Methods for Non-equilibrium Many Body Systems, 39–64.

1. Introduction

The past two decades have shown a growing interest in fundamental properties of concentrated colloidal suspensions. In this paper we present a summary of our contributions to this challenging field. Our aim was to develop a kinetic theory for the viscosity and diffusion in concentrated hard-sphere-like colloidal suspensions, based on the analogy of the transport properties of molecular fluids and colloidal suspensions. We have shown that it is possible to describe both the Newtonian and the dynamic, *i.e.*, frequency dependent viscosity as well as the self and the collective diffusion coefficients of concentrated colloidal suspensions from a single viewpoint with a molecular theory that takes into account the dominant physical microscopic processes at short and long times [1]. Here short-times refer to the range $\tau_B \sim t \ll \tau_P$ and long-times to the range $t \gg \tau_P$, where the Brownian time is defined as $\tau_B = m/\zeta_0 \sim 1$ ns (a measure for the average time in which the velocity of a Brownian particle relaxes to thermal equilibrium with the surrounding fluid) and the Péclet time as $\tau_P = \sigma^2/4D_0 \sim 1$ ms (a measure of the time-scale on which direct interactions between the Brownian particles become relevant), with m the mass and σ the diameter of the Brownian particle. The Stokes friction coefficient ζ_0 and the diffusion coefficient D_0 of an isolated Brownian particle are related through the Einstein relation $D_0 = k_B T/\zeta_0$, with T the temperature and k_B Boltzmann's constant.

For *short* times the dominant physical process is cage-diffusion: a quantitative interpretation of the short-time collective diffusion coefficient in terms of the characteristic escape rate of a particle out of the cage, formed around it by its nearest neighbors. The existence of the cage is directly related to the local ordering in the suspension as expressed in the first maximum of the static structure factor $S(k, \phi)$ at wavenumber $k = k^* \approx 2\pi/\sigma$, *i.e.*, an ordering of the particles on the length scale of their diameter σ with, typically, a fluid gap between the surfaces of two nearest neighbor particles of only $\approx \sigma/10$ at high volume fractions, $\phi > 0.4$. Here the volume fraction is defined as $\phi = \pi n \sigma^3/6$, with n the number density of the Brownian particles.

For *long* times the dominant process is a coupled relaxation mechanism as described by the mode-mode coupling theory, via two slowly decaying modes associated with conserved single-particle or collective dynamical variables. The lowest-order mode-mode coupling theory for the self-diffusion coefficient takes into account bilinear products of a self-diffusion mode and a cage-diffusion mode, while for the collective diffusion coefficient and the viscosity it takes into account bilinear products of two cage-diffusion modes. This coupled relaxation mechanism for the self-diffusion coefficient can be interpreted as a repetitive jumping from cage to cage. The diffusion inside a cage involves a self-diffusion mode, while the escape from one cage and entry into the next involves a cage-diffusion mode.

In this paper we focus on colloidal suspensions in which the Brownian particles behave effectively as hard-spheres. Examples of these are neutral suspensions like polymethylmethacrylate (PMMA) spheres or silica spheres in organic solvents and charged suspensions in which the added electrolyte concentration is so high that the Debye-sphere of counter-ions is collapsed to a very thin layer. Furthermore, we mainly consider monodisperse suspensions, in which all colloidal particles are of equal diameter. Only occasionally is the effect of polydispersity on the results discussed. Finally, we will not consider the shear-dependence of the viscosity. Our results for the viscosity are strictly valid only in the limit of vanishing shear-rate. Although there exists a vast amount of literature on the viscosity at finite shear rate and we certainly recognize the importance of this field, it is beyond the scope of this paper to discuss it even briefly.

In Section 2 a short summary of the basic theory is given, where we refer to the literature for a more extensive treatment. In Section 3 our expressions for the short-time transport coefficients are given and compared with experimental results. In Section 4 our theoretical results for the long-time wavenumber dependent self and collective diffusion coefficients and the Newtonian and the dynamic viscosity, obtained in the mode-mode coupling approximation, are presented and compared with experimental results on model hard-sphere suspensions. We conclude with a discussion of our results in Section 5.

2. Theoretical Background

2.1. DIFFUSION

The diffusive behavior in concentrated colloidal suspensions is connected to the decay of collective density fluctuations, described by the wavenumber k and time t dependent equilibrium intermediate scattering function $F(k, \phi, t)$, defined by

$$F(k, \phi, t) = \frac{1}{N} \left\langle \delta n(-\mathbf{k}) e^{\Omega t} \delta n(\mathbf{k}) \right\rangle, \tag{1}$$

where $\delta n(\mathbf{k})$ is a plane wave collective density fluctuation mode with wavevector $\mathbf{k}$,

$$\delta n(\mathbf{k}) = \sum_{i=1}^{N} e^{-i\mathbf{k}\cdot\mathbf{r}_i} - \left\langle \sum_{i=1}^{N} e^{-i\mathbf{k}\cdot\mathbf{r}_i} \right\rangle, \tag{2}$$

with $\mathbf{r}_i$ the position of particle i at time $t = 0$. The brackets $\langle\ \rangle$ denote the equilibrium ensemble average with the canonical distribution function and Ω is the adjoint Smoluchowski operator describing the time evolution of the system (*e.g.* [2, 3]),

$$\Omega = \sum_{i,j=1}^{N} [\nabla_i + \beta \mathbf{F}_i(\Gamma)] \cdot \mathbf{D}_{ij}(\Gamma) \cdot \nabla_j. \tag{3}$$

Here, $\nabla_i = \partial/\partial \mathbf{r}_i$, $\beta = 1/k_BT$, $\mathbf{F}_i(\Gamma) = -\nabla_i U(\Gamma)$ is the force on particle i due to direct interactions with all other particles in the configuration space $\Gamma = (\mathbf{r}_1, \cdots, \mathbf{r}_N)$ and $\mathbf{D}_{ij}(\Gamma)$ is the diffusion tensor which incorporates hydrodynamic interactions.

The corresponding definition of the equilibrium intermediate self scattering function $F_S(k, \phi, t)$ is

$$F_S(k, \phi, t) = \left\langle \delta n_1(-\mathbf{k}) e^{\Omega t} \delta n_1(\mathbf{k}) \right\rangle, \tag{4}$$

with $\delta n_1(\mathbf{k})$ a plane wave one-particle density fluctuation mode,

$$\delta n_1(\mathbf{k}) = e^{-i\mathbf{k}\cdot\mathbf{r}_1} - \left\langle e^{-i\mathbf{k}\cdot\mathbf{r}_1} \right\rangle. \tag{5}$$

Here, $\mathbf{r}_1$ is the tagged particle's position at time $t = 0$.

First, we consider the *short*-time *collective* and *self*-diffusion coefficients, $D^S(k, \phi)$ and $D^S_S(\phi)$, respectively. For short times, $\tau_B \sim t \ll \tau_P$, the equilibrium intermediate scattering function behaves as

$$F(k, \phi, t) = S(k, \phi) e^{-k^2 D^S(k,\phi) t}, \tag{6}$$

where $S(k, \phi) = F(k, \phi, t{=}0)$. From Eqs. (1) and (6) one obtains

$$D^S(k, \phi) = -\frac{\langle \delta n(-\mathbf{k}) \Omega \delta n(\mathbf{k}) \rangle}{N k^2 S(k, \phi)}, \tag{7}$$

or, using Eqs. (2) and (3),

$$D^S(k, \phi) = \frac{1}{N S(k, \phi)} \sum_{i,j=1}^{N} \left\langle \hat{\mathbf{k}} \cdot \mathbf{D}_{ij}(\Gamma) \cdot \hat{\mathbf{k}}\, e^{i\mathbf{k}\cdot(\mathbf{r}_i - \mathbf{r}_j)} \right\rangle, \tag{8}$$

with $\hat{\mathbf{k}} = \mathbf{k}/k$, so that $D^S(k, \phi)$ is expressed in terms of k-dependent averages over the diffusion tensor $\mathbf{D}_{ij}(\Gamma)$. The analogous expression for $F_S(k, \phi, t)$ is

$$F_S(k, \phi, t) = e^{-k^2 D^S_S(\phi) t}, \tag{9}$$

with the short-time self-diffusion coefficient given by (*cf.* Eqs. (3) to (5) and (9))

$$D^S_S(\phi) = -\frac{1}{k^2} \langle \delta n_1(-\mathbf{k}) \Omega \delta n_1(\mathbf{k}) \rangle = \left\langle \hat{\mathbf{k}} \cdot \mathbf{D}_{11}(\Gamma) \cdot \hat{\mathbf{k}} \right\rangle. \tag{10}$$

We remark that the averages in Eqs. (8) and (10) involve all particles 1 to N through the diffusion tensor $\mathbf{D}_{ij}(\Gamma)$. We give explicit expressions for $D^S(k,\phi)$ and $D_S^S(\phi)$ and compare them with experimental results in Section 3.

Next we consider the *long*-time, *i.e.*, $t \gg \tau_P$, *collective* diffusion coefficient $D^L(k,\phi)$, defined in terms of the Laplace transform $\tilde{F}(k,\phi,z) = \int_0^\infty dt \exp(-zt)F(k,\phi,t)$ of $F(k,\phi,t)$ for $z=0$, leading to (*e.g.* [2, 3])

$$D^L(k,\phi) = S(k,\phi)/\int_0^\infty dt k^2 F(k,\phi,t). \tag{11}$$

To develop a theory for $D^L(k,\phi)$ we start from Eqs. (1) and (11) and replace, like Brady [4], the diffusion tensor $\mathbf{D}_{ij}(\Gamma)$ in Eq. (3) by its mean-field average $\langle \mathbf{D}_{ij}(\Gamma) \rangle = \delta_{ij}\mathbf{1}D_S^S(\phi)$. Then, applying standard projection operator techniques to Eq. (1), and generalizing, for high concentrations, the procedures developed by Cichocki and Hess [5] for $\phi \to 0$, we obtain an expression for $D^L(k,\phi)$ for *all* ϕ [1, 7], *i.e.*,

$$D^L(k,\phi) = \frac{D_S^S(\phi)/S(k,\phi)}{1+\tilde{M}(k,\phi)/\left(k^2 D_S^S(\phi)\right)}, \tag{12}$$

with $\tilde{M}(k,\phi) = \int_0^\infty dt M(k,\phi,t)$. Here $\tilde{M}(k,\phi)$ is directly proportional to the generalized k-dependent longitudinal viscosity [5] and $M(k,\phi,t)$ is the longitudinal stress time-autocorrelation function,

$$M(k,\phi,t) = \frac{1}{N}\left(\beta k^2 D_S^S(\phi)\right)^2 \left\langle \sigma_{\parallel}(-\mathbf{k}) e^{\Omega_{MF}^{irr} t} \sigma_{\parallel}(\mathbf{k}) \right\rangle, \tag{13}$$

where the microscopic longitudinal stress is given by

$$\sigma_{\parallel}(\mathbf{k}) = -\sum_{i=1}^{N}\left[\frac{i\mathbf{k}\cdot\mathbf{F}_i}{k^2} + k_B T \frac{S(k,\phi)-1}{S(k,\phi)}\right] e^{-i\mathbf{k}\cdot\mathbf{r}_i}. \tag{14}$$

In Eq. (13), Ω_{MF}^{irr} is the mean-field expression for the irreducible adjoint Smoluchowski operator that generalizes, to high concentrations, the $\phi \to 0$ result of Cichocki and Hess [5] when use is made of the mean-field approximation for the diffusion tensor discussed above, *i.e*, by effectively replacing D_0 with $D_S^S(\phi)$.

The decay of $M(k,\phi,t)$ in Eq. (13) is determined in first approximation by the decay of *two* coupled collective density fluctuation modes only, since Ω_{MF}^{irr} keeps $\sigma_{\parallel}(\mathbf{k})$ orthogonal to all single collective density fluctuation modes when it evolves in time, hence the name mode-mode coupling approximation. This mode-mode coupling approximation to $M(k,\phi,t)$ follows straightforwardly by inserting in Eq. (13) a projection operator on bilinear products of two collective density fluctuation modes as described in [1, 2, 3, 6], with the final result

$$M(k,\phi,t) = \frac{1}{16\pi^3 n} \int d\mathbf{q} \left[\frac{V(\mathbf{k},\mathbf{q})}{S(q,\phi)S(|\mathbf{k}-\mathbf{q}|,\phi)}\right]^2 F(q,\phi,t)F(|\mathbf{k}-\mathbf{q}|,\phi,t). \tag{15}$$

Here the vertex function $V(\mathbf{k},\mathbf{q})$ represents the strength of the coupling between the microscopic longitudinal stress $\sigma_{||}(\mathbf{k})$ and the bilinear products $\delta n(\mathbf{q})\delta n(\mathbf{k}-\mathbf{q})$ and is given by

$$V(\mathbf{k},\mathbf{q}) = \frac{1}{N}\left(\beta k^2 D_S^S(\phi)\right)\left\langle \sigma_{||}(-\mathbf{k})\delta n(\mathbf{q})\delta n(\mathbf{k}-\mathbf{q})\right\rangle. \tag{16}$$

Using Eq. (14) for $\sigma_{||}(\mathbf{k})$ yields

$$V(\mathbf{k},\mathbf{q}) = D_S^S(\phi)\mathbf{k}\cdot\left[\frac{\mathbf{k}T(\mathbf{k},\mathbf{q})}{S(k,\phi)} - \mathbf{q}S(|\mathbf{k}-\mathbf{q}|,\phi) - (\mathbf{k}-\mathbf{q})S(q,\phi)\right], \tag{17}$$

where

$$T(\mathbf{k},\mathbf{q}) = \frac{1}{N}\langle \delta n(-\mathbf{k})\delta n(\mathbf{q})\delta n(\mathbf{k}-\mathbf{q})\rangle \tag{18}$$

is a wavenumber-dependent three-particle correlation function. One easily shows that $T(\mathbf{k},\mathbf{q}) = S(k,\phi)$ (k finite, q large), $T(\mathbf{k},\mathbf{q}) = S(q,\phi)$ (k large, q finite) and $T(\mathbf{k},\mathbf{q}) = S(|\mathbf{k}-\mathbf{q}|,\phi)$ (k large, q large and $|\mathbf{k}-\mathbf{q}|$ finite). For explicit calculations we use different approximations introduced in the literature, which satisfy these limiting properties. We then evaluate $M(k,\phi,t)$ of Eq. (15) for all times using the short-time expression (6) for $F(|\mathbf{k}-\mathbf{q}|,\phi,t)$ and $F(q,\phi,t)$. In Section 4.2, we compare $D^L(k,\phi)$ for different approximations of the three-particle correlation function with experimental results.

Finally, we consider the *long*-time, *i.e.*, $t \gg \tau_P$, *self*-diffusion coefficient $D_S^L(k,\phi)$, which can be obtained from the equilibrium intermediate self scattering function via

$$D_S^L(k,\phi) = 1/\int_0^\infty dt k^2 F_S(k,\phi,t). \tag{19}$$

Starting then from Eqs. (4) and (19), using the mean-field approximation for the diffusion tensor as discussed below Eq. (11), and proceeding as outlined above for $D^L(k,\phi)$ we obtain [1, 7]

$$D_S^L(k,\phi) = \frac{D_S^S(\phi)}{1+\tilde{M}_S(k,\phi)/\left(k^2 D_S^S(\phi)\right)}, \tag{20}$$

with $\tilde{M_S}(k,\phi) = \int_0^\infty dt M_S(k,\phi,t)$, where

$$M_S(k,\phi,t) = \frac{1}{8\pi^3 n} \int d\mathbf{q} \left[\frac{V_S(\mathbf{k},\mathbf{q})}{S(q,\phi)}\right]^2 F(q,\phi,t) F_S(|\mathbf{k}-\mathbf{q}|,\phi,t). \quad (21)$$

The self-vertex function $V_S(\mathbf{k},\mathbf{q})$ is defined as

$$V_S(\mathbf{k},\mathbf{q}) = D_S^S(\phi)(\mathbf{k}\cdot\mathbf{q})\,(S(q,\phi)-1)\,. \quad (22)$$

In section 4.1, we evaluate $M_S(k,\phi,t)$ for all times in terms of the short-time expressions for $F(q,\phi,t)$ and $F_S(|\mathbf{k}-\mathbf{q}|,\phi,t)$ of Eqs. (6) and (9) and compare the macroscopic self-diffusion coefficient, $D_S^L(\phi) = \lim_{k\to 0} D_S^L(k,\phi)$, with experimental results. We remark that the decay of $M_S(k,\phi,t)$ is determined in first approximation by the decay of one collective density fluctuation mode and one one-particle density fluctuation mode. Therefore, the projection operator for self-diffusion works on the bilinear product of cage-diffusion mode and a self-diffusion mode. This results in the appearance of the equilibrium intermediate self scattering function $F_S(|\mathbf{k}-\mathbf{q}|,\phi,t)$ in Eq. (21) (*cf.* Eq. (15)). Furthermore, we note that the lowest order mode-coupling approximation of the long-time self-diffusion coefficient does not involve the three-particle correlation function.

2.2. VISCOSITY

In previous publications we have shown that the long-time contribution to the viscosity can be obtained by directly solving the Smoluchowski equation for a colloidal suspension in the presence of an oscillating shear rate with vanishing amplitude [1, 8, 9]. We have shown that the final result is equivalent to the result obtained by applying the mode-mode coupling approximation to the orthogonal stress time-autocorrelation function [1, 8]. Here we summarize only the mode-mode coupling approach to show the analogy with the long-time self and collective diffusion coefficient. The dynamic viscosity $\eta(\phi,\omega)$ is obtained by adding the short and long-time contributions, *i.e.*,

$$\eta(\phi,\omega) = \eta_\infty(\phi) + \Delta\eta(\phi,\omega), \quad (23)$$

where $\eta_\infty(\phi)$ is the *short*-time or infinite frequency contribution, which contains corrections to the pure solvent viscosity η_0 due to hydrodynamic interactions between a Brownian particle and the solvent and between mutual Brownian particles via the solvent. The pure solvent viscosity η_0 is related directly to D_0 through the well-known Stokes-Einstein relation for an isolated spherical Brownian particle,

$$D_0 = \frac{k_B T}{3\pi\sigma\eta_0}, \quad (24)$$

where we have used that the friction coefficient ζ_0 of an isolated spherical Brownian particle is given by Stokes' formula $\zeta_0 = 3\phi\sigma\eta_0$.

The *long*-time contribution $\Delta\eta(\phi,\omega)$ is due to direct interactions between the Brownian particles and can be written in terms of the orthogonal stress time-autocorrelation function [1]

$$\Delta\eta(\phi,\omega) = \lim_{k\to 0} \frac{\beta}{V} \int_0^\infty dte^{i\omega t} \left\langle \sigma_\perp(-\mathbf{k}) e^{\Omega^{irr} t} \sigma_\perp(\mathbf{k}) \right\rangle, \tag{25}$$

where the microscopic orthogonal stress $\sigma_\perp(\mathbf{k})$ is given by

$$\sigma_\perp(\mathbf{k}) = -\sum_{i=1}^{N} \frac{ikF_{x,i}}{k^2} e^{-ikz_i} \tag{26}$$

with $\mathbf{k}$ chosen in the z-direction. Just as for the long-time diffusion, the irreducible Smoluchowski operator Ω^{irr} in Eq. (25) keeps $\sigma_\perp(\mathbf{k})$ orthogonal to all single collective density fluctuation modes when it evolves in time. Therefore, the contribution $\Delta\eta(\phi,\omega)$ to the dynamic viscosity can be evaluated in the mode-mode coupling approximation, where, just as for the long-time collective diffusion coefficient, the lowest-order mode-mode coupling approximation involves the projection on bilinear products of two collective density fluctuation modes, with the final result [1, 8]

$$\begin{aligned} \Delta\eta(\phi,\omega) &= \lim_{k\to 0} \frac{\beta}{16\pi^3} \int_0^\infty dte^{i\omega t} \int d\mathbf{q} \\ &\quad \left(\frac{V_\perp(\mathbf{k},\mathbf{q})}{S(q,\phi) S(|\mathbf{k}-\mathbf{q}|,\phi)} \right)^2 F(q,\phi,t) F(|\mathbf{k}-\mathbf{q}|,\phi,t). \end{aligned} \tag{27}$$

Here $V_\perp(\mathbf{k},\mathbf{q})$ represents the strength of the coupling between $\sigma_\perp(\mathbf{k})$ and the bilinear mode $\delta n(\mathbf{q})\delta n(\mathbf{k}-\mathbf{q})$ and is given by

$$\begin{aligned} V_\perp(\mathbf{k},\mathbf{q}) &= \frac{1}{N} \langle \sigma_\perp(-\mathbf{k}) \delta n(\mathbf{q}) \delta n(\mathbf{k}-\mathbf{q}) \rangle \\ &= \frac{q_x}{k\beta} \left(S(|\mathbf{k}-\mathbf{q}|,\phi) - S(q,\phi) \right). \end{aligned} \tag{28}$$

Using that, for $\mathbf{k}$ in the z-direction,

$$\lim_{k\to 0} \frac{1}{k} \left(S(|\mathbf{k}-\mathbf{q}|,\phi) - S(q,\phi) \right) = \frac{q_z}{q} \frac{\partial S(q,\phi)}{\partial q}, \tag{29}$$

the final result for $\Delta\eta(\phi,\omega)$ reads, after integration over the angular components of $\mathbf{q}$ (*cf.* Eqs. (27) to (29))

$$\Delta\eta(\phi,\omega) = \frac{k_B T}{60\pi^2} \int_0^\infty dte^{i\omega t} \int_0^\infty dq q^4 \left(\frac{S'(q,\phi)}{S(q,\phi)} \frac{F(q,\phi,t)}{S(q,\phi)} \right)^2, \tag{30}$$

with $S'(q,\phi) = \partial S(q,\phi)/\partial q$. We evaluated $\Delta\eta(\phi,\omega)$ for all times in terms of the short-time expression for $F(q,\phi,t)$ of Eq. (6) and compare the Newtonian viscosity, $\eta_N(\phi) = \eta(\phi,\omega = 0)$ and the dynamic viscosity $\eta(\phi,\omega)$ with experimental results in sections 4.3 and 4.4, respectively. We note that the decay of $\Delta\eta(\phi,\omega)$ is determined in first approximation by the decay of two cage-diffusion modes, just as for the long-time collective diffusion coefficient, but that the lowest order mode-mode coupling contribution to the viscosity does not involve the three-particle correlation function.

3. Short-Time Transport Coefficients

At short times, $\tau_B \sim t \ll \tau_P$, the dominant physical process involves the interaction of the Brownian particles with the fluid and with each other through the fluid, the so-called hydrodynamic interactions. These hydrodynamic interactions are complicated many-particle interactions for which exact evaluations exist only in limiting cases. To date there are only exact results for two isolated hard-sphere particles (a recent survey is presented by Dhont [3]). These results are valid up to second order in the volume fraction, thus clearly restricting the applicability to low volume fractions.

Approximate evaluations, based on series expansion of the hydrodynamic functions in powers of the inverse distance between the centers of the two particles were made by Beenakker and Mazur [10, 11]. They considered the hydrodynamical motion of one hard-sphere particle in a stationary field wherein all other particles are at rest, which, in effect, restricts the applicability to short-times. They obtained explicit expressions for the collective diffusion coefficient, the self-diffusion coefficient and the viscosity up to intermediate volume fractions $\phi \approx 0.4$. More recently, accurate numerical results were obtained for volume fractions up to $\phi = 0.45$, based on multipole expansion for the hydrodynamic interactions and lubrication theory for the singular forces near contact [12].

Up to now it seems that no *explicit* expressions are available in the literature for the transport properties of concentrated colloidal suspensions, *i.e.*, for high volume fractions, $\phi > 0.4$, and for which the complicated many-body hydrodynamic interactions are taken fully into account. Therefore, we propose to use a short-time collective diffusion coefficient $D^S(k,\phi)$, which is interpreted physically in terms of the characteristic cage-diffusion coefficient $D_C(k,\phi)$, in analogy with the result for concentrated molecular fluids from basic kinetic theory [13, 14, 15], *i.e.*,

$$D^S(k,\phi) \approx D_C(k,\phi) = \frac{D_S^S(\phi)d(k)}{S(k,\phi)}. \tag{31}$$

Here $d(k) = (1 - j_0(k\sigma) + 2j_2(k\sigma))^{-1}$, with $j_l(k\sigma)$ the spherical Bessel function of order l.

As presented, equation (31) is an adaptation of a very similar formula for the cage-diffusion coefficient in concentrated molecular fluids, where the short-time Enskog diffusion coefficient $D_E(\phi)$ is replaced by the short-time self-diffusion coefficient $D_S^S(\phi)$ of a colloidal suspension [16, 17]. Consistent with this analogy between $D_E(\phi)$ and $D_S^S(\phi)$, Cohen and de Schepper [18] proposed the (phenomenological) expression for the short-time self-diffusion coefficient of concentrated colloidal suspensions,

$$D_S^S(\phi) = \frac{D_0}{\chi(\phi)}, \tag{32}$$

where the low-density Boltzmann self-diffusion coefficient $D_B(\phi)$ is replaced consistently by its colloidal analogue, *i.e.*, the Stokes-Einstein diffusion coefficient D_0 of an isolated Brownian particle. Here $\chi(\phi)$, the equilibrium pair distribution function of two hard spheres at contact, is given in good approximation by the Carnahan-Starling expression

$$\chi(\phi) = \frac{1 - 0.5\phi}{(1-\phi)^3}. \tag{33}$$

From Eq. (32) an expression for the short-time viscosity $\eta_\infty(\phi)$ is deduced using a generalized Stokes-Einstein relation $D_S^S(\phi) = k_B T/3\pi\sigma\eta_\infty(\phi)$ (*cf.* Eq. (24)), so that

$$\eta_\infty(\phi) = \eta_0\chi(\phi). \tag{34}$$

This completes, for short times, the introduction of the transport properties in a colloidal suspension as expressed in $D^S(k,\phi)$, $D_S^S(\phi)$ and $\eta_\infty(\phi)$, which are explicitly known for all volume fractions ϕ.

Equations (31), (32) and (34) are compared with experimental results on neutral colloidal suspensions in Figs. 1, 2, and 3, respectively. Here the Carnahan-Starling approximation (33) is used and the static structure factor is calculated in the Percus-Yevick approximation, with a correction discussed by Henderson and Grundke [19]. The systematically lower result for $D^S(k,\phi)k^2\sigma^2/D_0$ as compared to the experimental data (*cf.* Fig. 1) is probably a result of the neglect of the polydispersity on the value of the maximum of the $S(k,\phi)$, since polydispersity lowers the first peak of the $S(k,\phi)$. The discrepancy can also be related to small inaccuracies in the determination of the effective hard-sphere diameter σ, as is illustrated in Fig. 2, bottom right. Here $D^S(k,\phi)k^2\sigma^2/D_0$ is calculated for two values of σ, corresponding to a surface layer of 9 nm and 14 nm [22]. The ensuing uncertainty in the determination of the volume fraction also causes difficulty in interpreting quantitatively the various experimental results [23, 24].

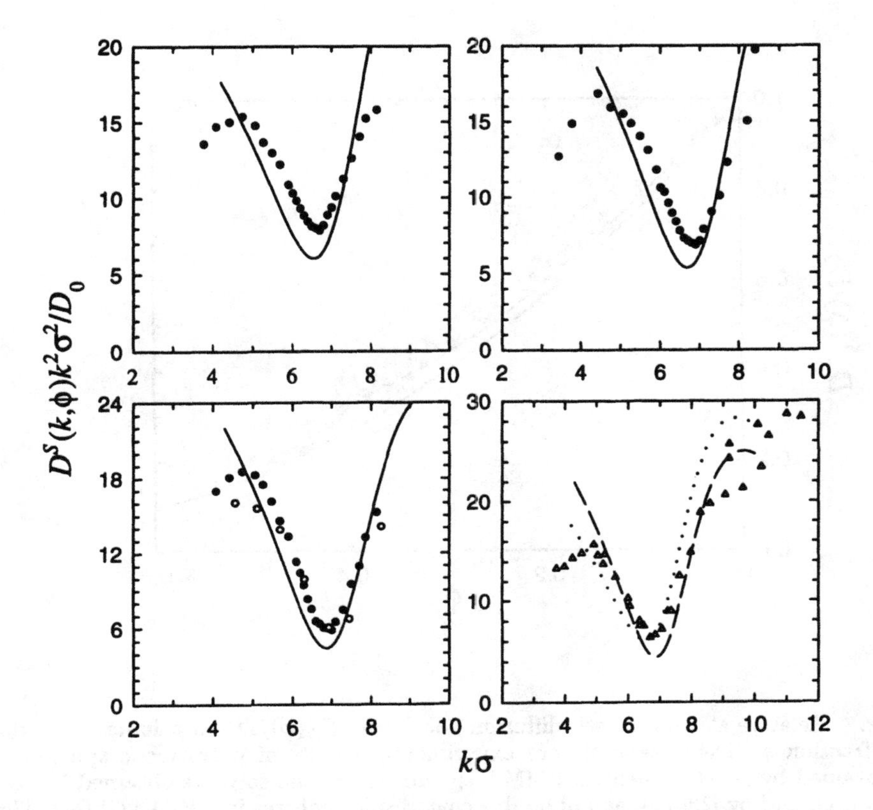

Figure 1. Reduced short-time collective diffusion coefficient $D^S(k,\phi)k^2\sigma^2/D_0$ as a function of $k\sigma$. The datapoints are experimental results for suspensions of neutral PMMA spheres obtained by [20] (•), [21] (○), and by [22] (△). The solid line represents Eq. (31) for the volume fractions $\phi = 0.443$ (top left), 0.465 (top right) and 0.494 (bottom left). In the bottom right, the dotted line corresponds to a surface layer of 9 nm ($\phi = 0.44$) and the dashed line to one of 14 nm ($\phi = 0.49$) [22].

In Figs. 2 and 3, the result of Beenakker and Mazur [10, 11], who obtained $D_S^S(\phi)$ and $\eta_\infty(\phi)$ for volume fractions up to $\phi = 0.45$, is also presented. Considering the small deviations between their result and that of Eqs. (32) and (34) and our primary interest in high volume fractions $\phi > 0.40$, equations (32) and (34) seem to be able to represent the experimental data over the entire fluid range, *i.e.*, for volume fractions $0 < \phi < 0.55$.

4. Long-Time Transport Coefficients

At long times, $t \gg \tau_P$, during and beyond which direct Brownian particle interactions take place, the dominant physical process is the increasing difficulty for a Brownian particle to diffuse out of the cage formed around

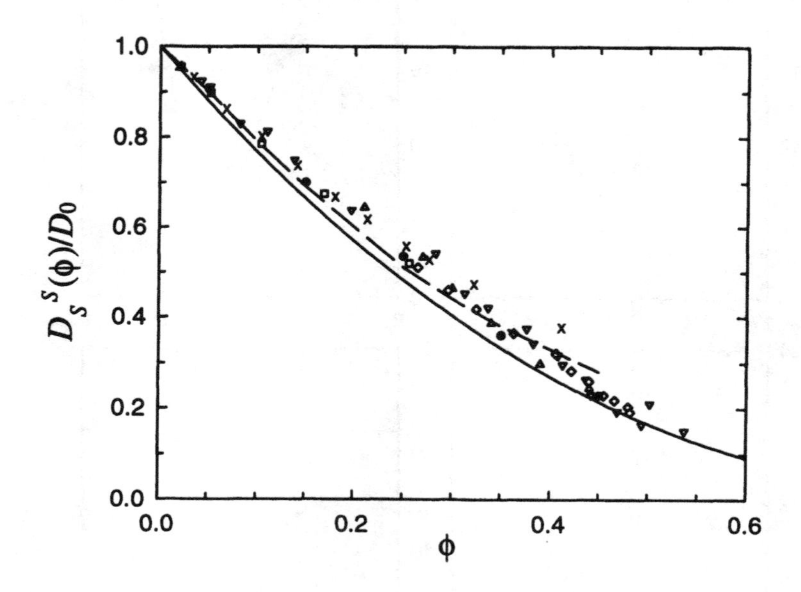

Figure 2. Relative short-time self-diffusion coefficient $D_S^S(\phi)/D_0$ as a function of the volume fraction ϕ. The datapoints are experimental results of polystyrene spheres in water obtained by [29] (□), neutral PMMA spheres in organic solvents obtained by [25] (▽), [26] (△), and by [20] (◇), and of double coated silica spheres in THFA [27] (×). The solid dots are computer simulation results [28]. The solid line represents Eq. (32) and the dashed line corresponds to the result of Beenakker and Mazur [11].

it by its neighbors, a process characterized by the cage-diffusion coefficient $D_C(k, \phi)$ (*cf.* Eq. (31)). Although hydrodynamical interactions are also present at this time-scale it seems that for high concentrations the increasing escape-difficulty is then the physical origin of the sharp increase (decrease) of the viscosity (diffusion) and that hydrodynamical interactions can be treated in good approximation in a mean-field-like manner.

In this section we compare our explicit expressions for the long-time transport properties in concentrated colloidal suspensions, obtained in Section 2, with experimental results. We consider the macroscopic self-diffusion coefficient $D_S^L(\phi) = \lim_{k\to 0} D_S^L(k, \phi)$, the long-time wavenumber dependent collective diffusion coefficient $D^L(k, \phi)$ and the dynamic, *i.e.*, frequency dependent, viscosity $\eta(\phi, \omega)$, which includes the Newtonian viscosity $\eta_N(\phi) = \eta(\phi, \omega = 0)$.

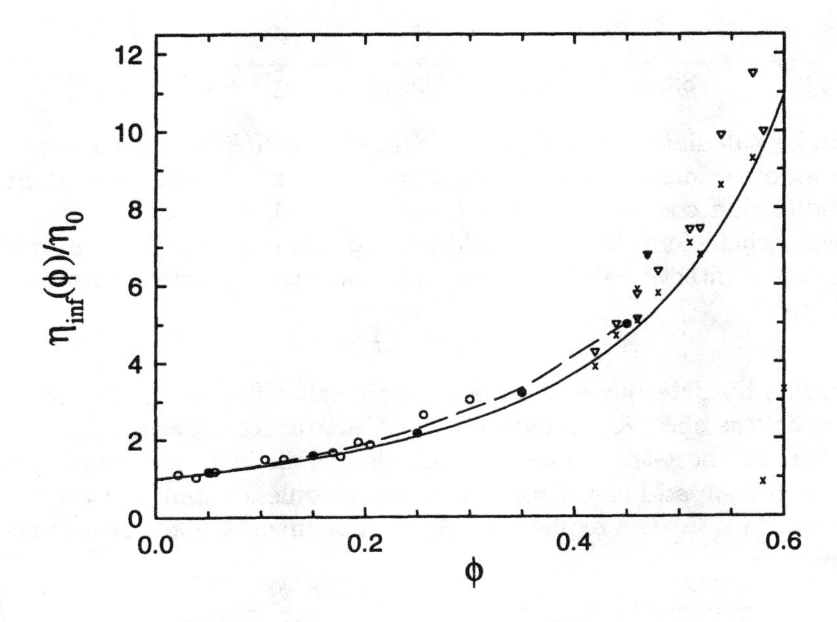

Figure 3. Relative infinite frequency viscosity $\eta_\infty(\phi)/\eta_0$ as a function of the volume fraction ϕ. The datapoints are experimental results obtained by Zhu *et al.* [29] (∘) and by van der Werff *et al.* [30] (▽) and computer simulation results obtained by Ladd [28] (•). The crosses are points obtained by Cichocki and Felderhof [31, 32] after an analysis of the data obtained by van der Werff *et al.* that is different from that of the authors. The solid line represents Eq. (34) and the dashed line the result of Beenakker [10].

4.1. LONG-TIME SELF-DIFFUSION COEFFICIENT

The mode-mode coupling contribution to the long-time self-diffusion coefficient is given in first approximation by the decay of the bilinear product of a cage-diffusion and a self-diffusion mode as discussed in Section 2.1. Restricting ourselves to lowest order mode-coupling theory implies that the decay of those two modes is determined in first approximation by the short-time expressions (6) for $F(q,\phi,t)$ and (9) for $F(|\mathbf{k}-\mathbf{q}|,\phi,t)$. This approximation has already been discussed in the application of the mode-mode coupling approximation to the diffusion and the viscosity of dense molecular fluids and appears to be quite good for volume fractions in the fluid range [33].

Then, the explicit result for the long-time wavenumber dependent self-diffusion coefficient $D_S^L(k,\phi)$ follows straightforwardly from Eqs. (20) to (22), *i.e.*,

$$\frac{D_S^S(\phi)}{D_S^L(k,\phi)} = 1 + \frac{D_S^S(\phi)}{8\pi^3 n}\int d\mathbf{q}\frac{\left[\hat{\mathbf{k}}\cdot\mathbf{q}\,(S(q,\phi)-1)\right]^2}{S(q,\phi)\,(D_S^S(\phi)|\mathbf{k}-\mathbf{q}|^2 + D^S(q,\phi)q^2)}, \quad (35)$$

which can be calculated once $D_S^S(\phi)$, $D^S(k,\phi)$ and $S(k,\phi)$ are known.

From an experimental point of view the long-time wavenumber dependent self-diffusion coefficient in the limit of $k = 0$, *i.e.*, the macroscopic self-diffusion coefficient $D_S^L(\phi) = D_S^L(k=0,\phi)$, is particularly important. This quantity is directly related to the long-time mean-square displacement according to

$$W(t) = 6D_S^L(\phi)t, \quad t \gg \tau_P, \quad (36)$$

and consequently determines the macroscopic self-diffusion coefficient of a tagged particle as observed in dynamic light scattering experiments. Using Eq. (31) for the short-time collective diffusion coefficients, performing the angular integration and changing to the dimensionless variable $x = q\sigma$, we obtain from Eq. (35) the explicit result for the macroscopic self-diffusion coefficient

$$\frac{D_S^S(\phi)}{D_S^L(\phi)} = 1 + \frac{1}{36\pi\phi}\int_0^\infty dx x^2 \frac{(S(x,\phi)-1)^2}{S(x,\phi)+d(x)}, \quad (37)$$

a result first published by Cohen and de Schepper [34]. In Fig. 4, this result is compared with experimental results of neutral colloidal suspensions obtained by several authors [25, 35, 36], where we used Eq. (32) for $D_S^S(\phi)$ and the same approximation for $S(k,\phi)$ as discussed above. Also given is the result obtained by Medina-Noyola [37], who solved approximately the generalized Langevin equation for the velocity of a tracer particle.

4.2. LONG-TIME COLLECTIVE DIFFUSION COEFFICIENT

The mode-coupling contribution to the long-time collective diffusion coefficient is given in first approximation by the decay of the bilinear product of two cage-diffusion modes as discussed in Section 2.1. Restricting ourselves again to lowest order mode-coupling theory, *i.e.*, using the short-time expression (6) for $F(q,\phi,t)$ and $F(|\mathbf{k}-\mathbf{q}|,\phi,t)$, we obtain an explicit result for the long-time wavenumber dependent collective diffusion coefficient $D^L(k,\phi)$,

$$\frac{D_S^S(\phi)/S(k,\phi)}{D^L(k,\phi)} = 1 + \frac{1}{16\pi^3 n k^2 D_S^S(\phi)}\int d\mathbf{q}\frac{[V(\mathbf{k},\mathbf{q})]^2/[S(q,\phi)S(|\mathbf{k}-\mathbf{q}|,\phi)]}{D^S(q,\phi)q^2 + D^S(|\mathbf{k}-\mathbf{q}|,\phi)|\mathbf{k}-\mathbf{q}|^2}, \quad (38)$$

with the vertex function $V(\mathbf{k},\mathbf{q})$ given by Eq. (17). The essential difference with expression (22) for the self-vertex function is the appearance of the

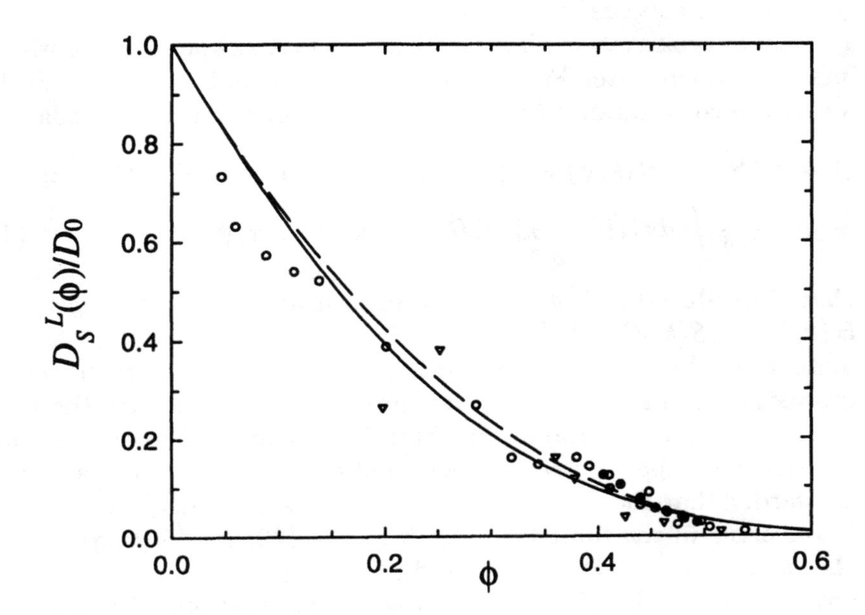

Figure 4. Relative long-time macroscopic self-diffusion coefficient $D_S^L(\phi)/D_0$ as a function of volume fraction ϕ. The datapoints are experimental results of neutral PMMA spheres in organic solvents by van Megen *et al.* [25] (∘) and by Segrè *et al.* [35] (•), and of neutral silica spheres in cyclohexane obtained by Kops-Werkhoven *et al.* [36] (▽). The solid line corresponds to the present theory (cf. Eq. (37)) and the dashed line to that of Medina-Noyola [37], who solved approximately the generalized Langevin equation for the velocity of a tracer particle.

poorly known three-particle correlation function $T(\mathbf{k}, \mathbf{q})$ in Eq. (17) for $V(\mathbf{k}, \mathbf{q})$. We have evaluated Eq. (38) using three approximations for $T(\mathbf{k}, \mathbf{q})$ obtained from the literature.

i) The first approximation, suggested by Hess and Klein [6], reads

$$T^{HK}(\mathbf{k}, \mathbf{q}) = S(k, \phi)\left(S(|\mathbf{k} - \mathbf{q}|, \phi) + S(q, \phi) - 1\right). \tag{39}$$

They used this approximation in a study of the collective diffusion coefficient of interacting charged colloidal suspensions, when hydrodynamic interactions are absent, and have calculated $\tilde{M}(k, \phi)$ of an overdamped one-component plasma for a Debye-Hückel potential.

ii) The second approximation, the so-called convolution approximation introduced by Jackson and Feenberg [38] in a study of elementary excitations in liquid helium, reads

$$T^{JF}(\mathbf{k}, \mathbf{q}) = S(k, \phi)S(|\mathbf{k} - \mathbf{q}|, \phi)S(q, \phi). \tag{40}$$

This approximation has recently been used by Baur *et al.* [39] in a study of charge-stabilized colloidal suspensions.

iii) The last approximation discussed here is the Kirkwood superposition approximation, which, after Fourier transformation and use of Parcival's theorem and the convolution property of the Fourier transform, reads

$$\begin{aligned} T^K(\mathbf{k},\mathbf{q}) &= n\left(S(q,\phi)H(k,\phi)+S(|\mathbf{k}-\mathbf{q}|,\phi)H(q,\phi)+S(k,\phi)H(|\mathbf{k}-\mathbf{q}|,\phi)\right) \\ &+ \frac{n^2}{8\pi^3}\int d\mathbf{x}H(|\mathbf{k}-\mathbf{x}|,\phi)H(|\mathbf{q}-\mathbf{x}|,\phi)H(x,\phi)+1. \end{aligned} \quad (41)$$

Here we have introduced $H(k,\phi)$, the Fourier transform of $h(r,\phi) = g(r,\phi) - 1$, *i.e.*, $H(k,\phi) = (S(k,\phi)-1)/n$.

We note that these three approximations do not exhaust the number of approximations found in the literature. However, they are the only closed results in **k**-space suitable for obtaining an explicit expression for $T(\mathbf{k},\mathbf{q})$ so that a comparison with experiment can be made. One can show straightforwardly that for all three approximations of $T(\mathbf{k},\mathbf{q})$, $V(\mathbf{k},\mathbf{q})$ of Eq. (17) converges in the limit for large k to Eq. (22) for $V_S(\mathbf{k},\mathbf{q})$. Thus, $\lim_{k\to\infty} D^L(k,\phi) = D_S^L(k,\phi)$, as expected physically.

We have evaluated Eq. (38) for the three approximations as a function of the wavenumber k for the volume fraction $\phi = 0.465$. Here we have used Eqs. (31) and (32) for $D^S(k,\phi)$ and $D_S^S(\phi)$, respectively. The results shown in Fig. 5 are compared to the recent experimental results of Segrè *et al.* [40]. We found that the Kirkwood approximation is unable to describe $D^L(k,\phi)$ for $k\sigma < 5$.

This is most probably due to the $S(k,\phi)$ in the first term on the right hand side of Eq. (17), since the $S(k,\phi)$ for hard spheres and high volume fractions is very small for a considerable range of k-values up to $k\sigma \approx 4$, resulting in a very high value of $V(\mathbf{k},\mathbf{q})$. This blow up at small-k is absent in the Hess and Klein approximation for the three-particle correlation function (*cf.* Eq. (39)), since it is linear in $S(k,\phi)$, thereby canceling the $S(k,\phi)$ in the first term on the right hand side of Eq. (17). However, this approximation is not symmetric in the mutual exchange of particles and therefore unphysical. Furthermore, it is approximately a factor 3 too large at intermediate values of k.

We found that the best results were obtained by using the convolution approximation of the triple correlation function. In Fig. 6, these results are given for three of the volume fractions for which experimental results were available. Also given in Fig. 6 is the result for the long-time wavenumber dependent self-diffusion coefficient (*cf.* Eq. (35)), around which $D^L(k,\phi)$ oscillates and to which it converges for large k. The intercept of $D_S^L(k,\phi)$ with the y-axis gives the macroscopic self-diffusion coefficient as presented in Fig. 4.

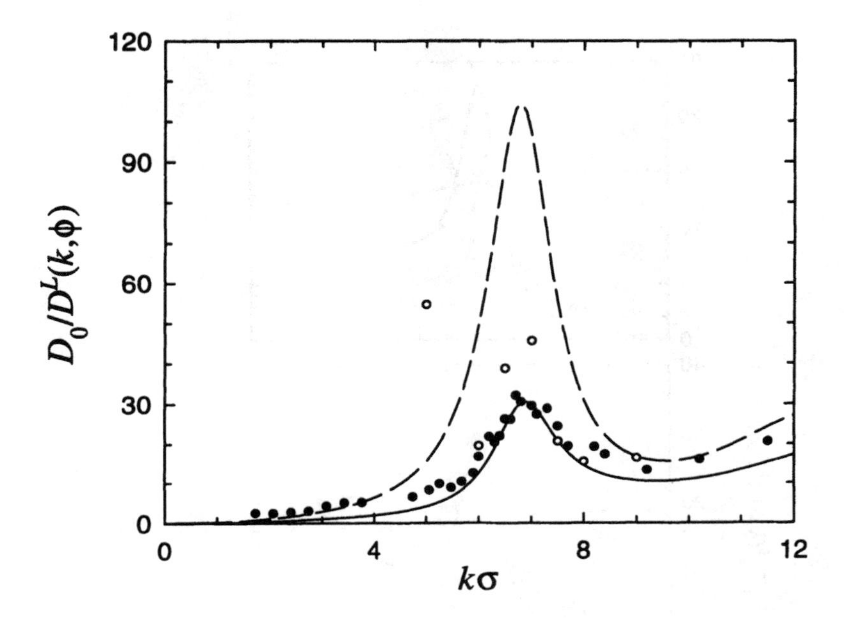

Figure 5. Inverse relative long-time wavenumber dependent collective diffusion coefficient $D_0/D^L(k,\phi)$ as a function of $k\sigma$ for various approximations of the three-particle correlation function at $\phi = 0.465$. The solid circles are experimental results of neutral PMMA spheres in an organic solvent [40]. The solid line corresponds to the present theory (cf. Eqs. (38) and (17)) in the convolution approximation (40), the dashed line to that in the Hess and Klein approximation (39) and the open circles to that in the Kirkwood approximation (41).

4.3. NEWTONIAN VISCOSITY

Setting $\omega = 0$ in Eqs. (23) and (30) and using Eqs. (6) and (31) one obtains a simple explicit expression for the Newtonian viscosity $\eta_N(\phi)$, *i.e.*,

$$\eta_N(\phi) = \eta_\infty(\phi) + \frac{\eta_0}{40\pi}\frac{D_0}{D_S^S(\phi)}\int_0^\infty dx x^2 \frac{[S'(x,\phi)]^2}{S(x,\phi)d(x)}, \tag{42}$$

where $x = q\sigma$, and the Stokes-Einstein relation (24) is used for D_0.

Although Eq. (42) has been derived for large ϕ ($0.3 < \phi < 0.55$), where cage-diffusion is the dominant finite-time contribution to the viscosity, it nevertheless appears to describe the ϕ-dependence of $\eta_N(\phi)$ for small and intermediate concentrations as well, due to the presence of the $\eta_\infty(\phi)$ term (*cf.* Fig. 7). Figure 7 also shows that the cage-diffusion describes well the very rapid increase of $\eta_N(\phi)$ with ϕ for $0.40 < \phi < 0.55$. Equation (42) has been evaluated using the short-time expressions (32) and (34) introduced in Section 3, the Carnahan-Starling approximation (33) for $\chi(\phi)$ and the

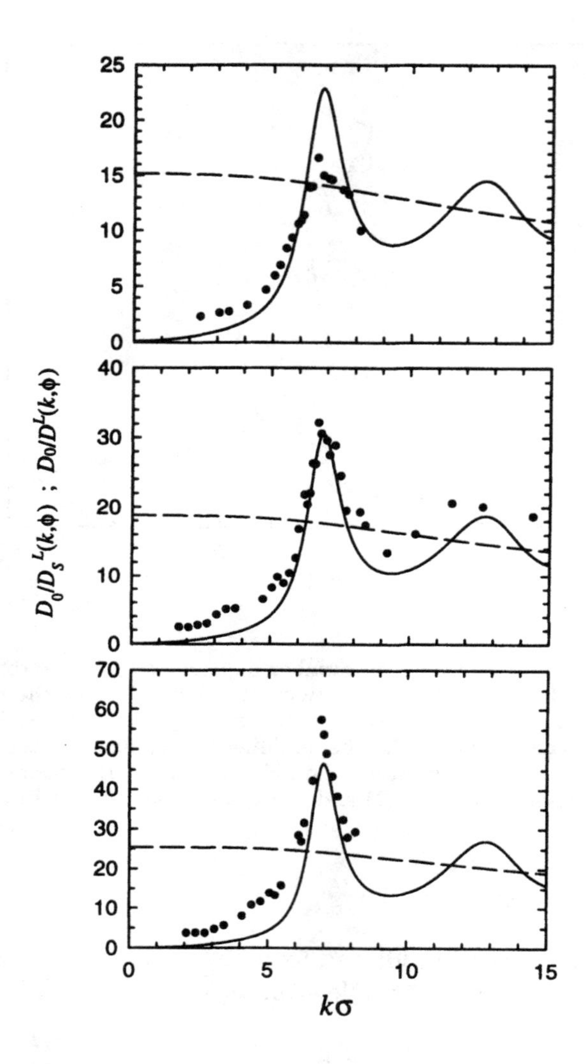

Figure 6. Inverse relative long-time wavenumber dependent collective diffusion coefficient $D_0/D^L(k,\phi)$ as a function of $k\sigma$ for three different volume fractions $\phi = 0.443$ (top), 0.465 (center) and 0.494 (bottom). The solid circles are experimental results of neutral PMMA spheres in an organic solvent [40]. The solid line represents the theoretical result in the convolution approximation (cf. Eqs. (38), (17) and (40)). The dashed line corresponds to the inverse relative long-time wavenumber dependent self-diffusion coefficient $D_0/D_S^L(k,\phi)$ (cf. Eq. (35)).

Henderson-Grundke correction [19] to the Percus-Yevick equation for the computation of the hard-sphere $S(x,\phi)$ and $S'(x,\phi)$. A convenient Padé

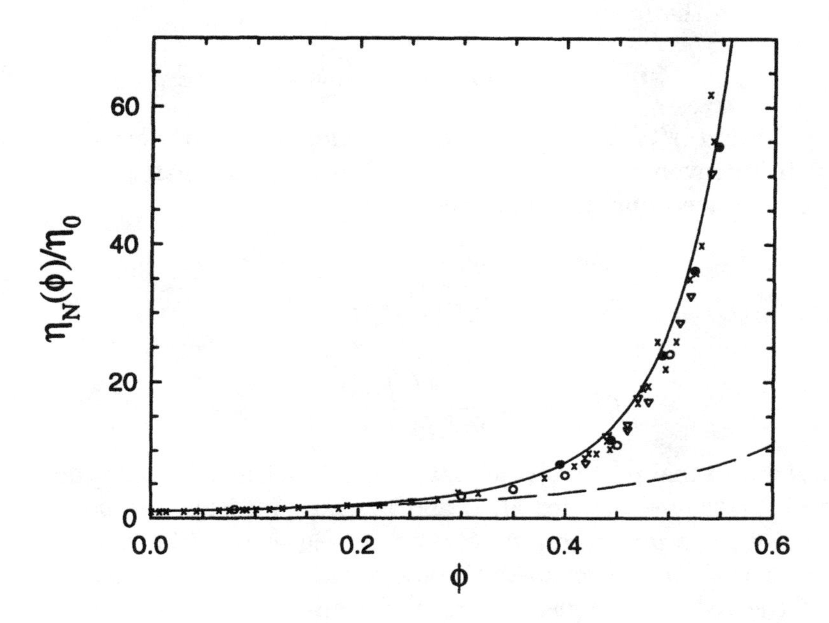

Figure 7. Relative Newtonian viscosity $\eta_N(\phi)/\eta_0$ as a function of the volume fraction ϕ. The datapoints are experimental results obtained by van der Werff and de Kruif [41] ($\times$), by van der Werff *et al.* [30] (Δ), by Jones *et al.* [42, 43] ($\bullet$) and by Papir and Krieger [44] ($\square$). The solid line represents Eq. (42) and the dashed line corresponds to $\eta_\infty(\phi)/\eta_0 = \chi(\phi)$ (cf. Eq. (34)).

approximation of $\eta_N(\phi)$ for practical use for all $0 < \phi < 0.55$ is given by

$$\eta_N(\phi) = \eta_0\chi(\phi)\left(1 + \frac{1.44\,(\phi\chi(\phi))^2}{1 - 0.1241\phi + 10.46\phi^2}\right) \tag{43}$$

with a relative accuracy that is better than 0.25%. This approximation yields for $\eta_N(\phi)$ the correct Einstein coefficient $\frac{5}{2}\phi$ as well as the same coefficient of $O(\phi^2)$ as Eq. (42).

4.4. DYNAMIC VISCOSITY

Performing the integration in time and using Eq. (6) for $F(q, \phi, t)$ one obtains from Eqs. (23) and (30) the final expression for the dynamic viscosity

$$\eta(\phi, \omega) = \eta_\infty(\phi) + \frac{k_B T}{60\pi^2}\int_0^\infty dq q^4 \left(\frac{S'(q,\phi)}{S(q,\phi)}\right)^2 \frac{1}{2q^2 D^S(k,\phi) - i\omega}, \tag{44}$$

where $D^S(k, \phi)$ and $\eta_\infty(\phi)$ are given by Eqs. (31) and (34), respectively. For $\omega \neq 0$, the dynamic viscosity $\eta(\phi, \omega)$ of Eq. (44) is complex, so that it can be written in the form

$$\eta(\phi,\omega) = \eta'(\phi,\omega) + i\eta''(\phi,\omega), \tag{45}$$

where $\eta'(\phi,\omega)$ and $\eta''(\phi,\omega)$ are the real and imaginary parts of $\eta(\phi,\omega)$, respectively. It is convenient and customary [30] to consider instead of $\eta'(\phi,\omega)$ and $\eta''(\phi,\omega)$ reduced quantities defined by:

$$\eta_R^*(\phi,\omega) = \frac{\eta'(\phi,\omega) - \eta_\infty(\phi)}{\eta_N(\phi) - \eta_\infty(\phi)} \tag{46}$$

and

$$\eta_I^*(\phi,\omega) = \frac{\eta''(\phi,\omega)}{\eta_N(\phi) - \eta_\infty(\phi)}, \tag{47}$$

where the reduced real part $\eta_R^*(\phi,\omega)$ varies as a function of ω between 1 (for $\omega \to 0$) and 0 (for $\omega \to \infty$) for all ϕ and $\eta_I^*(\phi,\omega)$ vanishes for $\omega \to 0$ and $\omega \to \infty$, exhibiting a maximum in between. In Fig. 8, $\eta_R^*(\phi,\omega)$ and $\eta_I^*(\phi,\omega)$ are compared with the experimental data of van der Werff *et al.* [30] as a function of the reduced frequency $\omega\tau_1(\phi)$ for all ϕ: $0.44 \le \phi \le 0.57$. Here the scaling of ω for the experimental data was performed in the same way as was done by van der Werff *et al.* [30] by fitting the data for large ω with the expression

$$\eta_R^*(\phi,\omega) = \eta_I^*(\phi,\omega) = \frac{3\sqrt{2}}{2\pi}\frac{1}{\sqrt{\omega\tau_1(\phi)}}, \tag{48}$$

where $\tau_1(\phi)$ is a phenomenological time.

Although the large experimental uncertainties in the data and the difficulty to obtain a reliable volume fraction from the experimental results complicate considerably a compelling detailed comparison of theory and experiment, a more detailed comparison of $\eta_{R,I}^*(\phi,\omega)$ as a function of ϕ can still be made. Examples are given in Fig. 9. In the same figure, the results are given of a general phenomenological description of the dynamic viscosity of colloidal suspensions due to Cichocki and Felderhof [31, 32]. Their description is based on a three-pole approximation in the complex $\sqrt{\omega}$-plane, where the locations of the poles are derived from the experimentally measured values $\eta_N^{exp}(\phi)$, $\eta_\infty^{exp}(\phi)$ and three additional parameters, one of which is a relaxation time. For the three concentrations $\phi = 0.44$, 0.46 and 0.53 for which their procedure could be implemented, $\eta'(\phi,\omega)$ and $\eta''(\phi,\omega)$ are consistent with our results within the experimental errors. As was shown by Cichocki and Felderhof, the strongly deviating cloud of points near $\omega\tau_1(\phi) \approx 1$ in the imaginary part of the reduced viscosity $\eta_I^*(\phi,\omega)$ (*cf.* Fig. 8 bottom) can be discarded, since they violate the Kramers-Kronig

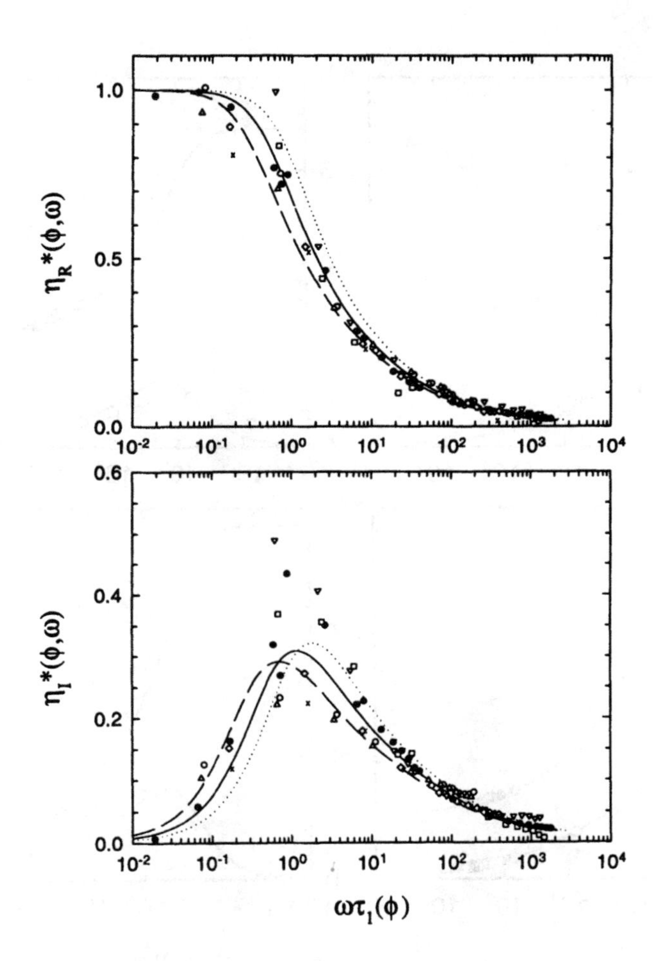

Figure 8. Real and imaginary part of the reduced viscosities $\eta_R^*(\phi,\omega)$ and $\eta_I^*(\phi,\omega)$, respectively, as a function of $\omega\tau_1(\phi)$. The data points are experimental results obtained by van der Werff *et al.* [30] for volume fractions $\phi = 0.44$ (o), 0.46 (⊙), 0.47 (□), 0.48 (△), 0.51 (▽), 0.52 (◇), 0.54 (×), and 0.57 (⊕). The lines represent the theoretical result (cf. Eqs. (44) to (47)) for volume fractions $\phi = 0.45$ (dotted line), 0.50 (solid line) and 0.55 (dashed line). The cloud of points (bottom) near $\omega\tau_1(\phi) = 1$ should be discarded since they do not satisfy the Kramers-Kronig relation [31, 32].

relations between the real and the imaginary part of $\eta(\phi,\omega)$ and must therefore be erroneous [31, 32].

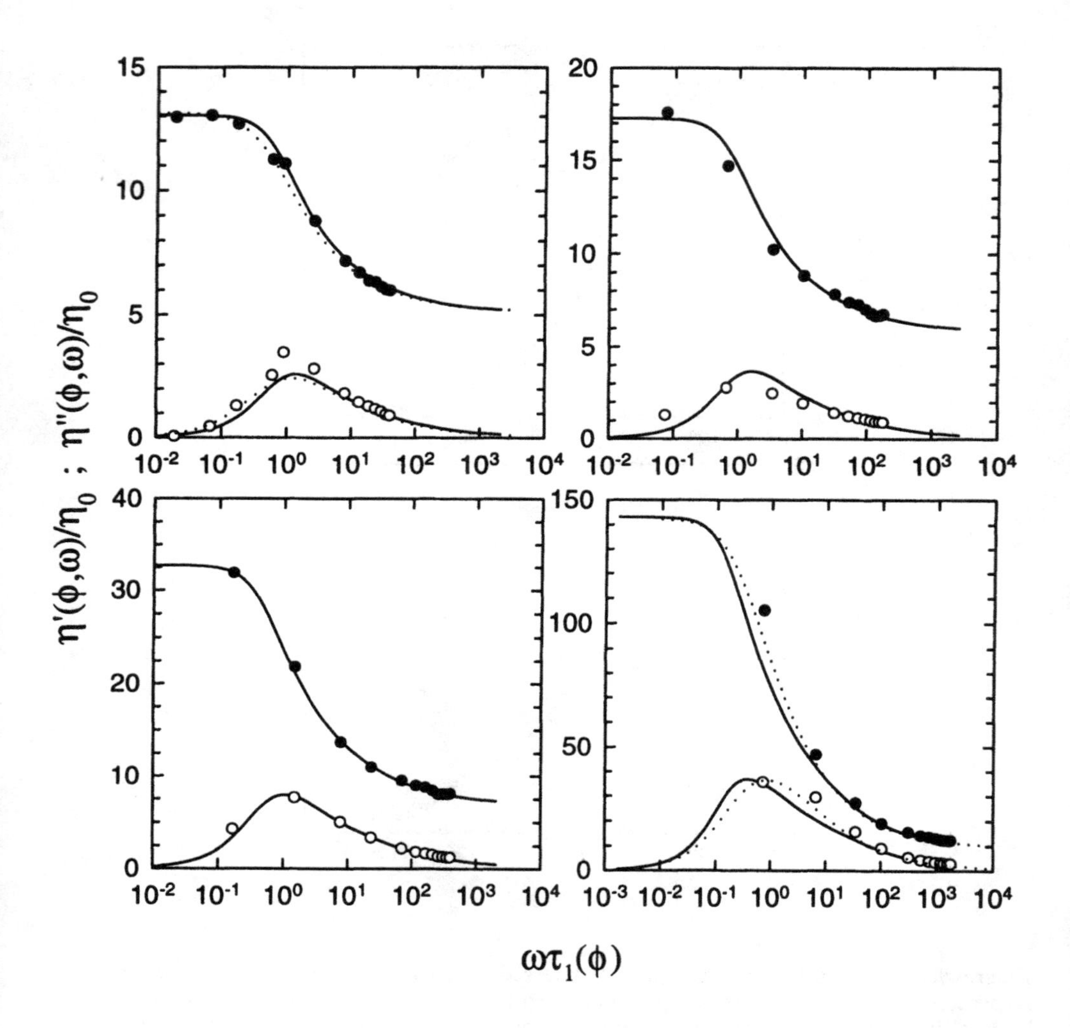

Figure 9. Relative real and imaginary parts of the dynamic viscosity, $\eta'(\phi,\omega)/\eta_0$ (•) and $\eta''(\phi,\omega)/\eta_0$ (∘), respectively, as a function of $\omega\tau_1(\phi)$, for four suspensions studied experimentally by van der Werff *et al.* [30], i.e., for $\phi = 0.46$ (top left), 0.48 (top right), 0.52 (bottom left), and 0.57 (bottom right). In order to make a fair and realistic comparison of the theory with experiment, keeping in mind the uncertainty in the determination of ϕ and the extreme sensitivity of the denominator of $\eta^*_{R,I}(\phi,\omega)$ on ϕ, as already pointed out by van der Werff *et al.* [30], an effective volume fraction ϕ^* is assigned to the experimental data, such that $(\eta_N^{theory}(\phi^*) - \eta_\infty^{theory}(\phi^*)) \equiv (\eta_N^{exp}(\phi) - \eta_\infty^{exp}(\phi))$, within the experimental uncertainty of ϕ. The solid line then represents the theoretical result from Eqs. (44) and (45) using $\phi = \phi^*$ [8]. The dotted lines correspond to the phenomenological results obtained by Cichocki and Felderhof [31, 32], which are only available for $\phi = 0.46$ and 0.57.

5. Discussion

In this paper we have presented a theoretical description of the viscosity and the diffusion of concentrated hard-sphere-like colloidal suspensions. We have shown that the dynamic viscosity, the Newtonian viscosity, the self-diffusion coefficient and the collective diffusion coefficient of concentrated colloidal suspensions can be described from one unifying viewpoint: the mode-coupling theory for long times and the microscopic cage-diffusion process for short times. We conclude with a number of remarks.

i) Figure 6 clearly shows that the theory predicts for $k\sigma < 5$ a larger value of $D^L(k,\phi)$ than was found experimentally, for all volume fractions. This deviation is most probably caused by the polydispersity of the PMMA particles as discussed more extensively in refs. [1] and [7]. In these references we have studied the result of a small polydispersity on the long-time wavenumber dependent collective diffusion coefficient $D^L(k,\phi)$ for a suspension of hard-sphere particles with a narrow size distribution.

ii) We have computed time-correlation functions $M(k,\phi,t)$, Eq. (15), and $M_S(k,\phi,t)$, Eq. (21), as well as the final expression (30) for $\Delta\eta(\phi,\omega)$ for *all* times using the short-time expressions (6) and (9) for $F(k,\phi,t)$ and $F_S(k,\phi,t)$, respectively. This inconsistency can be removed by using a self-consistent theory, where the behavior of $F(k,\phi,t)$ and $F_S(k,\phi,t)$ is consistently modified in concordance with the computation of the long-time transport coefficients $D^L(k,\phi)$, $D^L_S(k,\phi)$ and $\eta(\phi,\omega)$. This is done in calculations of $D^L(k=0,\phi)$ for undercooled fluids and the glass transition [45], but not needed in the normal fluid state considered here.

iii) Due to the high degree of order at very high volume fractions, $\phi > 0.55$, the Carnahan-Starling approximation (33), used to calculate expressions (32) and (34) for $D^S_S(\phi)$ and $\eta_\infty(\phi)$, respectively, is no longer adequate. However, it has been shown that $\eta_N(\phi)$ of Eq. (42) can still be used if, like Brady [4], an empirical expression is used for $\chi(\phi)$ with one pole at the volume fraction $\phi_p = \phi_{rcp} \approx 0.63$, incorporating the approach to a random close packed ordering at very high volume fractions $\phi > 0.55$ [46]. It has been shown [46] that this procedure can also be used to describe the Newtonian viscosity of the more ordered PMMA suspensions, with a polydispersity less then 5%, studied by Segrè *et al.* [35] and the highly ordered charged suspensions, with a high electrolyte concentration, studied by Imhof *et al.* [47].

iv) Our results for the long-time transport coefficients $D^L(k,\phi)$, $D^L_S(k,\phi)$ and $\eta(\phi,\omega)$ are based exclusively on the short-time transport coefficients $D^S(k,\phi)$, $D^S_S(\phi)$ and $\eta_\infty(\phi)$ and the cage-diffusion relaxation mechanism in the mode-mode coupling approximation, with the latter being almost solely responsible for the sharp increase of $\eta(\phi,\omega)$ and $\eta_N(\phi)$ and the sharp decrease of $D^L(k,\phi)$ and $D_S^L(k,\phi)$ for $\phi > 0.4$. From the agreement with experiment, it would seem that these two physical processes essentially suffice to understand the diffusion coefficient and the viscosity in the entire fluid range of hard-sphere-like colloidal suspensions. That this agreement occurs with consideration given only to the hydrodynamic interactions in a mean-field-like manner through $D_S^S(\phi)$ is rather puzzling. However, two observations in this respect are relevant. i) It is physically reasonable to assume that at high concentrations direct interactions are the main reason for the slowing down of the diffusion process and that hydrodynamic interactions in concentrated colloidal suspensions are suppressed or shielded to a great extent by the presence of many very close-by particles. Therefore, neglecting the complicated many-body effects and approximating the diffusion tensor by its mean-field average as discussed below Eq. (12) is a reasonable first approximation. ii) The theoretical results for $D^L(k,\phi)$, $D_S^L(k,\phi)$, $\eta(\phi,\omega)$ and $\eta_N(\phi)$ are given explicitly in terms of the short-time transport coefficients $D^S(k,\phi)$, $D_S^S(\phi)$ and $\eta_\infty(\phi)$ and the static structure factor $S(k,\phi)$. This impiies that any set of expressions for the short-time transport coefficients that describes the experimental results can be used to validate our expressions for the long-time transport coefficients. We used the analogy with molecular fluids to obtain explicit expressions for $D^S(k,\phi)$, $D_S^S(\phi)$ and $\eta_\infty(\phi)$ at high concentrations, which compare well with experimental results without specifically incorporating hydrodynamic interactions. It would be interesting to see how well our expressions for $D^L(k,\phi)$, $D_S^L(k,\phi)$, $\eta(\phi,\omega)$ and $\eta_N(\phi)$ compare with experiment if exact hydrodynamical expressions for $D^S(k,\phi)$, $D_S^S(\phi)$ and $\eta_\infty(\phi)$ were to be used.

v) Finally we note that recently Liu *et al.* [48, 49] have successfully adapted the Newtonian viscosity equation (42) to charged elliptical micelles and to neutral tri-block copolymer micelles. This indicates that the physics contained in our formalism is applicable to a wider class of suspensions than considered here.

Acknowledgement

R.V. gratefully acknowledges financial support from the Netherlands Foundation for Fundamental Research of Matter (Stichting FOM) and E.G.D.C. from the Engineering Research Program of the Office of Basic Energy Sciences of the Department of Energy under grant no. DE-FG02-88-ER13847.

References

1. R.Verberg, *Transport Properties in Concentrated Colloidal Suspensions*, Ph.D. thesis, (Delft University of Technology, 1998).
2. P.N. Pusey and R.J.A. Tough, Particle Interactions, in *Dynamic Light Scattering*,

R. Pecora, ed., (Plenum Press, New York, 1985).
3. J.K.G. Dhont, *An Introduction to Dynamics of Colloids*, (Elsevier, Amsterdam, 1996).
4. J.F. Brady, J. Chem. Phys. **99**, 567 (1993).
5. B. Cichocki and W. Hess, Physica A **141**, 475 (1987).
6. W. Hess and R. Klein, Adv. Phys. **32**, 173 (1983).
7. R. Verberg, I.M. de Schepper, and E.G.D. Cohen, Europhys. Lett. **48**, 397 (1999), and Phys. Rev. E, to appear (2000).
8. R. Verberg, E.G.D. Cohen, and I.M. de Schepper, Phys. Rev. E **55**, 3143 (1997).
9. I.M. de Schepper, H.E. Smorenburg, and E.G.D. Cohen, Phys. Rev. Lett. **70**, 2178 (1993).
10. C.W.J. Beenakker, Physica A **128**, 48 (1984).
11. C.W.J. Beenakker and P. Mazur, Physica A **126**, 349 (1984).
12. A.J.C. Ladd, J. Chem. Phys. **93**, 3484 (1990).
13. I.M. de Schepper, E.G.D. Cohen, and M.J. Zuilhof, Phys. Lett. A **101**, 399 (1984).
14. E.G.D. Cohen, I.M. de Schepper, and M.J. Zuilhof, Physica B **127**, 282 (1984).
15. E.G.D. Cohen, P. Westerhuijs, and I.M. de Schepper, Phys. Rev. Lett. **59**, 2872 (1987).
16. I.M. de Schepper, E.G.D. Cohen, P.N. Pusey, and H.N.W. Lekkerkerker, J. Phys.: Cond. Matter **1**, 6503 (1989).
17. P.N. Pusey, H.N.W. Lekkerkerker, E.G.D. Cohen, and I.M. de Schepper, Physica A **164**, 12 (1990).
18. E.G.D. Cohen and I.M. de Schepper, Phys. Rev. Lett. **75**, 2252 (1995).
19. D. Henderson and E.W. Grundke, J. Chem. Phys. **63**, 601 (1975).
20. P.N. Segrè, O.P. Behrend, and P.N. Pusey, Phys. Rev. E **52**, 5070 (1995).
21. P.N. Pusey and W. van Megen, Phys. Rev. Lett. **59**, 2083 (1987).
22. W. van Megen, R.H. Ottewil, S.M. Owens, and P.N. Pusey, J. Chem. Phys. **82**, 508 (1985).
23. I.M. de Schepper, E.G.D. Cohen, and R. Verberg, Phys. Rev. Lett. **77**, 584 (1996).
24. P.N. Segrè, S. P. Meeker, P.N. Pusey, and W.C.K. Poon, Phys. Rev. Lett. **77**, 585 (1996).
25. W. van Megen, S.M. Underwood, R.H. Ottewil, N.St.J. Williams, and P.N. Pusey, Far. Disc. Chem. Soc. **83**, 47 (1987).
26. P.N. Pusey and W. van Megen, J. Phys. **44**, 285 (1983).
27. A. van Veluwen, H.N.W. Lekkerkerker, C.G. de Kruif, and A. Vrij, J. Chem. Phys. **87**, 4873 (1987).
28. A.J.C. Ladd, Phys. Rev. Lett. **70**, 1339 (1993).
29. J.X. Zhu, D.J. Durian, J. Müller, D.A. Weitz, and D.J. Pine, Phys. Rev. Lett. **68**, 2559 (1992).
30. J.C. van der Werff, C.B. de Kruif, C. Blom, and J. Mellema, Phys. Rev. A **39**, 795 (1989).
31. B. Cichocki and B.U. Felderhof, Phys. Rev. A **46**, 7723 (1992).
32. B. Cichocki and B.U. Felderhof, J. Chem. Phys. **101**, 7850 (1994).
33. I.M. de Schepper, R. Verberg, and E.G.D. Cohen, Mol. Phys. **95**, 595 (1998).
34. E.G.D. Cohen and I.M. de Schepper, J. Stat. Phys. **63**, 241 (1991).
35. P.N. Segrè, S.P. Meeker, P.N. Pusey, and W.C.K. Poon, Phys. Rev. Lett. **75**, 958 (1995).
36. M.M. Kops-Werkhoven and H. M. Fijnaut, J. Chem. Phys. **77**, 2242 (1982).
37. M. Medina-Noyola, Phys. Rev. Lett. **60**, 2705 (1988).
38. H.W. Jackson and E. Feenberg, Rev. Mod. Phys. **34**, 686 (1962).
39. P. Baur, G. Nägele, and R. Klein, Phys. Rev. E **53**, 6224 (1996).
40. P.N. Segrè and P. N. Pusey, Phys. Rev. Lett. **77**, 771 (1996).
41. J.C. van der Werff and C.B. de Kruif, J. Rheol. **33**, 421 (1989).
42. D.A.R. Jones, B. Leary, and D.V. Boger, J. Colloid Interface Sci. **147**, 479 (1991).
43. D.A.R. Jones, B. Leary, and D.V. Boger, J. Colloid Interface Sci. **150**, 84 (1992).

44. Y.S. Papir and I.M Krieger, J. Colloid Interface Sci. **34**, 126 (1970).
45. W. Götze, Aspect of Structural Glass Transitions, in *Liquids, Freezing and the Glass Transition*, J.P. Hansen, D. Levesque, and J. Zinn-Justin, eds., (North-Holland, Amsterdam, 1991), p. 287.
46. E.G.D. Cohen, R. Verberg, and I.M. de Schepper, Physica A **251**, 251 (1998).
47. A. Imhof, A. van Blaaderen, G. Maret, J. Mellema, and J.K.G. Dhont, J. Chem. Phys. **100**, 2170 (1994).
48. Y.C. Liu and E.Y. Shue, E. Y., Phys. Rev. Lett. **76**, 700 (1996).
49. Y.C. Liu, S.H. Chen, and J.S. Huang, Phys. Rev. E **54**, 1698 (1996).

INTERACTING BROWNIAN PARTICLES

B. CICHOCKI

Institute of Theoretical Physics, Warsaw University
ul. Hoża 69, 00-681 Warsaw, Poland

1. Introduction

It is well known that colloidal suspensions are of great interest for their technological applications and in order to understand their properties it is necessary to use a statistical description [1, 2]. The latter is considered a difficult task. However, practical applications should not be the only reason for such an interest. As has already been pointed out, many phenomena of interest for statistical physics which are unobservable in atomic systems can be observed in colloidal suspensions due to much longer time and length scales [3]. That is an important experimental aspect of suspensions. There is also an equally important theoretical one which we would like to discuss in this short paper.

A correct model of colloidal dispersion, in the time scale characteristic of light scattering experiments or of macroscopic measurements, is a system of particles performing Brownian motion. If the dispersion is not very dilute, interactions between those particles must be taken into account. Application of the tools of kinetic theory to understand the relaxation processes in colloids is usually regarded as a very complicated problem, much more complicated than in the case of atomic fluids. Here we will try to show that the opposite is true and that the interacting Brownian particle systems are very convenient for tests of modern kinetic theory predictions. In various cases, corresponding expressions and calculations are much simpler for suspensions than for gases or liquids. To support this point of view, we will first introduce basic elements of statistical mechanics for the Brownian particle systems and next discuss results for the scattering function derived from the Enskog-type theory.

J. Karkheck (ed.),
Dynamics: Models and Kinetic Methods for Non-equilibrium Many Body Systems, 65–71.

2. Smoluchowski Dynamics

In order to demonstrate the structure of a statistical description for colloids let us consider a system of N spherical Brownian particles confined to a volume V. Let $\boldsymbol{R_i}$ denote the position of the center of the ith sphere. The system is described by a probability distribution $P(\boldsymbol{X}, t)$ for finding particles in the configuration $\boldsymbol{X} = (\boldsymbol{R}_1, \ldots, \boldsymbol{R}_N)$ at time t. It is not necessary to consider velocities in the scheme since their relaxation times are very small compared to the time scales related to the Brownian particles. The evolution of the distribution P is governed by the generalized Smoluchowski equation (GSE) [3]. In abbreviated form this reads

$$\frac{\partial P}{\partial t} = \mathcal{D}P, \tag{1}$$

where $\mathcal{D}$ is the Smoluchowski operator defined by

$$\mathcal{D}P = \frac{\partial}{\partial \boldsymbol{X}} \cdot \boldsymbol{D} \cdot \left(\frac{\partial P}{\partial \boldsymbol{X}} + \beta \frac{\partial \Phi}{\partial \boldsymbol{X}} P \right). \tag{2}$$

Here $\beta = 1/k_B T$ and the direct particle interactions (e.g. by electrostatic or hard-core forces) are incorporated in the potential energy $\Phi(\boldsymbol{X})$. Furthermore $\boldsymbol{D}(\boldsymbol{X})$, the diffusion matrix, depends on the configuration, due to hydrodynamic interactions, and is given by the generalized Einstein relation

$$\boldsymbol{D}(\boldsymbol{X}) = k_B T \, \boldsymbol{\mu}(\boldsymbol{X}), \tag{3}$$

where $\boldsymbol{\mu}(\boldsymbol{X})$ is the N-particle mobility matrix. The latter is found by solving the creeping flow equation for solvent response to particle motion [4]. In the last two decades there has been significant progress in understanding the hydrodynamic interactions, and efficient numerical algorithms for calculating the mobility matrix $\boldsymbol{\mu}(\boldsymbol{X})$ are available [4-7]. In the special case when the hydrodynamic interactions are neglected, the diffusion matrix is diagonal, i.e. $\boldsymbol{D}(\boldsymbol{X}) = D_0 \, \boldsymbol{I}$, where D_0 is a single-particle diffusion constant.

The GSE is a "Liouville" equation for Brownian particle systems. In order to get a proper description of the relaxation processes in these systems, it is necessary to apply to this equation kinetic theory methods. One then usually employs a formalism with basic objects such as collision operators or particle propagators. In the case of suspensions, when the Fourier transform is taken with respect to the configuration dependence, those objects are c-numbers. In the kinetic theory of fluids they are operators in velocity space. This is the first important simplification in favor of the Brownian particle systems.

However one must be careful when applying nonequilibrium statistical physics to suspensions. As an example, consider the intermediate scattering function which at wavenumber $k = | \boldsymbol{k} |$ is

$$F(k,t) = \frac{1}{N} \langle \hat{n}(-\boldsymbol{k},0)\, \hat{n}(\boldsymbol{k},t) \rangle, \tag{4}$$

where the angle brackets denote an average over the equilibrium distribution $P_{eq}(\boldsymbol{X}) \sim exp(-\beta\, \Phi(\boldsymbol{X}))$ and $\hat{n}(\boldsymbol{k},t)$ is the Fourier component of the particle-density fluctuation whose time dependence,

$$\hat{n}(\boldsymbol{k},t) = e^{\mathcal{L}t}\, \hat{n}(\boldsymbol{k}), \quad \hat{n}(\boldsymbol{k}) = \sum_{j=1}^{N} e^{-i\boldsymbol{k}\cdot\boldsymbol{R}_j} \tag{5}$$

is governed by the adjoint Smoluchowski operator $\mathcal{L}$ which in turn is defined by the relation

$$\mathcal{D}\, P_{eq} = P_{eq}\, \mathcal{L}. \tag{6}$$

The Laplace transform of the scattering function is

$$\hat{F}(k,z) = \int_0^{+\infty} e^{-z\,t}\, F(k,t)\, dt. \tag{7}$$

By applying the standard Mori-Zwanzig procedure with the projection

$$\mathcal{P} = \hat{n}(\boldsymbol{k})\,\rangle \frac{1}{NS(k)} \langle\, \hat{n}(-\boldsymbol{k}) \tag{8}$$

onto the density fluctuation subspace one gets the following relation

$$\hat{F}(k,z) = \frac{S(k)}{z + \frac{D_0\, k^2}{S(k)} \left(1 - \hat{M}(k,z)\right)}, \tag{9}$$

where $S(q)$ is the static structure factor and $\hat{M}(k,z)$ is the Laplace transform of the memory function, $M(k,t)$, given by

$$M(k,t) = \frac{S(k)}{D_0 k^2} \langle\, \hat{n}(-\boldsymbol{k})\, \mathcal{L}\mathcal{Q}\, e^{\mathcal{Q}\mathcal{L}\mathcal{Q}t} \mathcal{Q}\mathcal{L}\, \hat{n}(\boldsymbol{k})\,\rangle \tag{10}$$

with $\mathcal{Q} = 1 - \mathcal{P}$. Imagine now that one constructs an approximation for the function (10) according to kinetic theory recipes. This usually leads to very unpleasant results. For example, with the mode-mode coupling expression for the memory function one gets from Eq. (9) negative values for the effective diffusion coefficient [8]. In the case of Brownian particles, the Mori-Zwanzig procedure must be modified, as shown by Cichocki and Hess [9]. A technical reason is that the GSE is a second order differential equation.

An explanation of the problem based on physics can be found in Ref. [9]. This modified procedure leads to the additional relation

$$\hat{M}(k,z) = \frac{\hat{M}^{irr}(k,z)}{1+\hat{M}^{irr}(k,z)} \tag{11}$$

with the irreducible memory function M^{irr} given by the formula (10) with

$$\mathcal{L}' = \mathcal{L} - \mathcal{L}\,\hat{n}(\boldsymbol{k})\,\rangle \frac{1}{\langle\,\hat{n}(-\boldsymbol{k})\mathcal{L}\,\hat{n}(\boldsymbol{k})\,\rangle}\langle\,\hat{n}(-\boldsymbol{k})\mathcal{L} \tag{12}$$

instead of $\mathcal{L}$ in the exponential operator. It should be stressed that, in the case of suspensions, the irreducible memory function is a basic object appropriate for approximations since this function is an analogue of the collision operator introduced in the kinetic theory of atomic fluids. Recently, Kawasaki has generalized the results of Ref. [9] to dissipative systems [10].

Next, it is worth while to point out the second very important property of the Brownian particle system. The GSE is a master equation with the detailed balance. Hence time dependence of any correlation function related to the system can be expressed as a superposition of decaying exponentials. For the scattering function this gives

$$F(k,t) = \int_0^{+\infty} p(k,u)\,e^{-ut}\,du \tag{13}$$

with a positive spectral density $p(k,u)$. This property is very useful. Usually a theoretical analysis gives expressions in terms of the Laplace transforms. An inversion to time language is, in the case of atomic fluids, a very difficult task. Here, for example, by taking $z = -u + i\,\epsilon$ with infinitesimal and positive ϵ one gets the spectral density p as

$$p(k,u) = -\frac{1}{\pi} Im\,\hat{F}(k, z = -u + i\,\epsilon) \tag{14}$$

Then calculation of the scattering function $F(k,t)$ is a matter of a simple integration in Eq. (13).

3. Enskog Approximation

In order to illustrate how the scheme described in Sec.(2) works in practice, let us consider a system of Brownian hard spheres of radius a without hydrodynamic interactions. To first order in volume fraction, $\phi = 4\pi n a^3/3$, i.e. for low particle concentration $n = N/V$, an explicit expression for the irreducible memory function can be derived. To this end it is enough to solve the two-body GSE [11, 12]. The result is

$$\hat{M}_0^{irr}(k,z) = -24\,\phi \sum_{\substack{l=0\\ even}}^{\infty} (2l+1)\,[j_l'(y)]^2\, \frac{k_l(\alpha)}{\alpha\, k_l'(\alpha)}, \tag{15}$$

where

$$y = ka, \quad \alpha = \sqrt{k^2\,a^2 + \frac{2\,z\,a^2}{D_0}}. \tag{16}$$

In Eq. (15) j_l' and k_l' are the derivatives of the spherical Bessel function j_l and of the modified spherical Bessel function k_l, respectively [13].

For higher concentrations the above expression must be modified. The simplest proposition is the Enskog type approximation, analogous to that well-known result in the kinetic theory of atomic fluids, defined by the formula

$$\hat{M}_E^{irr}(k,z) = \chi\,\hat{M}_0^{irr}(k,z)\,, \tag{17}$$

where χ is the contact value of the equilibrium radial-distribution function. For short times, the irreducible memory function in the form (17) is exact [14]. Now one can use the Enskog approximation and calculates the scattering function. Appropriate steps are straightforward. Namely, by taking Eqs. (9) and (11) with $\hat{M}^{irr}$ in the form (17) and by performing the operation described above Eq. (14), one gets the spectral density $p(k,u)$. Then integration over u in Eq. (13) gives the function $F(k,t)$. Those rather simple calculations have been done by Cichocki and Felderhof in Ref. [14]. The final results have been compared with computer-simulation data for hard-sphere suspensions [15, 16]. Excellent agreement is found up to a volume fraction of about forty percent.

The results derived in [14] demonstrate also the existence of a collective mode for not very small wavenumbers. Namely, the spectral density $p(k,u)$ has a sharp peak located in the region of small inverse relaxation time u (see plots in [14]). This leads to an exponential decay for long times,

$$F(k,t) \sim e^{-u_0(k)\,t}, \tag{18}$$

where $u_0(k)$ is a position of the peak. Employing an analogue to the kinetic theory of dense atomic liquids, but without systematic analysis of the GSE, de Schepper et.al. [17, 18] have conjectured the above decay and the following formula

$$\frac{\chi\,S(k)}{D_0 k^2}\,u_0(k) = d(2ka), \tag{19}$$

where

$$d(x) = \frac{1}{1 - j_0(x) + 2j_2(x)}. \tag{20}$$

Having explicit expressions for the spectral density $p(k,u)$, Cichocki and Felderhof have been able to check this relation [14]. They have found that the wavenumber structure of the l.h.s. in Eq. (19) is similar to that of the function $d(2ka)$. However, there is a difference in magnitude up to a factor 2 for high particle concentrations. The stability of the structure suggests that the long-time decay may be used to measure the sphere radius.

4. Conclusions

The results presented in Sec.(3) support our statement that interacting Brownian particle systems are very convenient for checking results based on kinetic theory. The inverse relaxation spectra is the key tool which makes appropriate calculations relatively easy. The simple Enskog approximation works remarkably well for dilute and moderately dense suspensions. At high concentrations this is not enough. It is necessary to take into account more complicated dynamical events which correspond e.g. to the mode-mode coupling term in the irreducible memory function. Various propositions related to this problem have been checked by recent calculations [19].

References

1. W.B. Russel, D.A. Saville, and W.R. Schowalter, *Colloidal Dispersions*, (Cambridge University Press, Cambridge, 1989).
2. J.K.G. Dhont, *An Introduction to Dynamics of Colloids*, (Elsevier Science, Amsterdam, 1996).
3. P.N. Pusey, Colloidal suspensions, in *Liquids, Freezing and Glass Transition*, J.P. Hansen, D. Levesque, and J. Zinn-Justin, eds., (North-Holland, Amsterdam, 1991).
4. S. Kim and S.J. Karrila, *Microhydrodynamics: Principles and Selected Applications*, (Butterworth-Heinemann, Boston, 1991).
5. A.J.C. Ladd, J. Chem. Phys. **93**, 3484 (1990).
6. A.S. Sangani and G. Mo, Phys. Fluids **8**, 1990 (1996).
7. B. Cichocki, B.U. Felderhof, K. Hinsen, E. Wajnryb and J. Blawzdziewicz, J. Chem. Phys. **100**, 3780 (1994).
8. W. Hess and R. Klein, Adv. Physics **32**, 173 (1983).
9. B. Cichocki and W. Hess, Physica A **141**, 475 (1987).
10. K. Kawasaki, Physica A **215**, 61 (1995).
11. B.J. Ackerson and L. Fleishman, J. Chem. Phys. **76**, 2675 (1982).
12. B. Cichocki and B.U. Felderhof, J. Chem. Phys. **98**, 8186 (1993).
13. M. Abramowitz and I.A. Stegun (eds.), *Handbook of Mathematical Functions*, (Dover, New York, 1965).
14. B. Cichocki and B.U. Felderhof, Physica A **204**, 152 (1994).
15. B. Cichocki and K. Hinsen, Ber. Bunsenges. Phys. Chem. **94**, 243 (1990).
16. B. Cichocki and K. Hinsen, Physica A **166**, 473 (1990).
17. I.M. de Schepper, E.G.D. Cohen, P.N. Pusey, and H.N.W. Lekkerkerker, J. Phys.: Condens. Matter **1**, 6503 (1989).

18. P.N. Pusey, H.N.W. Lekkerkerker, E.G.D. Cohen, and I.M. de Schepper, Physica A **164**, 12 (1990).
19. B. Cichocki and P. Szymczak, *Dynamic scattering function of dense hard-sphere suspension* (to be published).

SPINODAL DECOMPOSITION KINETICS: THE INITIAL AND INTERMEDIATE STAGES

JAN K.G. DHONT
van 't Hoff Laborarory, Debye Research Institute
Padualaan 8, 3584 CH Utrecht, The Netherlands

Abstract. This text is a treatment of spinodal decomposition kinetics of colloidal systems in the initial and intermediate stages. When a stable, homogeneous system is quenched into an unstable or meta-stable state, density inhomogeneities will develop, which ultimately lead to complete phase separation, where two phases are in coexistence. The kinetics of the phase separation process can be described by analyzing equations of motion for the relevant order parameter. In the present text, the gas-liquid phase separation of colloids is considered, where the relevant quantity is the macroscopic density. Here, the "gas phase" is a colloidal fluid of low concentration, while the "liquid phase" is a more concentrated colloidal fluid.

The goal of this text is to develop a microscopic approach for spinodal decomposition kinetics, where the starting point is the Smoluchowski equation. This is an equation of motion for the probability density function of the phase space coordinates of the colloidal particles (the colloidal analogue of the Liouville equation for molecular systems). In section 1 the various stages that can be distinguished during spinodal decomposition, starting from a homogeneous system, are introduced. Spinodal decomposition kinetics in the initial stage, where density inhomogeneities have just started to develop, is discussed in section 2. First, the classic Cahn-Hilliard thermodynamic approach is considered, after which a microscopic rederivation of these results is given, based on the Smoluchowski equation. Section 2 concludes with a microscopic interpretation for the origin of the spinodal instability. In section 3 an alternative definition of the spinodal and binodal from a kinetic point of view is discussed, and the experimental relevance of the spinodal is considered. It turns out that the location of the theoretically well defined spinodal cannot be determined experimentally with arbitrary precision. Decomposition kinetics in the intermediate stage is treated in section 4. In the intermediate stage, density inhomogeneities are not small

J. Karkheck (ed.),
Dynamics: Models and Kinetic Methods for Non-equilibrium Many Body Systems, 73–120.

anymore, so that non-linear equations of motion must be considered. Due to the existence of a dominant length scale in the intermediate stage, dynamic scaling is expected. A dynamic scaling relation for the structure factor is derived, after which the Smoluchowski equation is analysed in the intermediate stage. The solution of the non-linear equation of motion for the density confirms the dynamic scaling relation for the structure factor, and leads to an explicit expression for the dynamic scaling function. Contrary to the thermodynamic type of approach, the effect of externally imposed shearing motion can be analysed in a quite straightforward manner, using the kinetic approach discussed in section 2. This is the subject of section 5. Theoretical predictions are compared to experimental findings in section 6. Finally, a few exercises are added, together with an overview of relevant literature (although this is by no means a complete overview).

1. The Various Stages during Spinodal Decomposition

Consider a homogeneous system that is unstable. In practice such a system may be prepared by suddenly cooling the system from a temperature in the stable region in the phase diagram to a temperature in the unstable part. Such a process is commonly referred to as *a quench*. Right after a quench the system starts to develop long ranged correlations. These correlations develop up to a point where they render the system unstable, from which time on the macroscopic density develops inhomogeneities. The evolution of the density is sketched in Fig. 1. As will be seen shortly, right after the quench, in the initial stages of the phase separation, one particular sinusoidal density fluctuation is amplified most rapidly.[1] There is one particular optimum wavelength for which the corresponding density waves grow most rapidly. This is depicted in Fig. 1a. In the *initial stage* of the phase separation, both the change $\delta\rho$ of the density and gradients of the density are small. The initial stage is also referred to as the *linear regime*, since equations of motion for the density may be linearized with respect to $\delta\rho$. Then there is the so-called *intermediate stage*, where $\delta\rho$ is not small so that linearization is no longer allowed. Gradients of the density are still small, as in the initial stage, due to the long wavelengths that demix. This stage is depicted in Fig. 1b. Subsequently, the decomposition reaches the *transition stage* where the lower and larger binodal concentrations ($\bar{\rho}_-$ and $\bar{\rho}_+$, respectively) are attained in various parts of the system, as sketched in Fig. 1c. At this stage, sharp interfaces between the regions with concentrations close to $\bar{\rho}_-$ and $\bar{\rho}_+$ exist. Inhomogeneities are now large, and higher

[1] Sinusoidally varying density profiles are also referred to as *density waves*.

order terms in gradients of the density come into play. In the *late, or final stage* of the phase separation the interfaces develop : concentration gradients sharpen and the interfacial curvatures change to ultimately establish coexistence (see Fig. 1d). We thus arrive at the following classification of the different stages during decomposition,

Initial stage : $\delta\rho/\bar{\rho}$ *is small,*
gradients are small ("diffuse interfaces"),

Intermediate stage : $\delta\rho/\bar{\rho}$ *is not small,*
gradients are small ("diffuse interfaces"),

Transition stage : $\delta\rho/\bar{\rho}$ *is large,*
gradients are not small ("sharp interfaces"),

Final stage : $\delta\rho/\bar{\rho}$ *is large,*
gradients are large ("very sharp interfaces").

Mathematically, "small" means that equations of motion can be linearized, "not small" means that the leading non-linear term must be taken into account, and "large" means that the full non-linear equation must be analyzed. Equations of motion for the density in the initial and intermediate stage can therefore be expanded to leading order with respect to gradients of the density, while the leading non-linear contribution in $\delta\rho/\bar{\rho}$ must be included in the intermediate stage. The first higher order terms in an expansion with respect to gradients of the density, which must be included in the transition stage, are referred to here as describing the dynamics of *sharp interfaces*, while even higher order terms describe the dynamics of *very sharp interfaces* in the final stage. These very sharp interfaces have a width of the order of a few particle diameters, except in case of quenches very close to the critical point, where the equilibrium interfacial thickness may be large.

2. Initial Spinodal Decomposition Kinetics

A homogeneous system that is quenched from the stable state into the unstable region of the phase diagram, by a sudden change of the temperature, develops inhomogeneities after the development of correlations that render the system unstable. Let $\bar{\rho} = N/V$ denote the number density of colloidal particles in the homogeneous state, before decomposition occurred, and let $\rho(\mathbf{r}, t)$ denote the macroscopic number density as a function of the position $\mathbf{r}$ in the system at time t after the system became unstable and started to demix. Define the change of the macroscopic density $\delta\rho(\mathbf{r}, t)$ relative to

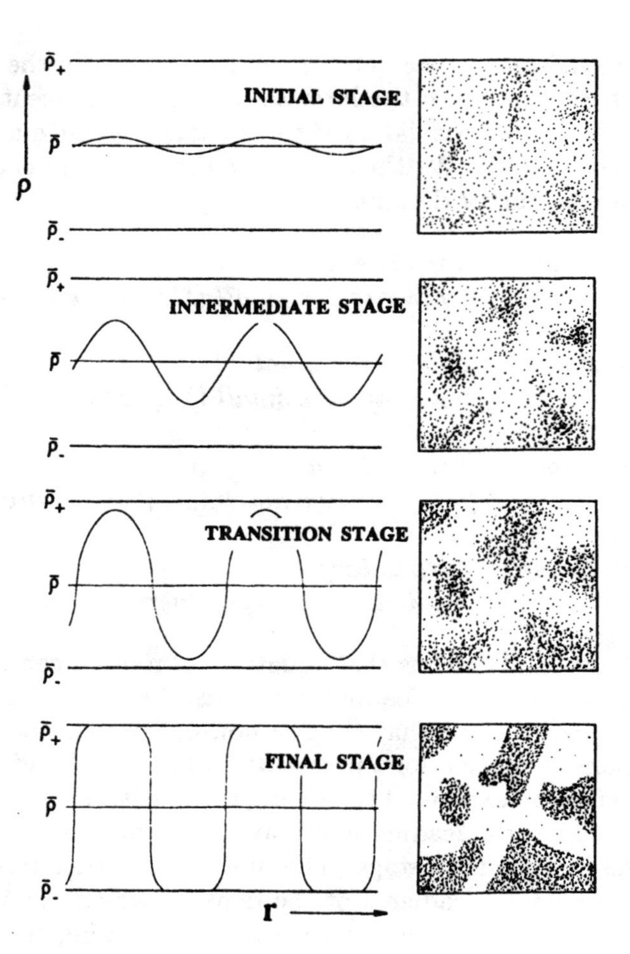

Figure 1. A sketch of the time development of the density after a quench in the unstable part of the phase diagram. Time increases from top to bottom. The left column of figures is a sketch of the density versus position, while the right column depicts the corresponding morphology of density variations in the system itself. The concentrations $\bar{\rho}_+$ and $\bar{\rho}_-$ are the binodal concentrations.

that in the homogeneous state as

$$\rho(\mathbf{r},t) \;=\; \bar{\rho} + \delta\rho(\mathbf{r},t)\,. \tag{1}$$

In the initial stage of the phase separation we have

$$\frac{|\;\delta\rho(\mathbf{r},t)\;|}{\bar{\rho}} \ll 1\,, \tag{2}$$

allowing linearization of equations of motion for the macroscopic density with respect to the change $\delta\rho$ of the density. In the present section we first describe the initial stage of phase separation on the basis of thermodynamic considerations, which approach is known as the Cahn-Hilliard theory, and subsequently on the basis of microscopic considerations.

2.1. THE CAHN-HILLIARD THEORY

In a thermodynamic type of approach, demixing can be described as transport of colloidal particles between volume elements which are internally in equilibrium. These volume elements are supposed to be so small that the wavelengths of density variations which are unstable are very much larger than the linear dimension of the volume elements. This is only possible when *large* wavelength density variations are unstable; this will indeed turn out to be the case. On the other hand these volume elements are supposed to be so large that they contain many colloidal particles and that the range of correlations between particles is small in comparison to its linear dimensions, in order to make a thermodynamic description of each volume element feasible. Furthermore, each volume element is supposed to be in thermal equilibrium at each instant during the initial stage of the phase separation. This can be achieved only when the rate of demixing is small in comparison to the relaxation time of density variations with a wavelength that fits many times into a volume element. That this is indeed the case is due to (i) the fact that it takes more time to displace colloidal particles over larger distances (for demixing) than over smaller distances (for internal equilibration) and (ii) the fact that the diffusion coefficient which describes the large wavelength demixing is very much smaller than the diffusion coefficient pertaining to relaxation of small wavelength density variations, as will be shown shortly. The latter is reminiscent of critical slowing down. The system is thus supposed to be in *local equilibrium*. The idea of fast-relaxing small-wavelength density fluctuations and slowly-growing large-wavelength density variations is depicted in Fig. 2.

If one is willing to accept the above assumptions, the Helmholtz free energy $A_V(\rho(\mathbf{r},t))$ per unit volume of a volume element at position $\mathbf{r}$, with a homogeneous density $\rho(\mathbf{r},t)$, is well defined and exhibits a "van der Waals loop-form" as a function of the local density. Such a form can be modelled by a fourth-order polynomial in the change $\delta\rho(\mathbf{r},t)$ of the local density relative to $\bar{\rho} = N/V$ as follows,

$$\begin{aligned} A_V(\rho(\mathbf{r},t)) &= A_V(\bar{\rho}) + a_1\delta\rho(\mathbf{r},t) + \frac{1}{2}a_2(\delta\rho(\mathbf{r},t))^2 \\ &\quad + \frac{1}{3}a_3(\delta\rho(\mathbf{r},t))^3 + \frac{1}{4}a_4(\delta\rho(\mathbf{r},t))^4 . \end{aligned} \tag{3}$$

The coefficient a_4 is positive, since for large densities the free energy is large and positive, while a_2 is negative to invoke instability. Notice that

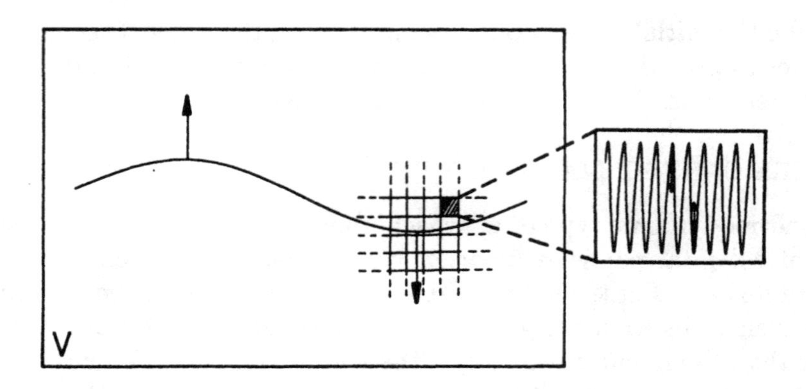

Figure 2. **Small volume elements are in internal equilibrium due to fast relaxation of small wavelength density fluctuations, while the large wavelength density variations grow slowly.**

this free energy is the free energy of a volume element in which fluctuations of the density with wavelengths larger than its own linear dimension are absent, since these simply do not "fit into" the volume. This is the free energy per unit volume of an infinitely large system (with a homogeneous density equal to $\bar{\rho}+\delta\rho$) in which the large wavelength density fluctuations are constrained by means of an external field. In a statistical mechanical calculation of the van der Waals form of the free energy, one should thus include an external field to account for the absence of density variations with wavelengths that are very much larger than the range of correlations between the colloidal particles.

The Helmholtz free energy of the total system is not simply equal to the sum of the free energies (3). There is also a contribution from the diffuse interfaces separating volume elements with different densities, that is, there is also a contribution arising from gradients in the density. Since we are only interested in the dynamics of the long wavelength density variations, leading to small gradients in the density, we may formally expand this contribution with respect to gradients in the density, and keep only the leading term. Since the free energy is invariant against inversion of the coordinate frame, that leading term is proportional to $\int d\mathbf{r}\ |\nabla\delta\rho(\mathbf{r},t)|^2$, where the integral ranges over the volume of the entire system under consideration. The proportionality constant is denoted here as $\frac{1}{2}\kappa$, where $\kappa > 0$ is referred to as *the Cahn-Hilliard square-gradient coefficient*. The total Helmholtz free energy is therefore the sum (read : integral) of the "bulk contributions" in eq.(3) and the above discussed "diffuse interface contribution",

$$A[\rho(\mathbf{r},t)] \quad = \quad A(\bar{\rho}) + \int d\mathbf{r}\ \left\{a_1\delta\rho(\mathbf{r},t) + \frac{1}{2}a_2(\delta\rho(\mathbf{r},t))^2 + \frac{1}{3}a_3(\delta\rho(\mathbf{r},t))^3\right.$$

$$+\frac{1}{4}a_4(\delta\rho(\mathbf{r},t))^4 + \frac{1}{2}\kappa\,|\nabla\delta\rho(\mathbf{r},t)|^2\Big\} \,. \tag{4}$$

The square brackets in $A[\rho(\mathbf{r},t)])$ denotes the functional dependence of A on $\rho(\mathbf{r},t)$.

The number density flux $\mathbf{j}(\mathbf{r},t)$ of colloidal particles is driven by gradients in the chemical potential. When these gradients are not too large, the number density flux is simply proportional to the gradient,

$$\mathbf{j}(\mathbf{r},t) \;=\; -D\nabla\mu(\mathbf{r},t)\,, \tag{5}$$

where D is a phenomenological transport coefficient. The chemical potential is in turn related to the functional derivative of the free energy with respect to the density,

$$\mu(\mathbf{r},t) \;=\; \frac{\delta A[\rho]}{\delta\rho(\mathbf{r},t)}\,. \tag{6}$$

Substitution of the expression (4) for the free energy gives

$$\mu(\mathbf{r},t) \;=\; a_1 + a_2\delta\rho(\mathbf{r},t) + a_3(\delta\rho(\mathbf{r},t))^2 + a_4(\delta\rho(\mathbf{r},t))^3 - \kappa\nabla^2\delta\rho(\mathbf{r},t)\,. \tag{7}$$

Now, conservation of the number of colloidal particles requires that

$$\frac{\partial}{\partial t}\rho(\mathbf{r},t) \;=\; -\nabla\cdot\mathbf{j}(\mathbf{r},t)\,. \tag{8}$$

Using eqs. (5, 6) in this equation of motion, linearization with respect to $\delta\rho(\mathbf{r},t)$ and Fourier transformation with respect to $\mathbf{r}$ finally yields

$$\frac{\partial}{\partial t}\delta\rho(\mathbf{k},t) \;=\; -D\,k^2\left[a_2 + \kappa\,k^2\right]\delta\rho(\mathbf{k},t)\,. \tag{9}$$

The solution of this equation is simply

$$\delta\rho(\mathbf{k},t) \;=\; \delta\rho(\mathbf{k},t=0)\,\exp\{-D^{eff}(k)\,k^2\,t\}\,, \tag{10}$$

where the *effective diffusion coefficient* is defined as

$$D^{eff}(k) \;=\; D\left[a_2 + \kappa\,k^2\right]. \tag{11}$$

Here, $\delta\rho(\mathbf{k},t=0)$ is the Fourier transform of the density at time $t=0$. At this instant of time correlations have developed to an extent that renders the system unstable, but phase separation has not yet occurred.

From eq.(10) it is clear that density waves with a wavelength $\Lambda = 2\pi/k$ are unstable when $D^{eff}(k) < 0$. For these *wavevectors* k, the amplitude

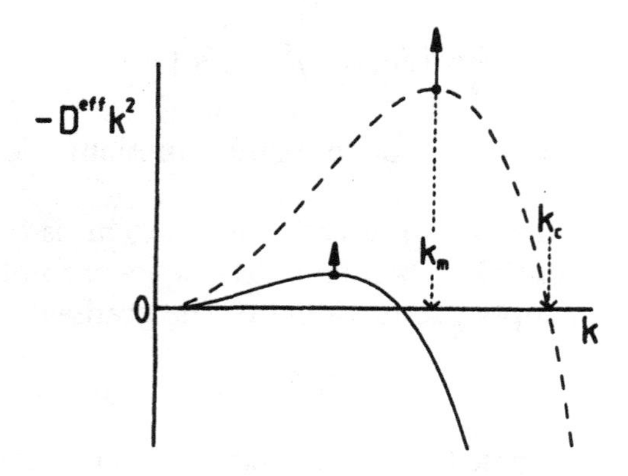

Figure 3. A sketch of the growth rate of sinusoidal density variations as a function of their wavevector. The dashed curve is for a deep quench, the solid line for a shallow quench.

$\delta\rho(\mathbf{k}, t)$ of the corresponding density wave grows exponentially with time. Since $\kappa > 0$, the effective diffusion coefficient can become negative for certain wavevectors only when $a_2 < 0$. From eq.(3) it follows that (differentiations are at constant N, and μ is the chemical potential of the colloidal particles)

$$a_2 = \frac{d^2(A/V)}{d\bar{\rho}^2} = \frac{d\,(\partial A/\partial N)}{d\bar{\rho}} = \frac{d\mu}{d\bar{\rho}} = \frac{1}{\bar{\rho}}\frac{dP}{d\bar{\rho}}\,, \tag{12}$$

so that negative values of the effective diffusion coefficient correspond to $dP/d\bar{\rho} < 0$, which is the thermodynamic criterion for instability. For density variations with a wavelength $\Lambda = 2\pi/k$ for which $D^{eff}(k) < 0$, diffusion occurs from regions of low density toward larger density. This phenomenon is often referred to as *uphill diffusion.*

The *growth rate* of a sinusoidal density variation is equal to $-D^{eff}(k)\,k^2$, and is sketched in Fig. 3. The wavevector k_m of the most rapidly growing density wave is easily found by straighforward differentiation,

$$k_m = \sqrt{-\frac{a_2}{2\kappa}}\,. \tag{13}$$

The so-called *critical wavevector* k_c is the wavevector beyond which density waves are stable. That is, for any $k > k_c$, $D^{eff}(k) > 0$. The critical wavevector is easily found to be equal to

$$k_c = \sqrt{-\frac{a_2}{\kappa}} = \sqrt{2}\,k_m\,. \tag{14}$$

Density variations with small wavevectors decompose slowly because it takes longer times to transport colloidal particles over large distances. Density variations with larger wavevectors decompose slowly because the driving force for uphill diffusion diminishes, as a result of the fact that less free energy is gained when larger density gradients are created.

Note that a deeper quench, where $-a_2$ is relatively large, results in a larger value for the most rapidly decomposing wavevector k_m.

2.2. SMOLUCHOWSKI EQUATION APPROACH

The description given in the previous subsection is based on thermodynamic arguments. A microscopic derivation of the Cahn-Hilliard result (10, 11) can be given on the basis of the Smoluchowski equation. The Smoluchowski equation is the equation of motion for the probability density function (pdf) $P \equiv P(\mathbf{r}_1, \mathbf{r}_2, \cdots, \mathbf{r}_N, t)$ of the position coordinates $\mathbf{r}_j$, $j = 1, 2, \cdots, N$, of all N colloidal particles in the system, and reads, with the neglect of hydrodynamic interaction,

$$\frac{\partial}{\partial t} P = D_0 \sum_{j=1}^{N} \nabla_{r_j} \cdot \left[\beta [\nabla_{r_j} \Phi] P + \nabla_{r_j} P\right], \tag{15}$$

where D_0 is the Stokes-Einstein diffusion coefficient, $\beta = 1/k_B T$ (with k_B Boltzmann's constant and T the temperature), and $\Phi \equiv \Phi(\mathbf{r}_1, \mathbf{r}_2, \cdots, \mathbf{r}_N)$ the potential energy of the assembly of colloidal particles. Since

$$\int d\mathbf{r}_2 \cdots \int d\mathbf{r}_N \, P(\mathbf{r}_1, \mathbf{r}_2, \cdots, \mathbf{r}_N, t) \equiv P_1(\mathbf{r}_1, t) = \frac{1}{N} \rho(\mathbf{r}_1, t), \tag{16}$$

with P_1 the probability density function for a single position coordinate, an equation of motion for the macroscopic density can be obtained from the Smoluchowski equation (15) by integration with respect to all the position coordinates, except for $\mathbf{r}_1$. In order to integrate the Smoluchowski equation, a pair-wise additive interaction potential is assumed, that is (with $r_{ij} = |\mathbf{r}_i - \mathbf{r}_j|$),

$$\Phi(\mathbf{r}_1, \mathbf{r}_2, \cdots, \mathbf{r}_N) = \sum_{i,j=1,\, i<j}^{N} V(r_{ij}), \tag{17}$$

with V the pair-interaction potential. Further, introducing the pair-correlation function g through[2]

$$\begin{aligned} P_2(\mathbf{r}_1, \mathbf{r}_2, t) &\equiv \int d\mathbf{r}_3 \cdots \int d\mathbf{r}_N \, P(\mathbf{r}_1, \mathbf{r}_2, \mathbf{r}_3, \cdots, \mathbf{r}_N, t) \\ &\equiv P_1(\mathbf{r}_1, t) \, P_1(\mathbf{r}_2, t) \, g(\mathbf{r}_1, \mathbf{r}_2, t), \end{aligned} \tag{18}$$

[2]The function g is called the "pair-correlation function" because it measures the correlation between two particles. Note that for statistically independent particles, $g \equiv 1$.

the integrated Smoluchowski equation reads, for identical Brownian particles (with $\mathbf{r}_1$ renamed as $\mathbf{r}$ and $\mathbf{r}_2$ as $\mathbf{r}'$),

$$\frac{\partial}{\partial t}\rho(\mathbf{r},t) = D_0\Big[\nabla^2\rho(\mathbf{r},t) + \beta\nabla\cdot\rho(\mathbf{r},t)\int d\mathbf{r}'\,[\nabla V(|\mathbf{r}-\mathbf{r}'|)]\,\rho(\mathbf{r}',t)\,g(\mathbf{r},\mathbf{r}',t)\Big]\,, \tag{19}$$

where ∇ is the gradient operator with respect to $\mathbf{r}$. There are two terms to be distinguished on the right hand-side : the first term between the square brackets describes the effect of Brownian motion, while the second term represents the effects of direct interactions. The combination

$$\mathbf{F}^{Int}(\mathbf{r},t) \equiv -\int d\mathbf{r}'\,[\nabla V(|\mathbf{r}-\mathbf{r}'|)]\,\rho(\mathbf{r}',t)\,g(\mathbf{r},\mathbf{r}',t)\,, \tag{20}$$

is the direct force on a colloidal particle at $\mathbf{r}$ due to particles in a volume element with position $\mathbf{r}'$, averaged with respect to the position of the latter. In the next subsection, we will come back to the role of these two contributions in rendering the system unstable.

Consider the initial stage of the phase separation, where the change $\delta\rho$ of the macroscopic density, as defined in eq. (1), is small. Let δg denote the accompanying change of the pair-correlation function,

$$g(\mathbf{r},\mathbf{r}',t) = g_0(|\mathbf{r}-\mathbf{r}'|) + \delta g(\mathbf{r},\mathbf{r}',t)\,. \tag{21}$$

Here, g_0 is the pair-correlation function after the quench, before phase separation occurred. Linearization of the Smoluchowski equation (19) with respect to these changes yields

$$\frac{\partial}{\partial t}\delta\rho(\mathbf{r},t) = D_0\,[\,\nabla^2\delta\rho(\mathbf{r},t) + \beta\bar{\rho}\nabla\cdot\int d\mathbf{r}'\,[\nabla V(|\mathbf{r}-\mathbf{r}'|)] \times (\delta\rho(\mathbf{r}',t)\,g_0(|\mathbf{r}-\mathbf{r}'|) + \bar{\rho}\,\delta g(\mathbf{r},\mathbf{r}',t))\,]\,. \tag{22}$$

To obtain a closed equation for $\delta\rho$, the change δg of the pair-correlation function must be expressed in terms of $\delta\rho$. Such a *closure relation* may be obtained as follows. The important feature is that the pair-correlation function in the integral in the Smoluchowski equation (22) is multiplied by the pair-force $\nabla V(|\,\mathbf{r}-\mathbf{r}'\,|)$, so that a closure relation is only needed for small distances $|\,\mathbf{r}-\mathbf{r}'\,|\leq R_V$, with R_V the range of the pair-interaction potential. Correlations over such small distances establish much faster than the demixing rates of the very long unstable wavelengths, simply because it takes more time to displace colloidal particles over larger distances. On a coarsened time scale that is much larger than relaxation times of density fluctuations with wavevectors $k \geq 2\pi/R_V$, but which still resolves the

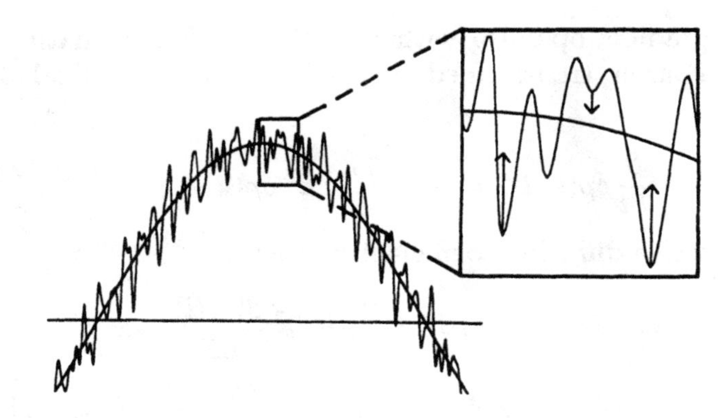

Figure 4. **The statistical local equilibrium assumption implies fast relaxation of short wavelength density fluctuations in comparison to the slowly demixing large wavelength density variations, rendering the pair-correlation function locally equal to the equilibrium pair-correlation function.**

phase separation process, the pair-correlation function in the integral in the Smoluchowski equation may therefore be replaced by the equilibrium pair-correlation function. This is the statistical equivalent of the thermodynamic local-equilibrium assumption made in the Cahn-Hilliard approach as described in the previous subsection. The statistical local-equilibrium assumption is illustrated in Fig. 4. The equilibrium pair-correlation function is to be evaluated at the instantaneous macroscopic density midway between the positions $\mathbf{r}$ and $\mathbf{r}'$. Hence, to first order in $\delta\rho$, and for $|\mathbf{r}-\mathbf{r}'| \leq R_V$,

$$\delta g(\mathbf{r},\mathbf{r}',t) = \delta g^{eq}(|\mathbf{r}-\mathbf{r}'|)\Big|_{density=\rho(\frac{\mathbf{r}+\mathbf{r}'}{2},t)} = \frac{dg^{eq}(|\mathbf{r}-\mathbf{r}'|)}{d\bar{\rho}}\,\delta\rho(\tfrac{\mathbf{r}+\mathbf{r}'}{2},t) \quad (23)$$

and

$$g_0(|\mathbf{r}-\mathbf{r}'|) = g^{eq}(|\mathbf{r}-\mathbf{r}'|)\,, \quad (24)$$

where g^{eq} is the equilibrium pair-correlation function for a homogeneous system with density $\bar{\rho}$ and the temperature after the quench. The two relations (23, 24) are certainly wrong for distances $|\mathbf{r}-\mathbf{r}'|$ comparable to the wavelengths of the unstable density variations. For such distances the system is far out of equilibrium. The validity of the relations (23, 24) is limited to small distances, where $|\mathbf{r}-\mathbf{r}'| \leq R_V$. Substitution of eqs. (23, 24) into the Smoluchowski equation (2) and renaming $\mathbf{R}=\mathbf{r}-\mathbf{r}'$ yields

$$\begin{aligned}\frac{\partial}{\partial t}\delta\rho(\mathbf{r},t) &= D_0\left[\nabla^2\delta\rho(\mathbf{r},t) + \beta\bar{\rho}\nabla\cdot\int d\mathbf{R}\,[\nabla_R V(R)]\right. \quad (25)\\ &\quad \left.\times\left(g^{eq}(R)\,\delta\rho(\mathbf{r}-\mathbf{R},t) + \bar{\rho}\,\frac{dg^{eq}(R)}{d\bar{\rho}}\,\delta\rho(\mathbf{r}-\tfrac{1}{2}\mathbf{R},t)\right)\right],\end{aligned}$$

with ∇_R the gradient operator with respect to $\mathbf{R}$. This equation of motion can now be Fourier transformed to yield (for mathematical details, see exercise 2)

$$\frac{\partial}{\partial t}\,\delta\rho(\mathbf{k},t) \;=\; -D^{eff}(k)\,k^2\,\delta\rho(\mathbf{k},t)\,, \tag{26}$$

where the effective diffusion coefficient is given by

$$\begin{aligned} D^{eff}(k) \;=\; & D_0\left[1+2\pi\beta\bar{\rho}\int_0^\infty dR\,R^3\,\frac{dV(R)}{dR}\right.\\ & \left.\times\left(2\,g^{eq}(R)\,\jmath(kR)+\bar{\rho}\frac{dg^{eq}(R)}{d\bar{\rho}}\,\jmath(\tfrac{1}{2}kR)\right)\right]. \end{aligned} \tag{27}$$

The $\jmath$-function is equal to

$$\jmath(x) \;=\; \frac{x\,\cos\{x\}-\sin\{x\}}{x^3}\,. \tag{28}$$

The equation of motion (26) is formally identical to the Cahn-Hilliard equation of motion (9), and its solution is given in eq.(10). The effective diffusion coefficient (27) may seem different from the Cahn-Hilliard diffusion coefficient (11) at first sight. However, since in the integrand in eq. (27) the factor $dV(R)/dR$ limits the integration range effectively to values $R \leq R_V$, and the wavevectors of interest are those for which $kR_V \ll 1$, the $\jmath$-functions may be expanded up to second order in their arguments. Taylor expansion of the sine and cosine functions in eq. (28) gives, $\jmath(x) = -1/3 + x^2/30 + O(x^4)$, so that the diffusion coefficient (27) is equal to

$$D^{eff}(k) \;=\; D_0\beta\left[\frac{dP}{d\bar{\rho}}+\Sigma\,k^2\right] \tag{29}$$

up to terms of order $D_0(kR_V)^4$, where

$$P \;=\; \bar{\rho}\,k_BT-\frac{2\pi}{3}\,\bar{\rho}^2\int_0^\infty dR\,R^3\,\frac{dV(R)}{dR}\,g^{eq}(R) \tag{30}$$

and

$$\Sigma \;=\; \frac{2\pi}{15}\,\bar{\rho}\int_0^\infty dR\,R^5\,\frac{dV(R)}{dR}\left(g^{eq}(R)+\frac{1}{8}\,\bar{\rho}\,\frac{dg^{eq}(R)}{d\bar{\rho}}\right)\,. \tag{31}$$

The expression in eq.(30) for P is precisely the osmotic pressure. Comparison of eq.(30) for the effective diffusion coefficient with the Cahn-Hilliard expression (11) identifies

$$\left.\begin{aligned} D\,a_2 \;&=\; D_0\,\beta\,\tfrac{dP}{d\bar{\rho}}\,,\\ D\,\kappa \;&=\; D_0\,\beta\,\Sigma\,. \end{aligned}\right\} \tag{32}$$

Using that $dP/d\bar{\rho} = \bar{\rho}\, d\mu/d\bar{\rho}$, and $a_2 = d\mu/d\bar{\rho}$ (see eq.(7)), with μ the chemical potential of the colloidal particles, the first of these equations reduces to

$$D/D_0 = \beta\,\bar{\rho}\,. \tag{33}$$

This identifies the transport coefficient D in the Cahn-Hilliard theory. The second of the above equations, together with the expression (31) for Σ, identifies the Cahn-Hilliard square gradient coefficient κ in terms of the microscopic quantities $V(R)$ and $g^{eq}(R)$. Note that eq.(13) for the wavevector of the most rapidly growing sinusoidal density component can be written as

$$k_m = \sqrt{-\frac{dP}{d\bar{\rho}}\,/\,2\Sigma}\,. \tag{34}$$

A deeper quench, where $-\beta dP/d\bar{\rho}$ is relatively large, results in a larger value for k_m.

2.3. THE MECHANISM THAT RENDERS A SYSTEM UNSTABLE

To understand on a microscopic level why a system can become thermodynamically unstable, let us rewrite the Smoluchowski equation (19) as

$$\frac{\partial}{\partial t}\rho(\mathbf{r},t) = -M\,\nabla\cdot\rho(\mathbf{r},t)\left[\mathbf{F}^{Br}(\mathbf{r},t)+\mathbf{F}^{Int}(\mathbf{r},t)\right]\,, \tag{35}$$

where $\mathbf{F}^{Int}$ is the direct force (20) and

$$\mathbf{F}^{Br}(\mathbf{r},t) = -k_BT\,\nabla\ln\{\rho(\mathbf{r},t)\}\,, \tag{36}$$

is the Brownian force on a colloidal particle at the position $\mathbf{r}$. Furthermore, $M = \beta\, D_0$ is a "mobility". Now consider a colloidal particle at $\mathbf{r}$ in an inhomogeneous environment, as sketched in Fig. 5. The inhomogeneous macroscopic density may be thought of as an instantaneous realization of the fluctuating density. A little thought shows that the Brownian force is always directed toward the region with lower concentration, as depicted in Fig. 5. Now suppose that the pair-interaction potential is purely attractive. The direct force $\mathbf{F}^{Int}$ is then directed in the opposite direction, toward the region with a larger density, since in that region there are more neighboring colloidal particles attracting the particle under consideration. This can also be seen formally from eq. (20) for the direct force, using a purely attractive pair-interaction potential. On lowering the temperature, the Brownian force diminishes, since that force is directly proportional to the temperature. The direct force, however, increases in magnitude, due to the fact

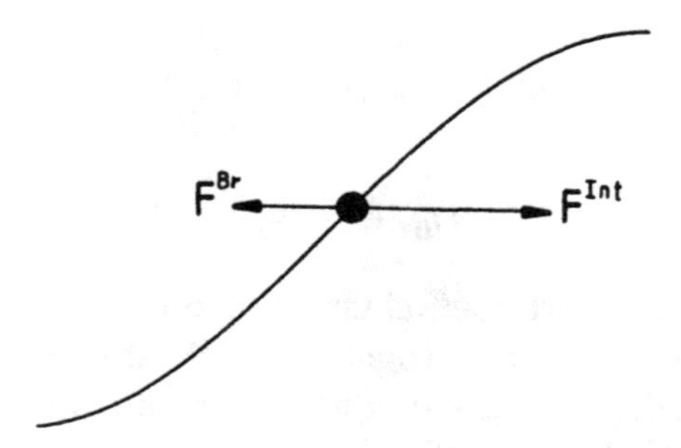

Figure 5. The direct and Brownian force on a colloidal particle, indicated by •, in an inhomogeneous system. An attractive direct force is directed toward the region with larger concentration, as sketched here.

that the pair-correlation function becomes more pronounced (to leading order in the density this follows from the expression $g = \exp\{-V/k_BT\}$, where $V < 0$ for an attractive pair-potential). At the temperature where $|\mathbf{F}^{Int}| > |\mathbf{F}^{Br}|$, the net force on the colloidal particle is directed toward the region with a larger density. This is the mechanism that is responsible for *uphill diffusion*, and leads to phase separation. The pair-potential is never purely attractive for real systems since there is always a hard-core repulsion, and there is a competition between the repulsive and attractive components of the direct force. For large densities one can imagine that hard-core repulsion becomes dominant, leading to stabilization. This causes the spinodal to shift to smaller temperatures on increasing the density at sufficiently large concentrations. At smaller concentrations the attractive force can be dominant, leading to an increase of the spinodal temperature on increasing the concentration.

3. Kinetic Definition of the Spinodal and Binodal and the Experimental Relevance of the Spinodal

The spinodal is theoretically perfectly well-defined as the set of temperatures and densities where the equation of motion for the macroscopic density becomes unstable. However, experimentally, the location of the spinodal can not be determined with arbitrary accuracy. The reason for this is discussed below.

The equation of motion for the macroscopic density is of the form

$$\frac{\partial \rho(\mathbf{r},t)}{\partial t} = F[\rho] \,, \tag{37}$$

where F is a non-linear functional of the macroscopic density. Let us again write

$$\rho(\mathbf{r},t) = \bar{\rho} + \delta\rho(\mathbf{r},t)\,, \tag{38}$$

where $\bar{\rho} = N/V$ is the particle number density of the homogeneous system. Let $L[\delta\rho]$ denote the linearized functional F with respect to $\delta\rho$, and $\Delta F = F - L$ its remainder, so that, $F[\delta\rho] = L[\delta\rho] + \Delta F[\delta\rho]$. The equation of motion (37) is thus rewritten as

$$\frac{\partial \rho(\mathbf{r},t)}{\partial t} = L[\delta\rho] + \Delta F[\delta\rho]\,. \tag{39}$$

The functional $\Delta F[\delta\rho]$ is at least $\mathcal{O}((\delta\rho)^2)$. Note that within the Smoluchowski equation approach as discussed in the previous section we found that, after Fourier transformation, $L[\delta\rho] \equiv -D^{eff}(k)k^2\delta\rho$. The spinodal was defined as the set of densities $\bar{\rho}$ and temperatures where the linearized equation of motion $\frac{\partial\rho(\mathbf{r},t)}{\partial t} = L[\delta\rho]$ becomes unstable. This is the set of $(\bar{\rho},T)$'s where the *homogeneous system* becomes *unstable* against density variations of an *infinitesimally small amplitude.* The spinodal is thus found from a linear stability analysis of the equation of motion for the macroscopic density.

Suppose that for a certain state $(\bar{\rho},T)$ of a homogeneous system the linearized equation of motion is stable, that is, density variations with infinitesimally small amplitude always relax to the homogeneous state with density $\bar{\rho}$. Then either the full non-linear equation of motion is stable against *any* density variation, of arbitrary form and amplitude, or there are types of density variations for which the equation of motion does *not* predict relaxation toward the homogeneous state, but rather predicts an increase of inhomogeneity. In the former case the system is *stable,* in the latter case the system is *metastable.* A metastable state is thus defined as a state where the homogeneous system is stable against density variations of infinitesimally small amplitude, but unstable against certain types of density variations with a *finite* amplitude. If a metastable system is unstable against a certain density variation $\delta\rho(\mathbf{r},t=0)$, then there is a finite number $0 < \lambda < 1$, such that the system is stable against the density variation $\lambda\delta\rho(\mathbf{r},t=0)$. For small enough, but finite λ the non-linear contribution $\Delta F \sim \lambda^2$ becomes insignificant relative to the stabilizing linear contribution $L[\delta\rho] \sim \lambda$. The binodal can now be defined as the set of $(\bar{\rho},T)$'s which marks the transition from a stable state to a metastable state. The density variations of finite amplitude against which a metastable system is unstable are possible candidates for nuclei. The probability for the occurence of a fluctuation giving rise to such a finite amplitude density variation determines the actually appearing nuclei. Phase separation proceeds only after a density variation occurs such that the non-linear contribution ΔF in eq.(39) "overrules" the stabilizing linear contribution. The lag-time for phase separation is related to the probability for such an unstable finite amplitude density variation to occur.

On approach to the spinodal from the metastable side, the minimum amplitude of density variations, which render the full non-linear equation of motion unstable, decreases. The reason for this is that the stabilizing linear term becomes smaller, rendering the non-linear contribution already important for relatively smaller amplitudes of density variations.

The experimental relevance of the binodal is clear. It marks the transition from states where the system remains homogeneous for ever, and those where phase separation occurs after a certain time lag that is determined by the probability for finite amplitude density fluctuations to occur which render the equation of motion unstable. The experimental relevance of the spinodal is less evident since density variations of finite amplitude are always present. On approaching the spinodal from the metastable side, a real system starts to behave as if it crossed the spinodal, that is, it phase separates without any time delay before the spinodal is actually reached; the density fluctuations with small yet finite amplitude, which are always present, are now the nuclei for phase separation from the metastable state. This happens when $| L[\delta\rho_{typ}] | \approx | \Delta F[\delta\rho_{typ}] |$, with $\delta\rho_{typ}(\mathbf{r}, t)$ a typical density fluctuation that has a large probability to occur.

In addition, in the unstable part of the phase diagram, just below the spinodal, the linear term L in the equation of motion (39) is small, and non-linear terms ΔF are important right from the start of the demixing process.

There is thus a gradual crossover from nucleation-dominated phase separation to spinodal-dominated phase separation on crossing the theoretical spinodal. This gradual crossover from nucleation to spinodal-like decomposition kinetics can be studied experimentally by means of time resolved turbidity experiments, after a temperature quench of a homogeneous system with a given density $\bar{\rho}$.[3] After a quench from a stable state into the metastable region of the phase diagram, away from the spinodal, there is a time lag before phase separation occurs, and the turbidity remains constant during that time. The initial slope of the turbidity versus time is thus zero for nucleation dominated phase separation. For spinodal-like dominated demixing, the initial slope is very large, since demixing occurs

[3]The turbidity τ measures the amount of light that is scattered by the system. It is related to the beam intensity I_0 that is incident on the sample and the intensity I of the beam after having passed the sample by the Lambert-Beer equation, $I/I_0 = \exp\{-\tau d\}$, with d the thickness of the sample. The turbidity increases as a function of time during demixing because the density inhomogeneities give rise to inhomogeneities in the refractive index, leading in turn to scattering of light.

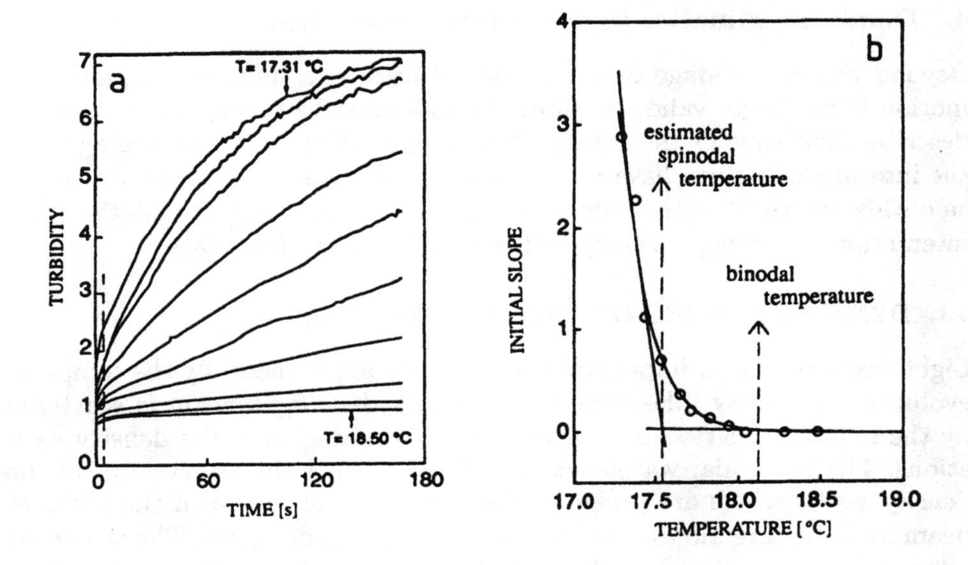

Figure 6. Results of turbidity measurements on a stearyl-silica colloid in benzene, at a volume fraction of 0.11. Data are taken from Verduin and Dhont (1995). (a) is a plot of the turbidity (in arbitrary units) as a function of time right after a quench from the stable state at 18.5 °C. The dotted line close to the vertical axis indicates the time required for the fluid and colloidal material to attain the new temperature : the slight increase in turbidity during the very first seconds after the quench is the result of an increased difference in refractive index between the colloidal particles and the fluid. The relevant initial slope is the slope after this short period of refractive index equilibration. The quench temperatures are : 17.31, 17.39, 17.46, 17.55, 17.60, 17.79, 17.85, 17.96, 18.06 and $18.50^0 C$. (b) is a plot of the initial slopes as obtained from Fig.a as a function of the quench temperature. The binodal temperature and the estimated location of the spinodal temperature are indicated. Notice the broad crossover from nucleation dominated to spinodal-decomposition dominated phase-separation kinetics.

without any time delay. The initial slope of the turbidity versus time right after a temperature quench is thus expected to change dramatically on crossing the spinodal. The temperature range where the crossover behavior for the initial slope occurs is indicative of the width of the region in the phase diagram where nucleation and spinodal decomposition mix up. The experimental relevance of the spinodal is that it locates the crossover region. An experimental example of time resolved turbidity measurements on a colloidal system is given in the Fig. 6. In Fig. 6a the time dependence of the turbidity is given for various quench depths and Fig. 6b is a plot of the initial slope versus the quench temperature; the estimated location of the spinodal is indicated in this figure. The crossover region is seen to be quite broad, extending far into the metastable region. Whether this is a generic feature, which also applies to molecular systems, is not known.

4. Demixing Kinetics in the Intermediate Stage

Beyond the initial stage of the phase separation, a linearized equation of motion is no longer valid. Non-linear terms must be taken into account to describe demixing in the intermediate stage. What the initial stage and the intermediate stage have in common, however, is that the density varies smoothly on the length scale of the order of the range R_V of the pair-interaction potential, contrary to the transition and late stage.

4.1. DYNAMIC SCALING OF THE STRUCTURE FACTOR

Light scattering is an important tool to study experimentally the temporal evolution of density inhomogeneities. An incident light beam is scattered by the inhomogeneities in refractive index resulting from the density variations. The particular wavelength $\Lambda = 2\pi/k$, with k the wavevector, of the density waves which are probed are set by the angle between the incident beam and the direction in which scattered light is detected. The scattered intensity is proportional to the so-called structure factor $S(\mathbf{k}, t)$, which at small wavevectors $\mathbf{k}$ is related to inhomogeneities of the macroscopic density as

$$S(\mathbf{k}, t) \;=\; \frac{1}{N} <|\rho(\mathbf{k}, t|^2>_{init} \,, \tag{40}$$

with $\rho(\mathbf{k}, t)$ the Fourier transform of the macroscopic density. The ensemble averaging $< \cdots >_{init}$ is over the initial realizations of the macroscopic density, before phase separation occurs. In the initial stage the structure factor immediately follows from eq.(26),

$$S(k, t) \;=\; S(k, t = 0) \, \exp\left\{-D^{eff}(k) k^2 t\right\} \,. \tag{41}$$

In the initial stage of spinodal demixing, a dominant length scale L develops, which length scale is related to the most rapidly growing wavevector in the initial stage (k_m in eq.(34)) as $L = 2\pi/k_m$. As will turn out, the dominant length scale becomes a function of time in the intermediate stage as a result of non-linear coupling. The dominant length scale is therefore denoted here as $L(t)$, which is now related to the wavevector $k_{ms}(t)$ where the structure factor peaks,

$$L(t) \;=\; 2\pi/k_{ms}(t) \,, \tag{42}$$

The dominance of a single length scale implies that distances can only be measured in units of that single length scale, so that

$$\frac{<\delta\rho(\mathbf{r}, t)\, \delta\rho(\mathbf{r}', t)>_{init}}{<\delta\rho^2(\mathbf{r}, t)>_{init}} \;\equiv\; F\left(\frac{|\mathbf{r} - \mathbf{r}'|}{L(t)}\right) \,. \tag{43}$$

The normalizing denominator on the left hand-side fixes the value of the *scaling function* $F(x)$ to unity at $x = 0$ for all times. It follows that (with $x = \mid \mathbf{r} - \mathbf{r}' \mid / L(t)$)

$$
\begin{aligned}
S(k,t) &= \frac{1}{N} <\mid \delta\rho(\mathbf{k},t) \mid^2>_{init} \\
&= \frac{1}{N} \int d\mathbf{r} \int d\mathbf{r}' \; <\delta\rho(\mathbf{r},t)\, \delta\rho(\mathbf{r}',t)>_{init} \exp\{i\mathbf{k}\cdot(\mathbf{r}-\mathbf{r}')\} \\
&= \frac{1}{N} <\delta\rho^2(\mathbf{r},t)>_{init} \int d\mathbf{r} \int d\mathbf{r}' \, F\Big(\frac{\mid \mathbf{r}-\mathbf{r}' \mid}{L(t)}\Big) \exp\{i\mathbf{k}\cdot(\mathbf{r}-\mathbf{r}')\} \\
&= \frac{4\pi}{\bar{\rho}} <\delta\rho^2(\mathbf{r},t)>_{init} \int_0^\infty d\mid\mathbf{r}-\mathbf{r}'\mid \mid\mathbf{r}-\mathbf{r}'\mid^2 F\Big(\frac{\mid\mathbf{r}-\mathbf{r}'\mid}{L(t)}\Big) \frac{\sin\{k\mid\mathbf{r}-\mathbf{r}'\mid\}}{k\mid\mathbf{r}-\mathbf{r}'\mid} \\
&= \frac{4\pi}{\bar{\rho}} L^3(t) \; <\delta\rho^2(\mathbf{r},t)>_{init} \int_0^\infty dx\, x\, F(x) \frac{\sin\{k\,L(t)\,x\}}{k\,L(t)} \, .
\end{aligned}
$$

In the fourth equation the integrations with respect to the spherical angular coordinates of $\mathbf{r} - \mathbf{r}'$ have been performed. Note that $< \delta\rho^2(\mathbf{r}, t) >_{init}$ is independent of $\mathbf{r}$ for initially translationally invariant systems. Integration of the structure factor (40) with respect to $\mathbf{k}$ yields (see also eq.(55))

$$
\frac{S(k,t)\, L^{-3}(t)}{\int_0^\infty dk'\, k'^2 S(k',t)} = \frac{2}{\pi} \int_0^\infty dx\, x\, F(x) \frac{\sin\{k\,L(t)\,x\}}{k\,L(t)} \, . \tag{44}
$$

The right hand-side of this *dynamic scaling relation* is a function of $k\,L(t) \sim k/k_{ms}(t)$ only. Therefore, plots of the quantity on the left hand-side of eq.(44) versus $k/k_{ms}(t)$ for various times must collapse onto a single curve.

Notice that it follows from the scaling equation (44), together with eq.(42) for the dominant length scale, that plots of $S(k,t)/S(k_{ms}(t),t)$ versus $k/k_{ms}(t)$ for various times should also collapse onto a single curve. This scaling means that the structure factor peaks have the same form, and differ only in the location of their maxima. One might call this scaling *dynamic similarity scaling.*

4.2. SOLUTION OF THE SMOLUCHOWSKI EQUATION

Let us first rederive the linearized equation of motion (26) in an alternative way that allows for the inclusion of non-linear terms. Instead of Fourier transforming the Smoluchowski equation eq.(25) with respect to $\mathbf{r}$, one may alternatively Taylor expand the changes $\delta\rho(\mathbf{r}-\mathbf{R},t)$ and $\delta\rho(\mathbf{r}-\frac{1}{2}\mathbf{R},t)$ of the density around $\mathbf{R} = \mathbf{0}$ in case the density is smooth on the length scale R_V, since the factor $\nabla_R V(R)$ in the integrand effectively limits the integration range to $R \leq R_V$. Substitution of the Taylor expansions

$$
\delta\rho(\mathbf{r}-\mathbf{R},t) = \delta\rho(\mathbf{r},t) - \mathbf{R}\cdot\nabla\delta\rho(\mathbf{r},t) \tag{45}
$$

$$+\frac{1}{2}\mathbf{RR}:\nabla\nabla\delta\rho(\mathbf{r},t)-\frac{1}{6}\mathbf{RRR}\,\vdots\,\nabla\nabla\nabla\rho(\mathbf{r},t)+\cdots,$$

and

$$\begin{aligned}\delta\rho(\mathbf{r}-\frac{1}{2}\mathbf{R},t) &= \delta\rho(\mathbf{r},t)-\frac{1}{2}\mathbf{R}\cdot\nabla\delta\rho(\mathbf{r},t) \\ &\quad +\frac{1}{8}\mathbf{RR}:\nabla\nabla\delta\rho(\mathbf{r},t)-\frac{1}{48}\mathbf{RRR}\,\vdots\,\nabla\nabla\nabla\rho(\mathbf{r},t)+\cdots,\end{aligned}\tag{46}$$

into eq.(25) and keeping only linear terms in $\delta\rho(\mathbf{r},t)$ yields

$$\begin{aligned}\frac{\partial}{\partial t}\delta\rho(\mathbf{r},t) &= D_0\Bigg[\nabla^2\delta\rho(\mathbf{r},t)-\beta\bar{\rho}\nabla\cdot\int d\mathbf{R}\,[\nabla_R V(R)] \\ &\quad\times\Bigg\{\mathbf{R}\cdot\nabla\delta\rho(\mathbf{r},t)\left(g^{eq}(R)+\frac{1}{2}\bar{\rho}\frac{dg^{eq}(R)}{d\bar{\rho}}\right) \\ &\quad+\mathbf{RRR}\,\vdots\,\nabla\nabla\nabla\delta\rho(\mathbf{r},t)\left(\frac{1}{6}g^{eq}(R)+\frac{1}{48}\bar{\rho}\frac{dg^{eq}(R)}{d\bar{\rho}}\right)\Bigg\}\Bigg].\end{aligned}\tag{47}$$

Since $\nabla_R V(R)$ is an odd function of $\mathbf{R}$ and $g^{eq}(R)$ an even function, integrals like

$$\int d\mathbf{R}\,[\nabla_R V(R)]g^{eq}(R) \quad\text{and}\quad \int d\mathbf{R}\,[\nabla_R V(R)]g^{eq}(R)\mathbf{RR}$$

are zero. Terms which are proportional to such integrals of odd functions are omitted in eq. (47). The spherical angular integrations can be performed after substitution of $\nabla_R V(R) = \hat{\mathbf{R}}\,dV(R)/dR$, with $\hat{\mathbf{R}} = \mathbf{R}/R$ the unit vector along $\mathbf{R}$, and using that

$$\int d\hat{\mathbf{R}}\,\hat{\mathbf{R}}\hat{\mathbf{R}} = \frac{4\pi}{3}\hat{\mathbf{I}}\,,\tag{48}$$

$$\int d\hat{\mathbf{R}}\,\hat{R}_i\hat{R}_j\hat{R}_k\hat{R}_l = \frac{4\pi}{15}\left[\delta_{ij}\delta_{kl}+\delta_{ik}\delta_{jl}+\delta_{il}\delta_{jk}\right],\tag{49}$$

where the integration ranges over the unit spherical surface and where δ_{ij} is the Kronecker delta ($\delta_{ij}=0$ for $i\neq j$, and $\delta_{ij}=1$ for $i=j$). The equation of motion now reduces to

$$\frac{\partial}{\partial t}\delta\rho(\mathbf{r},t) = D_0\beta\left[\frac{dP}{d\bar{\rho}}\nabla^2\delta\rho(\mathbf{r},t)-\Sigma\,\nabla^2\nabla^2\delta\rho(\mathbf{r},t)\right],\tag{50}$$

with P and Σ given in eqs. (30, 31), respectively. Fourier transformation reproduces eqs. (26, 29).

The above procedure can be applied to include higher order terms in $\delta\rho(\mathbf{r},t)$. The equation of motion for the structure factor can be found from that of the macroscopic density as follows,

$$\begin{aligned}\frac{\partial}{\partial t}S(k,t) &= \frac{1}{N}\frac{\partial}{\partial t}\int d\mathbf{r}\int d\mathbf{r}' < \delta\rho(\mathbf{r},t)\,\delta\rho(\mathbf{r}',t) >_{init} \exp\{i\mathbf{k}\cdot(\mathbf{r}-\mathbf{r}')\} \\ &= 2\frac{1}{N}\int d\mathbf{r}\int d\mathbf{r}' < \delta\rho(\mathbf{r}',t)\,\frac{\partial\delta\rho(\mathbf{r},t)}{\partial t} >_{init} \exp\{i\mathbf{k}\cdot(\mathbf{r}-\mathbf{r}')\}\,. \qquad (51)\end{aligned}$$

The last equation follows from

$$\begin{aligned}\int d\mathbf{r}\int d\mathbf{r}' < \delta\rho(\mathbf{r}',t)\,\frac{\partial\delta\rho(\mathbf{r},t)}{\partial t} >_{init} \exp\{i\mathbf{k}\cdot(\mathbf{r}-\mathbf{r}')\} &= \\ \int d\mathbf{r}'\int d\mathbf{r} < \delta\rho(\mathbf{r},t)\,\frac{\partial\delta\rho(\mathbf{r}',t)}{\partial t} >_{init} \exp\{i\mathbf{k}\cdot(\mathbf{r}-\mathbf{r}')\}\,,\end{aligned}$$

which in turn follows from inversion invariance of the ensemble averages, meaning that these do not change under the transformation $\mathbf{r}\to-\mathbf{r}$ and $\mathbf{r}'\to-\mathbf{r}'$.

The equation of motion (19) is now substituted into eq. (51) and subsequently expanded with respect to $\delta\rho(\mathbf{r},t)$ and $\delta\rho(\mathbf{r}',t)$, as discussed in the first part of this subsection, but now including higher order terms. We will assume here that $\delta\rho(\mathbf{r},t)$ for a fixed position and time is approximately a Gaussian variable. This is certainly wrong in the transition and late stage, where the probability density function (pdf) of the density is peaked around two concentrations, which ultimately become equal to the two binodal concentrations. In the initial and intermediate stage such a splitting of the pdf is assumed not to occur, and the pdf is approximately "bell-shaped" like a Gaussian variable. When one is willing to accept the Gaussian character of the macroscopic density, averages $<\cdots>_{init}$ of odd products of changes in the density are zero, while averages of products of four density changes can be written as a sum of products containing only two density changes. Hence, in the expansion of the integrand in eq. (51) with respect to $\delta\rho$, only even products need be considered, and averages of products of four density changes can be reduced to products of two densities with the help of Wick's theorem. Furthermore, as discussed above, in the intermediate stage there is no need to take higher order spatial derivatives than fourth order into account. Extending the Taylor expansion (23) to third order (with $g^{eq}\equiv g^{eq}(|\mathbf{r}-\mathbf{r}'|)$),

$$\delta g(\mathbf{r},\mathbf{r}',t) = \frac{dg^{eq}}{d\bar{\rho}}\delta\rho(\tfrac{\mathbf{r}+\mathbf{r}'}{2},t) + \frac{1}{2}\frac{d^2g^{eq}}{d\bar{\rho}^2}\delta\rho^2(\tfrac{\mathbf{r}+\mathbf{r}'}{2},t) + \frac{1}{6}\frac{d^3g^{eq}}{d\bar{\rho}^3}\delta\rho^3(\tfrac{\mathbf{r}+\mathbf{r}'}{2},t),$$

one obtains, after a considerable effort (see the appendix for mathematical details),

$$\begin{aligned}\frac{\partial}{\partial t} S(k,t) &= -2D_0\beta k^2 S(k,t)\left[\frac{dP}{d\bar{\rho}} + \Sigma k^2\right] \\ &\quad - D_0\beta k^2 S(k,t)\left[\frac{d^3P}{d\bar{\rho}^3} + \frac{d^2\Sigma}{d\bar{\rho}^2}k^2\right] < \delta\rho^2(\mathbf{r},t) >_{init} \qquad (52)\\ &+2D_0\beta k^2 S(k,t)\left[\Sigma^\circ < \delta\rho(\mathbf{r},t)\nabla^2\delta\rho(\mathbf{r},t) >_{init} + \Sigma^\bullet <|\nabla\delta\rho(\mathbf{r},t)|^2>_{init}\right],\end{aligned}$$

where

$$\Sigma^\circ = \frac{4\pi}{15}\int_0^\infty dR\, R^5 \frac{dV(R)}{dR}\left(\frac{5}{8}\frac{dg^{eq}(R)}{d\bar{\rho}} + \frac{5}{6}\bar{\rho}\frac{d^2g^{eq}(R)}{d\bar{\rho}^2} + \frac{5}{48}\bar{\rho}^2\frac{d^3g^{eq}(R)}{d\bar{\rho}^3}\right) \qquad (53)$$

and

$$\Sigma^\bullet = \frac{4\pi}{15}\int_0^\infty dR\, R^5 \frac{dV(R)}{dR}\left(\frac{5}{8}\bar{\rho}\frac{d^2g^{eq}(R)}{d\bar{\rho}^2} + \frac{1}{16}\bar{\rho}^2\frac{d^3g^{eq}(R)}{d\bar{\rho}^3}\right). \qquad (54)$$

Notice that averages like $< \delta\rho^2(\mathbf{r},t) >_{init}$ are independent of position, but are still time dependent.

4.2.1. *Evaluation of the ensemble averages in terms of the structure factor* First, consider the average $< \delta\rho^2(\mathbf{r},t) >_{init}$. For isotropic systems, integration of the structure factor (40) with respect to $\mathbf{k}$ yields

$$\begin{aligned}\int d\mathbf{k}\, S(k,t) &= \frac{1}{N}\int d\mathbf{r}\int d\mathbf{r}'\int d\mathbf{k} < \delta\rho(\mathbf{r},t)\,\delta\rho(\mathbf{r}',t) >_{init} \exp\{i\mathbf{k}\cdot(\mathbf{r}-\mathbf{r}')\} \\ &= (2\pi)^3\frac{1}{N}\int d\mathbf{r}\int d\mathbf{r}'\,\delta(\mathbf{r}-\mathbf{r}') < \delta\rho(\mathbf{r},t)\,\delta\rho(\mathbf{r}',t) >_{init} \\ &= (2\pi)^3\frac{1}{N}\int d\mathbf{r} < \delta\rho^2(\mathbf{r},t) >_{init},\end{aligned}$$

where it is used that $\int d\mathbf{k}\exp\{i\mathbf{k}\cdot(\mathbf{r}-\mathbf{r}')\} = (2\pi)^3\delta(\mathbf{r}-\mathbf{r}')$, with $\delta(\mathbf{r}-\mathbf{r}')$ the 3-dimensional delta distribution. Since there is no preferred position in the system on average, the ensemble average with respect to initial conditions is independent of position. It is thus found that

$$< \delta\rho^2(\mathbf{r},t) >_{init} = \frac{1}{(2\pi)^3}\bar{\rho}\int d\mathbf{k}\, S(k,t) = \frac{1}{2\pi^2}\bar{\rho}\int_0^\infty dk\, k^2 S(k,t). \qquad (55)$$

The average $< \delta\rho(\mathbf{r},t)\nabla_r^2\delta\rho(\mathbf{r},t) >_{init}$ is calculated as follows. Using Green's second integral theorem, with the neglect of surface integrals, yields similarly

$$\int d\mathbf{k}\, k^2\, S(k,t) = \frac{-1}{N}\int d\mathbf{r}\int d\mathbf{r}'\int d\mathbf{k} < \delta\rho(\mathbf{r},t)\,\delta\rho(\mathbf{r}',t) >_{init} \nabla_r^2 \exp\{i\mathbf{k}\cdot(\mathbf{r}-\mathbf{r}')\}$$

$$= \frac{-1}{N}\int d\mathbf{r}\int d\mathbf{r}'\int d\mathbf{k} < [\nabla_r^2\delta\rho(\mathbf{r},t)]\,\delta\rho(\mathbf{r}',t) >_{init} \exp\{i\mathbf{k}\cdot(\mathbf{r}-\mathbf{r}')\}$$
$$= \frac{-(2\pi)^3}{N}\int d\mathbf{r} < [\nabla_r^2\delta\rho(\mathbf{r},t)]\,\delta\rho(\mathbf{r},t) >_{init} .$$

Since the ensemble average is position independent it follows that

$$< \delta\rho(\mathbf{r},t)\nabla_r^2\delta\rho(\mathbf{r},t) >_{init} = -\frac{1}{2\pi^2}\,\bar{\rho}\int_0^\infty dk\, k^4 S(k,t)\,. \tag{56}$$

The neglect of surface integrals in Green's integral theorem means that the influence of the walls of the container of the system on the decomposition process is not considered. Similarly,

$$\int d\mathbf{k}\, k^2\, S(k,t) = \frac{1}{N}\int d\mathbf{r}\int d\mathbf{r}'\int d\mathbf{k} < \delta\rho(\mathbf{r},t)\,\delta\rho(\mathbf{r}',t) >_{init} \nabla_r\cdot\nabla_{r'} \exp\{i\mathbf{k}\cdot(\mathbf{r}-\mathbf{r}')\}$$

$$= \frac{1}{N}\int d\mathbf{r}\int d\mathbf{r}'\int d\mathbf{k} < [\nabla_r\delta\rho(\mathbf{r},t)]\cdot[\nabla_{r'}\delta\rho(\mathbf{r}',t)] >_{init} \exp\{i\mathbf{k}\cdot(\mathbf{r}-\mathbf{r}')\}$$

$$= \frac{(2\pi)^3}{N}\int d\mathbf{r} < [\nabla_r\delta\rho(\mathbf{r},t)]\cdot[\nabla_r\delta\rho(\mathbf{r},t)] >_{init} ,$$

so that

$$<|\nabla_r\delta\rho(\mathbf{r},t)|^2>_{init} = \frac{1}{2\pi^2}\bar{\rho}\int_0^\infty dk\, k^4 S(k,t) = -<\delta\rho(\mathbf{r},t)\nabla_r^2\delta\rho(\mathbf{r},t)>_{init} . \tag{57}$$

It is important to note that **the structure factor that is integrated with respect to the wavevector in the above equations is only that part of the structure factor that relates to the demixing process, and is given in eq. (40)**. The integration therefore does not extend to infinity, but really goes up to some finite wavevector of the order of a few times k_{ms}, where the demixing peak of the structure factor attains its maximum value. The "molecular contribution" to the structure factor is understood not to be included in any of the above equations. In an experiment, the integrals over the structure factor in the above equations can be obtained by numerically integrating the intensity peak at small scattering angles that emerges during demixing.

The explicit non-linear equation of motion for the structure factor is now obtained from eq. (52) by substitution of eqs. (55-57), yielding

$$\frac{\partial}{\partial t} S(k,t) \;=\; -2\,D_0\,\beta k^2 S(k,t)\left[\frac{dP}{d\bar{\rho}} + \Sigma k^2\right]$$

$$- \; D_0 \, \beta k^2 S(k,t) \left[\frac{d^3 P}{d\bar{\rho}^3} + \frac{d^2 \Sigma}{d\bar{\rho}^2} k^2 \right] \frac{\bar{\rho}}{2\pi^2} \int_0^\infty dk' \, k'^2 S(k',t)$$

$$- 2 \, D_0 \beta k^2 S(k,t) \, \Sigma^{\circ - \bullet} \, \frac{\bar{\rho}}{2\pi^2} \int_0^\infty dk' \, k'^4 S(k',t) \,, \qquad (58)$$

with $\Sigma^{\circ-\bullet} = \Sigma^\circ - \Sigma^\bullet$. Keeping only the first term on the right hand-side in the above equation of motion reproduces the linear theory result (to see this, multiply both sides of eq.(26) by $\delta\rho^*(\mathbf{k}, t)$ and average with respect to initial conditions).

4.2.2. *Simplification of the equation of motion*

Not all terms on the right hand-side of the equation of motion (58) are equally important. Neglect of the irrelevant terms simplifies the equation of motion considerably and reduces the number of independent parameters.

The wavevector dependent contribution $\sim \Sigma k^2$ in the very first term on the right hand-side of eq. (58) is essential, even thought the wavevectors of interest are small. This is due to the fact that near the spinodal $dP/d\bar{\rho}$ is small and negative. The wavevector dependent contribution $\sim d^2\Sigma/d\bar{\rho}^2 \, k^2$ to the second term, however, is not essential, since $d^3P/d\bar{\rho}^3$ is not small, except possibly for quenches close to the critical point. For the small wavevectors under consideration one may neglect the contribution $\sim d^2\Sigma/d\bar{\rho}^2 \, k^2$ in the second term on the right hand-side in eq. (58). Physically, this means that the local density dependence of the contribution of gradients in the density to the Helmholtz free energy is neglected, that is, the density dependence of the Cahn-Hilliard square gradient coefficient is neglected.

Furthermore, the dimensionless numbers $\beta\bar{\rho}^2 d^3P/d\bar{\rho}^3$ and $\beta\bar{\rho}^2\Sigma^{\circ-\bullet}/R_V^2$ are probably not of a different order of magnitude. The ratio of the third and second term on the right hand-side of eq. (58) is thus of the order

$$\frac{\textit{third term}}{\textit{second term}} = O\Big(\int_0^\infty dk' \, k'^2 (k' R_V)^2 S(k',t) \; / \; \int_0^\infty dk' \, k'^2 S(k',t) \Big) \, .$$

This ratio is small since $k' R_V \ll 1$, so that the third term may be neglected against the second term.

The equation of motion (58) thus reduces to

$$\frac{\partial}{\partial t} S(k,t) \;=\; -2 \, D_0 \, \beta k^2 S(k,t) \left[\frac{dP}{d\bar{\rho}} + \Sigma k^2 \right] \qquad (59)$$

$$- \; D_0 \, \beta k^2 S(k,t) \, \frac{d^3 P}{d\bar{\rho}^3} \, \frac{\bar{\rho}}{2\pi^2} \int_0^\infty dk' \, k'^2 S(k',t) \, .$$

This equation contains the relevant features of phase separation in the intermediate stage, for quenches away from the critical point.

4.2.3. *Shift of $k_m(t)$ and $k_c(t)$ with time*
During the initial stage the wavevector of the most rapidly growing density wave is independent of time, and coincides with the maximum of the structure factor (see eq. (34)). This is no longer true beyond the initial stage. The wavevector of the relatively most rapidly growing sinusoidal density variation is easily found from eq.(59),

$$k_m(t) = \sqrt{-\left[\frac{dP}{d\bar{\rho}} + \frac{d^3P}{d\bar{\rho}^3}\frac{\bar{\rho}}{4\pi^2}\int_0^\infty dk'\, k'^2 S(k',t)\right] / \, 2\Sigma} \,. \qquad (60)$$

Since $d^3P/d\bar{\rho}^3 > 0$, and $\int_0^\infty dk'\, k'^2 S(k',t) = \frac{2\pi^2}{\bar{\rho}} < \delta^2\rho(\mathbf{r},t) >_{init}$ evidently increases with time, $k_m(t)$ shifts to smaller wavevectors as time proceeds. This means that the regions of lower and higher density increase their size.

The wavevector where the structure factor peaks, which is denoted as $k_{ms}(t)$, does not coincide with the most rapidly growing wavevector $k_m(t)$ beyond the linear stage. Since the maximum of the structure factor shifts to lower wavevectors we must have that $k_m(t) < k_{ms}(t)$ beyond the initial stage.

The critical wavevector $k_c(t)$, beyond which density waves are stable, is easily seen to be equal to

$$k_c(t) = \sqrt{2}\, k_m(t) \,, \qquad (61)$$

just as in the initial stage.

Note it follows from eq. (55) and the Gaussian character of the density that eq. (60) can also be written as

$$k_m(t) = \sqrt{- < \frac{dP(\bar{\rho} + \delta\rho(\mathbf{r},t))}{d\bar{\rho}} >_{init} / \, 2\Sigma} \,, \qquad (62)$$

which expression is valid up to second order in $\delta\rho(\mathbf{r},t)$. This expression reduces to eq. (34) for k_m during the initial stage where $\delta\rho(\mathbf{r},t)$ is small compared to $\bar{\rho}$.

4.2.4. *The dimensionless equation of motion*
For numerical purposes and to reduce the number of parameters, the equation of motion (59) is rewritten in dimensionless form. First, using eq. (60) for $k_m(t)$, it is found that eq. (59) can be written as

$$\frac{\partial}{\partial t} S(k,t) = 4D_0\beta\Sigma\, k_m^4(t) \left(\frac{k}{k_m(t)}\right)^2 \left[1 - \frac{1}{2}\left(\frac{k}{k_m(t)}\right)^2\right] S(k,t) \,. \qquad (63)$$

Let us now introduce the dimensionless wavevector K and time τ,

$$\begin{aligned} K &= k/k_{m,0}\,, & (64)\\ \tau &= -2D_0\beta\frac{dP}{d\bar{\rho}}\,k_{m,0}^2\,t\,, & (65)\end{aligned}$$

where $k_{m,0} = k_m(t=0)$ is the wavevector of the most rapidly growing density wave during the initial stage, which is given in eq. (60).[4] The dimensionless variable τ is the time in units of the time that a particle with an effective diffusion coefficient $-D_0\,\beta dP/d\bar{\rho}$ requires for diffusion over a distance $\sim k_{m,0}^{-1}$. The equation of motion (59) in the desired dimensionless form reads

$$\frac{\partial}{\partial\tau}\,S(K,\tau) = \left[\left(\frac{k_m(\tau)}{k_{m,0}}\right)^2 K^2 - \frac{1}{2}K^4\right] S(K,\tau)\,. \qquad (66)$$

The ratio $k_m(\tau)/k_{m,0}$ is similarly written in dimensionless form, using eq. (60),

$$\left(\frac{k_m(\tau)}{k_{m,0}}\right)^2 = 1 - C\int_0^\infty dK'\,K'^2 S(K',\tau)\,, \qquad (67)$$

with

$$C = \sqrt{-\frac{dP/d\bar{\rho}}{2\Sigma}\,\frac{d^3P/d\bar{\rho}^3}{2\Sigma}\,\frac{\bar{\rho}}{4\pi^2}} > 0 \qquad (68)$$

and $K' = k'/k_{m,0}$. The number of parameters is thus reduced to the single dimensionless constant C.

4.3. SOLUTION OF THE EQUATION OF MOTION

The equation of motion (66, 67) is easily solved numerically, where the wavevector integration extends up to the non-zero wavevector where the actual structure factor becomes equal to the initial structure factor. Fig. 7a shows the numerical solution with $C = 0.01$. The insert is a blow up for earlier times, corresponding to the initial stage.

As can be seen from this figure, the maximum in the structure factor shifts to smaller wavevectors. The wavevector $k_m(t)$ of the most rapidly growing density wave is therefore smaller than the wavevector $k_{ms}(t)$ where the structure factor peaks. The critical wavevector $k_c(t)$ is quite close to $k_{ms}(t)$, resulting in a decrease of the structure factor just beyond its maximum.

[4]The assumption here is that the integral $\int_0^\infty dk'\,k'^2 S(k',t=0)$ is of no significance.

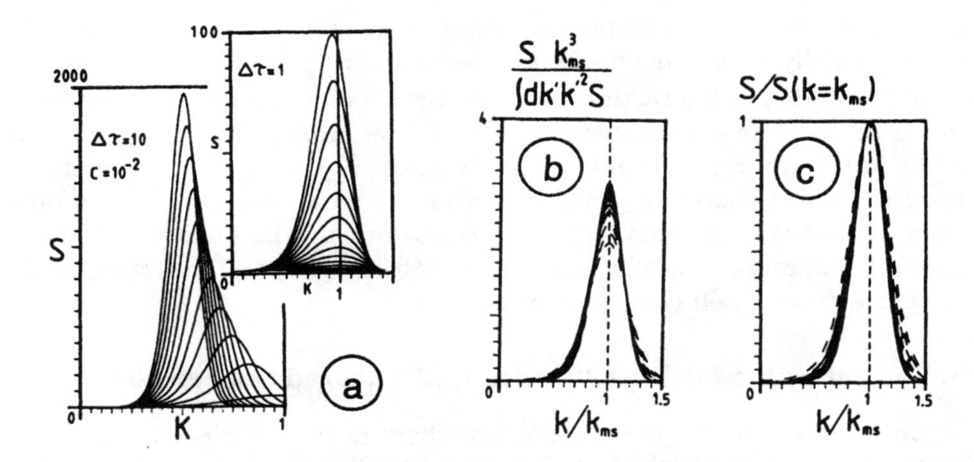

Figure 7. (a) The numerical solution of the equation of motion (66, 67) for $C = 0.01$. The initial structure factor $S(K, t = 0)$ is simply taken equal to 1. The inserts show the structure factor for early times. (b) A test of the scaling relation (44). (c) A test of dynamic similarity scaling. In Figs. b, c all curves in Fig. a where $S > 100$ are included.

Scaling is always approximate since there is never rigorously a single length scale, but only dominance of a single length scale. As can be seen from Figs. 7b, c, scaling becomes more accurate as time proceeds, and ultimately all curves converge to the thick solid curve in these figures. These thick solid lines are the *dynamic scaling functions.* The universality of these scaling functions admit a direct experimental verification of the ideas developed in this section.[5]

It should be noted that the structure-factor scaling functions are in principle dependent on the initial state of the density and the value of the parameter C, and possibly on the particular manner in which the quench is realized. However, it is found from numerical calculations that **there is remarkably little variation of the scaling functions on varying the initial ensemble average $<\mid \delta\rho(\mathbf{k}, t = 0) \mid^2>_{init}$ and the value of the parameter C. This property of the equation of motion (66, 67) makes the dynamic scaling functions universal in the sense that they are independent of initial conditions and quench**

[5]The dynamic similarity scaling function in Fig. 7c is almost perfectly described by the simple function,

$$\frac{S(k,t)}{S(k = k_{ms}, t)} = \exp\left\{-30\left(\frac{k}{k_{ms}} - 1\right)^2 - 25\left(\frac{k}{k_{ms}} - 1\right)^3\right\}.$$

characteristics.[6] The scaling functions as given in Figs. 7b, c, apply for any physically reasonable choice of these quantities.[7]

Hydrodynamic interactions between the colloidal particles can be taken into account in an approximate way (see Dhont (1996, 1997)), leading to an additional term in the equation of motion (66, 67), with an additional constant C', which, like C, depends on the quench parameters. The remarkable thing is that the dynamic scaling functions remain the same when hydrodynamic interactions are included, although the temporal evolution of the structure factor itself changes considerably.

5. Initial Spinodal Decomposition of Sheared Suspensions

In this section we analyze the effect of shear flow on the evolution of the macroscopic density in the initial stage, where linearization with respect to changes of the density with time is allowed. Being based on thermodynamic reasoning, the Cahn-Hilliard approach is not easily extended to include effects of shear flow. As yet, it is not known how to extend thermodynamics to include shear flow, if possible at all. Within the Smoluchowski approach, however, it is rather straightforward to include a shear flow.

The Smoluchowski equation (15) for a sheared system, with neglect of the hydrodynamic interaction, reads

$$\frac{\partial}{\partial t} P = D_0 \sum_{j=1}^{N} \nabla_{r_j} \cdot \left[\beta[\nabla_{r_j} \Phi]\, P + \nabla_{r_j} P\right] - \sum_{j=1}^{N} \nabla_{r_j} \cdot [\mathbf{\Gamma} \cdot \mathbf{r}_j \, P] \,. \tag{69}$$

P is again the probability density function (pdf) of the position coordinates of all colloidal particles. The *velocity gradient matrix* $\mathbf{\Gamma}$ defines the externally imposed shear flow velocity $\mathbf{u}(\mathbf{r}) = \mathbf{\Gamma} \cdot \mathbf{r}$. The velocity of a colloidal particle with position coordinate $\mathbf{r}_j$ due to the externally imposed shearing motion of the system is equal to $\mathbf{\Gamma} \cdot \mathbf{r}_j$. We will use

$$\mathbf{\Gamma} = \dot{\gamma} \begin{pmatrix} 0 & 1 & 0 \\ 0 & 0 & 0 \\ 0 & 0 & 0 \end{pmatrix} , \tag{70}$$

[6]Provided that the quench is deep enough. The equations of motion derived here are valid for quenches not too close to the spinodal where $\beta\, dP/d\bar{\rho}$ and $\beta\bar{\rho}^2 d^3P/d\bar{\rho}^3$ are not very small.

[7]"Physically reasonable" is any choice where $k_{ms}(t)/k_{m,0}$ smoothly evolves from 1 to smaller values. That is, any choice of $S(K, \tau = 0)$ and C for which the non-linear terms in the equation of motion (66, 67) are insignificant at zero time are termed "physically reasonable". Non-linear terms should thus become important solely due to the growth of the structure factor.

with $\dot{\gamma}$ the *shear rate*. This matrix corresponds to a flow velocity $\mathbf{\Gamma} \cdot \mathbf{r}$ in the x-direction, with its gradient in the y-direction. The x-, y- and z-directions are referred to as the *flow*, *gradient* and *vorticity* directions, respectively.

The analysis of the Smoluchowski equation (69) is much the same as for the unsheared case considered before. There is one essential difference with the unsheared case, however, concerning the closure relation for the pair-correlation function. Short-wavelength density fluctuations relax fast, as for the unsheared system, rendering the short-ranged behavior of the pair-correlation function equal to the pair-correlation function g^{stat} of a stable homogeneous sheared system (the superscript "stat" refers to "stationary"). This is not the equilibrium pair-correlation function, since the shear flow may affect short-ranged correlations. For a zero shear rate g^{stat} becomes equal to the equilibrium pair-correlation function. For the sheared system, the closure relation (23) is replaced by

$$\delta g(\mathbf{r}, \mathbf{r}', t \,|\, \dot{\gamma}) = \frac{d g^{stat}(\mathbf{r} - \mathbf{r}' \,|\, \dot{\gamma})}{d\bar{\rho}} \, \delta\rho(\tfrac{\mathbf{r}+\mathbf{r}'}{2}, t \,|\, \dot{\gamma}) \ , \textit{for} \ |\mathbf{r} - \mathbf{r}'| < R_V . \quad (71)$$

In addition, the pair-correlation function at time $t = 0$, after the quench when the system became unstable before significant phase separation occurred, is equal to g^{stat}. Notice that the shear flow renders g^{stat} anisotropic, that is, it is a function of the vector $\mathbf{r} - \mathbf{r}'$, not just of its length $|\mathbf{r} - \mathbf{r}'|$.

The analysis of subsection 2.2 can be copied to the present case, except that all pdf's are shear-rate dependent, to obtain the following equation of motion for the Fourier transform of the macroscopic density,

$$\left(\frac{\partial}{\partial t} - \dot{\gamma} \, k_1 \, \frac{\partial}{\partial k_2} \right) \delta\rho(\mathbf{k}, t \,|\, \dot{\gamma}) \;=\; -D(\mathbf{k} \,|\, \dot{\gamma}) \, k^2 \, \delta\rho(\mathbf{k}, t \,|\, \dot{\gamma}) \,, \quad (72)$$

where k_j is the jth component of $\mathbf{k}$ and where $D(\mathbf{k}|\dot{\gamma})$ is a diffusion coefficient given by

$$\begin{aligned} D(\mathbf{k} \,|\, \dot{\gamma}) \;=\; & D_0 \left[1 - \frac{1}{2} \beta\bar{\rho} \int d\mathbf{R} \, R \frac{dV(R)}{dR} (\hat{\mathbf{k}} \cdot \hat{\mathbf{R}})^2 \right. \quad (73) \\ & \left. \times \left(2 \, g^{stat}(\mathbf{R} \,|\, \dot{\gamma}) \, \frac{\sin\{\mathbf{k} \cdot \mathbf{R}\}}{\mathbf{k} \cdot \mathbf{R}} + \bar{\rho} \, \frac{d g^{stat}(\mathbf{R} \,|\, \dot{\gamma})}{d\bar{\rho}} \, \frac{\sin\{\frac{1}{2}\mathbf{k} \cdot \mathbf{R}\}}{\frac{1}{2}\mathbf{k} \cdot \mathbf{R}} \right) \right] , \end{aligned}$$

where $\hat{\mathbf{k}} = \mathbf{k}/k$ and $\hat{\mathbf{R}} = \mathbf{R}/R$ are unit vectors. The shear rate dependence of the density is denoted explicitly.

The unstable density waves have a wavelength that is much larger than the range R_V of the pair-interaction potential. The effect of the shear flow is therefore much more pronounced for the demixing density variations than for the short-ranged part of the pair-correlation function. In fact, the shear-rate dependence of the pair-correlation function for distances of the order R_V and smaller may be neglected when the *bare Peclet number* $Pe^0 = \dot{\gamma} R_V^2 / 2D_0$ is not too large. This dimensionless number measures the distortion of structures with linear dimensions of at most R_V. The distortion of the large scale structures with linear dimensions Λ formed during spinodal decomposition is measured, roughly, by the Peclet number $\dot{\gamma}\Lambda^2/2D_0$. Since $\Lambda \gg R_V$, severe effects of shear flow on the decomposition kinetics are observed even for small bare Peclet numbers. We may therefore replace $g^{stat}(\mathbf{R} \mid \dot{\gamma})$ by the equilibrium pair-correlation function $g^{eq}(R)$, provided that $O(Pe^0)$ is small. The spherical angular integrations in eq.(73) can now be performed (see exercise 3 for mathematical details) to obtain, not surprisingly, the effective diffusion coefficient (29) for the unsheared system,

$$D(\mathbf{k} \mid \dot{\gamma}) = D^{eff}(k) , \quad \text{up to} \quad D_0 \times O\left(Pe^0\right) . \tag{74}$$

For larger shear rates, where Pe^0 is not small, spinodal decomposition is affected by the distortion of short-ranged correlations. In the sequel these short-ranged distortions are neglected.

The solution of the Smoluchowski equation (72), with $D(\mathbf{k} \mid \dot{\gamma})$ replaced by $D^{eff}(k)$ in eq.(29), reads

$$\delta\rho(\mathbf{k}, t \mid \dot{\gamma}) = \delta\rho(\mathbf{k} = (k_1, k_2 + \dot{\gamma} k_1 t, k_3), t = 0 \mid \dot{\gamma}) \exp\left\{-D^{eff}(\mathbf{k}, t \mid \dot{\gamma}) k^2 t\right\} , \tag{75}$$

where the time and shear-rate dependent effective diffusion coefficient is equal to

$$D^{eff}(\mathbf{k}, t \mid \dot{\gamma}) = \frac{1}{\dot{\gamma} k_1 t} \int_{k_2}^{k_2 + \dot{\gamma} k_1 t} dx \, D^{eff}\left(\sqrt{k_1^2 + x^2 + k_3^2}\right) \frac{k_1^2 + x^2 + k_3^2}{k^2} . \tag{76}$$

For shear rates for which Pe^0 is large, $D^{eff}(k)$ must be replaced here by $D(\mathbf{k} \mid \dot{\gamma})$.

Notice that there is a time dependence in the exponential prefactor in eq.(75). Hence, besides the exponential function, the wavevector dependence of the initial density variation contributes to the time evolution of the density.

The integral (76) for the effective diffusion coefficient can be done explicitly, with a little effort, after substitution of the small wavevector expansion (29) for $D^{eff}(k)$, to yield

$$D^{eff}(\mathbf{k}, t \mid \dot{\gamma}) = D_0 \left[\beta \frac{dP}{d\bar{\rho}} \left\{ 1 + \frac{K_1 K_2 \dot{\gamma} t + \frac{1}{3} K_1^2 (\dot{\gamma} t)^2}{K^2} \right\} \right.$$

$$+\left(\beta\Sigma/R_V^2\right)\left\{(K_1^2+K_3^2)\left(1+\frac{K_2^2+2K_1K_2\dot{\gamma}t+\frac{2}{3}K_1^2(\dot{\gamma}t)^2}{K^2}\right)\right. \tag{77}$$
$$\left.\left. + \frac{K_2^4+2K_1K_2^3\dot{\gamma}t+2K_1^2K_2^2(\dot{\gamma}t)^2+K_1^3K_2(\dot{\gamma}t)^3+\frac{1}{5}K_1^4(\dot{\gamma}t)^4}{K^2}\right\}\right] .$$

Here, $\mathbf{K} = \mathbf{k}\,R_V$ is a dimensionless wavevector.

The sheared system is unstable when there is a wavevector for which $D^{eff}(\mathbf{k}, t = 0\,|\,\dot{\gamma}) < 0$. From eq.(77), however, it follows that

$$D^{eff}(\mathbf{k}, t = 0\,|\,\dot{\gamma}) \;=\; D^{eff}(k)\,, \tag{78}$$

so that a sheared system is unstable if, and only if, the unsheared system is unstable. The spinodal is therefore not shifted by applying a shear flow. This is true only within the approximation (74), where the shear rate dependence of $D(\mathbf{k}\,|\,\dot{\gamma})$ is omitted. This omission is correct up to $O(Pe^0)$. The conclusion is thus that **the spinodal is shifted only slightly for not too large values of this bare Peclet number** (see Dhont (1996) for a quantitative prediction of how the location of the spinodal changes as a function of Pe^0). Since large effects on the demixing kinetics should be observed already for small values of Pe^0, meaningful experiments can be performed where the shift of the spinodal may be neglected.

Since the time always appears in eq. (77) as a product with k_1, it follows that in directions where $k_1 = 0$ there is no effect of shear flow,

$$D^{eff}(\mathbf{k}, t\,|\,\dot{\gamma})\Big|_{k_1=0} \;=\; D^{eff}(k)\Big|_{k_1=0} \,. \tag{79}$$

Density variations in the (y, z)-plane, that is the gradient-vorticity plane where $k_1 = 0$, are therefore not affected by shear flow.

Apart from the exponential prefactor in eq. (75), the growth rate of density variations is proportional to $-D^{eff}(\mathbf{k}, t\,|\,\dot{\gamma})\,k^2$. This is an anisotropic function, that is, a function of the vector $\mathbf{k}$, not just of its length $k = |\,\mathbf{k}\,|$. Moreover, the anisotropy changes as time proceeds. A plot of the anisotropic growth rates at various values of $\dot{\gamma}t$ is given in Fig. 8. The spherically symmetric growth rates become ellipsoidal for small times (see the figures for $\dot{\gamma}t = 1$ and 2). In the velocity-gradient plane, where $K_3 = 0$, the ellipsoid makes an angle with the K_1 and K_2 axes, while in the velocity-vorticity plane, where $K_2 = 0$, the long axis of the ellipsoid is along the line where $K_1 = 0$. The angle of the major axis of the ellipsoidal distortion in the (K_1, K_2)-plane with the K_2-axis is seen to decrease for larger values of $\dot{\gamma}t$ (note that in the two bottom figures in Fig. 8 the K_1-scale is expanded by a factor of 10 relative to the other figures). For somewhat larger values of $\dot{\gamma}t$, the growth rates along the major and minor axes of the ellipsoid in the

(K_1, K_2)-plane diminish, while in the (K_1, K_3)-plane the growth rate along the minor axis diminishes. According to eq. (79), growth rates in directions where $K_1 = 0$ are not affected by shear flow, so that the corresponding cross sections of all the figures in Fig. 8 are identical. For larger values of $\dot{\gamma}t$ the only remaining unstable modes are those where K_1 is small. Shear flow thus stabilizes density variations in directions along the flow direction, resulting in a "two-dimensional demixing" in planes perpendicular to the flow direction. The resulting structure, in which regions of different concentration form, extends in the flow direction. The extent of these elongated structures is equal to $2\pi/k_{1m}$, with k_{1m} the value of k_1 where the growth rate is maximal.

6. Experiments on Spinodal Decomposition

During the linear regime of spinodal decomposition, eq. (34) predicts a time independent location of the wavevector k_m of the most rapidly growing density wave. Moreover, plots of $\ln\{S(\mathbf{k}, t)\}/k^2t$ versus k^2 should be time independent straight lines with a slope equal to $D\kappa = D_0\beta\Sigma$ and an intercept $Da_2 = D_0\beta dP/d\bar{\rho}$. Sometimes these characteristics of the initial stage are indeed observed, but in most experiments they are not observed. For quenches in the direct vicinity of the spinodal non-linear terms are important right from the start, leading to non-Cahn-Hiliard like decomposition. Furthermore, it may well be that in some experiments the decomposition is so fast that a first meaningful measurement can be performed only beyond the initial stage. The scattering peak emerging at small wavevectors is always observed, together with the displacement of its maximum to smaller scattering angles due to non-linear coupling.

Figure 9a shows experimental scattering curves for a spinodally decomposing microemulsion system in the intermediate stage. These experimental curves are much like the theoretical curves in Fig. 7a. In particular, the shift of the maximum of the structure factor peak towards smaller wavevectors is indeed observed. Moreover, dynamic similarity scaling is seen to apply in Fig. 9c, and is in reasonable agreement with the theoretically predicted scaling function (the dashed curve in these figures). There is some deviation for the larger wavevectors. These deviations are due to scattering by sharp interfaces which are beginning to form. Mathematically, these contributions are neglected in our theory through the neglect of terms of $O(K^6)$ in the equation of motion (19). These sharp interface contributions to the experimental intensities cause the experimental value of the integral $\int dk'\, k'^2 S(k', t)$ to be large because they occur at large wavevectors which weigh more heavily in the integrand. This is the reason why the scaling relation (44) is not verified by experiments, as shown in Fig. 9b. Due

Figure 8. The anisotropic growth rates $-D^{eff}(\mathbf{k}, t \mid \dot{\gamma})K^2$, with $\mathbf{K} = \mathbf{k}\, R_V$ a dimensionless wavevector, for various times $\dot{\gamma}t$ (see eq.(77)). The ratio of the two dimensionless numbers $\beta dP/d\bar{\rho}$ and $\beta\Sigma/R_V^2$ is taken equal to $-1/10$ here. Negative values for the growth rate, corresponding to stable fluctuations, are not shown. The left column of figures is for $K_3 = 0$ the right column for $K_2 = 0$. The vertical scales are the same for all figures. For the two lower figures the wavevector scale in the K_1-direction is 10 times as small as compared to the other plots.

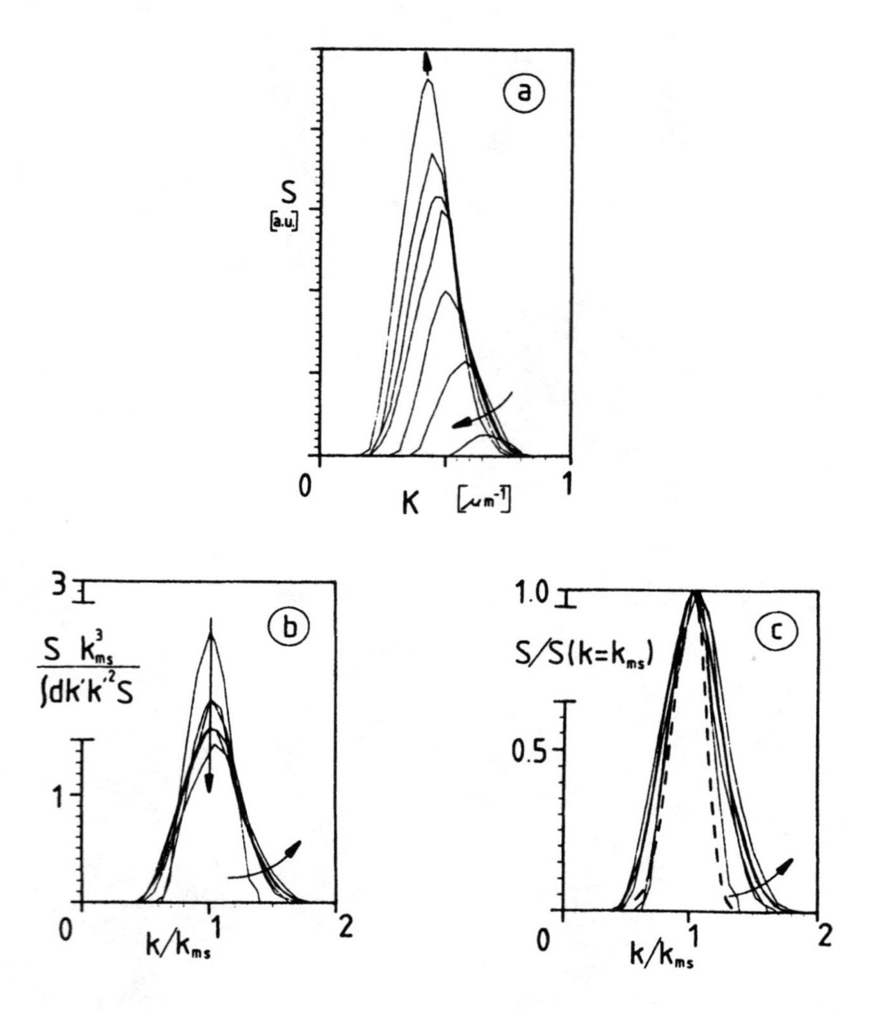

Figure 9. (a) Scattering peaks during decomposition of an AOT/water/decane microemulsion. (b) A test of the dynamic scaling relation (44). (c) Dynamic similarity scaling. The dashed curve is the theoretical prediction (see Fig. 7c). The arrows indicate changes as time proceeds. Experimental data are taken from Mallamace et. al. (1995).

to the already developing sharp interfaces the above mentioned integral is grossly overestimated, giving rise to experimental scaling functions with a diminishing amplitude (the arrows in Fig. 9 indicate trends with increasing time).

A well known empirical scaling relation for the structure factor is due to Furukawa (1984). This scaling function is much broader than the scaling function that we found for the intermediate stage, and applies in the final stage. One of the features of sharp (and very sharp) interfaces is a decay of the structure factor at larger wavevectors like $\sim k^{-4}$. This so-called *Porod behavior* is one of the ingredients for constructing the Furukawa scaling function. Such behavior is absent in the intermediate stage where sharp interfaces are just beginning to form.

Spinodal decomposition of sheared systems can be studied by means of light scattering, just as for unsheared systems. According to eq. (75) the scattering intensity of a decomposing sheared system is proportional to

$$\begin{aligned} S(\mathbf{k},t\,|\,\dot{\gamma}) &= \frac{1}{N} <|\,\delta\rho(\mathbf{k},t\,|\,\dot{\gamma})\,|^2> \\ &= <|\,\delta\rho(\mathbf{k}=(k_1,k_2+\dot{\gamma}t,k_3),t=0\,|\,\dot{\gamma})\,|^2> \exp\left\{-2D^{eff}(\mathbf{k},t\,|\,\dot{\gamma})k^2t\right\}. \end{aligned}$$

Apart from the exponential prefactor, which in principle also contributes to the time and wavevector dependence of the scattered intensity, the scattering patterns should resemble the time and wavevector dependence of the anisotropic growth rate $-D^{eff}(\mathbf{k},t\,|\,\dot{\gamma})k^2$ as depicted in Fig. 8. So far, no scattering experiments of this kind have been performed for colloidal systems. Experiments on binary fluids are reported by Chan et. al. (1988,1991), Perrot et. al. (1989) and Baumberger et. al. (1991). Scattering patterns taken from Baumberger et. al. (1991) are given in Fig. 10 (see also Chan et. al. (1988)). There is a striking resemblance between these experimental results and our predictions in Fig. 8. In the (k_1, k_2)-plane the ellipsoidal scattering pattern is rotated relative to both axes, contrary to the patterns in the (k_1, k_3)-plane, where the major axis of the ellipsoid is oriented parallel to the k_3-axis. Furthermore, the predicted decrease of the angle between the major axis of the ellipsoidal scattering pattern in the (k_1, k_2)-plane and the k_2-axis at later times is observed. This reorientation of the ellipsoidal scattering pattern in the (k_1, k_2)-plane at larger values of $\dot{\gamma}t$ towards alignment along the k_2-axis ultimately leads to quasi two-dimensional growth.

The prediction (74) that shear has no effect in directions where $k_1 = 0$ is also found experimentally. Perrot et al. (1989) state that "$\cdots$ in the direction perpendicular to the flow and to the shear, the characteristic length is nearly insensitive to the shear and is identical to that obtained without shear flow," and Chan et. al. (1991) state that "in the direction perpendicular to the flow and the shear, shear seems to have little effect on the growth." In fact, the effect of shear in these directions is expected to be of order Pe^0.

Also mentioned by Perrot et. al. (1989) and Baumberger et. al. (1991) is the diminishing intensity along the major axis of the ellipsoid in the (k_1, k_2)-plane. This feature seems to be in accord with the theoretical predictions in Fig. 8.

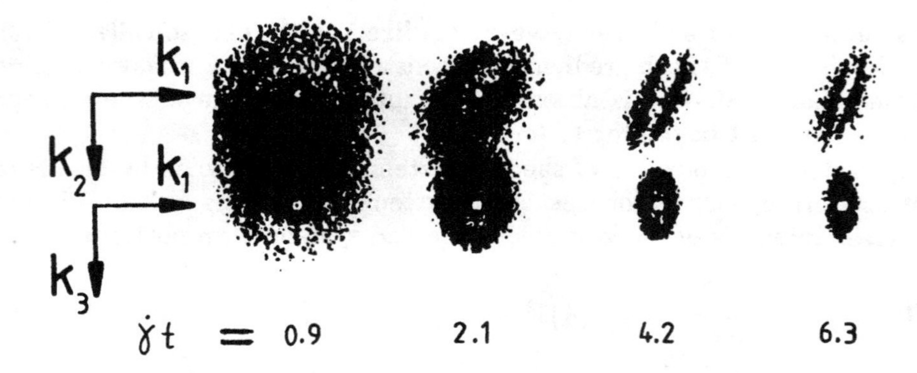

Figure 10. Scattering patterns of a sheared demixing binary fluid (isobutyric acid and water) in the (k_1, k_2)-plane (top figures) and the (k_1, k_3)-plane (bottom figures). These figures are taken from Baumberger et. al. (1991).

As can be seen from eq. (77) for the effective diffusion coefficient, growth rates scale with $\dot{\gamma}t$. Such a scaling is observed by Baumberger et. al. (1991), who state that "$\cdots$ varying $\dot{\gamma}$ at constant t is similar to varying t at constant $\dot{\gamma}$."

As we have seen in section 5, the only unstable density waves for large shear rates are those where the component k_1 of the wavevector along the flow direction is small, leading to a kind of two-dimensional growth. Such two-dimensional growth has been observed in experiments on polymer systems (Hashimoto et. al. (1995)) and binary fluids (Perrot et. al. (1989)). The title of the latter reference refers explicitly to this phenomenon: "Spinodal Decomposition under Shear: Towards a Two-Dimensional Growth?". Such a "Dimensional Reduction in Phase-Separating Critical Fluids under Shear Flow" was first predicted theoretically by Imaeda and Kawasaki (1985).

Appendix

The derivation of the equation of motion eq. (52) requires a considerable effort. The mathematical treatment of one of the terms encountered in the derivation is discussed in this appendix. Other terms are treated similarly.

One of the typical terms encountered is

$$\begin{aligned} I &\equiv \int d\mathbf{r} \int d\mathbf{r}' < \delta\rho^2(\mathbf{r},t)\,[\nabla_r^2 \delta\rho(\mathbf{r},t)]\,\delta\rho(\mathbf{r}',t) >_{init} \exp\{i\mathbf{k}\cdot(\mathbf{r}-\mathbf{r}')\} \\ &= \nabla_{r''}^2 \int d\mathbf{r} \int d\mathbf{r}' < \delta\rho^2(\mathbf{r},t)\,\delta\rho(\mathbf{r}'',t)\,\delta\rho(\mathbf{r}',t) >_{init} \exp\{i\mathbf{k}\cdot(\mathbf{r}-\mathbf{r}')\}\big|_{\mathbf{r}''=\mathbf{r}} . \end{aligned}$$

In the last line, $\mathbf{r}''$ is to be taken equal to $\mathbf{r}$ after the differentiation is performed. Assuming Gaussian statistics one obtains

$$< \delta\rho^2(\mathbf{r},t)\,\delta\rho(\mathbf{r}'',t)\,\delta\rho(\mathbf{r}',t) >_{init} = \\ < \delta\rho^2(\mathbf{r},t) >_{init} < \delta\rho(\mathbf{r}'',t)\,\delta\rho(\mathbf{r}',t) >_{init} \\ + 2 < \delta\rho(\mathbf{r},t)\,\delta\rho(\mathbf{r}',t) >_{init} < \delta\rho(\mathbf{r},t)\,\delta\rho(\mathbf{r}'',t) >_{init} . \tag{80}$$

The first term on the right hand-side contributes

$$< \delta\rho^2(\mathbf{r},t) >_{init} \nabla^2_{r''} \int d\mathbf{r} \int d\mathbf{r}' < \delta\rho(\mathbf{r}'',t)\,\delta\rho(\mathbf{r}',t) >_{init} \exp\{i\mathbf{k}\cdot(\mathbf{r}-\mathbf{r}')\}_{|\mathbf{r}''=\mathbf{r}}$$
$$= < \delta\rho^2(\mathbf{r},t) >_{init} \int d\mathbf{r} \int d\mathbf{r}' < [\nabla^2_r \delta\rho(\mathbf{r},t)]\,\delta\rho(\mathbf{r}',t) >_{init} \exp\{i\mathbf{k}\cdot(\mathbf{r}-\mathbf{r}')\}$$
$$= -k^2 < \delta\rho^2(\mathbf{r},t) >_{init} \int d\mathbf{r} \int d\mathbf{r}' < \delta\rho(\mathbf{r},t)\,\delta\rho(\mathbf{r}',t) >_{init} \exp\{i\mathbf{k}\cdot(\mathbf{r}-\mathbf{r}')\}$$
$$= -k^2 < \delta\rho^2(\mathbf{r},t) >_{init} N\,S(k,t) .$$

In the third line, Green's second integral theorem is used, with the omission of surface integrals. The second term on the right hand-side of eq. (80) contributes

$$2\nabla^2_{r''} \int d\mathbf{r} \int d\mathbf{r}' < \delta\rho(\mathbf{r},t)\delta\rho(\mathbf{r}',t) >_{init} \times \\ < \delta\rho(\mathbf{r},t)\delta\rho(\mathbf{r}'',t) >_{init} \exp\{i\mathbf{k}\cdot(\mathbf{r}-\mathbf{r}')\}_{|\mathbf{r}''=\mathbf{r}}$$
$$= 2 < \delta\rho(\mathbf{r},t)\nabla^2\delta\rho(\mathbf{r},t) >_{init} \int d\mathbf{r} \int d\mathbf{r}' < \delta\rho(\mathbf{r},t)\delta\rho(\mathbf{r}',t) >_{init} \exp\{i\mathbf{k}\cdot(\mathbf{r}-\mathbf{r}')\}$$
$$= 2 < \delta\rho(\mathbf{r},t)\nabla^2\delta\rho(\mathbf{r},t) >_{init} NS(k,t).$$

The term under consideration here is thus equal to

$$I = N\,S(k,t)\left[-k^2 < \delta\rho^2(\mathbf{r},t) >_{init} + 2 < \delta\rho(\mathbf{r},t)\nabla^2_r\delta\rho(\mathbf{r},t) >_{init}\right] .$$

The averages with respect to initial conditions are independent of position, since there is no preferred position on average.

Averages like $< \delta\rho(\mathbf{r},t)\nabla_r\nabla^2_r\delta\rho(\mathbf{r},t) >_{init}$ are zero, since each component of the vector $\nabla_r\nabla^2_r\delta\rho(\mathbf{r},t)$ is equally likely to be positive and negative, independent of the local value of $\delta\rho(\mathbf{r},t)$.

Exercises

1) *Stability and decomposition kinetics of a van der Waals fluid*

A van der Waals fluid is defined as a one-component fluid (or a suspension of monodisperse colloidal particles) with a hard-core repulsion and an additional attractive pair-interaction potential w of infinite range. Subdivide the entire system into little volume elements as was done in subsection 2.1 on the Cahn-Hilliard theory. These volume elements are now so small that the additional pair-interaction potential is a constant over distances equal to the linear dimensions of the volume elements, but at the same time so large that they contain many particles. Such a long-ranged pair-interaction potential is not realistic, but it allows for an analysis of thermodynamic behavior and phase-separation kinetics. Despite the unrealistic nature of the pair-interaction potential, the equation of state of a van der Waals fluid exhibits all features that one expects for gasses/fluids. The equation of state is analyzed in (a), thermodynamic stability is considered in (b) and decomposition kinetics in (c).

Let us first derive an expression for the free energy of a van der Waals system (this derivation is taken from van Kampen (1964)). The canonical configurational partition function is equal to

$$\begin{aligned} Q_N &= \frac{1}{N!}\int d\mathbf{r}_1 \cdots \int d\mathbf{r}_N \exp\{-\beta\Phi(\mathbf{r}_1, \cdots, \mathbf{r}_N)\} \\ &= \frac{1}{N!}\int d\mathbf{r}_1 \cdots \int d\mathbf{r}_N\, \chi(\mathbf{r}_1, \cdots, \mathbf{r}_N) \exp\left\{-\beta \sum_{n>m=1}^{N} w(|\mathbf{r}_n - \mathbf{r}_m|)\right\}, \end{aligned}$$

where the so-called "characteristic function" χ is 0 when two or more hard cores overlap and 1 otherwise. The characteristic function enters through the hard-core part of the interaction potential Φ, which is infinite when two or more hard cores overlap and 0 otherwise. Let N_j denote the number of particles in the jth volume element. The partition sum is now rewritten in terms of a sum of all possible realizations $\{N_j\}$ of these so-called occupation numbers. Since the additional pair potential w is supposed to be constant within each volume element, the partition function can be written as (n is the number of volume elements)

$$\begin{aligned} Q_N = \frac{1}{N!}\sum_{\{N_j\}} \frac{N!}{\prod_j N_j!} \underbrace{\int d\mathbf{r}_1 \cdots \int d\mathbf{r}_{N_1}}_{N_1\ in\ V_1} \cdots \cdots \underbrace{\int d\mathbf{r}_{N+1-N_n} \cdots \int d\mathbf{r}_N}_{N_n\ in\ V_n} \\ \times\, \chi(\mathbf{r}_1, \cdots, \mathbf{r}_N) \exp\left\{-\beta \sum_{i>j} w_{ij} N_i N_j\right\}. \end{aligned}$$

Here, w_{ij} is the long ranged pair potential evaluated at the distance between the volume elements i and j. Each of the integrals pertaining to a single volume element renders the average volume available to a single particle, taking into account that part of the total volume is excluded due to the presence of the other particles. This *free volume* is approximately equal to

$\Delta - N_j\delta$, with Δ the volume of a volume element and δ being a measure for the core size of the particles. Hence,

$$Q_N = \frac{1}{N!}\sum_{\{N_j\}} \frac{N!}{\prod_j N_j!} \prod_j (\Delta - N_j\delta)^{N_j} \exp\left\{-\beta \sum_{i>j} w_{ij} N_i N_j\right\}.$$

This result can also be written as

$$Q_N = \sum_{\{N_j\}} \exp\{-\beta\Psi(N_1, \cdots, N_n)\},$$

with

$$\Psi(N_1, \cdots, N_n) = -k_B T \sum_j (N_j \ln\{\Delta - N_j\delta\} - N_j \ln\{N_j\} + N_j) + \sum_{i>j} w_{ij} N_i N_j. \tag{81}$$

Stirling's approximation $\ln\{N_j!\} = N_j \ln\{N_j\} - N_j$ is used here. The canonical partition function is related to the Helmholtz free energy A as $A = -k_B T \ln\{Q_N\}$. For large N's, Ψ is sharply peaked around its minimum value, and positive and large otherwise. There is therefore a dominant term in the above sum that defines the partition function, pertaining to the occupation numbers where Ψ attains its minimum value. Hence

$$A = \Psi(N_1, \cdots, N_n), \tag{82}$$

where the occupation numbers are those for which Ψ attains its minimum value.[8]

(a) Assume that the density is homogeneous, that is, assume that

$$N_j = N\frac{\Delta}{V}, \quad \textit{for all } j,$$

[8]Notice that the minimization of Ψ is constrained by the condition that the total number of particles in the canonical ensemble is a constant, that is,

$$N = \sum_j N_j = \textit{constant}.$$

The actual function that one should minimize is therefore

$$\Psi^{\dagger}(N_1, \cdots, N_n) = \Psi(N_1, \cdots, N_n) - \lambda \sum_j N_j,$$

where λ is a *Lagrange multiplier*, which can be determined after minimization. In this way van Kampen (1964) constructs, quite elegantly, the two-phase equilibrium states. We do not go into this matter here.

where V is the volume of the entire system under consideration. Show from eqs. (81, 82) that the free energy is now equal to (note that $n = V/\Delta$)

$$A = -k_B T \left(N \ln \left\{ \frac{V - N\delta}{N} \right\} + N \right) - \frac{1}{2} w_0 \frac{N^2}{V} ,$$

where

$$w_0 = -\frac{1}{V} \sum_{i \neq j} w_{ij} \Delta^2 = -\frac{1}{V} \int_{r>d} d\mathbf{r} \int_{r>d} d\mathbf{r}' \, w(|\mathbf{r} - \mathbf{r}'|) = -4\pi \int_d^\infty dr \, r^2 w(r) .$$

Since w is defined only outside the hard cores, the integration ranges do not include distances smaller than the diameter d of the cores. Notice that for an attractive additional pair-potential w the parameter w_0 is positive. Now use that the osmotic pressure is equal to $P = -\partial A/\partial V_{|N,T}$ to show that (with $\bar{\rho} = N/V$)

$$P = \frac{\bar{\rho} k_B T}{1 - \bar{\rho}\delta} - \frac{1}{2} w_0 \bar{\rho}^2 .$$

This is the *van der Waals equation of state.*

(b) Show from the instability criterion $dP/d\bar{\rho} < 0$ that the homogeneous state with density $\bar{\rho}$ is unstable when

$$\frac{\beta w_0}{\delta} > \frac{1}{\bar{\rho}\delta \, (1 - \bar{\rho}\delta)^2} . \tag{83}$$

Verify that the the minimum value for the function $1/x(1-x)^2$ is $27/4$ which is attained for $x = 1/3$. Conclude that there is no unstable homogeneous state when $\beta w_0/\delta < 27/4$, and that the critical temperature is given by $T_c = \frac{4}{27} \frac{w_0}{k_B \delta}$.

(c) Equations (81, 82) allow for the construction of the Helmholtz free energy functional of the density for an inhomogeneous state. To this end, the summations over volume elements in eq. (81) are to be replaced by volume integrals. This can be done as follows. Instead of working with number densities, it is more convenient here to work with a quantity that is proportional to the volume fraction of colloidal particles,

$$\varphi_j \equiv N_j \delta / \Delta .$$

When δ is taken equal to four times the core volume of a particle, this is four times the volume fraction in the jth volume element. According to eqs. (81, 82), the free energy can be written in terms of this concentration parameter as

$$A = -\frac{k_B T \Delta}{\delta} \sum_j \left(\varphi_j \left[\ln \left\{ \frac{1 - \varphi_j}{\varphi_j} \right\} + \ln\{\delta\} + 1 \right] \right) + \frac{1}{2} \left(\frac{\Delta}{\delta} \right)^2 \sum_{i \neq j} w_{ij} \varphi_i \varphi_j .$$

The summations can be identified as integrals as follows,

$$\sum_j \Delta(\cdots)_j \equiv \int d\mathbf{r}\,(\cdots)(\mathbf{r})\,.$$

Verify that the free energy can now be written as

$$\begin{aligned} A[\varphi(\mathbf{r})] &= -\frac{k_B T}{\delta}\int d\mathbf{r}\,\left(\varphi(\mathbf{r})\left[\ln\left\{\frac{1-\varphi(\mathbf{r})}{\varphi(\mathbf{r})}\right\}+\ln\{\delta\}+1\right]\right) \\ &\quad + \frac{1}{2\delta^2}\int d\mathbf{r}\int d\mathbf{r}'\, w(|\,\mathbf{r}-\mathbf{r}'\,|)\varphi(\mathbf{r}')\varphi(\mathbf{r})\,. \end{aligned}$$

The functional dependence of A on $\varphi(\mathbf{r})$ is denoted as usual by the square brackets. Show by functional differentiation that

$$\begin{aligned} \frac{\delta A[\varphi]}{\delta\varphi(\mathbf{r})} &= -\frac{k_B T}{\delta}\left[\ln\left\{\frac{1-\varphi(\mathbf{r})}{\varphi(\mathbf{r})}\right\}+\ln\{\delta\}+1-\frac{1}{1-\varphi(\mathbf{r})}\right] \\ &\quad + \frac{1}{\delta^2}\int d\mathbf{r}'\, w(|\,\mathbf{r}-\mathbf{r}'\,|)\varphi(\mathbf{r}')\,. \end{aligned}$$

The chemical potential is equal to $\mu(\mathbf{r}) = \frac{\delta A[\rho]}{\delta\rho(\mathbf{r})} = \frac{\delta A[\varphi]}{\delta\varphi(\mathbf{r})} \times \delta$. Verify that the particle current density is equal to

$$\mathbf{j}(\mathbf{r}) = -D\nabla\mu(\mathbf{r}) = -\frac{k_B T D}{\varphi(\mathbf{r})(1-\varphi(\mathbf{r}))^2}\nabla\varphi(\mathbf{r}) - \frac{D}{\delta}\int d\mathbf{r}'\,[\nabla w(|\,\mathbf{r}-\mathbf{r}'\,|)]\varphi(\mathbf{r}').$$

Apply Gauss's integral theorem to arrive at the following equation of motion

$$\begin{aligned} \frac{\partial}{\partial t}\varphi(\mathbf{r},t) &= -k_B T D\delta\,\frac{1-4\varphi(\mathbf{r},t)+3\varphi^2(\mathbf{r},t)}{\varphi^2(\mathbf{r},t)\,(1-\varphi(\mathbf{r},t))^4}\,|\,\nabla\varphi(\mathbf{r},t)\,|^2 \\ &\quad + \frac{k_B T D\delta}{\varphi(\mathbf{r},t)\,(1-\varphi(\mathbf{r},t))^2}\nabla^2\varphi(\mathbf{r},t) + D\int d\mathbf{r}'\, w(|\,\mathbf{r}-\mathbf{r}'\,|)\nabla'^2\varphi(\mathbf{r}',t)\,, \end{aligned}$$

where the time dependence of φ is now denoted explicitly. Linearize with respect to $\delta\varphi(\mathbf{r},t) = \varphi(\mathbf{r},t) - \bar{\varphi}$, with $\bar{\varphi} = \bar{\rho}\delta$, and show that

$$\frac{\partial}{\partial t}\delta\varphi(\mathbf{r},t) = \frac{k_B T D\delta}{\bar{\varphi}\,(1-\bar{\varphi})^2}\nabla^2\delta\varphi(\mathbf{r},t) + D\int d\mathbf{r}'\, w(|\,\mathbf{r}-\mathbf{r}'\,|)\nabla'^2\delta\varphi(\mathbf{r}',t)\,.$$

Fourier transform this equation of motion with respect to the position coordinate $\mathbf{r}$ with the help of the convolution theorem to show that

$$\delta\rho(\mathbf{k},t) = \delta\rho(\mathbf{k},t=0)\exp\left\{-D^{eff}(k)\,k^2\,t\right\}\,,$$

where the effective diffusion coefficient is equal to

$$D^{eff}(k) = \frac{k_B T D \delta}{\bar{\varphi}(1-\bar{\varphi})^2} + D \underbrace{\int_{r>d} d\mathbf{r} \exp\{-i\mathbf{k}\cdot\mathbf{r}\} w(r)}_{\equiv w(k)} .$$

Expand the Fourier transform $w(k)$ up to "$O(k^2)$" to show that

$$D^{eff}(k) = \frac{k_B T D \delta}{\bar{\varphi}(1-\bar{\varphi})^2} - D\left[w_0 - k^2 w_2\right] .$$

The parameter w_0 is defined in exercise (b), while

$$w_2 = -\frac{2\pi}{3}\int_d^\infty dr\, r^4 w(r) .$$

This is the standard form of the Cahn-Hilliard diffusion coefficient. Verify that $D^{eff}(k=0) < 0$ whenever the instability criterion is satisfied. Use that $D/D_0 = \beta\bar{\rho}$ (see eq. (33)) and the van der Waals equation of state in (a) to show that $D^{eff}(k=0) = D_0 \beta dP/d\bar{\rho}$, in accordance with our general expression (29) for the effective diffusion coefficient. Derive an expression for Σ (see eq. (31)) in terms of the interaction parameter w_2, and verify that $\Sigma > 0$ for an attractive long ranged pair-interaction potential w.

2) Fourier transformation of eq. (25) with respect to $\mathbf{r}$ yields integrals of the type

$$I(\mathbf{k}) \equiv i\mathbf{k}\cdot \int d\mathbf{r} \int d\mathbf{R}\, [\nabla_R V(R)] f(R) \delta\rho(\mathbf{r}-\alpha\mathbf{R},t) \exp\{-i\mathbf{k}\cdot\mathbf{r}\} ,$$

where α is either 1 or 1/2. Verify each of the following mathematical steps which lead to an expression for the integral in terms of the Fourier transform $\delta\rho(\mathbf{k},t)$.

$$\begin{aligned} I(\mathbf{k}) &= i\mathbf{k}\cdot \int d\mathbf{r} \int d\mathbf{R}[\nabla_R V(R)] f(R) \delta\rho(\mathbf{r}-\alpha\mathbf{R},t) \times \\ &\qquad \exp\{-i\mathbf{k}\cdot(\mathbf{r}-\alpha\mathbf{R})\} \exp\{-i\alpha\mathbf{k}\cdot\mathbf{R}\} \\ &= i\mathbf{k}\cdot \int d\mathbf{R}[\nabla_R V(R)] f(R) \exp\{-i\alpha\mathbf{k}\cdot\mathbf{R}\} \times \\ &\qquad \int d(\mathbf{r}-\alpha\mathbf{R}) \delta\rho(\mathbf{r}-\alpha\mathbf{R},t) \exp\{-i\mathbf{k}\cdot(\mathbf{r}-\alpha\mathbf{R})\} \\ &= \delta\rho(\mathbf{k},t) i\mathbf{k}\cdot \int d\mathbf{R}[\nabla_R V(R)] f(R) \exp\{-i\alpha\mathbf{k}\cdot\mathbf{R}\}. \end{aligned}$$

Now use that $\nabla_R V(R) = \hat{\mathbf{R}}\, dV(R)/dR$, with $\hat{\mathbf{R}} = \mathbf{R}/R$, and verify that ($\nabla_k$ is the gradient operator with respect to $\mathbf{k}$)

$$
\begin{aligned}
& i\mathbf{k} \cdot \int d\mathbf{R}\, [\nabla_R V(R)] f(\mathbf{R}) \exp\{-i\mathbf{k}\cdot\mathbf{R}\} \\
&\quad = i\mathbf{k} \cdot \int_0^\infty dR\, R^2 \frac{dV(R)}{dR} f(R) \oint d\hat{\mathbf{R}}\, \hat{\mathbf{R}} \exp\{-i\alpha\mathbf{k}\cdot\hat{\mathbf{R}}R\} \\
&\quad = i\mathbf{k} \cdot \int_0^\infty dR\, R^2 \frac{dV(R)}{dR} f(R) \frac{1}{-i\alpha R} \nabla_k \oint d\hat{\mathbf{R}} \exp\{-i\alpha\mathbf{k}\cdot\hat{\mathbf{R}}R\} \\
&\quad = i\mathbf{k} \cdot \int_0^\infty dR\, R^2 \frac{dV(R)}{dR} f(R) \frac{1}{-i\alpha R} \nabla_k\, 4\pi \frac{\sin\{\alpha k R\}}{\alpha k R} \\
&\quad = i\mathbf{k} \cdot \int_0^\infty dR\, R^2 \frac{dV(R)}{dR} f(R) \frac{1}{-i\alpha R} \frac{\mathbf{k}}{k} \frac{d}{dk}\, 4\pi \frac{\sin\{\alpha k R\}}{\alpha k R} \\
&\quad = i\mathbf{k} \cdot \int_0^\infty dR\, R^2 \frac{dV(R)}{dR} f(R) \frac{1}{-i\alpha R} 4\pi\alpha^2 R^2\, \mathbf{k}\, \jmath(\alpha k R)\,.
\end{aligned}
$$

In the third equation it is used that

$$
\oint d\hat{\mathbf{R}} \exp\{\pm i\alpha\mathbf{k}\cdot\mathbf{R}\} = 4\pi \frac{\sin\{\alpha k R\}}{\alpha k R}\,. \tag{84}
$$

The $\jmath$-function is defined in eq. (28). Conclude that

$$
I(\mathbf{k}) \;=\; -\delta\rho(\mathbf{k},t)\, 4\pi\alpha k^2 \int_0^\infty dR\, R^3 \frac{dV(R)}{dR} f(R)\, \jmath(\alpha k R)\,.
$$

Use this result to verify eqs. (26, 27).

3) To obtain eq. (74) for the diffusion coefficient defined in eq. (73), integrals of the kind

$$
I \equiv f \oint d\hat{\mathbf{R}}\, (\hat{\mathbf{k}}\cdot\hat{\mathbf{R}})^2 \frac{\sin\{\mathbf{k}\cdot\mathbf{R}\}}{\mathbf{k}\cdot\mathbf{R}}
$$

must be evaluated, where $\oint d\hat{\mathbf{R}}$ is the integration with respect to spherical angular coordinates ranging over the unit sphere. Show that this integral is equal to

$$
I \;=\; -\frac{1}{2\,(KR)^2} \frac{\partial}{\partial\alpha} \oint d\hat{\mathbf{R}}\; [\, \exp\{i\alpha\mathbf{k}\cdot\mathbf{R}\} + \exp\{-i\alpha\mathbf{k}\cdot\mathbf{R}\}\,]\Big|_{\alpha=1}\,,
$$

where α is to be set equal to 1 after the differentiation is performed. Use eq. (84) to show that

$$
I \;=\; -4\pi\, \jmath(kR)\,,
$$

with the j-function defined in eq. (28). Verify eq. (74).

4) ***Stability and demixing of confined suspensions***

We have considered systems of infinite extent, where density waves of infinite wavelength become unstable on the spinodal. Suppose now that the suspension in contained in a cube with sides of length L. The maximum wavelength of density waves is now L, corresponding to wavevectors $2\pi/L$. Suppose that the container is still large enough to neglect the influence of the walls of the container. Show that the spinodal is now given by

$$\frac{dP}{d\bar{\rho}} = -\Sigma \left(\frac{2\pi}{L}\right)^2 .$$

At a given density the spinodal temperature is thus lower than for a system of infinite extent.

Consider a rectangular geometry with two small equal sides of length l and a large length $L: \; L \gg l$. Argue that upon cooling, density waves with wavevectors along the long side will become unstable first. The demixing process will then have a one-dimensional character.

In a realistic description of the shift of the spinodal due to a confining geometry, the effects of the walls on the microstructure of the suspension should be taken into account, which is not a simple matter.

Further Reading

The turbidity data in Fig. 6 are taken from,

· H. Verduin and J.K.G. Dhont, J. Coll. Int. Sci. **172**, 425 (1995).

A few of the original papers on the Cahn-Hilliard theory are:

· J.W. Cahn and J.E. Hilliard, J. Chem. Phys. **28**, 258 (1958); **31**, 688 (1959).

· M. Hillert, Acta Metallica **9**, 525 (1961).

· J.W. Cahn, Acta Metallica **9**, 795 (1961).

· J.W. Cahn, J. Chem. Phys. **42**, 93 (1965).

· J.W. Cahn, Trans. Metall. Soc. Aime **242**, 166 (1968).

· H.E Cook, Acta Metallica **18**, 297 (1970).

· J.E. Hilliard, in *Phase Transformations*, H.J. Aronson, ed., (American Society for Metals, Metals Park OH, 1970), ch. 12.

Extensions of the Cahn-Hilliard theory, including computer simulations, are:

· J.S. Langer, Annals of Physics **65**, 53 (1971).

· J.S. Langer and M. Bar-on, Annals of Physics **78**, 421(1973).

· J.S. Langer, M. Bar-on, and H.D. Miller, Phys. Rev. A **11**, 1417 (1975).

· K. Kawasaki, Prog. Theor. Phys. **57**, 826 (1977).

· K. Kawasaki and T. Ohta, Prog. Theor. Phys. **59**, 362, 1406 (1978).

· R. Evans and M.M. Telo da Gama, Mol. Phys. **38**, 687 (1979).
· K. Binder, J. Chem. Phys. **79**, 6387 (1983).
· K. Binder, Coll. Pol. Sci. **265**, 273 (1987).
· C. Billotet and K. Binder, Z. Phys. B **32**, 195 (1979).
· G.F. Mazenko, Phys. Rev. B **42**, 4487 (1990).
· A. Sariban and K. Binder, Macromolecules **24**, 578 (1991).
· P. Fratzl, J.L. Lebowitz, O. Penrose, and J. Amar, Phys. Rev. B **44**, 4794 (1991).
· A. Shinozaki and Y. Oono, Phys. Rev. Lett. **66**, 173 (1991).
· J.A. Alexander, S. Chen, and D.W. Grunau, Phys. Rev. B **48**, 634 (1993).
· T. Koga and K. Kawasaki, Physica A **196**, 389 (1993).

In the 1975 paper of Langer, Bar-on and Miller, an expression for the time dependence of $k_m(t)$ is found for molecular systems that is similar to eq. (62). They also derive the identification in eq. (55). A few of the above papers start from equations of motion for the density, and solve these (numerically), including the late stage. It turns out that this is not realistic. Scaling behavior is predicted in a more reliable way from heuristic considerations about the driving mechanisms during the transition and late stage. See

· K. Binder and D. Stauffer, Phys. Rev. Lett. **33**, 1006 (1974).
· E.D. Siggia, Phys. Rev. A **20**, 595 (1979).

Scaling functions for the transition and late stages are constructed in:

· H. Furukawa, Physica A **123**, 497 (1984).

A Smoluchowski equation approach to spinodal decomposition for rigid rod like Brownian particles, where correlations are neglected (that is, where the pair-correlation function is taken equal to 1), can be found in:

· T. Shimada, M. Doi, and K. Okano, J. Chem. Phys. **88**, 7181 (1988).

The Smoluchowski approach as discussed in subsection 2.2 is taken from:

· J.K.G. Dhont, A.F.H. Duyndam, and B.J. Ackerson, Physica A **189**, 503 (1992).
· J.K.G. Dhont, A.F.H. Duyndam, and B.J. Ackerson, Langmuir **8**, 2907 (1992).

Demixing kinetics in the intermediate stage, including effects of hydrodynamic interactions between the colloidal particles, is described in:

· J.K.G. Dhont, J. Chem. Phys. **105**, 5112 (1996).
· J.K.G. Dhont, Progr. Coll. Polym. Sci. **104**, 66 (1997).

Theory on the effect of shear flow on decomposition kinetics can be found in:

· T. Imaeda, A. Onuki, and K. Kawasaki, Prog. Theor. Phys. **71**, 16 (1984).
· T. Imaeda and K. Kawasaki, Prog. Theor. Phys. **73**, 559 (1985).

· A. Onuki, Physica A **140**, 204 (1986).

· J.K.G. Dhont and A.F.H. Duyndam, Physica A **189**, 532 (1992).

· J. Lai and G.G. Fuller, J. Pol. Sci.: part B: Pol. Physics **32**, 2461 (1994).

In most of these papers the tendency for concentration fluctuations to acquire two-dimensional character as time proceeds is explicitly mentioned, in accordance with the results of section 5. The approach developed in section 5 is taken from the paper by Dhont and Duyndam.

The displacement of the spinodal of colloidal systems due to shear flow is discussed in:

· J.K.G. Dhont, Phys. Rev. Lett. **76**, 4269 (1996).

Experiments on spinodal decomposition in binary fluids are reported in:

· P. Guenoun, R. Gastaud, F. Perrot, and D. Beysens, Phys. Rev. A **36**, 4876 (1987).

· A. Cumming, P. Wiltzius, F.S. Bates, and J.H. Rosendale, Phys. Rev. A **45**, 885 (1992).

· N. Kuwahara, K. Kubota, M. Sakazume, H. Eda, and K. Takiwaki, Phys. Rev. A **45**, 8324 (1992).

· K. Kubota, N. Kuwahara, H. Eda, M. Sakazume, and K. Takiwaki, J. Chem. Phys. **97**, 9291 (1992).

· A.E. Bailey and D.S. Cannell, Phys. Rev. Lett. **70**, 2110 (1993).

Experiments on polymer systems can be found in:

· C.A. Smolders, J.J. van Aartsen, and A. Steenbergen, Kolloid-Z.u.Z. Polymere **243**, 14 (1971).

· I.G. Voigt-Martin, K.-H. Leister, R. Rosenau, and R. Koningsveld, J. Pol. Sci.: Part B: Pol. Phys. **24**, 723 (1986).

· P. Wiltzius, F.S. Bates, and W.R. Heffner, Phys. Rev. Lett. **60**, 1538 (1988).

· F.S. Bates and P. Wiltzius, J. Chem. Phys. **91**, 3258 (1989).

· H. Lee, T. Kyu, A. Gadkari, and J.P. Kennedy, Macromolecules **24**, 4852 (1991).

· M. Takenaka and T. Hashimoto, J. Chem. Phys. **96**, 6177 (1992).

· N. Kuwahara, H. Sato, and K. Kubota, J. Chem. Phys. **97**, 5905 (1992); Phys. Rev. E **47**, 1132 (1993).

· M. Takenaka and T. Hashimoto, Macromolecules **27**, 6117 (1994).

· C.C. Lin, H.S. Jeon, and N.P. Balsara, J. Chem. Phys. **103**, 1957 (1995).

Spinodal decomposition in other systems, like alloys (Komura) and surfactant systems (Mallamace et al.) is discussed in:

· S. Komura, K. Osamura, H. Fujii, and T. Takeda, Phys. Rev. B **31**, 1278 (1985).

· F. Mallamace, N. Micali, S. Trusso, and S.H. Chen, Phys. Rev. E **51**, 5818 (1995).

The data in Fig. 9 are taken from Malamace et al. (1995).

Experiments on the effect of steady and oscillatory shear flow on the spinodal decomposition kinetics of binary fluids can be found in:

· D. Beysens, M. Gbadamassi, and L. Boyer, Phys. Rev. Lett. **43**, 1253 (1979).

· D. Beysens, M. Gbadamassi, and B. Moncef-Bouanz, Phys. Rev. A **28**, 2491 (1983).

· D. Beysens and F. Perrot, J. Physique-Lettres **45**, 31 (1984).

· C.K. Chan, F. Perrot, and D. Beysens, Phys. Rev. Lett. **61**, 412 (1988).

· F. Perrot, C.K. Chan, and D. Beysens, Europhysics Lett. **9**, 65 (1989).

· T. Baumberger, F. Perrot, and D. Beysens, Physica A **174**, 31 (1991).

· C.K. Chan, F. Perrot, and D. Beysens, Phys. Rev. A **43**, 1826 (1991).

· T. Baumberger, F. Perrot, and D. Beysens, Phys. Rev. A **46**, 7636 (1992).

Similar experiments on polymer systems are reported in:

· T. Hashimoto, T. Takebe, and K. Fujioka, in *Dynamics and Patterns in Complex Fluids*, A. Onuki and K. Kawasaki, eds., Springer Proceedings in Physics vol. 52, (Springer Verlag, Berlin, Heidelberg, 1990).

· T. Hashimoto, T. Takebe, and K. Asakawa, Physica A **194**, 338 (1993).

· T. Hashimoto, K. Matsuzaka, E. Moses, and A. Onuki, Phys. Rev. Lett. **74**, 126 (1995).

Overview articles, where in some cases nucleation is also discussed, and which contain additional references, are:

· K. Binder, Rep. Prog. Phys. **50**, 783 (1987).

· W.I. Goldburg, in *Scattering Techniques Applied to Supramolecular and Nonequilibrium Systems*, S.H. Chen et. al., eds., (Plenum Press, New York, 1981), p. 383.

· J.D. Gunton, M. San Miquel, and P.S. Sahni, in *Phase Transitions and Critical Phenomena*, vol. 8, C. Domb and J.L. Lebowitz, eds., (Academic Press, New York, 1983), p. 267.

· S.W. Koch, in *Dynamics of First-order Phase Transitions in Equilibrium and Nonequilibrium Systems*, Lecture Notes in Physics, H. Araki et. al., eds., (Springer Verlag, Berlin, 1984).

· K. Binder and D.W. Heermann, in *Scaling Phenomena in Disordered Systems*, R. Pynn and A. Skjeltorp, eds., (Plenum Press, New York, 1985), p. 207.

· H. Furukawa, Adv. Phys. **34**, 703 (1985).

· P. Guyot and J.P. Simon, Journal de Chim. Phys. **83**, 703 (1986).

The derivation of the free energy functional of a van der Waals fluid, used in exercise 1, and a description of two-phase equilibrium can be found in:

· N.G. van Kampen, Phys. Rev. **135**, A362 (1964).

FIRST
INTERLUDE

SEVENTEENTH CENTURY DUTCH ART: A BRIEF GUIDE TO THE MAURITSHUIS, THE HAGUE, AND TO THE RIJKSMUSEUM, AMSTERDAM

J.R. DORFMAN
Department of Physics and Institute for Physical Science and Technology, University of Maryland
College Park, MD 20742, USA

1. Introduction

The Netherlands gained its independence from Spain in 1648 at the conclusion of the Eighty Years War, after a number of bloody battles. This led to the formation of the confederation of the Dutch provinces with a Protestant leadership, under the house of Orange. Both the years that preceded and those that followed the Peace of Münster saw the growth of a large and well-to-do middle class, with considerable business interests in the newly colonized regions of the Americas and East Indies.

Along with the growth of the middle class and the decline of the power of the Catholic Church, this period was also characterized by a strong interest in developing a national identity, and by a strong intellectual life. There was a natural identification of the Dutch national history with that of the Jews of the Old Testament, with Spain taking the role of Egypt as the oppressor. Further, an interest in the natural sciences developed, and the science of optics was pursued with spectacular achievements. The names of Huygens and van Leeuwenhoek are still familiar today, after more than three centuries. Fifty years before the establishment of the Protestant government in the Netherlands, and following the Sack of Antwerp, there was also a migration of Flemish Protestants northward to escape Spanish domination, and many of these immigrants brought with them a highly developed cultural and artistic education. Moreover, the Dutch proved to be reasonably tolerant for that time, for not only were Catholics able to continue their religious practices, at least in the privacy of their homes, a relatively large Jewish community lived in the Netherlands without serious persecution.

J. Karkheck (ed.),
Dynamics: Models and Kinetic Methods for Non-equilibrium Many Body Systems, 123–129.

None of this is meant to suggest that the Netherlands was a utopia, only to indicate that the country was open to new enterprises in business, in science and in the arts. Many families had enough money to indulge themselves in collecting art and interesting objects, including shells, musical instruments, scientific instruments, etc. Dutch society also provided a ready market for art of all types. No longer was art purchased almost exclusively by churches and the noble classes, but people of the merchant class were interested in owning paintings and the artistic community grew to meet the demand. Paintings, etchings, and drawings showing landscapes, scenes of everyday life, now known generally as Dutch genre art, cityscapes, still life art, were generally available and inexpensive enough that ordinary people would be able to buy them, typical prices being on the order of a few guilders for the cheapest commercial paintings. Commissioned paintings were somewhat more expensive, of course, as was the output of the best artists of the time. For example one could buy a Vermeer at an auction in 1696 for as little as 17 Guilders, or, for THE VIEW Of DELFT, 200 Guilders. A particular Rembrandt etching was sufficiently expensive that it was called THE HUNDRED GUILDER PRINT (1647), and it is known by that title even today. [1]

The character of Dutch art was quite different in spirit from that of the Catholic countries to the south. While history, biblical and mythological paintings were still quite common and regarded as the highest form of art, at least by the art theorists of the time, the religious messages conveyed by the paintings were often more subtle than those in earlier paintings, and the Virgin Mary appeared only in paintings commissioned by Catholics, and never in Dutch Protestant art. However, even genre art and still lifes were often packed with religious meaning, of a Calvinist sort, that probably any reasonably educated person of the time could understand. Art historians have written a great deal on how this art was understood by viewers in the seventeenth century, and it will not come as any surprise that there continue to be serious controversies over many points raging in art circles, as there are in physics circles, too.

A very common theme in Dutch art of the period is the idea of the vanity of this life, or *vanitas*. That is, earthly pleasures are very transitory, and the only lasting happiness will come in the next life, and then only to those who have led lives of moderation and industriousness, and devoted themselves to good works. This idea is often expressed in symbolic form through the appearance of skulls and bones in paintings, or through more subtle symbols such as cut flowers, clocks, hour glasses, broken objects, bubbles, or reflections in mirrors. Also common are scenes illustrating moral

[1] Apparently it was Rembrandt himself who paid this price for a copy of his own work.

lessons. Books of drawings and short poems illustrating moral themes were quite common, and the symbols associated with them were generally known. Common sayings like "a fool and his money are soon parted" often appear in visual form in scenes where an old man is being robbed by a prostitute. Many of the Dutch artists had, and their successors still have, a ribald sense of humor, so bordello scenes, phallic objects, sexual illusions of subtle and overt types are quite common in their art.

A number of Dutch cities are especially associated with the artists who lived there. Rembrandt grew up in Leiden, studied with Pieter Lastman, and moved to Amsterdam at age 25, (1631), which was the main artistic center. However the Leiden school, with Jan Lievens, the fijnschilders (fine painters), founded by a Rembrandt student, Gerrit Dou, and Jan Steen, was a major center of art, as was Utrecht, the Utrecht school being largely composed of followers of an Italian painter, Caravaggio. Of course, Delft with Jan Vermeer, Haarlem with Hals and Ruisdael, Kampen with Avercamp, only illustrate the wide spread of painters of genius over the Netherlands at the time.

When looking at a painting for the first time you should just let the beauty of the painting sink in for a while. The artists were, after all, incredible technicians with paint. They could make clothing, people, animals, minerals, and vegetables look three dimensional and almost real, just using paint on a two dimensional surface. You will notice that they liked to paint gold and silver objects with light falling on them, peeled lemons, reflections of objects on metal or glass surfaces, and so on, to exhibit their skills as artists. Then you should look more closely at the paintings, examining the small details and always asking why the painter took this approach to the subject, and why the various objects appear in the paintings. You will soon learn that empty barrels mean that somebody in the painting makes a loud noise, but is essentially empty, that oysters are aphrodisiacs, that dogs can represent fidelity or loose and amoral sexual behavior, that broken eggs can indicate a vanitas theme or a loss of virginity, flowers or leaves eaten by insects represent the transient nature of worldly beauty, and so on.

It is generally agreed that the greatest genius of this period was Rembrandt. However there were quite a number of painters of very real genius. Among these, my own favorites include Jan Vermeer, Jan Steen, Frans Hals, Hendrik Avercamp, Jacob van Ruisdael, Frans van Mieris, Gerrit Dou, Pieter Saenredam, Judith Leyster, Gerrit van Honthorst, ...!

2. The Mauritshuis, The Hague

With this general introduction, we turn now to to a brief mention of a few of the great paintings in museums easily accessible from Leiden by a short train ride. I should certainly mention the Lakenhal Museum in Leiden as an interesting nearby place to visit. You will see a very early Rembrandt, PALAMEDES BEFORE AGAMEMNON, which looks a lot like a painting by his teacher, Pieter Lastman, but nothing at all like a "real" Rembrandt. Like any good graduate student, Rembrandt did what his advisor told him to do, but after 'getting his degree', went on to do his own stuff, and largely abandoned Lastman's style.

In the Mauritshuis, I suggest that you take a careful look at:

- GIRL OFFERING OYSTERS, by Jan Steen, (1658-1660): This little gem is one of the masterpieces of Seventeenth Century art. It may be the sexiest painting of all time, and says a lot about oysters, too. Notice the lion head chair. They are very popular in Dutch art.

- VASE WITH FLOWERS, by Ambrosius Bosschaert, the Elder, (ca. 1620). Bosschaert was one of the Flemish artists who emigrated after the fall of Antwerp to the Spanish. Notice the cut flowers, the shells, the insects, and the leaves with holes. This is not just some pretty flowers, it is a moral lesson about the temporary nature of earthly beauty, and the landscape in the background is certainly not Dutch. Not only that, the flowers could not possibly have been arranged that way since they bloom at many different times of the year.

- THE ANATOMY LESSON OF DR. TULP, by Rembrandt van Rijn, (1632). This was one of Rembrandt's first major commissions. If you look closely, you will see numbers around the painting to identify all of the people in it. Even the name of the corpse is known. He was an executed criminal. He stole a coat. The people in the painting look very much alive, if somewhat posed, except for the corpse, who really does look dead. This painting, too, conveys a moral, since it is not accidental that Rembrandt chose to illustrate a hand which was considered to be one of God's most wonderful creations. The even more intricate marvels of molecular biology still had a few hundred years to wait for their discovery.

- THE VIEW OF DELFT, by Jan Vermeer, (ca. 1660-1661). (Figure) I can talk for hours about this painting. Look at the textures of the paint across the painting. Try to find Vermeer's monogram on the painting, and notice that some of the shadows on the water are not quite what one would expect from the physics of light. Vermeer has also rearranged the buildings in Delft a bit. Take your time to look at this painting and the other paintings by Vermeer in the room. You are truly in heaven. Marcel Proust thought so, too. He was a great admirer of Vermeer and made special trips to visit the Mauritshuis from France. Both the VIEW OF DELFT and the GIRL WITH THE PEARL EAR-

RING have just been restored. Two of Vermeer's paintings, THE ASTRONOMER (Paris) and THE GEOGRAPHER (Frankfurt) depict van Leeuwenhoek who was a trustee for Vermeer's estate. [2]

- THE YOUNG MOTHER, by Gerrit Dou, (1658). This is a painting by the student of Rembrandt who founded the Leiden School of fijnschilders. The crystalline clarity of the painting is amazing. Dou is reputed to have taken several days or more just to paint a broom. If this painting conveys a profound moral, no one seems to know just what it is.
- VIEW OF HAARLEM WITH BLEACHING GROUNDS, by Jacob van Ruisdael, (ca. 1670). This magnificent landscape shows Haarlem, with its central buildings around the St. Bavo Church, which is the (overly) large building in the background. Look at the clouds, the sunlight, the windmills, and the lines of the painting. Let your eye follow along from the front of the painting to the back.

3. The Rijksmuseum, Amsterdam

Along with the Mauritshuis, the Rijksmuseum in Amsterdam houses one of the world's great collections of art from the Golden Age of Dutch art. Here we mention only a very few of these paintings.

- WEDDING PORTRAIT OF ISAAC MASSA AND BEATRIX VAN DER LAEN, by Frans Hals, (1622). This wonderfully informal portrait of a bride and groom is full of symbols of steadfast love and fidelity, including the ivy, the vines, and the thistle next to Isaac. There seems to be a garden of love in the background. One nice feature of this painting is that the bride is as important to the painting as the groom is. Isaac seems about to say something. Massa was a good friend of Hals and is the subject of other portraits as well.
- EMBLEMATIC STILL LIFE, by Johannes Torrentius, (1614). This is a real vanitas painting. Everything suggests moderation, the horse bridle, the water to mix with wine, and the measured notes of the music. Even the bridle is painted to look like a skull. The water and wine was a well known emblem of the time, symbolizing moderation. This is the only known painting of Torrentius, who was principally known at the time for blasphemy, and for painting erotic nudes, none of which survive.

[2]The idea behind the pairing of the Astronomer and the Geographer can easily be traced to the first line of the book of Genesis, God having created the heavens and the earth.

- WINTER LANDSCAPE WITH ICE SKATERS, by Hendrik Avercamp, (1608). No one paints the misty nature of Dutch winter life on canals like Avercamp. It is miraculous how he captures the mist in paint. Look carefully at all of the figures, you will see all kinds of funny stuff going on, and the same people show up in many of Avercamp's canal scenes. The perspective of this painting is reminiscent of a winter scene of hunters by Breugel, now in Vienna.
- THE LITTLE STREET, by Jan Vermeer, (1658). Like THE VIEW OF DELFT, this is a city scene, and a very realistic looking one at that. A very careful examination reveals that the realism is actually illusory, and the scene is actually fanciful. Note that a woman on a chair has been painted out in the entrance to the passageway. You can still see a bit of her. To see some of the illusions that Vermeer uses to fool our eyes, take a good look at the shutters on the house at the right, for example. This is a great painting, full of Vermeer's tricks.
- THE MERRY FAMILY, by Jan Steen, (1668). This in one of the several paintings where Steen illustrates a moral lesson of the time, in this case, "the young ones chirp as the old ones used to sing". Look carefully at what the children are doing. Steen liked to put his family in his paintings. Here his father is at the left, sitting next to his sister, dressed as a religious woman. His children are in the painting, too.
- THE NIGHT WATCH, otherwise known as THE COMPANY OF CAPTAIN FRANS BANNING COCQ AND LIEUTENANT WILLEM VAN RUYTENBURCH, by Rembrandt van Rijn, (1642). This is probably the most well known painting of the Golden Age. It is truly spectacular. It is hard not to notice the Captain in black and the lieutenant in the brilliant yellow suit. But notice the little girl in the yellow dress on the other side of the captain, and follow all of the colors around the painting. A number of the group are demonstrating various aspects of musketry.

Further Reading

A very good general introduction to the art of this period is to be found in:

Seymour Slive, *Dutch Painting 1600-1800*, (Yale University Press, New Haven, 1995).

Arthur Wheelock has written beautifully on the art of the Golden Age. I can recommend his book:

Arthur K. Wheelock, Jr., *Vermeer and the Art of Painting*, (Yale University Press, New Haven, 1995), as a perfect example of clarity of thought and explanation. After reading this book, you will understand what is required to really look at a painting. No doubt you will run to the nearest museum to see if you can look this closely, too.

A wonderful collection of essays on Dutch art of this period has been written by the Polish poet Zbigniew Herbert. The book is: Zbigniew Herbert, *Still Life With a Bridle*, (The Ecco Press, New York, 1991). The title essay discusses the life and work of Torrentius, whose only extant painting, described above, is in the Amsterdam Rijksmuseum.

Along with journal articles and monographs, scholars of art history write essays for exhibition catalogs. These are a very valuable source of information, and you might want to check the numerous catalogs of recent exhibitions of Dutch art in all its aspects. Try the catalogs of the Vermeer, Steen, Hals, and Rembrandt exhibitions that took place within the last few years.

End Notes

These notes were prepared for the scientists attending the NATO-Advanced Study Institute on "Dynamics: Models and Kinetic Methods for Non-equilibrium Many-Body Systems", Leiden, the Netherlands, July - August, 1998.

I would like to thank Prof. Arthur Wheelock, Jr, of the National Gallery of Art, Washington D.C., and the University of Maryland, as well as Ms. Esmee Quodbach for valuable advice, corrections and suggestions. I also would like to thank Ms. Sarah Miller and Ms. Phoebe Avery for their gracious help with slide preparation.

KINETIC THEORY OF DYNAMICAL SYSTEMS

R. VAN ZON AND H. VAN BEIJEREN
Institute for Theoretical Physics, University of Utrecht
Princetonplein 5, 3584 CC Utrecht, The Netherlands

AND

J.R. DORFMAN
Institute for Physical Science and Technology and
Department of Physics, University of Maryland
College Park, MD 20742, USA

Abstract. It is generally believed that the dynamics of simple fluids can be considered to be chaotic, at least to the extent that they can be modeled as classical systems of particles interacting with short range, repulsive forces. Here we give a brief introduction to those parts of chaos theory that are relevant for understanding some features of non-equilibrium processes in fluids. We introduce the notions of Lyapunov exponents, Kolmogorov-Sinai entropy and related quantities using some simple low-dimensional systems as "toy" models of the more complicated systems encountered in the study of fluids. We then show how familiar methods used in the kinetic theory of gases can be employed for explicit, analytical calculations of the largest Lyapunov exponent and KS entropy for dilute gases composed of hard spheres in d dimensions. We conclude with a brief discussion of interesting, open problems.

1. Introduction

We consider here a classical many-particle system, a gas of hard spheres or of hard disks. Our principal concern will be to develop methods by means of which we can understand and calculate the properties of such gases as chaotic dynamical systems. It is, of course, well known that to describe the macroscopic, equilibrium properties of such gases, we can easily dispense with any knowledge of most of the dynamical properties of the particles of

J. Karkheck (ed.),
Dynamics: Models and Kinetic Methods for Non-equilibrium Many Body Systems, 131–167.

which the gas is composed. That is, one can use thermodynamics and equilibrium statistical mechanics, i.e. statistical thermodynamics, to describe the relevant equilibrium properties of the gas. All of the relevant microscopic properties of the system needed for statistical thermodynamics are contained in the partition sum, which is defined in terms of the Hamiltonian of the system. The partition function is based on a probability measure on phase space. The macroscopic properties are simply related to averages of certain microscopic expressions with respect to this measure. Of course it is far from trivial to compute these averages for anything like a real physical system.

Our interest here, though, is to consider a gas as a mechanical system and to understand its behavior in time, rather than its equilibrium properties, and to try to make quantitative statements about the motion of the trajectories of the phase points that describe the gas, in the usual $2Nd$-dimensional phase-space, Γ-space, of the system, where N is the number of particles, d the number of the spatial dimensions of the system, $d = 2$ or 3, and the phase-space has dN spatial coordinates and dN momentum coordinates. We take the particles to be identical hard spheres or hard disks, each of mass m and diameter σ. When we wish to describe the typical or average properties of the system, we must start with the specification of some useful probability measure, with respect to which averages can be defined. Any dynamical system, therefore, consists of: 1) a space Γ, 2) a measure $\mu(A)$, $A \subset \Gamma$, and 3) a transformation $S : \Gamma \to \Gamma$. We will see that the dynamical viewpoint can explain some features of macroscopic systems from their microscopic behavior. The explanations can be followed most easily in dynamical systems of very low dimensionality. However, even in simple low dimensional systems, dynamics may become so complicated that it is effectively impossible to follow the dynamics for long times, starting from a typical initial point, and we will be forced to consider typical behaviors using some appropriate probability measure. Our interest will be focused on chaotic systems which have the property that any uncertainty in the specification of the exact initial state of the system will grow exponentially in time, to the point where the future of a phase-space point can no longer be predicted to within a reasonable accuracy [1]. But we can still say something about probabilities.

It turns out that there is a close connection between the chaoticity of the system and issues like irreversibility on the macroscopic level and, for a gas of particles that interact with short-range forces, the validity of kinetic theory [2]. This connection will first be outlined in section 2, for low dimensional systems. A more extensive treatment can be found in Ref. [3]. In section 3 we return to a high dimensional system in the form of a hard-sphere gas in equilibrium. At low densities we can use kinetic theory to calculate a measure of chaoticity called the largest Lyapunov exponent. In section 4 another chaotic characteristic of this system is calculated using kinetic theory: the Kolmogorov-Sinai entropy. In section 5 we make some concluding remarks and present some open questions.

2. Dynamical Systems

The standard approaches to the theory of non-equilibrium processes in fluids are based on three foundational pillars: (1) The identification of the macroscopic quantities of physical interest as averages of microscopic quantities over an appropriate ensemble of similarly prepared systems; (2) The use of the Liouville equation, either in its classical or in its quantum mechanical version, to compute the time evolution of the ensemble distribution function; and (3) The utilization of some kind of physically reasonable factorization assumption for the ensemble distribution function in order to transform the Liouville equation into a tractable equation whose solution can be used to make quantitative statements about the macroscopic quantities. Such a procedure is followed in the derivation of the Navier-Stokes fluid dynamic equations from the Liouville equation [4] for general fluids, and in the derivation of the Boltzmann transport equation, and its extensions to higher densities, from the Liouville equation for dilute and moderately dense gases. More phenomenological approaches to irreversible behavior in fluids often depend on explicit stochastic assumptions about the underlying dynamical processes taking place in the fluid [5].

While they are of the highest importance for the development of theories of irreversible processes in fluids, both of these approaches to irreversible behavior leave the answers to some fundamental questions obscure. In particular, these approaches offer only qualitative insights into the reasons for the validity of the stochastic assumptions imbedded in these various procedures – either through factorization assumptions, which in essence, are statements about correlations and probabilities – or through the replacement of the exact dynamics by a stochastic, Langevin-type, dynamics. Further, while the approaches outlined above do predict an approach to an equilibrium state, under the proper physical conditions, and do provide experimentally verifiable statements about the approach to equilibrium, they do not give a complete picture of why the system approaches an equilibrium state, based upon the underlying microscopic dynamics. The general arguments for the use of stochastic methods are based upon the randomness of the microscopic motions of the particles but are not much more specific than that. The picture that we have of the approach to equilibrium is generally based on the idea that the local averages of conserved mechanical quantities, such as mass, momentum, and energy, change very slowly in time compared to the local averages of nonconserved quantities. Thus the macroscopic behavior will be dominated by the slowest variables in the system, the local conserved quantities, and the equilibrium state will be achieved when these quantities have reached steady, homogeneous values. This picture, suggested by solutions of the Boltzmann equation, has led to important advances in the theory of fluids, among others, to mode-coupling theory. What is missing from it is a basic understanding of the necessary (or sufficient) properties of the intermolecular potential for the system to approach an equilibrium state, as well as an understanding of the properties of the trajectories of the system, and the evolution of measures in Γ-space, that are responsible for the approach to equilibrium states and, perhaps, under different boundary conditions, to more complicated, but interesting non-equilibrium steady states.

The application of ideas from dynamical systems theory to non- equilibrium statistical mechanics allows us to make some progress in resolving the issues described above. The application of ideas from chaos theory, in particular, enables us to make some quantitative statements about the type and degree of randomness of a dynamical system, even of large systems typically treated by statistical mechanics. It also allows us to describe equilibrium and non-equilibrium states of a system in terms of probability measures defined in Γ-space, and in terms of the time evolution of these measures. Moreover, there are interesting and unexpected connections between the macroscopic transport coefficients that describe the approach of a fluid to equilibrium, and microscopic quantities that describe the chaotic behavior of a fluid, considered as a large, dynamical system. In this section we will outline some of these rather new ideas, and illustrate their applications to statistical mechanics by seeing how they work for systems of low dimensions and then generalizing them, when possible, to higher dimensional systems. We begin with a very simple two-dimensional reversible system, the baker's map, which exhibits many of the features we would like to see in more general, higher dimensional systems.

2.1. THE BAKER'S MAP

The simplest example of a reversible system with chaotic dynamics is probably the baker's map. Here we consider a two-dimensional phase space on a unit square. That is, $\Gamma = (x, y); 0 \leq x, y \leq 1$. The map, $\mathbf{B}$, operates only at discrete time steps, and moves points (x, y) to $\mathbf{B}(x, y) = (x', y')$ given by

$$\mathbf{B}\begin{pmatrix} x \\ y \end{pmatrix} = \begin{pmatrix} x' \\ y' \end{pmatrix} = \begin{pmatrix} 2x \\ y/2 \end{pmatrix} \quad \text{for } 0 \leq x < 1/2; \text{ and}$$

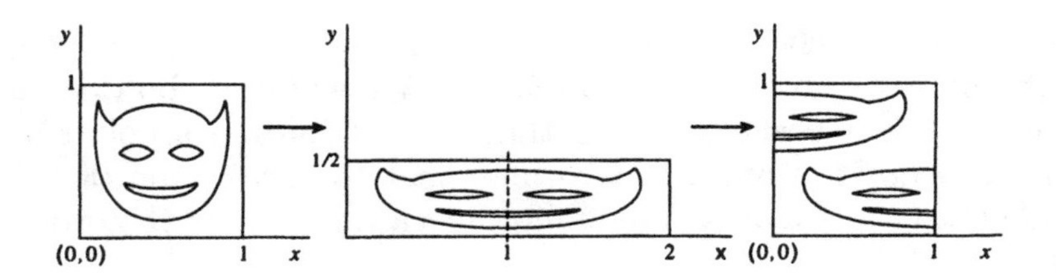

Figure 1. The baker's transformation.

$$= \begin{pmatrix} 2x-1 \\ (y+1)/2 \end{pmatrix} \quad \text{for } 1/2 \leq x < 1. \tag{1}$$

This map is illustrated in Fig. 1. It is immediately clear that this map possesses an inverse, $\mathbf{B}^{-1}$, given by

$$\mathbf{B}^{-1}\begin{pmatrix} x \\ y \end{pmatrix} = \begin{pmatrix} x/2 \\ 2y \end{pmatrix} \quad \text{for } 0 \leq y < 1/2 \text{ and}$$
$$= \begin{pmatrix} (x+1)/2 \\ 2y-1 \end{pmatrix} \quad \text{for } 1/2 \leq y < 1. \tag{2}$$

The baker's map is clearly area-preserving, and it is time-reversible in the sense that the transformation $\mathbf{T} : (x, y) \to (1-y, 1-x)$ serves as a time reversal transformation for this map, such that $\mathbf{T} \circ \mathbf{B} \circ \mathbf{T} = \mathbf{B}^{-1}$, and $\mathbf{T} \circ \mathbf{T} = \mathbf{1}$, where $\mathbf{1}$ is the unit operator.

Now we regard the unit square as a "toy" phase-space. The dynamics of the baker's map in this phase-space has the following properties: 1) Consider two infinitesimally separated points. Unless they have exactly the same x-coordinates, or the same y-coordinates, the images of these two points under several applications, called iterations, of the map $\mathbf{B}$ will cause the x-components to separate *exponentially* with the number of applications, with an exponent of $\lambda_+ = \ln 2$, whereas the y-components will converge exponentially to a common value with an exponent of $\lambda_- = -\ln 2$. The exponents, $\lambda_\pm$, characterizing exponential separation or convergence of points in phase-space are called *Lyapunov exponents.* The directions in which the points converge exponentially, in this case just the y-direction, are called *stable* directions, and the directions in which they separate exponentially, in this case the x-direction, are called *unstable* directions. The fact that the Lyapunov exponents sum to zero is a simple consequence of the area preserving property of the baker's map, as can easily be seen by considering the evolution, with successive iterations, of the small rectangle with corners at $(x, y), (x + \delta x, y), (x, y + \delta y), (x + \delta x, y + \delta y)$ under the baker map, $\mathbf{B}$. This rectangle has constant area, but grows exponentially long in the x-direction, and exponentially thin in the y-direction. Sooner or later it gets stretched and folded in such a way that on a coarse grained scale, the unit square is covered uniformly. We will describe this behavior on a coarse-grained scale, by saying that the distribution of points becomes weakly uniform. It is important to note that the projection of the small rectangle onto the x-axis will be uniform in a time, n_u on the order of

$$n_u \sim \frac{-\ln \delta x}{\lambda_+}, \tag{3}$$

where $\lambda_+ = \ln 2$ is the positive Lyapunov exponent for the baker's map. That is, the projection of the small rectangle on the unstable direction becomes uniform much sooner than the distribution of points on the entire unit square becomes weakly uniform.

2.2. THE ARNOLD CAT MAP AND HYPERBOLIC SYSTEMS

A different map with a similar dynamical behavior, i.e., area preserving, with exponentially separating and converging trajectories characterized by positive and negative Lyapunov exponents, is provided by the Arnold cat map, $\mathbf{T}(x, y)$, illustrated in Fig. 2, and given by

$$\begin{pmatrix} x' \\ y' \end{pmatrix} = \mathbf{T}\begin{pmatrix} x \\ y \end{pmatrix} = \begin{pmatrix} 2 & 1 \\ 1 & 1 \end{pmatrix}\begin{pmatrix} x \\ y \end{pmatrix} \mod 1. \tag{4}$$

The area preserving property is guaranteed by the fact that the matrix representation of $\mathbf{T}$ has unit determinant, the integer coefficients of the matrix together with the mod 1 condition implies that the unit square, or more properly, the unit torus, is mapped smoothly onto itself by $\mathbf{T}$. Such area preserving maps with integer coefficients are called *toral automorphisms*. The Arnold cat map also has stable and unstable directions associated with positive and negative Lyapunov exponents given by $\lambda_\pm = \ln[(3 \pm \sqrt{5})/2]$. In a way much like the baker's map, a small region of the unit square will be stretched and squeezed under the iterated action of the cat map, with projections on both the x and y-axes becoming uniform on a similar time scale as in Eq. (3), and with the distribution of points becoming weakly uniform on a longer time scale.

The baker's map and the Arnold cat map are two simple examples of what are called *hyperbolic* dynamical systems. Briefly, and somewhat loosely stated, hyperbolic dynamical systems are defined by the action of some dynamical transformation, $\mathbf{S}$, on a phase-space Γ, such that: (a) one can identify stable and unstable directions in Γ, under the action of $\mathbf{S}$, with negative and positive Lyapunov exponents, all bounded away from zero; (b) the stable and unstable manifolds (lines, surfaces, etc.) are continuous functions of the variables that define the phase space, and when the two

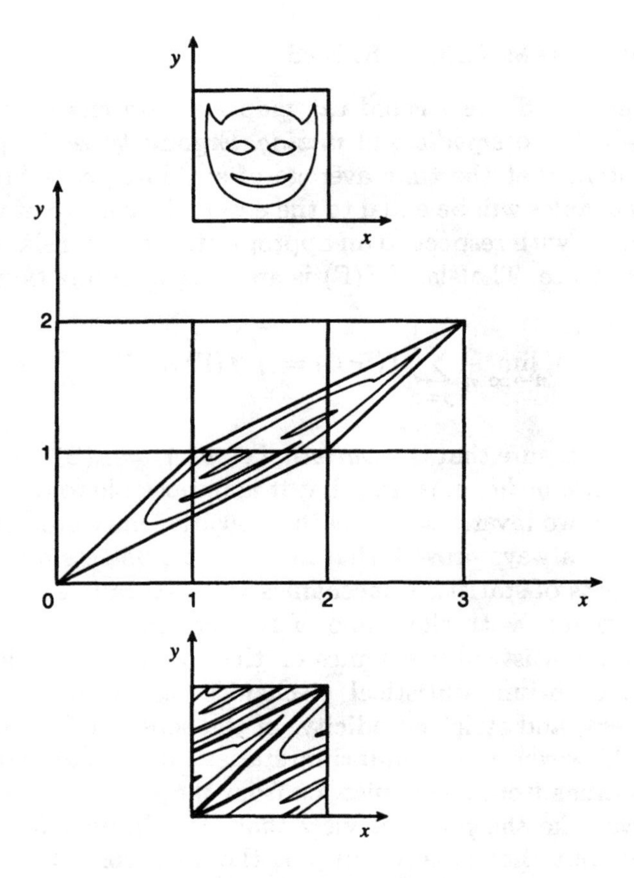

Figure 2. The Arnold cat map.

manifolds intersect, they do so transversely; (c) the system is *transitive*, i.e., there exists some trajectory in the phase-space that is dense on the phase-space; and (d) for maps, i.e., for dynamical systems where **S** acts only at discrete times, there are no directions in phase-space with a Lyapunov exponent of zero, while for "flows", i.e. systems where **S** depends upon a continuous time parameter, the only directions in Γ with zero Lyapunov exponents are the direction of the flow itself and those directions perpendicular to the surfaces of the prescribed constants of the motion, such as total energy and, where appropriate, total momentum and total angular momentum. Clearly the baker's map and the Arnold cat map are hyperbolic maps. A typical flow that one might examine for hyperbolicity is the motion of a phase point on surfaces of constant energy for a system of interacting particles.

2.3. ERGODIC AND MIXING SYSTEMS

The baker's map and the Arnold cat map are also examples of dynamical systems which are *ergodic* and *mixing.* Ergodicity is the property of a dynamical system that the time average of any integrable function of the phase-space variables will be equal to the ensemble average of this function, the average taken with respect to an appropriate time translation invariant, equilibrium measure. That is, if $f(\Gamma)$ is an integrable function, then

$$\lim_{n\to\infty} \frac{1}{n} \sum_{j=o}^{n-1} f(\mathbf{S}^j\Gamma) = \int f(\Gamma)\, \mu(d\Gamma), \tag{5}$$

where $\mu(A)$ is a measure that is *invariant*, i.e., $\mu(A) = \mu(\mathbf{S}^{-1}A)$ for any nontrivial set A, and *ergodic*, meaning that it is impossible to divide the whole phase-space into two invariant sets, each of positive measure.[1] It is generally assumed, but not always proved, that our systems possess a unique ergodic measure. Students of statistical mechanics will naturally associate the idea of an ergodic system with the name of Boltzmann who used this idea to base equilibrium statistical mechanics on the laws of mechanics.[2]

Much of equilibrium statistical mechanics can be based on the laws of large numbers, and strict ergodicity, in the sense of Boltzmann, is not that essential. However, non-equilibrium statistical mechanics requires some deep underpinnings from mechanics, or from the theory of stochastic processes. Here we take the point of view that Hamiltonian mechanics is all that is needed, but that is certainly not the only possible point of view. For non-equilibrium statistical mechanics, it is useful to explore an idea of Gibbs, which is called the mixing property of a dynamical system. Mixing systems are always ergodic, but the reverse is not always true. To define a mixing system, we consider two arbitrary sets in the phase-space, A and B, say, both of nonzero measure, and the evolution of the set A in time. Suppose after n iterations of the map $\mathbf{S}$ the set A has moved to $\mathbf{S}^n A$, then the system is mixing if

$$\lim_{n\to\infty} \frac{\mu(B \cap \mathbf{S}^n A)}{\mu(B)} = \frac{\mu(A)}{\mu(\Gamma)}, \tag{6}$$

where $\mu(\Gamma)$ is the measure of the entire phase-space, such as the unit square for the baker's map or the cat map, or the constant energy surface for a more general system. The mixing condition simply means that the time

[1]Eq. (5) doesn't have to hold for all points Γ, as long as the set of points violating it has measure zero, with respect to the measure in the definition of the dynamical system.

[2]Traditionally, statistical mechanical systems were called ergodic if they are ergodic under Hamiltonian flow with the Liouville measure on the energy shell.

evolution of a set in phase-space is such that, in a coarse grained sense, it gets uniformly distributed, with respect to the measure μ, over the entire phase-space. It can be proved rather easily that for a mixing system, non-equilibrium averages of integrable functions f will approach their equilibrium values in the course of time.

2.4. THE APPROACH TO EQUILIBRIUM

A nice illustration of the approach to equilibrium, as provided by baker or cat maps, is to consider the behavior in time of reduced distribution functions. That is, if we think of the unit square, again, as a phase space, then we can define a phase-space distribution function, $\rho_n(x, y)$, as a function of the number, n, of iterations of the map and the coordinates x and y. The phase-space distribution function satisfies a discrete-time version of the Liouville equation, which is a form of the Frobenius-Perron equation for area preserving maps. The appropriate equation for the baker map is

$$\rho_n(x, y) = \rho_{n-1}(\mathbf{B}^{-1}(x, y)), \tag{7}$$

which, written out in full detail, becomes

$$\begin{aligned} \rho_n(x, y) &= \rho_{n-1}(x/2, 2y) \text{ for } 0 \leq y < 1/2 \\ &= \rho_{n-1}((x+1)/2, 2y-1) \text{ for } 1/2 \leq y < 1. \end{aligned} \tag{8}$$

A similar, but somewhat more complicated equation could be given for the Arnold cat map, but we will not use it here.

For these simple two-dimensional models, a reduced distribution function is obtained by integrating the distribution function over one of the two phase-space variables, x or y. This integration is motivated by the fact that for a system of N particles, we are not particularly interested in the full N-particle distribution function, but rather in the one or two-particle distribution functions that can be used to evaluate the macroscopic quantities of interest, such as mass, momentum, and energy densities. Since our simple maps have only two coordinates, we can only consider the very simple case where a reduced distribution function is obtained by integrating over one of the coordinates. For the baker's map we will construct the distribution function for the density of points in the x-direction, for reasons that will become clear as we proceed. That is, we define a reduced distribution function $W_n(x)$ by

$$W_n(x) = \int_0^1 dy\, \rho_n(x, y). \tag{9}$$

Using Eq. (8), we can easily obtain a difference equation for $W_n(x)$, as

$$W_n(x) = \frac{1}{2}\left[W_{n-1}(\frac{x}{2}) + W_{n-1}(\frac{x+1}{2})\right]. \tag{10}$$

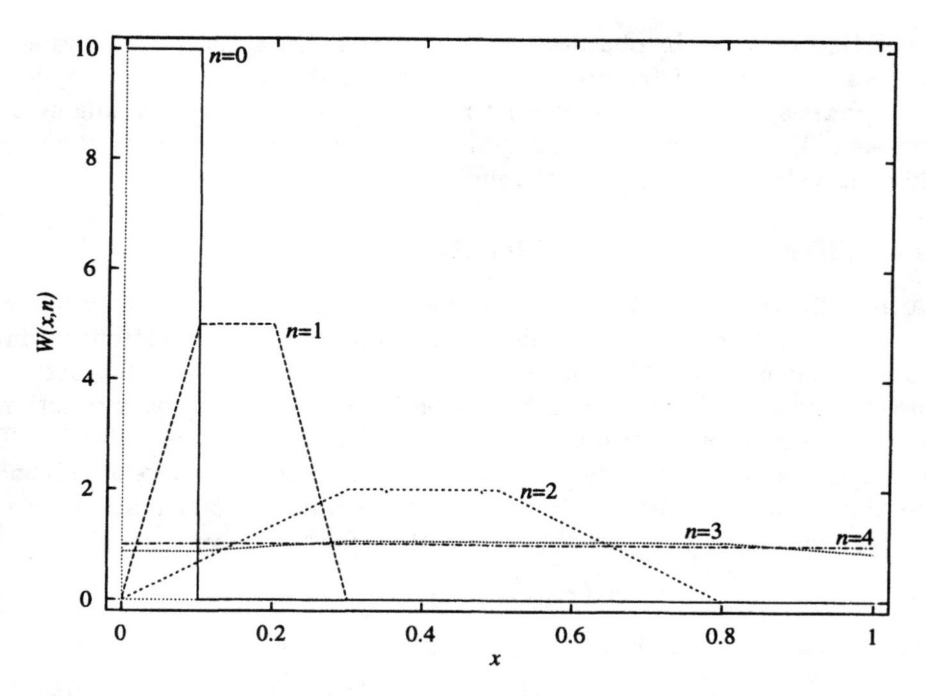

Figure 3. The projection, $W(x, n)$, onto the x-coordinate, of the phase-space distribution function for the Arnold cat map.

This equation is, among other things, the Frobenius-Perron equation for the one-dimensional map, $x' = 2x \pmod 1$, on the interval $(0, 1)$. What is more important here, though, is that except for very special initial values for $W_0(x)$, such as Dirac delta functions on the periodic points of the map, $W_n(x)$ approaches a constant, independent of x, as $n \to \infty$. This may be proved in a number of ways, but may be understood most simply by just drawing some possible functional forms for $W_0(x)$ and follow what happens to them after a few iterations of Eq. (10). A standard procedure is to make a Fourier expansion of $W_0(x)$, and to notice that only the constant term remains as the number of iterations gets large. The approach to equilibrium in this simple system can be associated with the properties of the expanding manifold in our simple two-dimensional phase-space. Because of the stretching of regions in phase-space in the unstable directions, functions defined on the unstable manifold will get "smoothed out" in the course of time, much the same way that a ball of dough gets smoother and smoother along the direction that the baker stretches it. The initial wrinkles in the phase-space distribution function will not get smoothed out along the stable direction, on the contrary, they typically will get more and more wrinkled as the system evolves. From these considerations we can see that the inte-

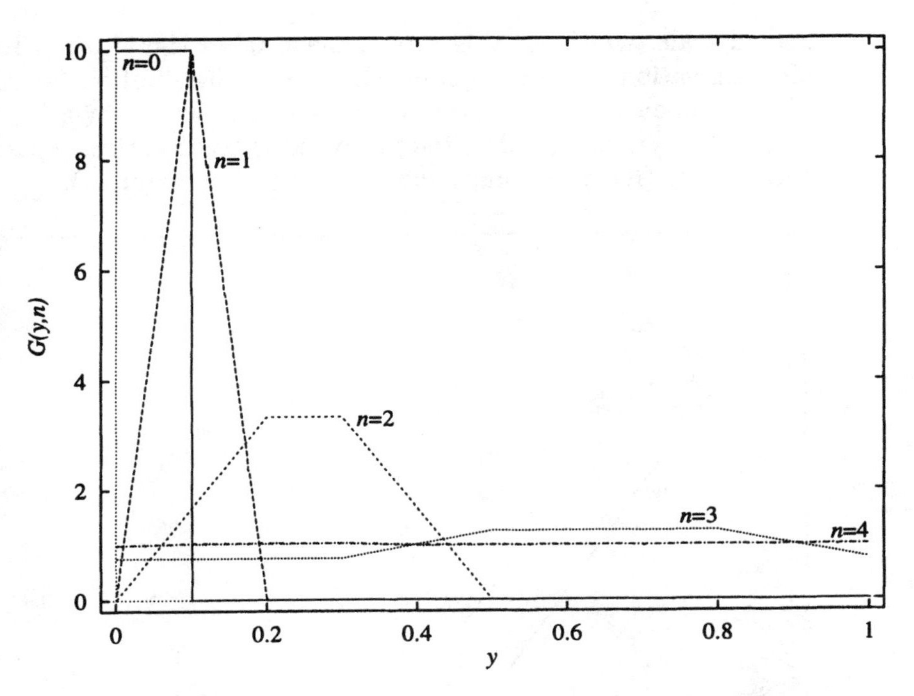

Figure 4. The projection, $G(y,n)$, onto the y-coordinate, of the phase-space distribution function for the Arnold cat map.

gration of the phase-space distribution over the stable direction in Eq. (9) was not chosen accidentally; had we integrated over x instead, we would not have obtained an equation with a nice equilibrium solution as $n \to \infty$. In fact, a typical initial distribution will become smooth in the expanding directions but very striated in the contracting directions. However, eventually it will look uniform on a coarse grained scale, consistent with the mixing behavior of the baker's map.

The connection between the approach to equilibrium and the expanding direction of a measure preserving map can be further explored by considering the Arnold cat map. Here, the expanding direction is along a line that is not aligned along either of the coordinate axes. One would expect that for this model a projection of the phase-space distribution function along either the x-axis, or the y-axis, would approach an equilibrium value. That this is so can be seen from a simple computer calculation. We start with a phase-space distribution that is concentrated in a small region $0 \leq x, y \leq 0.1$. We then follow the evolution in time of x and y projections of the distribution function. In Figs. 3 and 4 we can easily see that these distributions approach constant values after three or four iterations of the map, much before the entire phase-space distribution is smooth, even on a reasonably coarse grained scale, which for this arrangement takes eight to ten iterations. This observation may be generalized: reduced distribution functions on some lower dimensional projection of phase-space will always become smooth under the dynamics, unless the projected space is entirely spanned by stable directions (hence is some subset of the stable manifold).

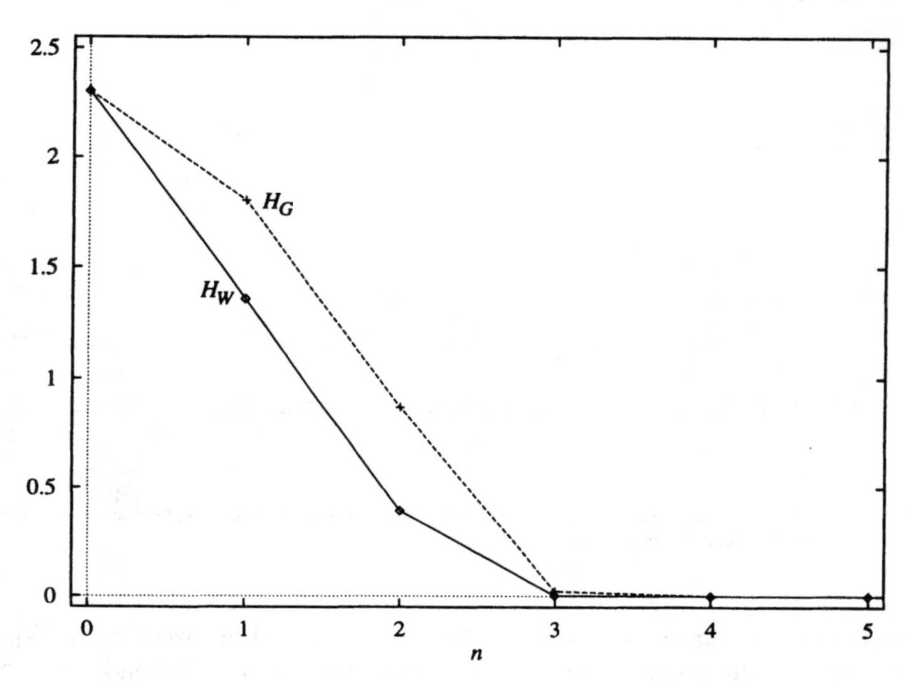

Figure 5. The Boltzmann H-functions $H_W(n)$ and $H_G(n)$, obtained by using $W(x,n)$ and $G(y,n)$ respectively.

Not only does one see an approach to an equilibrium distribution for the projected distribution functions for these maps, one also sees that a suitably defined Boltzmann H-function decreases monotonically as the number of iterations increases. This is illustrated in Fig. 5 for the Arnold cat map, for both projected distribution functions, starting from the same initial state as described above. The figure shows the H-function for both projections, as calculated on a computer. For the baker's map, we can easily show the monotonic decrease in the H function analytically. To do this we define the H-function by

$$H(n) = \int_0^1 dx\, W_n(x)\, \ln W_n(x). \tag{11}$$

If we now use the recursion relation for $W_n(x)$, Eq. (10), we find that

$$H(n+1) = \int_0^1 dx\, W_{n+1}(x)\, \ln W_{n+1}(x)$$

$$
\begin{aligned}
&= \int_0^1 dx \frac{1}{2}\left[W_n(\frac{x}{2}) + W_n(\frac{x+1}{2})\right] \ln\left\{\frac{1}{2}\left[[W_n(\frac{x}{2}) + W_n(\frac{x+1}{2})]\right]\right\} \\
&\leq \frac{1}{2}\int_0^1 dx \left[W_n(\frac{x}{2}) \ln W_n(\frac{x}{2}) + W_n(\frac{x+1}{2}) \ln W_n(\frac{x+1}{2})\right] \\
&= H(n). \qquad (12)
\end{aligned}
$$

The inequality in Eq. (12) follows from the fact that $f[(a+b)/2] \leq [f(a)+f(b)]/2$ if $f(x) = x \ln x$. That is, a chord connecting two points on the curve $x \ln x$ lies above the curve. Thus, we see that the H function decreases with time until $W_n(x)$ becomes a constant.

We conclude this discussion with a few remarks. For the baker's map and the Arnold cat map, admittedly "toy" models, highly simplified versions of N particle systems, with almost trivial phase-spaces, we have been able to derive irreversible equations and H-theorems with a minimum of assumptions. We have associated the approach to equilibrium of projected distribution functions with the existence of unstable manifolds for the dynamics in phase-space, and the fact that the projection is not orthogonal to the unstable directions. In a more general context, such as a corresponding, but not yet possible, dynamical derivation of the Boltzmann transport equation, we would expect that the approach to equilibrium, seen here for baker and Arnold cat maps, would correspond to the approach to a local equilibrium state in the fluid. In such a local equilibrium state, the system has equilibrium values for density, temperature, and local mean velocity over distances of a few mean free paths. Then much slower hydrodynamic processes with a different kind of dynamics govern the approach to an overall equilibrium state for the entire fluid.[3]

2.5. THE KOLMOGOROV-SINAI ENTROPY

We next turn to a brief discussion of an important quantity that characterizes both deterministic chaos of the type we have been studying, as well as Markov, stochastic processes. This quantity is called the Kolmogorov-Sinai (KS) entropy. For a deterministic system it measures the rate at which information is gained about the initial state of a system. That is, suppose that we know that the initial phase point of our system is in some small region of Γ-space of dimension ε on a side, and that we cannot resolve the

[3]We wish to insert one word of caution about this picture. It seems clear from an examination of diffusion in some non-chaotic systems, such as the famous wind-tree model, that chaoticity is sufficient, but not always necessary for understanding the approach to equilibrium of systems of many particles. However, in such non-chaotic systems there often is a non-dynamical source of randomness, as in the random locations of scatterers in the wind-tree model. This non-dynamical source of randomness is not needed to explain the approach to equilibrium in chaotic systems.

location in Γ-space to any better precision. Consider now the evolution of this small volume element in Γ-space. For a system with non-zero Lyapunov exponents, this small volume will get exponentially stretched along the expanding directions. After some time this stretching will make some sides exponentially longer than the initial value ε, typically of length $\varepsilon \exp(\lambda_i t)$, where λ_i is one of the positive Lyapunov exponents. Since we can resolve points in Γ-space to within a distance of ε in any direction, we can now determine which of many small regions, of dimension ε on a side, our system is in at time t. Then by inferring where this region came from in the initial volume, we learn more about the initial location of the phase point. Although it requires some careful analysis to prove, it is not difficult to imagine that there is some direct relation between the positive Lyapunov exponents and the KS entropy, h_{KS}. In fact for a closed, hyperbolic system, Pesin has proved [6] that the relation is as direct as our discussion above would imply, namely,

$$h_{KS} = \sum_{\lambda_i > 0} \lambda_i. \tag{13}$$

A deep and interesting fact is that at least one way to prove Pesin's result (which we will not do here) depends on the fact that hyperbolic systems can be mapped onto Markov stochastic processes. That is, for such systems, with baker and Arnold cat maps as simple examples, one can represent the dynamics to within any arbitrary degree of precision, as a Markov process. Such Markov processes have a measure of their own information entropy, a quantity which measures the degree of uncertainty in the next outcome of the stochastic process. One can show that the information entropy, suitably defined, of the stochastic representation of a hyperbolic dynamical system is equal to the KS entropy of the system. This is one of the deep results of dynamical systems theory, which provides a firm mathematical basis for the correspondence of hyperbolic dynamical systems with Markov processes. It expresses the fact that from the point of view of mathematical analysis, at least, there is no real difference between a Newtonian, hyperbolic dynamical system with a finite KS entropy, and a Markov stochastic process with equal values of the information entropy. From the point of view of physics there is a big difference, of course. The central idea of the physical approach we take is to show that Newtonian dynamical systems are sufficiently hyperbolic to behave as if they were Markov stochastic systems and in consequence all important properties of Markov systems apply to dynamical systems, too.

An important first step is to show that some useful models of physical systems have positive Lyapunov exponents and a finite, positive KS entropy per particle. In the next sections we will show how methods of statistical mechanics can be used to calculate Lyapunov exponents and KS entropies of some simple many-particle systems—gases of hard spheres in d dimensions, where $d = 2, 3$. Before doing so, however, we briefly turn our attention to an application of the ideas of this section to non-equilibrium statistical mechanics.

2.6. THE ESCAPE-RATE FORMALISM FOR TRANSPORT COEFFICIENTS

We conclude this section with a discussion of a formal relation between the transport coefficients that characterize hydrodynamic processes in fluid systems, and the chaotic properties, such as Lyapunov exponents and KS entropies that characterize the underlying dynamical behavior of the fluid. The relation of interest here is called the "escape-rate" formula for transport coefficients and is due to Gaspard, Nicolis, and co-workers [7, 8]. It applies to those fluids which can be considered to be classical, transitive, hyperbolic dynamical systems. While we do not know with certainty if any models of fluid systems satisfy this hyperbolicity requirement, we can suppose, as a working hypothesis, that generic fluid systems are well described as transitive hyperbolic systems, and then explore the consequences that result. This hypothesis has been assumed by almost all workers in this field, but its most satisfactory articulation was provided by Cohen and Gallavotti in their study of non-equilibrium fluctuations in thermostatted systems [9].[4]

To illustrate the escape-rate formula, we consider only the case of particle diffusion in an array of fixed scatterers, and refer to the literature for more general cases [8, 10]. We suppose that a collection of particles is moving in a region R in space, and that there is also a collection of fixed scatterers placed in R, as illustrated in Fig. (6). We suppose that the size of R is characterized by a length L which is much larger than the typical mean free path of the particles moving in R. For simplicity we suppose that the moving particles do not interact with each other but only with the scatterers, and that the scatterers do not "trap" the moving particles in regions of microscopic or macroscopic size. In order to provide a non-equilibrium situation which would exhibit the diffusive properties of this arrangement, we suppose that R is surrounded by absorbing boundaries such that if a particle crosses the boundary of R from the interior, it is absorbed and lost to the system.

Under most circumstances,[5] the classical, macroscopic dynamics of this

[4]It is customary to use the term *Anosov* system to describe a transitive hyperbolic system without singularities, such as the Arnold cat map. The class of Anosov systems does not include baker maps or hard-sphere systems, since discontinuities of the map or flow, present in these systems, are not allowed by the Anosov condition. For this reason we prefer to generalize the chaotic hypothesis to include transitive, hyperbolic systems.

[5]That is, when the surface area to volume of R scales to zero as $L \to \infty$.

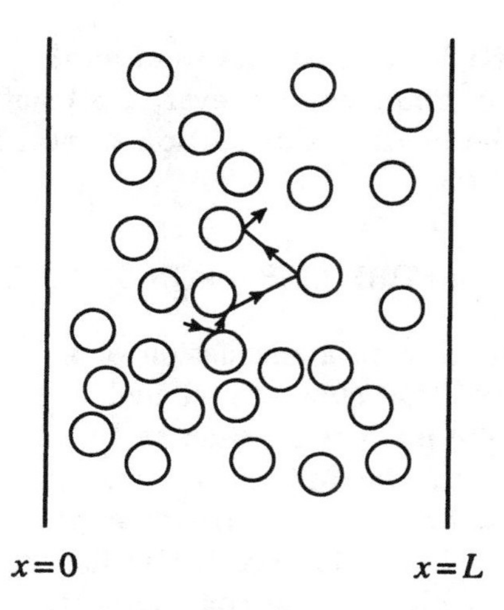

Figure 6. **A slab geometry for diffusion in a system of moving particles is an array of fixed scatterers, with absorbing boundaries.**

process is described by the diffusion equation

$$\frac{\partial n(\vec{r},t)}{\partial t} = D\nabla^2 n(\vec{r},t), \tag{14}$$

where $n(\vec{r},t)$ is the density of moving particles at time t at point $\vec{r}$, and D is the coefficient of diffusion of the moving particles for this system. The absorbing boundary conditions require that $n(\vec{r},t) = 0$ on the boundaries. While the exact solution of Eq. (14) depends on the geometry of the system, one can easily see that for long times, the total number of particles in the system decays with time as

$$N(t) = \int_R d\vec{r}\, n(\vec{r},t) \simeq N(t=0)\exp[-DA\,t/L^2], \tag{15}$$

where A is a factor of order unity that depends on the geometry of the region R, D is the diffusion coefficient, and L is the characteristic size of R. We see that this macroscopic process is characterized by an exponential escape of particles from R, with an escape-rate, $\gamma_{mac} = DA/L^2$. There is a corresponding *microscopic* description of the escape process based upon the properties of the trajectories of the particles moving in the array of scatterers. There is a set of initial points in the phase space for the moving particles which are associated with trajectories that never leave the system for either the forward or time reversed motion. This set of points is called a *repeller* and denoted by $\mathcal{R}$, and it is invariant, in the sense that any time-translation of this set of points is the set $\mathcal{R}$ itself. Further, the repeller usually forms a fractal set in phase space with zero Lebesgue measure.

Simple examples of repellers can be found in most books on chaos theory [1, 3]. It is possible to consider the dynamical properties of the trajectories on the repeller, and to define the Lyapunov exponents, $\lambda_i(\mathcal{R})$, and the KS entropy, $h_{KS}(\mathcal{R})$, of such trajectories [10, 11, 12]. For transitive hyperbolic systems, one finds that the sum of the positive Lyapunov exponents on the repeller is not equal to the KS entropy as would be the case for closed systems according to Pesin's theorem, but that the two quantities differ by an amount equal to the rate at which the other trajectories escape from the system, which we denote by γ_{mic}. That is,

$$\gamma_{mic} = \sum_{\lambda_i > 0} \lambda_i(\mathcal{R}) - h_{KS}(\mathcal{R}). \tag{16}$$

Now we make the reasonable conjecture that the microscopic and the macroscopic escape-rates are equal, which leads to an expression for the diffusion coefficient D given by

$$D = \lim_{L \to \infty} \frac{L^2}{A} \left[\sum_{\lambda_i > 0} \lambda_i(\mathcal{R}) - h_{KS}(\mathcal{R}) \right], \tag{17}$$

where we have taken the large system limit in order to remove terms of higher order in $1/L$ resulting from deviations of the actual dynamics from the diffusion law. In the event that the limit on the right hand side of Eq. (17) exists, one has an expression for a macroscopic transport coefficient in terms of microscopic dynamical quantities. The escape-rate formalism has been applied by Gaspard and Baras [13] to determine the diffusion coefficient of a particle moving in a dense array of hard-disk scatterers, where the centers of the scatterers are placed at the vertices of a triangular lattice, and by van Beijeren, Dorfman, and Latz, to determine the KS entropy on the repeller of a dilute, random Lorentz gas with hard-disk or hard-sphere scatterers [14].

3. Largest Lyapunov Exponent of a Gas of Hard Spheres at Low Density

3.1. THE HARD-SPHERE GAS

Often the calculation of chaotic characteristics of a system can only be done numerically. It would be preferable if one could at least find approximate values for such quantities using more analytical methods, and thus gain some insight into the relevant physical processes. For the Lorentz Gas at low densities, this was done by Dorfman, Van Beijeren and others [16, 14, 17, 18]. Here we present a calculation of the largest Lyapunov exponent for a system that is closer to a real gas, namely a dilute hard-sphere gas. A brief presentation of this calculation can be found in Ref. [19].

We take N hard spheres in a volume V, in d dimensions. The diameter of the hard spheres is σ, the reduced density is defined as the dimensionless number $\tilde{n} = N\sigma^d/V$. We work in the thermodynamic limit $N, V \to \infty$, keeping $\tilde{n}$ fixed (but small). In this limit we need not be concerned with boundary conditions, but one may think of periodic boundary conditions, which have been used in Molecular Dynamics simulations to which we will eventually compare our results.

The phase space of the hard-sphere gas consists of the positions $\{\vec{r}_i\}$ and velocities $\{\vec{v}_i\}$ of all N particles. To calculate the largest Lyapunov exponent, denoted here by λ_+, we need to consider two infinitesimally close trajectories in phase-space, $\Gamma = (\vec{r}_1, \vec{v}_1, \ldots, \vec{r}_N, \vec{v}_N)$ and $\Gamma + \delta\Gamma = \Gamma + (\delta\vec{r}_1, \delta\vec{v}_1, \ldots, \delta\vec{r}_N, \delta\vec{v}_N)$. The dynamics of the $\delta\vec{r}_i$ and $\delta\vec{v}_i$ is found from linearizing the dynamics of $\vec{r}_i$ and $\vec{v}_i$, which consists of a sequence of free flights and binary collisions. In free flight there are continuous changes,

$$\begin{aligned}
\dot{\vec{r}}_i &= \vec{v}_i \\
\dot{\vec{v}}_i &= 0 \\
\delta\dot{\vec{r}}_i &= \delta\vec{v}_i \\
\delta\dot{\vec{v}}_i &= 0,
\end{aligned} \tag{18}$$

and at collisions the values for the two particles i and j change discontinuously according to:

$$\begin{aligned}
\vec{r}_i' &= \vec{r}_i \\
\vec{v}_i' &= \vec{v}_i - (\vec{v}_{ij} \cdot \hat{\sigma})\hat{\sigma} \\
\delta\vec{r}_i' &= \delta\vec{r}_i - (\delta\vec{r}_{ij} \cdot \hat{\sigma})\hat{\sigma} \\
\delta\vec{v}_i' &= \delta\vec{v}_i - (\delta\vec{v}_{ij} \cdot \hat{\sigma})\hat{\sigma} - \boldsymbol{Q} \cdot (\delta\vec{r}_i - \delta\vec{r}_j).
\end{aligned} \tag{19}$$

For particle j, interchange i and j. Primes are used to denote values right after the collision. The $\vec{v}_{ij} = \vec{v}_i - \vec{v}_j$ is the relative velocity and $\hat{\sigma} = (\vec{r}_i - \vec{r}_j)/\sigma$ is the collision parameter. $\boldsymbol{Q}$ is the matrix [19]

$$\boldsymbol{Q} = \frac{[(\hat{\sigma} \cdot \vec{v}_{ij})\mathbf{1} + \hat{\sigma}\vec{v}_{ij}] \cdot [(\hat{\sigma} \cdot \vec{v}_{ij})\mathbf{1} - \vec{v}_{ij}\hat{\sigma}]}{\sigma(\hat{\sigma} \cdot \vec{v}_{ij})}. \tag{20}$$

The non-dotted products of vectors are dyadic products and $\mathbf{1}$ is the identity matrix.

Our approach will be based on kinetic theory. We are concerned with the distribution of $(\vec{r}, \vec{v}, \delta\vec{r}, \delta\vec{v})$ as a function of time. For low densities, the evolution of the distribution function f is described by a kinetic equation

[20]. This equation is based on the assumption that two colliding particles are uncorrelated, so that the probability of a collision between a particle with $(\vec{r}_1, \vec{v}_1, \delta\vec{r}_1, \delta\vec{v}_1)$ and one with $(\vec{r}_2, \vec{v}_2, \delta\vec{r}_2, \delta\vec{v}_2)$ is proportional to the product $f(\vec{r}_1, \vec{v}_1, \delta\vec{r}_1, \delta\vec{v}_1)\, f(\vec{r}_2, \vec{v}_2, \delta\vec{r}_2, \delta\vec{v}_2)$.[6]

The kinetic equation can, unlike the ordinary Boltzmann equation, be expanded in powers of $1/|\ln \tilde{n}|$ to get the low density behavior of f, and thus of λ_+. We will take a different but roughly equivalent approach, we will derive the effective dynamics of the $\delta\vec{r}_i$ and $\delta\vec{v}_i$ for low densities, and use that to write down a kinetic equation.

3.2. LOW DENSITY DYNAMICS – CLOCK MODEL

The main characteristics of the low density region is the typically long free flight time τ of an individual particle between collisions, compared to the time it would take two transparent hard spheres to cross each other. If v_0 is the typical thermal velocity, the latter is σ/v_0, while $\tau \approx 1/(v_0\sigma^{d-1}N/V) = \sigma/(v_0\tilde{n})$. Thus, $\tilde{n}$ is the small parameter.

Just before a collision at time t the $\delta\vec{r}_i(t)$ will be

$$\delta\vec{r}_i(t) = \delta\vec{r}_i(t_0) + \delta\vec{v}_i(t_0)\tau_i\,, \tag{21}$$

if t_0 is the time of the previous collision and τ_i is the (large) free flight time $t - t_0$. Suppose that initially $\delta\vec{r}_i(t_0)/\sigma$ and $\delta\vec{v}_i(t_0)/v_0$ are of the same order, then just before collision,

$$\delta\vec{r}_i = \tau_i\left[\delta\vec{v}_i + \mathcal{O}(\tilde{n})\right].$$

We insert this into the collision rules and neglect the terms of relative order $\mathcal{O}(\tilde{n})$ to obtain

$$\begin{aligned} \delta\vec{r}_j^{\,\prime} &\approx \tau_j\delta\vec{v}_j + \{(\tau_i\delta\vec{v}_i - \tau_j\delta\vec{v}_j)\cdot\hat{\sigma}\}\hat{\sigma} \\ \delta\vec{v}_j^{\,\prime} &\approx \boldsymbol{Q}\cdot(\tau_i\delta\vec{v}_i - \tau_j\delta\vec{v}_j). \end{aligned} \tag{22}$$

Using Eq. (20), we see that $\delta\vec{r}_i^{\,\prime}/\sigma$ and $\delta\vec{v}_i^{\,\prime}/v_0$ are both of order $(\delta\vec{v}_i - \delta\vec{v}_j)/v_0\tilde{n}$, and, relative to these quantities before collision, are of the order of the ratio of the mean free time to the time it takes to traverse a particle diameter. For a dilute gas, this ratio is large and terms of relative order $\tilde{n}$ can be neglected. In this way we have eliminated the $\delta\vec{r}_i$ from the $\delta\vec{v}_i$ dynamics.

The neglected terms were of relative order $\tilde{n}$, relative to either $\tau_i\delta\vec{v}_i$ or $\tau_j\delta\vec{v}_j$. These two are not necessarily of the same order. Now, if one of them is one or more orders of $\tilde{n}$ higher than the other, we should also neglect it.

[6] Of course one also has to demand that the particles should be a distance σ apart.

If they are both of the same order, we should keep both. To know which terms to neglect, we have to keep track of the orders of $\tilde{n}$ in $\delta\vec{v}_i$. For that purpose, we define

$$\delta\vec{v}_i = v_0(\tilde{n})^{-k_i}\hat{e}_i, \tag{23}$$

where $\|\hat{e}_i\| = 1$. The number k_i counts the number of orders of $\tilde{n}$ and we will call this the *clock value* of particle i. The clock values are real numbers at this point, but later will be approximated by integers. Inserting Eq. (23) into Eq. (22), we can get the clock values k_i' and k_j' after collision. Since, to leading order in density, $\delta\vec{v}_i'$ and $\delta\vec{v}_j'$ differ only in sign, $k_i' = k_j' = k'$, with

$$k' = \frac{1}{-\ln\tilde{n}} \ln\|\boldsymbol{Q}\cdot(\tau_i\delta\vec{v}_i - \tau_j\delta\vec{v}_j)\|.$$

Both τ_i and τ_j are typically of order $\sigma/(v_0\tilde{n})$. This means that if $k_i > k_j$, we should neglect the term with $\delta\vec{v}_j$, and if $k_i < k_j$, we should neglect the other term. This yields

$$k' = k_D + \frac{1}{-\ln\tilde{n}} \ln\|\boldsymbol{Q}\cdot\tau_D\hat{e}_D\|,$$

where $D = i$ if $k_i > k_j$, and $D = j$ otherwise. Particle D is called the dominant particle. Using the property $\tau_D = \mathcal{O}(\sigma/(v_0\tilde{n}))$ and the explicit form of $\boldsymbol{Q}$ from Eq. (20), one gets

$$k' = k_D + 1 + \mathcal{O}(\frac{1}{\ln\tilde{n}}).$$

So far we have ignored the possibility that $k_i = k_j$. But the resulting correction in fact only contributes to the $\mathcal{O}(1/\ln\tilde{n})$ part.

Differences in the number of collisions suffered by different particles cause the clock values to not be all the same, even if they were so initially. They are indispensible for determining the magnitudes of postcollisional velocity deviations. But in fact no more is needed! For if we know how fast the clock values grow, we know the linear growth of $\ln\|\delta\vec{v}\|$, which is precisely the largest Lyapunov exponent. So we define the clock speed as

$$w = \lim_{t\to\infty} \frac{< k(t) >}{\bar{\nu}t},$$

in which $< k >$ is the average clock value and $\bar{\nu}$ is the average collision frequency.[7] Because we extracted $\bar{\nu}$, which is $\mathcal{O}(\tilde{n})$, this clock speed is of order 1.

[7]In a hard-sphere gas in d dimensions, $\bar{\nu} = \frac{2\pi^{\frac{d-1}{2}}}{\Gamma(\frac{d}{2})}\sqrt{\frac{k_BT}{m\sigma^2}}\,\tilde{n}$.

The Lyapunov exponent is related to w via

$$\lambda_+ = -w\bar{\nu}\ln\tilde{n}.$$

The clock speed w will be calculated in an expansion in $1/|\ln\tilde{n}|$. The leading order for low density, calculated in the next section, is a non-trivial constant. This behavior of λ_+ as a function of density had been conjectured already by Krylov [21], however without a nontrivial prefactor w. A first estimate of w was made by Stoddard and Ford [22], who got $w = \ln N$. This would mean that there is no thermodynamic limit for the Lyapunov exponent. Some numerical simulations [23] have been interpreted as supporting the logarithmic divergence with N, but with a prefactor much smaller than 1. However, we will find a finite thermodynamic limit for w and show that, even in a mean-field approach that fully ignores local density fluctuations, it approaches this thermodynamic value so slowly that in the range of particle numbers accessible to simulations one could not distinguish between saturation or slow but steady increase.

3.3. KINETIC APPROACH

For low densities, we may describe the effective dynamics of the clock values by

$$k'_i = k'_j = \max(k_i, k_j) + 1, \tag{24}$$

where the collision pairs (i, j) are chosen completely randomly with Poisson distributed collision times. The model with this dynamics we will call the clock model. For simplicity we will consider integer clock values only, though this restriction is by no means necessary or important. The clock speed w found in this model gives the leading term in the density expansion of the Lyapunov exponent:

$$\lambda_+ = w\bar{\nu}\left[-\ln\tilde{n} + \mathcal{O}(1)\right]. \tag{25}$$

A distribution function $f_k(t)$ will denote the fraction of particles having clock value k at time t. From the dynamics specified above we can derive an equation for the distribution function $f_k(t)$ of clock values. We expect the clock values to grow linearly with time. If they all grow at the same rate, we have

$$f_k(t) = g(k - w\bar{\nu}t),$$

with w as defined before, because

$$\lim_{t\to\infty}\frac{1}{\bar{\nu}t}\sum_{k=-\infty}^{\infty} g(k - w\bar{\nu}t)k = \lim_{t\to\infty}\frac{1}{\bar{\nu}t}\sum_{x=-\infty}^{\infty} g(x)(x + w\bar{\nu}t) = w.$$

So once we have a kinetic equation for $f_k(t)$, we will look for these *propagating solutions.*

Consider the contributions to $\frac{\partial f_k}{\partial t}$ from collisions in which a clock value k is lost. In each collision where a particle with k enters, the k gets lost, so the fraction of particles with k decreases at a rate $\bar{\nu} f_k(t)$ due to these processes. There are also processes which increase the fraction of particles having k. For these the larger incoming clock value should be $k-1$. We have to distinguish collisions with equal incoming clock values, both $k-1$, from ones with different incoming clock values $k-1$ and l. The rate for the latter type of collisions is $\bar{\nu} f_{k-1} \sum_{l=-\infty}^{k-2} f_l$. For the former ones the rate is only $1/2\ \bar{\nu} f_{k-1}^2$, because the two particles are drawn from the same fraction f_{k-1}, and it doesn't matter in which order they are picked. In either case the number of particles with clock value k increases by two, so we get the kinetic equation

$$\frac{\partial f_k(t)}{\partial \bar{\nu} t} = -f_k(t) + f_{k-1}^2 + 2 f_{k-1} \sum_{l=-\infty}^{k-2} f_l.$$

We scale out $\bar{\nu}$ by defining a new time variable $\tau = \bar{\nu} t$. The equation is simplified further by replacing the fraction f_k of particles having clock value k with the fraction C_k of particles that have clock values k *or less*:

$$C_k(\tau) = \sum_{l=-\infty}^{k} f_l(\tau).$$

The kinetic equation then takes the short form

$$\frac{\partial C_k}{\partial \tau} = -C_k + C_{k-1}^2. \tag{26}$$

3.4. FRONT PROPAGATION

Let us investigate the solutions of Eq. (26). We see that, given C_{k-1} and the initial value of C_k, we can solve for C_k,

$$C_k(\tau) = e^{-\tau} \left[C_k(0) + \int_0^{\tau} e^{\tau'} C_{k-1}^2(\tau') d\tau' \right].$$

If $C_k = 0$ initially, it will remain zero, because $0 \le C_{k-1} \le C_k$. We take initial conditions such that $C_{k<1}(0) = 0$ and $C_{k>0}(0) = 1$, i.e. all particles have $k = 1$. The solutions C_k are all polynomials in $\exp(-\tau)$, but with increasing k the order grows exponentially. Nonetheless, we calculated the C_k up to

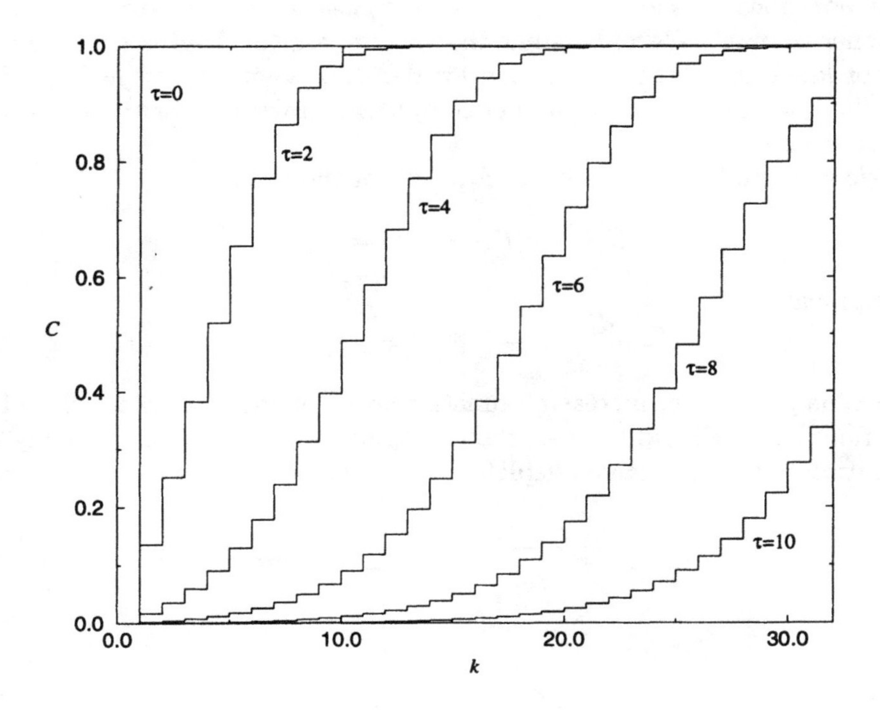

Figure 7. Time evolution of the clock value distribution. All initial clock values are 0.

$k = 32$ using a computer program to handle the analytic manipulations. The C_k's are plotted in Fig. 7 for $\tau = 0, 2, 4, 6$ and 10.

We see in Fig. 7 that the initial distribution changes to some smoother shape, and moves to the right. After a while the shape seems to stay constant. We can now view Eq. (26) as describing a propagating front: $C \equiv 0$ is a stable phase and $C \equiv 1$ is an unstable phase. On the left we have the stable phase, on the right, the unstable phase and in between is the intermediate region called *the front* , that propagates to the right, into the unstable phase. The velocity at which it moves to the right is w. From Fig. 7 we see that for $\tau = 10$, the speed is still increasing, and is about 3.8.

Front propagation into an unstable phase comes in two flavors [24]. In general, the instability of the unstable phase sets a velocity w^* by which small perturbations propagate to the right. This velocity w^* is determined from the linearized equation describing the front propagation around the unstable phase. For *pulled* fronts w^* is the asymptotic velocity of any solution with an initial shape that is sufficiently steep. If, however, no solution of the full non-linear front equation with this velocity exists, or if it is unstable with respect to some nonlinear perturbation, the velocity is set by these non-linearities and the front is called *pushed.* In that case the velocity is higher than w^*. We will assume that in our case the front is *pulled.* As we do not know of a general criterion for deciding whether a front is pushed or pulled, we will use the results of computer simulations for the validation of our assumption.

We want to find a solution to Eq. (26) of the form

$$C_k(\tau) = F(k - w\tau) = F(x),$$

which means

$$-w\frac{dF(x)}{dx} = -F(x) + F^2(x-1). \tag{27}$$

Now C is a positive, increasing function of k, bounded between 0 and 1. The function F should have all these properties too. So we are looking for a solution of a form such as depicted in Fig. 8.

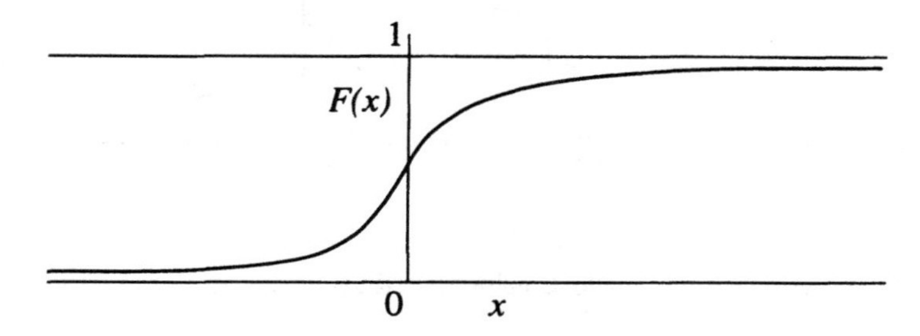

Figure 8. Clock value distribution as a propagating front

For a pulled front, we have to investigate the Eq. (27) linearized around the unstable phase. Writing $F = 1 - \Delta$, we get

$$-w\frac{d\Delta(x)}{dx} = -\Delta(x) + 2\Delta(x-1) + \mathcal{O}(\Delta^2). \tag{28}$$

Given w, the asymptotic solution of Eq. (28) is given by a sum of exponentials [25]

$$\Delta(x) = \sum_i A_i e^{-\gamma_i x}. \tag{29}$$

The possible values of γ_i are found by inserting $\exp(-\gamma x)$ into the linearized equation (in case of degeneracies, A_i should be replaced by a polynomial in x). This gives w as a function of γ,

$$w(\gamma) = (2e^\gamma - 1)/\gamma. \tag{30}$$

The γ_i may be complex, and, for given w, there are infinitely many of them. However, we know that Δ should be monotonic, so the most slowly decaying term in the sum should have a real positive γ.

From the plot of Eq. (30) in Fig. 9, we see that there are three cases: If w is larger than some critical w^*, there are two γ's, which are real and positive. If $w = w^*$ these two become degenerate and for $w < w^*$ they become complex. But, for the slowest term to have complex γ was not allowed by the condition of monotonicity of F: asymptotically the function would oscillate. So we conclude that, for the the Ansatz of a propagating front to work, the velocity should be at least w^*, the minimal w from Eq. (30) for positive real γ. This value can be expressed in terms of Lambert's W function,[8]

$$w^* = \frac{-1}{W(\frac{-1}{2e})} \approx 4.31107\ldots,$$

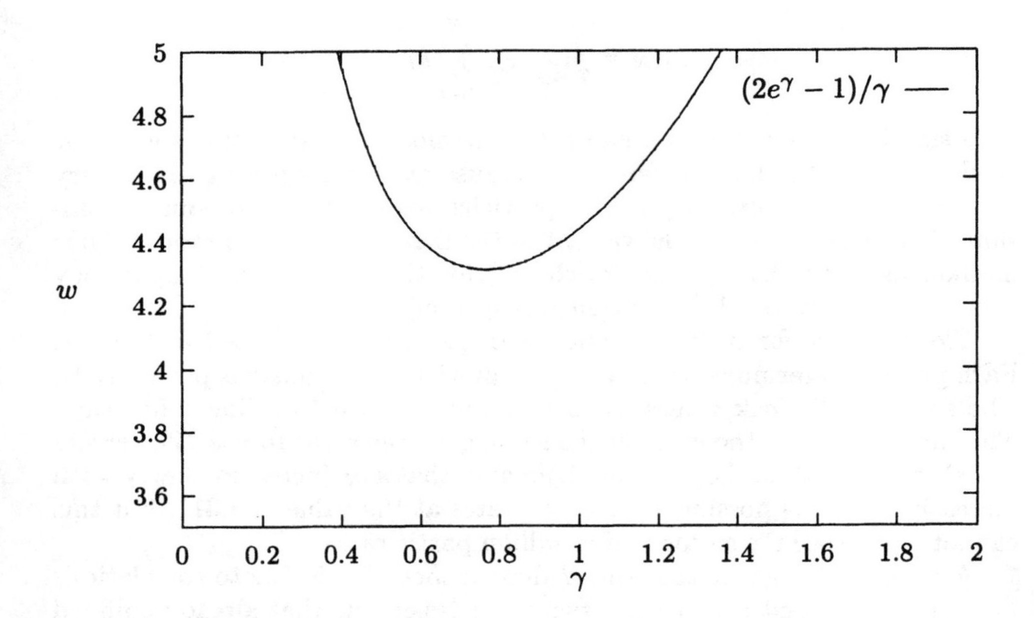

Figure 9. Velocity versus asymptotic decay rate γ.

and it is the velocity set by the instability that we mentioned before. So it *is* the velocity w we were after. Its value is compatible with the estimate 3.8 from Fig. 7 (a lower value than 3.8 would not), which supports the assumption of a pulled front. This concludes the calculation of the leading order in Eq. (25) of the largest Lyapunov exponent in the infinite system

[8]Defined as $W(x)\exp[W(x)] = x$, where the branch analytic in 0 is meant.

limit within the framework of our clock model, but we still have to consider the effects of a finite number of particles.

3.5. LARGE FINITE N EFFECTS

The kinetic equation works well in the thermodynamic limit, but to compare our results with simulations we need to correct for the effects of the finiteness of the number of particles. The clock model is very suitable for simple simulations. We take a set of N integer numbers $\{k_i\}$. In each time step two of them are picked at random, and the collision rule in Eq. (24) is applied to them. We do this T times. The average number of collisions per particle is then $2T/N$, because in each of the T collisions two particles are involved. An estimate for the clock speed is the average clock value $\sum_i k_i/N$ divided by the average number of collisions per particle. For large T this approaches the clock speed w_N, so

$$w_N = \lim_{T\to\infty} \frac{1}{2T} \sum_{i=1}^{N} k_i.$$

This simulation is not even a bad chacterization of what happens with the clock values in the hard-sphere gas, because at low densities there is very little correlation between pairs of particles involved in subsequent collisions. It is very similar to the variant of the Direct Simulation Monte Carlo method used by Dellago and Posch [26] for the calculation of Lyapunov exponents in a "spatially homogeneous system".

Clock speeds for different numbers of particles are plotted in Fig. 11. Each point is determined from a single run with 2000 collisions per particle. The sums of all clock values at 20 times were fitted to a linear function. The slope gave w_N, the error in the slope gave the error in w_N. The errors are always less than 0.5 percent. One sees that w_N increases slowly with increasing N. It is possible that it saturates at the value of 4.311, but this cannot yet be seen even for half a million particles.

A natural thought is that the N dependence of w is due to correlations between subsequent collisions: if two particles collide that already collided just before, or that had their clock values reset shortly before by particles that were roughly synchronized already, the gain in clock value of the "slower" particle will be less than average. Hence the average clock speed is reduced. However, as we will see, the reduction in clock speed can be explained entirely on the basis of the linearized equation plus some simple bounds on its region of validity and this explanation does not seem to require any effects of correlated collisions.

In the head of the distribution, the finiteness of the number of particles becomes important when $f_k = \mathcal{O}(1/N)$. For this reason, Brunet and Derrida

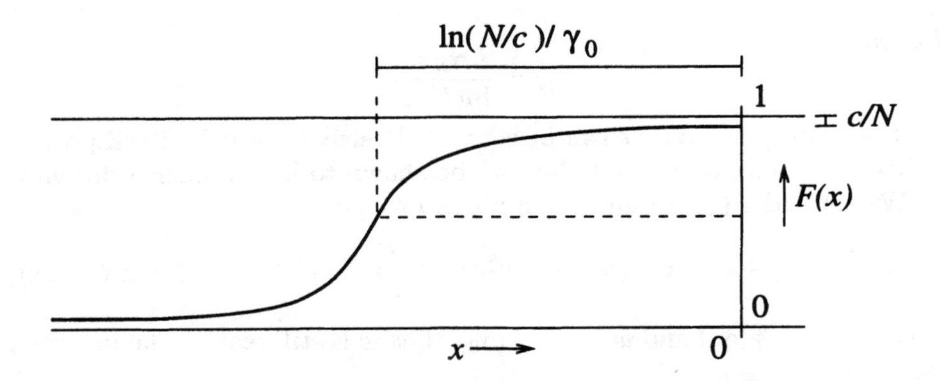

Figure 10. **The cutoff at the head of the distribution and the linear regime.**

[27] treat the finite N effects by the introduction of a cutoff $\varepsilon = 1/N$ in the equation, i.e. they modify the equation as soon as $f_k < \varepsilon$. They distinguish three regions: one where the non-linear behavior is dominant, one where the linear equation holds, and one where the cutoff is effective. By glueing the solutions in these regions together, one obtains a new ε-dependent front velocity.

The introduction of a cutoff in something that is supposed to be a distribution function, which is an average over realizations, seems hard to justify. A more satisfactory way to obtain it, is to shift all clockvalues in each realization such that the particle with the largest clock value has $x = 0$, and then average the distribution function (this was also suggested by Kessler *et. al.* [28]). Then the cutoff occurs naturally.

For $x < 0$ there is a region where we can use the linearized equation. We consider the leading term in Eq. (29):

$$\Delta \sim \frac{c}{N} e^{-\gamma_0 x}.$$

The prefactor c/N is obtained from the fact that $\Delta = \mathcal{O}(1/N)$ at $x = 0$ (c is order 1). The linear regime ends when Δ becomes of order 1, so for $x = -\ln(\frac{N}{c})/\gamma_0$. This is illustrated in Fig. 10.

In contrast with the infinite system, now we only have to demand monotonicity and positivity from the solution of the linearized equation in this interval of width $\ln(\frac{N}{c})/\gamma_0$. This means that a small imaginary value is allowed for the γ in Eq. (29) with the smallest real part. We denote $\gamma = \gamma_R + i\gamma_I$. The consequential oscillations,

$$\Delta = a_0 \, e^{-\gamma_R x} \cos(\gamma_I x + \phi),$$

should not cause sign changes in the function or its derivative. The derivative can be sign definite if at most half a wavelength fits in the interval, i.e.

if at most

$$\gamma_I = \frac{\gamma_R \pi}{\ln(N/c)}.$$

For the leading behavior c can be set to 1. Positivity of Δ by itself poses an additional bound on γ_I, but this can be shown to be a higher order effect.

We expand $w(\gamma)$ around its minimum at γ_0:

$$w(\gamma) = w(\gamma_0 + \delta\gamma) \approx w(\gamma_0) + \frac{1}{2}w''\delta\gamma^2 + \frac{1}{6}w'''\delta\gamma^3 + \text{h.o.(higher orders)},$$

where $w'' = \frac{d^2w}{d\gamma^2}(\gamma_0)$ and $w''' = \frac{d^3w}{d\gamma^3}(\gamma_0)$. The w is still real, so the imaginary part gives

$$0 = \delta\gamma_I \left\{ w''\delta\gamma_R + \frac{1}{6}w'''[3\delta\gamma_R^2 - \delta\gamma_I^2] + \text{h.o.} \right\},$$

where $\delta\gamma = \delta\gamma_R + i\delta\gamma_I$. To first order this says that $w''\delta\gamma_R = \frac{1}{6}w'''\delta\gamma_I^2$: the shift in the real part of γ is higher order compared to the imaginary part. The new velocity is written as $w = w_0 + \delta w$, and δw is obtained from the real part of equation Eq. (30) expanded to first order:

$$\begin{aligned} \delta w &= \frac{1}{2}w''\left\{\delta\gamma_R^2 - \delta\gamma_I^2\right\} + \frac{1}{6}w'''\left\{\delta\gamma_R^3 - \delta\gamma_R\delta\gamma_I^2\right\} + \text{h.o.} \\ &= -\frac{1}{2}w''\delta\gamma_I^2 + \text{h.o.} = -\frac{w''\pi^2\gamma_0^2}{2\ln^2(N/c)} + \text{h.o.}, \end{aligned} \tag{31}$$

which coincides with Brunet and Derrida's result (when c is set to 1). However, they needed to consider how the linear region connects to the others, while we only need that Δ is of order 1 at the border with the non-linear region. From Eq. (30), one can show that $w''\gamma_0^2 = (w_0 - 1)$, so we find the result

$$w_N = w_0 - \frac{(w_0 - 1)\pi^2}{2\ln^2(N/c)}. \tag{32}$$

In Fig. 11 the simulation results are plotted together with a fit. We calculate $\delta w^{-1/2} = 1/\sqrt{w_0 - w_N}$, with w_N taken from the simulations, which should be a linear function of $\ln N$ for large N,

$$\delta w^{-1/2} \overset{N\to\infty}{\longrightarrow} a + b\ln N. \tag{33}$$

Indeed this behavior is seen in the inset of Fig. 11. According to Eq. (32), $a = -\ln c$ and $b = \pi^{-1}\sqrt{2/(w_0 - 1)} = 0.247\ldots$. The fit to the data[9] shown in the inset yields $b = 0.23 \pm 0.01$, consistent with the theoretical prediction[10], and $a = -0.07 \pm 0.07$. The value of c corresponding to this a is $c \approx 1.07$, so it is of order 1 as expected.

[9]As the prediction is for large N, only points for $N > 256$ are used in the fit.

[10]By allowing changes in a and b simultaneously, one finds an appreciably larger range of acceptable values for b than the error in b indicates.

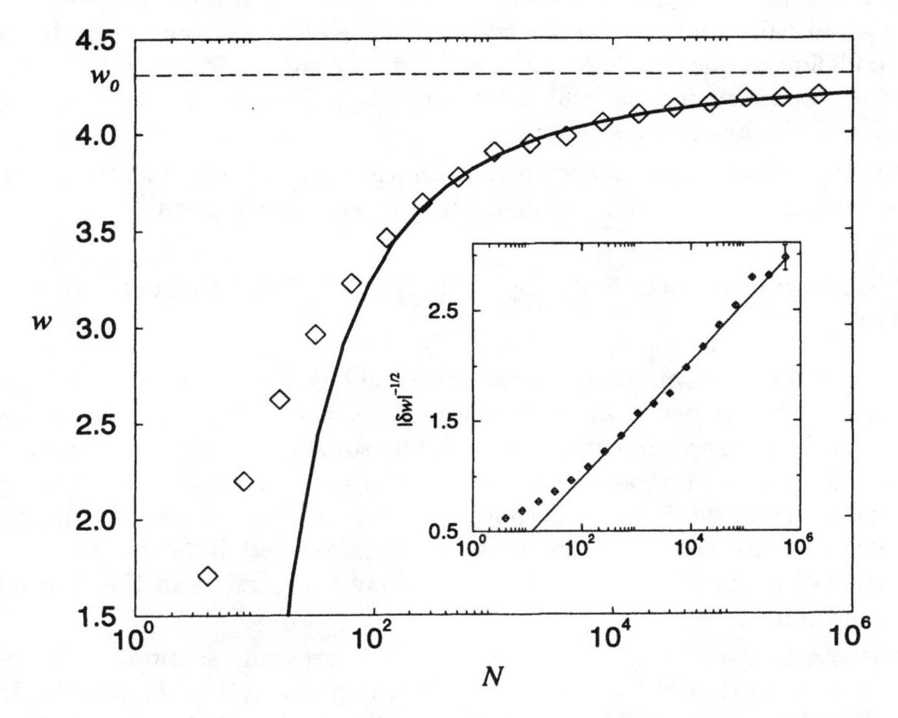

Figure 11. Effect of finite N on the clock speed. Diamonds are the simulations; the dashed line is the asymptotic value 4.311.... The inset shows the linear behavior of $1/\sqrt{w_0 - w_N}$ (plotted with error bars) as a function of $\ln N$. The solid line in both is the fit to the form in Eq. (33).

The values from our simulations have been compared in Ref. [19] to molecular dynamics simulations of hard spheres, from which w is found from a fit of λ_+ to the form Eq. (25). The values agreed very well with the simulations.

3.6. FURTHER REFINEMENTS

Comparing the results from high precision simulations of the clock model on the one hand and of hard-disk systems with exactly the same number of particles and equal collision frequency on the other hand, one finds that the dimensionless clock speed w in the latter is significantly higher than in the former; for a system of 10,000 particles the clock model gives a w of 4.05 ± 0.01, whereas the corresponding hard-disk value is 4.47 ± 0.02. The cause of this difference is the velocity dependence of the collision frequency, which is an increasing function of speed. As a result particles in the head of the distribution tend to have a higher speed than average (a higher collision frequency enhances the clock value), resulting in a clock speed that is systematically higher than it would be in the case of a velocity independent collision frequency. The clock model does have a velocity independent collision frequency, indeed, which explains the difference in w between this model and the hard-disk system.

An explicit calculation accounting for these effects and producing improved estimates for w in hard-disk systems will appear soon [29].

4. Kolmogorov-Sinai Entropy of a Gas of Hard Spheres at Low Density

In the previous section we reviewed the calculation by Van Zon and Van Beijeren of the largest Lyapunov exponent for a gas of hard disks at low densities. Here we will outline a related, but somewhat simpler calculation of the KS entropy of a gas of hard disks or of hard spheres at low density. This calculation, while not rigorous in a mathematical sense, is a strong indication of the chaotic behavior of such gases, and indicates that the chaoticity of a dilute hard-disk or hard-sphere gas persists in the thermodynamic limit.

The starting point is the same as in the previous section. It is important to note that the dynamics of a hard-sphere system is exactly described as free motion of the particles, punctuated by instantaneous, binary collisions between some pair of particles. To describe this motion and the quantities we need for the KS-entropy, we consider the dynamical behavior of the positions and coordinates of all the particles in the gas, $(\vec{r}_1, \vec{v}_1, \vec{r}_2, \vec{v}_2, \ldots, \vec{r}_N, \vec{v}_N)$, as well as a set of deviation vectors which describe the motion of a pencil of nearby trajectories in phase space, $(\delta\vec{r}_1, \delta\vec{v}_1, \ldots, \delta\vec{r}_N, \delta\vec{v}_N)$. The equations of motion for these quantities between the binary collisions are given by Eq. (18) and the changes of these quantities at a collision between particles i and j are given in Eqs. (19,20).

To proceed with the calculation of the KS-entropy, we will suppose that the Pesin formula holds, i.e., that the KS-entropy is the sum of the positive Lyapunov exponents. This sum can be obtained by considering the growth in time of the volume of the dN-dimensional projection of an arbitrary, infinitesimal volume in the full $2dN$-dimensional phase space. This observation requires some explanation, as follows. The typical rate of separation of two arbitrary, but infinitesimally close, points in phase-space, for a hyperbolic system, will be exponential and the rate will be given by the largest Lyapunov exponent. If we consider a typical two-dimensional, infinitesimal area in phase-space, then this area will grow exponentially with a rate determined by the sum of the two largest Lyapunov exponents. In other words, the exponential growth of a typical infinitesimal n-dimensional subvolume in phase space is determined by the sum of the n largest Lyapunov exponents. Further, for a Hamiltonian system, the Lyapunov exponents come in plus-minus conjugate pairs, so that the sum of a conjugate pair of exponents is always zero. Consequently, the growth of an infinitesimal dN-dimensional subvolume is determined by the sum of the dN non-negative Lyapunov exponents, and the volume of an infinitesimal $2dN$-dimensional volume remains constant, in accord with Liouville's theorem.

For the typical dN dimensional subvolume, we consider a volume formed by the projection of $2dN$ infinitesimal displacement vectors on velocity space, $(\delta\vec{v}_1, \delta\vec{v}_2, \ldots, \delta\vec{v}_N)$. Given initial values for each of these vectors, as well as for all of the positions, $\vec{r}_i$, velocities, $\vec{v}_i$, and position deviation vectors, $\delta\vec{r}_i$, we can follow their evolution in time and, in principle at least, determine the time dependence of an infinitesimal volume element in velocity space, which we denote as $\delta\mathcal{V}(t)$. Then

$$\begin{aligned} h_{KS} = \sum_{\lambda_i \geq 0} \lambda_i &= \lim_{t\to\infty} \frac{1}{t} \ln \frac{\delta\mathcal{V}(t)}{\delta\mathcal{V}(0)} \\ &= \lim_{t\to\infty} \frac{1}{t} \int_0^t d\tau \frac{d \ln \delta\mathcal{V}(\tau)}{d\tau} \\ &= \left\langle \frac{d \ln \delta\mathcal{V}(t)}{dt} \right\rangle. \end{aligned} \tag{34}$$

The last line of Eq. (34) is based on the assumption (still unproven) that a gas of hard spheres is ergodic, so that time averages can be replaced by equilibrium averages taken with respect to a microcanonical ensemble. Here this average is denoted by angular brackets. Since we are considering a volume element in velocity space, we can use the fact that the velocity displacement vectors do not change during the free flight motion of the particles between the collisions, but do change at a collision. Under these circumstances, one can use elementary kinetic theory considerations to show that the final term on the right hand side of Eq. (34) is

$$\left\langle \frac{d \ln \delta\mathcal{V}(t)}{dt} \right\rangle = \left\langle \sum_{i<j} \mathcal{T}_{i,j} \ln \delta\mathcal{V} \right\rangle = \frac{N(N-1)}{2} \left\langle \mathcal{T}_{12} \ln \delta\mathcal{V} \right\rangle. \tag{35}$$

Here $\mathcal{T}_{12}$ is a binary collision operator, discussed in some detail in Refs. [30, 31], given by

$$\mathcal{T}_{12} = \sigma^{d-1} \int_{\vec{v}_{12}\cdot\hat{\sigma}<0} d\hat{\sigma} |\vec{v}_{12} \cdot \hat{\sigma}| \delta(\vec{r}_{12} - \sigma\hat{\sigma}) [\mathcal{P}_{\hat{\sigma}}(1,2) - 1]. \tag{36}$$

In Eq. (36), d is the number of spatial dimensions of the system, σ again is the diameter of the spheres, $\vec{r}_{12} = \vec{r}_1 - \vec{r}_2; \vec{v}_{12} = \vec{v}_1 - \vec{v}_2$, and the operator $\mathcal{P}_{\hat{\sigma}}(1,2)$ is a substitution operator which replaces the precollision values, $\vec{r}_1, \vec{v}_1, \vec{r}_2, \vec{v}_2, \delta\vec{r}_1, \delta\vec{v}_1, \delta\vec{r}_2, \delta\vec{v}_2$, by their post collision values, denoted with primes, given by Eqs. (19, 20). The unit vector $\hat{\sigma}$ is an impact parameter, running in the direction of the line connecting the centers at collision and is integrated over a hemisphere corresponding to all allowed directions.

At this point it is useful to express the precollision quantities $\delta\vec{r}_i$ as

$$\delta\vec{r}_i = \delta\vec{r}_i(0) + \tau_i \delta\vec{v}_i, \tag{37}$$

where $\delta\vec{r}_i(0)$ is the position displacement of particle i just after its previous collision, and τ_i is the time between the previous collision of particle i with some other particle, and the next collision involving particle i. To further simplify the expression for the KS-entropy, we now neglect, as in section 3, the initial displacement vectors, $\delta\vec{r}_i(0)$, when we calculate the change of the infinitesimal volume in velocity space at the $(1,2)$ collision in Eq. (35). This turns out to be a serious approximation. It leads to the correct value for the leading density term in h_{KS}, at low density, but the first order correction to this term is obtained incorrectly in this approximation. This can be repaired, but at the cost of a much longer and intricate calculation which we will present elsewhere.

If we insert $\delta\vec{r}_1 = \tau_1\delta\vec{v}_1$ and $\delta\vec{r}_2 = \tau_2\delta\vec{v}_2$ into the expression, Eq. (19), for the post collision velocity deviations for particles 1 and 2, we find that

$$[\mathcal{P}_{\hat{\sigma}}(1,2) - 1] \ln \delta\mathcal{V} = \ln \frac{\delta\mathcal{V}'}{\delta\mathcal{V}} = \ln |\det \mathbf{M}_{12}|, \tag{38}$$

where

$$\mathbf{M}_{12} = \mathbf{1} - 2\hat{\sigma}\hat{\sigma} - \frac{2T_{12}}{\sigma}\left[(\vec{v}_{12}\cdot\hat{\sigma})\mathbf{1} - \vec{v}_{12}\hat{\sigma} + \hat{\sigma}\vec{v}_{12} - \frac{(\vec{v}_{12})^2}{(\vec{v}_{12}\cdot\hat{\sigma})}\hat{\sigma}\hat{\sigma}\right]. \tag{39}$$

In Eq. (39), $T_{12} = (\tau_1 + \tau_2)/2$ and $\mathbf{1}$ is the unit matrix. The determinants are easily evaluated. For $d = 2$, one finds

$$|\det \mathbf{M}_{12}| = 1 + \frac{2T_{12}|\vec{v}_{12}|}{\sigma\cos\phi}, \tag{40}$$

where ϕ is the angle of incidence in the $1,2$ collision and ranges over the values $-\pi/2 \leq \phi \leq \pi/2$. A similar calculation for $d = 3$ shows that

$$|\det \mathbf{M}_{12}| = 1 + \frac{2T_{12}|\vec{v}_{12}|}{\sigma\cos\phi}(\cos^2\phi + 1) + \left(\frac{2T_{12}|\vec{v}_{12}|}{\sigma}\right)^2. \tag{41}$$

To obtain the leading term in the density, for low density gases, we keep the highest power of the time in each of the expressions for the determinant. Further, at low densities we can compute the ensemble averages appearing in Eq. (34) by ignoring possible pre-collision correlations between particles 1 and 2, and using equilibrium values for the single-particle distribution functions appearing in the ensemble averages. In this way we find for $d = 2$

$$\begin{aligned} h_{KS}/N &= \frac{a}{2n}\int d\vec{v}_1 \int d\vec{v}_2 \int d\tau_1 \int d\tau_2 \int_{\vec{v}_{12}\cdot\hat{\sigma}<0} d\hat{\sigma}|\vec{v}_{12}\cdot\hat{\sigma}| \times \\ &\times F_1(\vec{v}_1,\tau_1)F_1(\vec{v}_2,\tau_2)\ln T_{12} + \cdots, \end{aligned} \tag{42}$$

where the normalized equilibrium single-particle distribution functions $F_1(\vec{v}_i, \tau_i)$ are given, in d dimensions, by

$$F_1(\vec{v}_i,\tau_i) = n(\frac{\beta m}{2\pi})^{d/2}\nu(\vec{v}_i)e^{-\beta m\vec{v}_i^2/2}e^{-\nu(\vec{v}_i)\tau_i}. \tag{43}$$

Here n is the number density of the gas, $\beta = (k_B T)^{-1}$, where T is the gas temperature, and k_B is Boltzmann's constant, $\nu(\vec{v}_i)$ is the equilibrium collision frequency for a particle with velocity $\vec{v}_i$. For two dimensions, the evaluation of the integrals [32] leads directly to

$$h_{KS}/N = \frac{\nu}{2}[-\ln(n\sigma^2) + \cdots], \tag{44}$$

where $\nu = [(2\pi^{1/2}n\sigma)/(\beta m)^{1/2}]$ is the average collision frequency at equilibrium for a two-dimensional gas of hard disks. The terms left out are of higher order in the density.

For three-dimensional gases, a parallel calculation [32] leads to

$$h_{KS}/N = \nu[-\ln(\pi n\sigma^3) + \cdots], \tag{45}$$

where for a gas of hard spheres ($d = 3$) the average collision frequency $\nu = [(4\pi^{1/2}n\sigma^2)/(\beta m)^{1/2}]$. These results are in excellent agreement with the numerical simulations of Dellago and Posch [26]. The higher order terms take, as mentioned earlier, considerably more work, and are discussed elsewhere [33].

5. Conclusions and Outlook

In the previous sections we have reviewed some of the ideas that motivate the interest in the chaotic foundations of non-equilibrium processes in fluids. We have provided an elementary discussion of transitive, hyperbolic dynamical systems such as the baker and the Arnold cat maps to illustrate some of the central notions and dynamical quantities. We then turned to the applications of kinetic theory to compute the largest Lyapunov exponent and the KS entropy for a dilute gas of hard disks or hard spheres. The explicit results obtained by these methods are in good agreement with the results of computer simulations, and, apart from the corrections due to the velocity dependence of the collision frequency referred to at the end of section 3, represent the present state of the art in the analytical calculation of chaotic quantities for dilute gases with short range, repulsive forces. There are, however, still many open problems which need solving. Here we mention a few of them:

1. We have a theory for the leading density behavior of the largest Lyapunov exponent for a dilute gas of disks or spheres. We also have some understanding of the number dependence of this quantity when we are not quite in the thermodynamic limit. However we do not know much about the higher density corrections to this exponent, nor do we know anything about the rest of the Lyapunov spectrum, other than the KS entropy per particle (in the thermodynamic limit). The determination of the complete spectrum would be quite an accomplishment.
2. Recent results of Rom-Kedar and Turaev [34] imply that systems with short-range repulsive forces, other than hard disks or spheres, may not be totally hyperbolic. Instead their phase spaces may have elliptic islands where the motion is not chaotic. It would be interesting to know first of all if there are any experimental or theoretical consequences of the existence of these elliptic regions for non-equilibrium processes in real fluid systems and also whether such elliptic islands will persist for arbitrarily large energies.
3. One important application of the methods described here is to the determination of the chaotic properties of thermostatted, driven systems for which a non-equilibrium steady state is reached and maintained. The general properties of such systems are described in the books of Hoover [35] and of Evans and Morriss [36], and a clear mathematical description has recently been given by Ruelle [37]. Of special interest is the perturbation of the Lyapunov spectrum produced by the thermostatted driving field. For dilute, random Lorentz gases it has been possible to use kinetic theory methods to determine the spectrum when the field is small [17]. It would be worthwhile to extend these results to larger fields and to gases where all of the particles are moving.
4. The escape-rate formalism described here has two drawbacks: (a) It is not at all easy to describe the fractal repeller that forms in the phase-space of a system with many degrees of freedom. (b) Even in those cases where the sum of the positive Lyapunov exponents on the repeller can be calculated analytically, the KS entropy is not yet directly accessible to analytic methods. Instead one has to use the transport

coefficients and the sum of the positive Lyapunov exponents to infer the KS entropy of trajectories on the repeller. It would be very helpful to have a better understanding of the properties of high-dimensional repellers and to have an independent analytical means to compute the KS entropy of trajectories on the repeller.

5. One of the main goals of current research in this area is to obtain, if possible, some deeper understanding of the dynamical basis of the laws of irreversible thermodynamics. For two-dimensional diffusive models based upon the baker map, it has been possible to show that the laws of irreversible thermodynamics result from a careful analysis of the fractal structures that appear in the relevant phase spaces of these models when the systems are in non-equilibrium steady states [38, 39, 40]. The main physical idea is that entropy is produced by the irreversible loss of information when changes are taking place in a system on very fine scales, beyond experimental resolution. However, as has been emphasized by other authors [41], these models may be too simple and/or the thermostats considered may be too special to allow for any general conclusions to be drawn. This area of research is active and many issues remain to be understood.
6. Our description of fluid systems as composed of classical particles interacting through repulsive, short-range forces is certainly incomplete. Typical fluid systems are better modeled by short-range forces with both attractive and repulsive regions. We do not yet know what effects a more careful analysis of interparticle forces will have on our picture of the chaotic behavior of fluids. An even more serious problem is connected with our use of classical mechanics to describe systems which are quantum mechanical in nature. We have almost no understanding of how to correctly obtain a quantum version of the classical chaotic picture of fluids, or even know for sure if such a thing is possible.

Acknowledgements

We thank A. Latz, P. Gaspard, M. H. Ernst, and E. G. D. Cohen for many helpful conversations. JRD acknowledges support by the National Science Foundation (USA) under grant NSF PHY-96-00428. HvB and RvZ are supported by FOM, SMC and by the NWO Priority Program Non-Linear Systems, which are financially supported by the "Nederlandse Organisatie voor Wetenschappelijk Onderzoek (NWO)".

References

1. E. Ott, *Chaos in Dynamical Systems*, (Cambridge University Press, Cambridge, 1993).

2. J. R. Dorfman, Phys. Rep. **301**, 151 (1998).
3. J. R. Dorfman, *An Introduction to Chaos in Non-equilibrium Statistical Mechanics*, (Cambridge University Press, Cambridge, 1999).
4. M. H. Ernst and J. R. Dorfman, J. Stat. Phys. **12**, 311 (1975).
5. N. Wax, *Noise and Stochastic Processes*, (Dover Publishing Co., New York, 1954).
6. Ja. B. Pesin, Sov. Math. Doklady **17**, 196 (1976), reprinted in R. S. MacKay and J. D. Meiss, *Hamiltonian Dynamical Systems: A Reprint Selection* (Adam Hilger, Bristol, 1987).
7. P. Gaspard and G. Nicolis, Phys. Rev. Lett. **65**, 1693 (1990).
8. J. R. Dorfman and P. Gaspard, Phys. Rev. E **51**, 28 (1995).
9. G. Gallavotti and E. G. D. Cohen, J. Stat. Phys. **80**, 931 (1995); Phys. Rev. Lett. **74**, 2694 (1995).
10. P. Gaspard, *Chaos, Scattering Theory and Statistical Mechanics*, (Cambridge University Press, Cambridge, 1998).
11. N. Chernov and R. Markarian, Bol. de Soc. Brasil. Mat. **28**, 271 (1997).
12. D. Ruelle and J.-P. Eckmann, Rev. Mod. Phys. **57**, 617 (1985).
13. P. Gaspard and F. Baras, Phys. Rev. E **51**, 5332 (1995).
14. H. van Beijeren and J. R. Dorfman, Phys. Rev. Lett. **74**, 4412 (1995); erratum **76**, 3238 (1996); see also H. van Beijeren, A. Latz, and J. R. Dorfman (to be published).
15. Special issue on the *Proceedings of the Euroconference on The Microscopic Approach to Complexity in Non-Equilibrium Molecular Simulations. CECAM at ENS-Lyon, 1996*, M. Mareschal, ed., Physica (Amsterdam) **240A** (1997).
16. J. R. Dorfman and H. van Beijeren, in Ref. [15], pp. 12–42.
17. H. van Beijeren, J. R. Dorfman, E. G. D. Cohen, Ch. Dellago, and H. A. Posch, Phys. Rev. Lett. **77**, 1974 (1996).
18. A. Latz, H. van Beijeren, and J. R. Dorfman, Phys. Rev. Lett. **78**, 207 (1997).
19. R. van Zon, H. van Beijeren, and Ch. Dellago, Phys. Rev. Lett. **80**, 2035 (1998).
20. J. R. Dorfman and H. van Beijeren, in *Statistical Mechanics* Part B, B. Berne, ed., (Plenum Publishing Co., New York 1977).
21. N. S. Krylov, *Works on the Foundations of Statistical Physics*, (Princeton University Press, Princeton, 1979).
22. S. D. Stoddard and J. Ford, Phys. Rev. A **8**, 1504 (1973).
23. D. J. Searles, D. J. Evans, and D. J. Isbister, in Ref. [15], pp. 96–104.
24. W. van Saarloos, Phys. Rep. **301**, 9 (1998); U. Ebert and W. van Saarloos, Phys. Rev. Lett. **80**, 1650 (1998); U. Ebert and W. van Saarloos, *Front propagation into unstable states: Universal algebraic convergence towards uniformly translating pulled fronts*, Physica D (to appear).
25. R. E. Bellman and K. L. Cooke, *Differential-Difference Equations*, (Academic Press, New York/London, 1963).
26. Ch. Dellago and H. A. Posch, in Ref. [15], pp. 68–83.
27. E. Brunet and B. Derrida, Phys. Rev. E **56**, 2597 (1997).
28. D. A. Kessler, Z. Ner, and L. M. Sander, Phys. Rev. E **58**, 107 (1998).
29. R. van Zon and H. van Beijeren, in preparation.
30. M. H. Ernst, J. R. Dorfman, W. Hoegy, and J. M. J. van Leeuwen, Physica **45**, 127 (1969).
31. M. H. Ernst and J. R. Dorfman, J. Stat. Phys. **57**, 581 (1989).
32. H. van Beijeren, J. R. Dorfman, H. A. Posch, and Ch. Dellago, Phys. Rev. E **56**, 5272 (1997).
33. J. R. Dorfman, A. Latz, and H. van Beijeren (to be published).
34. V. Rom-Kedar and D. Turaev, Physica D **130**, 187 (1999).
35. W. Hoover, *Computational Statistical Mechanics*, (Elsevier Science Publishers, Amsterdam, 1991).
36. D. J. Evans and G. P. Morriss, *Statistical Mechanics of Non-Equilibrium Liquids*, (Academic Press, London, 1990); now available from the web site `http://rsc.anu.edu.au/~evans`.

37. D. Ruelle, J. Stat. Phys. **95**, 393 (1999).
38. P. Gaspard, J. Stat. Phys. **89**, 1215 (1997).
39. J. Vollmer, T. Tél, and W. Breymann, Phys. Rev. E **58**, 1672 (1998), and references contained therein.
40. T. Gilbert and J. R. Dorfman, J. Stat. Phys. **96**, 225 (1999).
41. L. Rondoni and E. G. D. Cohen (to be published).

MULTIFRACTAL PHASE-SPACE DISTRIBUTIONS FOR STATIONARY NONEQUILIBRIUM SYSTEMS

H. A. POSCH AND R. HIRSCHL
Institute for Experimental Physics, University of Vienna, Boltzmanngasse 5, A-1090 Vienna, Austria

AND

WM. G. HOOVER
Department of Applied Science, University of California at Davis-Livermore and Lawrence Livermore National Laboratory Livermore, CA 94551-7808, USA

Abstract. The phase-space density of stationary nonequilibrium particle systems is known to be a multifractal object with an information dimension smaller than the phase-space dimension. The rate of heat flowing through the system, divided by the Boltzmann constant and the kinetic temperature, is equal to the sum of the Lyapunov exponents. The reduction in dimensionality is determined from the spectrum of Lyapunov exponents. We show here that also many-body systems in nonequilibrium states with *stochastic thermostats* can be found that have similar properties and support fractal structures in phase space. We study two two-dimensional examples: first, color conductivity for a system of hard disks, which are thermostated by a stochastic map which affects the momenta of randomly chosen particles; second, color conductivity of a system of soft disks which are subjected to a stochastic force and perform Brownian motion. Full Lyapunov spectra were computed for both models, and the information dimensions of their underlying attractors determined.

1. Introduction

The phase-space collapse and the existence of multifractal strange attractors and repellers in the phase space is a well-established and natural property of any deterministically-thermostated, stationary, nonequilibrium sys-

J. Karkheck (ed.),
Dynamics: Models and Kinetic Methods for Non-equilibrium Many Body Systems, 169–189.

tem [1, 2, 3, 4]. See also Refs. [5, 6]. Although highly successful for the computation of transport properties in nonequilibrium steady states [7, 8, 9, 10], these models are still viewed as somewhat artificial and unphysical by some researchers [11, 12], and they make up only a small subclass of all possible nonequilibrium flows. Thus, in a more general setting, the usefulness of the concept of multifractal attractors in the phase-space of many-body systems is still in some doubt. The question therefore arises whether this concept carries over also to other forms of thermostats such as *stochastic boundaries.* There are theoretical arguments suggesting that the phase-space distribution of a nonequilibrium system with stochastic walls, maintaining a constant heat flow, is absolutely continuous [13, 14] implying that no fractal attractor exists for such a case. However, in a recent computer simulation of a field-driven Galton board, also known as the driven periodic Lorentz gas, we have found numerical evidence for the existence of a multifractal structure with a partially stochastic boundary [15]. Also, the Lyapunov spectrum of a one-dimensional particle chain supporting heat conduction between two stochastic boundaries exhibits all the properties which point to the existence of a multifractal attractor in the nonequilibrium steady state [16]. The sum of the Lyapunov exponents is negative, indicating a shrinkage of an arbitrary phase space volume, and is proportional to the heat current through the system. The information dimension D_1, which is computed from the Lyapunov spectrum with the help of the Kaplan-Yorke formula [17], is strictly smaller than the phase-space dimension for steady nonequilibrium states.

Lyapunov exponents describe the exponential growth or shrinkage of infinitesimal perturbations in the phase space of chaotic systems. A reference state $\mathbf{\Gamma} = \{\mathbf{q}_1, \mathbf{q}_2, \dots, \mathbf{q}_N, \mathbf{p}_1, \mathbf{p}_2, \dots, \mathbf{p}_N\}$ of a many-body system evolves in time along its phase trajectory as dictated by the equations of motion of the system. Any perturbation or tangent vector $\delta\mathbf{\Gamma} = \{\delta\mathbf{q}_1, \delta\mathbf{q}_2, \dots, \delta\mathbf{q}_N, \delta\mathbf{p}_1, \delta\mathbf{p}_2, \dots, \delta\mathbf{p}_N\}$ evolves according to the linearized equations of motion. The Lyapunov exponents are defined by

$$\lambda_l = \lim_{t\to\infty} \frac{1}{t} \log \frac{|\delta\mathbf{\Gamma}_l\,(t)|}{|\delta\mathbf{\Gamma}_l\,(0)|}. \tag{1}$$

The multiplicative ergodic theorem of Oseledec [18, 19] assures that for ergodic systems with phase space dimension $2dN$ there are as many orthonormal initial vectors $\delta\mathbf{\Gamma}_l\,(0)$, which yield a set of $2dN$ exponents $\{\lambda_l\}$. The ordered set $\{\lambda_1 \geq \lambda_2 \geq \cdots \geq \lambda_L\}$ is referred to as the Lyapunov spectrum of the system. The l is an index which numbers the exponents. The λ_l are independent of the metric and of the initial conditions of the reference trajectory.

For numerical implementation, the reference trajectory and a complete set of tangent-vector trajectories must be computed simultaneously. The difficulties associated with the choice of the unknown initial vectors $\{\delta\Gamma_l(0)\}$, taken to be orthonormal and with arbitrary orientation, and with the rounding errors of the computer are overcome by periodic reorthonormalization of the offset vectors during the simulation [20, 21]. The Lyapunov exponents are obtained from the time-averaged logarithms of the corresponding normalizing factors.

Our aim is the computation of the full Lyapunov spectrum for a many-body system in a stationary nonequilibrium state, because it is the only practical means of detecting multifractal attractors in a multidimensional phase space, and of computing their fractal dimension. As an example we consider here the color-conductivity problem in a two-dimensional model fluid with a current generated by an externally applied field. We approach the problem from three different starting points, which differ mainly by the way the dissipated energy, which is continuously produced by the current, is removed from the system to achieve a stationary state. In Section 2 we apply the by-now classical Gaussian thermostat to constrain, by way of feedback, the kinetic energy of the system [7, 8, 9, 10]. A dynamical friction term is introduced into the equations of motion, which remain time reversible and deterministic. In Section 3 this dynamical and time-reversible thermostat is replaced by a stochastic scheme, through which the momenta of randomly selected particles are replaced by stochastic momenta obtained from an equilibrium distribution belonging to a lower temperature. Finally, in Section 4 we consider heavy soft disks in two dimensions, which perform Brownian motion and release their excess kinetic energy into the heat bath. In all three cases it is possible to compute the Lyapunov spectrum for the system in a stationary conducting state. We discuss these results in Section 5.

2. Color Conductivity with Gaussian Thermostat

Color conductivity is, arguably, the simplest stationary nonequilibrium system. It was studied by computer simulation before [22, 3, 7, 23], and Lyapunov spectra for soft-particle interactions were obtained for stationary conducting states with dynamic-thermostat control [3, 23]. A method for the computation of the Lyapunov spectrum of hard-body systems in two and three dimensions was developed only recently [24, 25]. To set the stage and for later reference, we summarize this method below.

Our two-dimensional system consists of N hard disks with diameter σ and mass m, located at the positions $\mathbf{q}_j, j = 1 \ldots N$, and carrying color charges $c_j = \pm c$. The system is assumed to be "neutral", $\sum_j c_j = 0$. Between collisions the particles stream continuously. They are accelerated by a homogeneous and constant external field $\mathbf{E}$ of strength E according to the equations of motion

$$\dot{\mathbf{q}}_j = \mathbf{p}_j/m, \quad \dot{\mathbf{p}}_j = c_j\mathbf{E} - \zeta\mathbf{p}_j, \tag{2}$$

where

$$\zeta = \sum_{j=1}^{N} c_j(\mathbf{p}_j \cdot \mathbf{E}) \Big/ \sum_{j=1}^{N} \mathbf{p}_j^2 \tag{3}$$

is a Gaussian-thermostat variable, which keeps the kinetic energy $K = \sum_j \mathbf{p}_j^2/2m$ of the disks exactly constant. The continuous streaming is interrupted when two particles, say k and l, collide. Their position is not modified by the collision, but their momenta change according to the collision map

$$\mathbf{p}_k^f = \mathbf{p}_k^i + \left(\mathbf{p}^i \cdot \mathbf{q}^i\right) \mathbf{q}^i/\sigma^2, \quad \mathbf{p}_l^f = \mathbf{p}_l^i - \left(\mathbf{p}^i \cdot \mathbf{q}^i\right) \mathbf{q}^i/\sigma^2, \tag{4}$$

where $\mathbf{q}^i \equiv \mathbf{q}_l^i - \mathbf{q}_k^i$ and $\mathbf{p}^i \equiv \mathbf{p}_l^i - \mathbf{p}_k^i$ are the relative position and momentum immediately before the collision, respectively. The superscripts i and f refer to the initial and final states of the collision.

For computation of the Lyapunov exponents, the linearized equations of motion for the streaming, and the linearized collision map, are required for the construction of the tangent-vector trajectories. From (2) and (3) the former is readily obtained:

$$\begin{aligned} \delta\dot{\mathbf{q}}_j &= \delta\mathbf{p}_j/m, \\ \delta\dot{\mathbf{p}}_j &= -\zeta\delta\mathbf{p}_j - \mathbf{p}_j \sum_{k=1}^{N} (c_k\mathbf{E} - 2\mathbf{p}_k\zeta) \cdot \delta\mathbf{p}_k \Big/ \sum_{k=1}^{N} (\mathbf{p}_k \cdot \mathbf{p}_k) \end{aligned} \tag{5}$$

The linearization of the collision map (4) is more complicated, and we refer to Ref. [24] for details. It yields for the particles $j \neq k, l$ not partaking in the collision

$$\begin{aligned} \delta\mathbf{q}_j^f &= \delta\mathbf{q}_j^i, \\ \delta\mathbf{p}_j^f &= \delta\mathbf{p}_j^i + \Delta\zeta\mathbf{p}_j^i\delta\tau_c, \end{aligned} \tag{6}$$

and for the colliding particles k, l

$$\begin{aligned} \delta\mathbf{q}_k^f &= \delta\mathbf{q}_k^i + \left(\delta\mathbf{q}^i \cdot \mathbf{q}^i\right) \mathbf{q}^i/\sigma^2, \\ \delta\mathbf{q}_l^f &= \delta\mathbf{q}_l^i - \left(\delta\mathbf{q}^i \cdot \mathbf{q}^i\right) \mathbf{q}^i/\sigma^2, \end{aligned}$$

$$\begin{aligned}
\delta\mathbf{p}_k^f &= \delta\mathbf{p}_k^i + \left(\delta\mathbf{p}^i\cdot\mathbf{q}^i\right)\mathbf{q}^i/\sigma^2 + \frac{1}{\sigma^2}\left[\left(\mathbf{p}^i\cdot\delta\mathbf{q}_c\right)\mathbf{q}^i + \left(\mathbf{p}^i\cdot\mathbf{q}^i\right)\delta\mathbf{q}_c\right] \\
&\quad + \left(\Delta\zeta\mathbf{p}_k^i + \frac{c}{\sigma^2}\left(\mathbf{E}\cdot\mathbf{q}^i\right)\mathbf{q}^i\right)\delta\tau_c, \\
\delta\mathbf{p}_l^f &= \delta\mathbf{p}_l^i - \left(\delta\mathbf{p}^i\cdot\mathbf{q}^i\right)\mathbf{q}^i/\sigma^2 - \frac{1}{\sigma^2}\left[\left(\mathbf{p}^i\cdot\delta\mathbf{q}_c\right)\mathbf{q}^i + \left(\mathbf{p}^i\cdot\mathbf{q}^i\right)\delta\mathbf{q}_c\right], \\
&\quad + \left(\Delta\zeta\mathbf{p}_l^i - \frac{c}{\sigma^2}\left(\mathbf{E}\cdot\mathbf{q}^i\right)\mathbf{q}^i\right)\delta\tau_c,
\end{aligned} \tag{7}$$

where $\delta\mathbf{q}^i \equiv \delta\mathbf{q}_l^i - \delta\mathbf{q}_k^i$ and $\delta\mathbf{p}^i \equiv \delta\mathbf{p}_l^i - \delta\mathbf{p}_k^i$ are the relative position and momentum displacements before the collision. The vector $\delta\mathbf{q}_c = \delta\mathbf{q}^i + \mathbf{p}^i\delta\tau_c$ denotes the infinitesimal displacement of the collision points of the perturbed trajectory from the reference trajectory, and $\delta\tau_c = -\left(\delta\mathbf{q}^i\cdot\mathbf{q}^i\right)/\left(\mathbf{p}^i\cdot\mathbf{q}^i\right)$ is the delay time between the collisions of the reference and the perturbed trajectories. It may be positive or negative. Furthermore,

$$\Delta\zeta = -c\left(\mathbf{p}^i\cdot\mathbf{q}^i\right)\left(\mathbf{E}\cdot\mathbf{q}^i\right) \Big/ \left(\sigma^2\sum_{j=1}^{N}\mathbf{p}_j^i\cdot\mathbf{p}_j^i\right) \tag{8}$$

is the change of ζ due to the collision between k and l, and $c = (c_l - c_k)$ is their charge difference. $\Delta\zeta$ affects also the collision rules for the non-colliding particles, but it vanishes if c is zero. The Eqs. (2-4) suffice to compute the reference trajectory, the Eqs. (5-8) are required to determine the tangent-vector dynamics. In two dimensions there are $4N$ Lyapunov exponents, and as many tangent vectors need to be followed for the computation of the complete spectrum. Since the number of equations to be solved increases with N^2, the number of particles is restricted to about one thousand with present-day workstations.

In our numerical work reported in detail in Ref. [24] reduced units are used for which the particle mass m, the disk diameter σ, the kinetic energy per particle K/N, and the Boltzmann constant k_B are unity. $K = \sum\mathbf{p}^2/2m$ is the total kinetic energy. With this choice the unit of time is $(m\sigma^2 N/K)^{1/2}$. The density of the gas is defined by $\rho = N/V$, where V is the volume of the simulation box, which has an aspect ratio of $2/\sqrt{3}$.

For comparison, some of the results of Ref. [24] for densities of $0.6\sigma^{-2}$ and $0.8\sigma^{-2}$ are reproduced in Fig. 1 and in Table 1. The shift of the Lyapunov spectrum to more negative values is hardly visible for a field of $E = 1$ on the scale of the figure, but it is most noticeable in the sum of conjugate exponents $z_l = \lambda_l + \lambda_{4N-l+1}$, which is shown as the dashed line and refers to the expanded vertical scale on the right-hand side of the figure. Within experimental errors, z_l is found to agree with $-\zeta$, independent of l. This is a formulation of the conjugate pairing rule [26, 27], which obviously holds

ρ	E	$\langle\zeta\rangle$	λ_1	$\sum_l \lambda_l$	h_{KS}/N	κ	ΔD_1
0.4	0.0	0.0	2.91	0.0	2.68	-	0
0.4	0.5	0.0566	2.86	-7.19	2.46	0.453	2.54
0.4	1.0	0.2246	2.73	-28.52	2.14	0.449	10.89
0.6	0.0	0.0	4.28	0.0	4.58	-	0
0.6	0.5	0.0235	4.25	-2.99	4.44	0.188	0.699
0.6	1.0	0.1003	4.23	-12.73	4.16	0.200	3.046
0.8	0.0	0.0	6.51	0.0	7.85	-	0
0.8	0.5	0.0068	6.51	-0.868	7.79	0.055	0.133
0.8	1.0	0.0271	6.45	-3.435	7.57	0.054	0.531

TABLE 1. Results for color-conductivity simulations of 64 hard disks with a Gaussian thermostat. ρ is the density, and E is the strength of the applied field. The Lyapunov spectra corresponding to these data are shown in Fig. 10 of Ref. [24]. ζ is the Gaussian friction, λ_1 is the maximum Lyapunov exponent, and h_{KS} is the Kolmogorov-Sinai entropy, which is equal to the sum of the positive exponents. κ is the color conductivity. ΔD_1 is the dimensionality reduction of the phase space density in the stationary nonequilibrium state. All quantities are given in the reduced units introduced in Section 2.

for these systems [24]. The nonequilibrium version of Liouville's Theorem provides the link between the rate of heat loss, $\dot{Q}$, with the rate of shrinkage of an arbitrary (differentially small) volume element δV, co-moving with the flow in phase space. The following well-known chain of equalities holds for our color-conductivity model with a Gaussian thermostat,

$$\langle d \ln \delta V/dt \rangle = -(2N-1)\langle\zeta\rangle = \langle dQ/dt\rangle/kT = \sum_{l=1}^{4N} \lambda_l \leq 0, \tag{9}$$

where $\langle\cdots\rangle$ denotes a time average. The equality sign is valid only in equilibrium. As usual, k is the Boltzmann constant, and T is the kinetic temperature. This shows that a negative sum of Lyapunov exponents is a clear indication of the existence of a fractal attractor in phase space for stationary nonequilibrium systems of this kind. It provides the motivation for our attempts to compute Lyapunov spectra for nonequilibrium systems with stochastic thermostats in the following sections. These relations are well obeyed by the simulation results of our Gauss-thermostated conducting flows summarized in Table 1. The respective spectra are shown in full in Fig.10 of Ref. [24], and partly in Fig. 1 of this work.

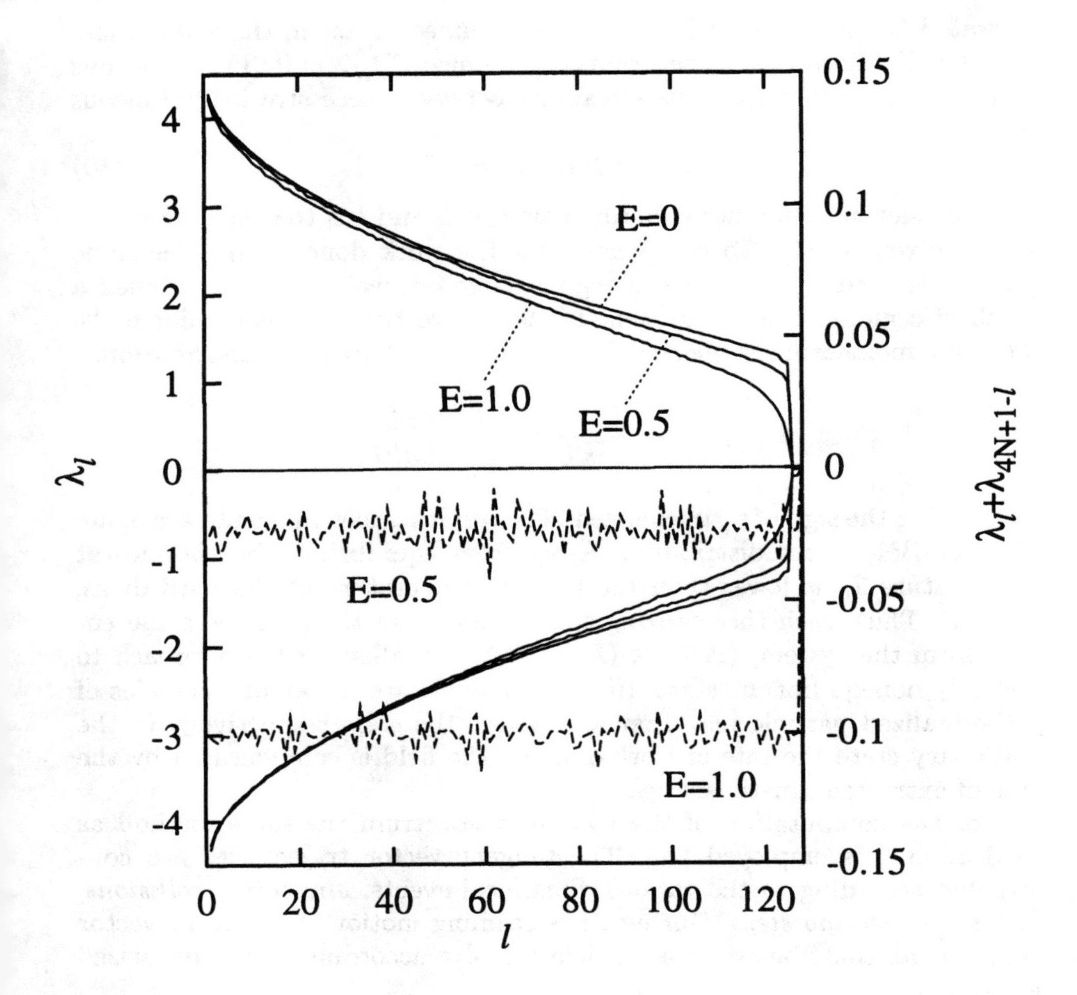

Figure 1. Lyapunov spectra of a system of 64 Gauss-thermostated hard disks for a density of $\rho = 0.6\sigma^{-2}$. E is the applied field. The dashed curves show the sum of conjugate exponents, $\lambda_l + \lambda_{4N-l+1}$, on an expanded scale on the right-hand side of the figure. The Lyapunov spectra are given in units of $(K/Nm\sigma^2)^{1/2}$. The Lyapunov spectra are only defined for integer values of the index l on the abscissa, which labels pairs of exponents.

3. Color Conductivity with Stochastic Thermostat

We study in this section the same conductivity problem for a two-dimensional hard-disk system as before, but a stochastic map replaces the deterministic thermostat. The same notation is used as in the previous section.

Between collisions the particles carrying color charges $c_i = \pm c$ are accelerated by the external field which is assumed to act in the x direction, $\mathbf{E} = (E, 0)$. Total charge neutrality is assumed, $\sum_i c_i = 0$. The equations of motion for the continuous streaming between successive instantaneous events are

$$\dot{\mathbf{q}}_i = \mathbf{p}_i/m, \quad \dot{\mathbf{p}}_i = c_i\mathbf{E}. \tag{10}$$

The collision map for two colliding particles k and l is the same as before and is given by (4). To compensate for the work done by the field, one particle is randomly selected at regular time intervals τ and is assigned a stochastic momentum component in the y direction, perpendicular to $\mathbf{E}$. The new momentum is selected from a Maxwell-Boltzmann distribution,

$$w(p_y) = \frac{1}{\sqrt{2\pi mkT_0}} \exp\left\{-\frac{p_y^2}{2mkT_0}\right\}, \tag{11}$$

but leaving the sign of p_y unchanged. The resulting new momenta are again Maxwell-Boltzmann distributed. Away from equilibrium the thermostat temperature T_0 is lower than the kinetic temperature of the hard disks, $T_0 < T$. Thus, each thermalizing step removes, on the average, some energy from the system, $\langle \Delta K \rangle \equiv \langle K_f - K_i \rangle < 0$, allowing the approach to a steady nonequilibrium state. Here, K_f and K_i are the kinetic energies of a thermalized particle just after and before the step, respectively. In the stationary state the rate of work done by the field is compensated by the rate of extracted kinetic energy.

For the computation of the Lyapunov spectrum the same method as in Section 2 is employed [24]. The tangent vector trajectories are constructed according to the various dynamical events, *streaming, collisions*, and *thermostating steps*. During the streaming motion the tangent vector components contributed by a particle j evolve according to the linearized equations

$$\dot{\delta\mathbf{q}}_j = \delta\mathbf{p}_j/m, \quad \dot{\delta\mathbf{p}}_j = \mathbf{0}. \tag{12}$$

The linearized collision map is given by (6) and (7), but is simplified by the fact that there is no friction term in the streaming equations (10). Consequently, $\Delta\zeta = 0$ in (6) and (7). Only the thermostating step requires more thought. Guided by the results for dynamically thermostated systems in Eq. (9) connecting the heat transfer through the system with the change in the phase space volume, we require that the logarithm of an infinitesimal volume changes according to [16]

$$\ln \delta V^f - \ln \delta V^i = \Delta K/kT_0 \tag{13}$$

for the thermostating step. Since only the p_y-component of a single particle is involved in such an event, it follows for the respective tangent vector component that

$$\delta p_y^f = \delta p_y^i e^{\Delta K/kT_0}. \tag{14}$$

The other components are unchanged. All necessary information is now available to follow the time evolution of the tangent vectors and to compute the Lyapunov spectrum.

We carried out simulations for color-conducting hard-sphere systems with this method, where various thermostat intervals τ were tested. The systems contained up to 128 particles. To present our results we use reduced units for which the hard-disk diameter σ, the thermostat energy kT_0, the disk mass m, and the charge c are unity. The kT_0 determines the width of the momentum distribution for the thermostating steps. These reduced units are identical to those in Section 2, since $kT_0 = K/N$ for dynamically thermostated systems with feedback. As always, temperatures are measured in units for which the Boltzmann constant is unity. The aspect ratio of the simulation box with periodic boundary conditions is $A = \sqrt{3}/2$. Its area is denoted by V. An event-based algorithm is used, stepping from one instantaneous event to the next, where the collisions are determined to machine precision. Since the streaming motion of the particles is on parabolic curves, this requires solving a fourth-order polynomial. Although finding the roots is rather time-consuming, it preserves the advantages of an event-based algorithm and is more accurate than a numerical integration.

Our results for various fields E and number densities $\rho \equiv N/V$ are summarized in Table 2. For comparison we note that the close-packed density $\rho_c = 2/\sqrt{3}\sigma^{-2}$. As expected, the temperatures for the nonequilibrium steady states ($E > 0$) are significantly higher than in equilibrium. Since in hard-body systems all dynamical events scale strictly with the velocity, we account for this temperature increase by introducing rescaled Lyapunov exponents

$$\lambda_l^* = \sqrt{K_0/K}\lambda_l, \tag{15}$$

and a rescaled Kolmogorov-Sinai entropy. The latter, according to Pesin [28], is obtained as the sum of the positive exponents, $h_{KS}^* = \sum_{\lambda_l^*>0} \lambda_l^*$. Here, $K_0/N = kT_0$ is the unit kinetic energy of the stochastic bath. One deduces from Table 2 that the fluid temperature T is strongly affected by the thermostating interval τ. After rescaling, however, the respective Lyapunov spectra $\{\lambda_l^*; l = 1, \ldots, 4N\}$ are found to coincide to within the numerical accuracy of about 0.3%. Thus, τ is an irrelevant parameter and will not be considered further.

In Fig. 2 the spectra for a 64-particle system at a density $\rho = 0.8\sigma^{-2}$ are shown in equilibrium, and for a field $E = 1$. The sum of the exponents is clearly shifted to more negative values, as is inferred from Table

N	ρ	E	τ	$\langle K\rangle$	λ_1^*	$\sum_l \lambda_l$	κ	h_{KS}^*/N	ΔD_1
64	0.8	0.0	∞	64.0	6.50	0.000	0.000	7.85	0
64	0.8	0.0	0.1	65.7	6.51	0.00	0.001	7.79	0
64	0.8	0.5	0.05	71.1	6.60	-0.918	0.059	7.74	0.13
64	0.8	0.5	0.1	76.8	6.50	-0.872	0.056	7.72	0.12
64	0.8	0.5	0.2	82.9	6.47	-0.789	0.051	7.73	0.11
64	0.8	1.0	0.1	104.9	6.50	-2.95	0.048	7.66	0.34
64	0.4	1.0	0.1	256.0	2.90	-14.19	0.227	2.60	1.90
128	0.5	0.0	0.05	128.9	3.60	0.00	-0.001	3.50	0
128	0.5	1.0	0.05	376.8	3.69	-16.42	0.130	3.36	2.34

TABLE 2. Simulation results for N hard disks in two dimensions accelerated by an external color field of strength E and subjected to a stochastic thermostat with interval τ as introduced in Section 3. The Lyapunov exponents λ_l are reduced according to $\lambda_l^* = (T_0/T)^{1/2}\lambda_l$ to account for the increase in temperature, and the (reduced) Kolmogorov-Sinai entropy is obtained from the sum of the positive exponents, $h_{KS}^* \sum_{\lambda_l^*>0} \lambda_l^*$. κ is the color conductivity, and $\Delta D_1 = 4N - D_1$ is the reduction in dimensionality of the phase space density, where D_1 is the information dimension. All quantities are given in the reduced units defined in Section 3.

2, but the shift of the individual exponents is weak and is essentially restricted to the most negative exponents. This can be clearly seen from the dashed horizontal line in Fig. 2, which refers to the expanded vertical scale on the right-hand side and which represents the sum of conjugate exponents, $z_l \equiv \lambda_l^* + \lambda_{4N-l+1}^*$. The conjugate pairing rule does not hold, which would require a uniform shift of this pair sum for all Lyapunov exponent pairs. This behavior is even much more pronounced for hard-disk gases with smaller density as is shown in Fig. 3. There, two spectra, normalized by their respective maximum exponent λ_1, are shown for the same color field ($E = 1$) but for different densities. The spectrum of the lower-density gas is very asymmetric. A similar behavior was already found for an inhomogeneously-driven model of planar Couette flow with a dynamic Nosé-Hoover thermostat [9]. Interestingly, the most-negative exponents for both spectra in Fig. 3 are almost equal, as is shown in the inset, but we do not know an explanation.

The conductivity coefficient is given by $\kappa = (\langle \mathbf{J}\rangle \cdot \mathbf{E})/NE^2$, where $\mathbf{J} = \sum_{i=1}^N c_i \mathbf{p}_i/m$ is the total color current. According to the equalities in (9) for a dynamically thermostated system, it is simply related to the sum of

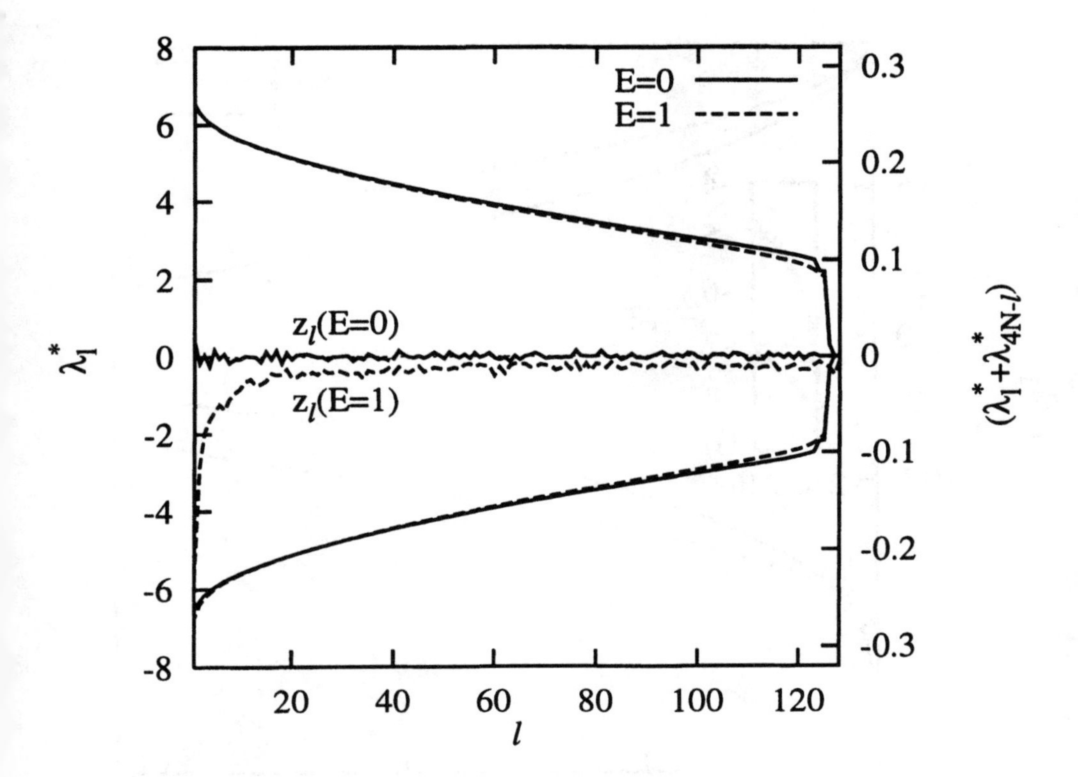

Figure 2. Rescaled Lyapunov spectra of 64 hard disks at a density $\rho = 0.8\sigma^{-2}$, subjected to an external color field E. The system has a stochastic thermostat described in Section 3. The dashed curves show the sum $z_l = \lambda_l^* + \lambda_{4N-l+1}^*$ of conjugate exponents on an expanded vertical scale (on the right-hand side). The rescaled Lyapunov spectra are given in units of $(K_0/Nm\sigma^2)^{1/2}$, where $K_0/N = kT_0$ is the kinetic energy per particle of the thermostat. The Lyapunov spectra are only defined for integer values of the index l.

the Lyapunov exponents according to

$$-\sum_{l=1}^{4N} \lambda_l = \frac{\kappa N^2 E^2}{\langle K \rangle}. \tag{16}$$

This relation is also reasonably-well obeyed by the stochastically thermostated systems treated here. The most important aspect of our results, however, is the fact that the Lyapunov spectra yield an information dimension D_1 which is strictly smaller than the phase space dimension $4N$. In Table 2 the dimensionality reduction $4N - D_1$ is listed, where D_1 was estimated from the Kaplan-Yorke formula [17].

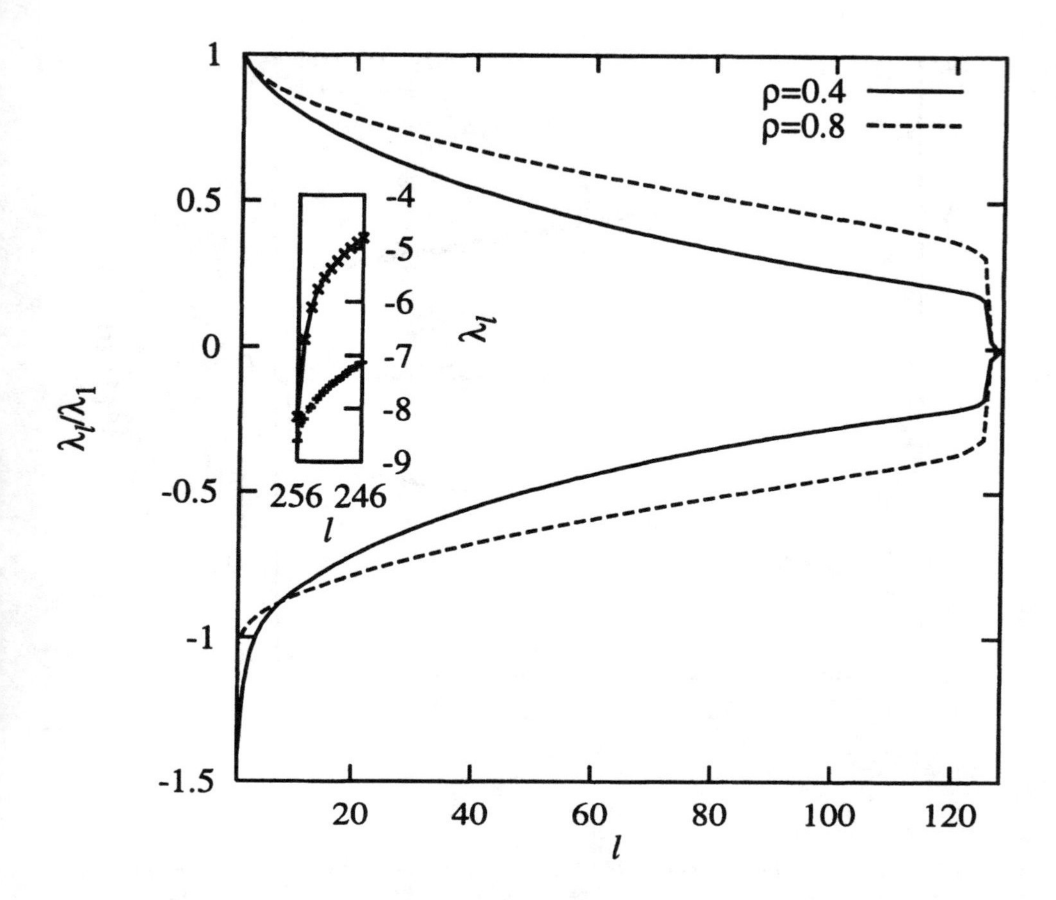

Figure 3. Lyapunov spectra of 64 hard disks for the densities $\rho = 0.8\sigma^{-2}$ and $0.4\sigma^{-2}$, subjected to an external color field $E = 1$. The spectra are normalized by their maximum exponent λ_1. The inset shows the unnormalized value of the most-negative exponents for the two spectra. The Lyapunov spectra are only defined for integer values of the index l.

4. Color Conductivity and Langevin Dynamics

Another and conceptually very appealing way for the removal of the irreversibly generated heat from the system is by treating the color particles as heavy Brownian particles of mass m diffusing in a hypothetical gas of particles with negligible mass $\mu \ll m$. The color particles interact with each other through a smooth short-ranged potential $\phi_{i,j}$, and the total potential energy is taken pairwise additive, $\Phi = \sum\sum \phi_{i<j}$. For simplicity, we neglect in the following hydrodynamic interactions mediated by the gas, such as the Oseen tensor [29, 30, 31], and assume that the dynamics of the Brownian

particles is described by a single-particle Langevin equation

$$\begin{aligned}\dot{\mathbf{q}}_i &= \mathbf{p}_i/m, \\ \dot{\mathbf{p}}_i &= \mathbf{F}_i(\{\mathbf{q}\}) + c_i\mathbf{E} - \zeta\mathbf{p}_i + \mathbf{A}_i(t),\end{aligned} \tag{17}$$

where $\mathbf{F}_i = -\partial\Phi/\partial\mathbf{q}_i$. As before, $c_i = \pm c$ is the color charge coupling to the field $\mathbf{E}$, and total charge neutrality is assumed, $\sum_i c_i = 0$. The stochastic forces $\mathbf{A}_i(t)$ are δ-correlated Gaussian random variables, for which

$$\langle \mathbf{A}_i(0) \cdot \mathbf{A}_i(t) \rangle = 2d\zeta kT_0\delta(t) \tag{18}$$

holds. In these expressions ζ is a friction constant which is related to the viscosity of the solvent, and d is the dimension. Integrating the last equation yields $\langle \mathbf{A}_i^2 \rangle = d\zeta kT_0$, which is the detailed-balance expression relating the fluctuating forces to the friction. The k is the Boltzmann constant, and T_0 is the fixed temperature of the gas. Strictly speaking, these relations are true in the linear-response limit of vanishing fields, in which case the averages $\langle \cdots \rangle$ are averages over an equilibrium ensemble.

If Eq. (17) is integrated over one timestep Δt of the simulation,

$$\mathbf{p}_i(t + \Delta t) = \mathbf{p}_i(t) + \int_t^{t+\Delta t} (\mathbf{F}_i + c_i\mathbf{E} - \zeta\mathbf{p}_i)\, d\tau + \Delta t\bar{\mathbf{A}}_i, \tag{19}$$

the time averaged stochastic force for particle i,

$$\bar{\mathbf{A}}_i \equiv \frac{1}{\Delta t}\int_t^{t+\Delta t} \mathbf{A}_i(\tau)d\tau, \tag{20}$$

is a vector with components $\bar{A}_{i,\alpha}$ which are Gaussian random variables and are distributed according to [32]

$$w(A_{i,\alpha}) = (2\pi\langle \bar{A}_{i,\alpha}^2 \rangle)^{-1/2} \exp(-\bar{A}_{i,\alpha}^2/2\langle \bar{A}_{i,\alpha}^2 \rangle). \tag{21}$$

Furthermore, $\langle \bar{A}_{i,\alpha} \rangle = 0$, and $\langle \bar{A}_{i,\alpha}^2 \rangle = 2m\zeta kT_0/\Delta t$. We note that (21) is a consequence of the central-limit theorem and will always hold, as long as the correlation time of the noise is much shorter than the timestep Δt. Finally, the equations of motion (17) we consider in the following become

$$\begin{aligned}\dot{\mathbf{q}}_i &= \mathbf{p}_i/m, \\ \dot{\mathbf{p}}_i &= \mathbf{F}_i(\{\mathbf{q}\}) + c_i\mathbf{E} - \zeta\mathbf{p}_i + \bar{\mathbf{A}}_i,\end{aligned} \tag{22}$$

with constant $\bar{\mathbf{A}}_i$ during a timestep.

The energy of the Brownian particles, $H = K + \Phi = \sum_i \mathbf{p}_i^2/2m + \Phi$, changes according to

$$\begin{aligned} \dot{H} &= \sum_i \frac{\mathbf{p}_i \cdot \dot{\mathbf{p}}_i}{m} - \sum_i \mathbf{F}_i \cdot \dot{\mathbf{q}}_i \\ &= \sum_i c_i \frac{\mathbf{p}_i}{m} \cdot \mathbf{E} - \zeta \sum_i \frac{\mathbf{p}_i^2}{m} + \sum_i \frac{\mathbf{p}_i}{m} \cdot \bar{\mathbf{A}}_i, \end{aligned} \tag{23}$$

where we have used (22). Since $\mathbf{J} = \sum_i c_i \mathbf{p}_i/m$ is the total color current, the first term of (23) is readily identified as the rate of work done by the external field,

$$\dot{W} = \mathbf{J} \cdot \mathbf{E} \geq 0, \tag{24}$$

where the equal sign only applies if the field vanishes. The two remaining terms, the second negative and the third positive when averaged over time, are a consequence of the heat bath. Writing

$$\dot{Q} = 2K\zeta - \sum_i \frac{\mathbf{p}_i}{m} \cdot \bar{\mathbf{A}}_i, \tag{25}$$

we may identify $\dot{Q}$ as the rate of heat extracted by the bath. Here, $K = \sum_i \mathbf{p}_i^2/2m$ is the kinetic energy of the Brownian particles. It is equal to $dNkT/2$ when averaged over time, where T is the Brownian particles' kinetic temperature. T is generally larger than the fixed temperature T_0 of the bath, and $T = T_0$ holds only in equilibrium. In the nonequilibrium steady state to which the system eventually reverts for nonvanishing fields, one has $\langle \dot{H} \rangle = \langle \dot{W} \rangle - \langle \dot{Q} \rangle = 0$, and $\langle \dot{Q} \rangle \geq 0$, where the equal sign only applies in equilibrium, and where $\langle \cdots \rangle$ denotes a time average. The heat bath not only slows down the Brownian particles with the constant friction ζ, but accelerates them also randomly, through $\bar{\mathbf{A}}_i$ and prevents freezing. The second term of (25) does not vanish when averaged over time, in spite of the fact that the stochastic force is a random variable. During a timestep Δt, during which $\bar{\mathbf{A}}_i$ is constant, a particle i is particularly accelerated in the direction of $\bar{\mathbf{A}}_i$ such that the average over many consecutive timesteps of $(1/\Delta t) \int_t^{t+\Delta t} (\mathbf{p}_i \cdot \bar{\mathbf{A}}_i)/m$ is positive.

For the computation of the Lyapunov spectrum of our system we recall that the logarithmic rate of change of a phase space volume δV is related to the rate of heat transfer $\dot{Q}$ and to the sum of the Lyapunov exponents according to

$$\langle d \ln \delta V/dt \rangle = -\frac{\langle \dot{Q} \rangle}{kT} = \sum_{l=1}^{L} \lambda_l. \tag{26}$$

As was sketched in Section 2, the Lyapunov exponents are determined by following the motion of a set of tangent vectors representing infinitesimal perturbations of the reference trajectory which expand or shrink exponentially on the average. At first sight the motion of these vectors seems to be given by the motion equations obtained by linearizing the Eqs. (22):

$$\begin{aligned} \delta\dot{\mathbf{q}}_i &= \delta\mathbf{p}_i/m, \\ \delta\dot{\mathbf{p}}_i &= \sum_j \frac{\partial\mathbf{F}_i}{\partial\mathbf{q}_j}\cdot\delta\mathbf{q}_j - \zeta\delta\mathbf{p}_i \end{aligned} \tag{27}$$

However, this linearized equation is not complete as it is. It does not explicitly involve the stochastic force $\bar{\mathbf{A}}_i$, although Eq. (26) indicates that $\dot{Q}$, defined in (25), determines the sum of the exponents. The first term in the motion equation (27) for $\delta\mathbf{p}_i$ does not make any contributions to the Lyapunov-exponent sum, and the second term containing the constant ζ is responsible for a uniform shift of the spectrum to more negative values. Taken by themselves, the linearized equations (27) would give a Lyapunov spectrum whose sum of exponents is $-dN\zeta$, which is strictly negative *also for a vanishing field.* This is clearly a nonsensical result. Strictly speaking, the original motion equations (22) are not autonomous because of the periodically changed random force, and the $\bar{\mathbf{A}}_i$ contribute although they are constant during a time step.

To find the missing term we note that an infinitesimal phase-space volume element may be written as $\delta V = \prod_{i,\alpha}\delta q_{i,\alpha}\delta p_{i,\alpha}$; $\alpha = 1,\ldots,d$; $i = 1,\ldots,N$, where $\delta q_{i,\alpha}, \delta p_{i,\alpha}$ are the tangent-vector components associated with a given Lyapunov exponent. If we follow Eq. (26) and equate $d\ln\delta V/dt$ with $\dot{Q}/kT$ obtained from (25), we may identify the α-component contribution of particle i according to

$$\frac{\delta\dot{p}_{i,\alpha}}{\delta p_{i,\alpha}} = \frac{1}{mkT}(-\zeta p_{i,\alpha}^2 + p_{i,\alpha}\bar{A}_{i,\alpha}). \tag{28}$$

The first term involving ζ is a consequence of the analogous term in the motion equation (22) for the reference trajectory. If we replace $2K$ in (25) by its time average $2\langle K\rangle = dNkT$, as is appropriate for large systems, this term simply reduces to $-\zeta$ as in Eq. (27). But since we include the fluctuations, which may be significant for small systems, this term is slightly modified as indicated in (28). The second term containing $\bar{A}_{i,\alpha}$ is the missing contribution and must be added to Eq. (27). The complete linearized equations of motions finally become

$$\begin{aligned} \delta\dot{q}_{i,\alpha} &= \delta p_{i,\alpha}/m, \\ \delta\dot{p}_{i,\alpha} &= \sum_{j=1}^{N}\sum_{\beta=1}^{d}\frac{\partial F_{i,\alpha}}{\partial q_{j,\beta}}\delta q_{j,\beta} - \frac{N}{\sum_{i=1}^{N} p_{i,\alpha}^2}(p_{i,\alpha}^2\zeta - p_{i,\alpha}\bar{A}_{i,\alpha})\delta p_{i,\alpha}, \end{aligned} \tag{29}$$

E	ζ	T	$\langle K \rangle$	$\langle \Phi \rangle$	λ_1	$\sum_l \lambda_l$	$\langle \dot{Q} \rangle$	$\langle \dot{W} \rangle$	$\langle J \rangle$	κ	ΔD_1
0	0.5	1.0	64.1	1.72	2.02	0.00	0.01	0.00	0.01	-	0.00
0.125	0.5	1.0	64.0	1.74	1.98	-0.79	0.77	0.76	6.10	0.76	0.39
0.25	0.5	1.04	66.6	1.82	2.00	-2.95	3.10	3.09	12.4	0.77	1.5
0.5	0.5	1.19	76.1	2.17	2.01	-9.94	12.3	12.3	24.7	0.77	5.0
1	0.5	1.73	110.6	3.43	2.19	-25.7	47.4	47.4	47.4	0.74	12.1
1.0	1.0	1.25	80.2	2.44	2.24	-24.8	33.9	33.9	33.9	0.53	11.5

TABLE 3. Simulation results for 64 Brownian disks in two dimensions immersed in a bath with constant friction ζ and accelerated by an external color field E. The particle density of the disks is $\rho = 0.2$. The random stochastic force is constant for timesteps $\Delta t = 0.01$ and is sampled from a Gaussian distribution with a bath temperature $T_0 = 1$. $\langle J \rangle = \langle |\mathbf{J}| \rangle$ is the time-averaged magnitude of the color current. All other quantities are defined in the main text. They are given in the reduced units for which the potential parameters ϵ and σ, the Brownian particle mass m, the color charge c, and the Boltzmann constant k are unity.

where we have further replaced kT in (28) by $\sum_{i=1}^N p_{i,\alpha}^2/Nm$, twice the kinetic energies per particle associated with the various components α. In practical applications they may be different parallel or perpendicular to the field, far from equilibrium. These equations of motion are our final results and replace (27).

We have used (29) to simulate systems containing 64 soft disks in two dimension, $d = 2$, and interacting with the potential

$$\phi_{i,j} = 100\epsilon \left(1 - \left(\frac{r_{ij}}{\sigma}\right)^2\right)^4, \tag{30}$$

where $r_{ij} = |\mathbf{r}_j - \mathbf{r}_i|$ is the interparticle distance. The simulation box is square, and periodic boundaries are employed. Reduced units are used for which the potential parameters ϵ and σ, the mass m, and the charge c are unity. As usual, the temperature is measured in units for which the Boltzmann constant k is unity as well. All numbers in Fig. 4 and in the Table 3 are given in these units. The number density of the Brownian particles is $\rho = 0.2\sigma^{-2}$ for all cases. The strength of the field $|\mathbf{E}|$ is denoted by E, and its units are $\epsilon/(\sigma c)$. The Lyapunov exponents λ_l are given in units of $(\epsilon/(m\sigma^2))^{1/2}$. Relevant simulation results are collected in Table 3.

In Fig. 4 the Lyapunov spectra for equilibrium ($E = 0$) and for $E = 1$ are shown. The dashed line near the abscissa is the arithmetic mean of the positive and negative branches of the spectra. Clearly, in equilibrium the spectrum is symmetric around the abscissa, and the sum of exponents

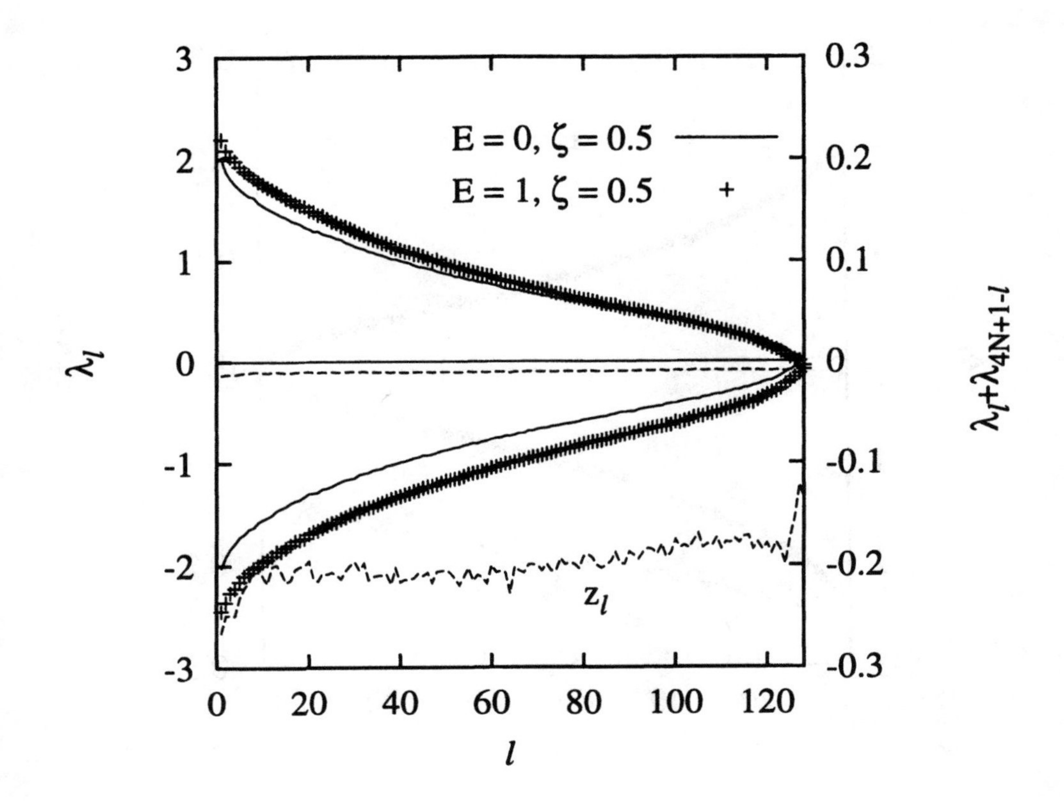

Figure 4. Lyapunov spectra for 64 Brownian soft disks of unit mass m and a particle density $\rho = 0.2\sigma^{-2}$ immersed in a bath with friction coefficient $\zeta = 0.5$. E is the strength of the applied field. The arithmetic mean of conjugate Lyapunov exponents for $E = 1$ is shown by the dashed line parallel to the abscissa. Twice that value, $z_l \equiv \lambda_l + \lambda_{4N-l+1}$, is also shown on an expanded vertical scale on the right-hand side. The Lyapunov exponents are given in units of $(\epsilon/(m\sigma^2))^{1/2}$, and fields in units of $\epsilon/(\sigma c)$. The Lyapunov spectra are only defined for integer values of the index l.

vanishes, as expected. The temperature of the particles in the stationary nonequilibrium state is significantly higher. The sum of conjugate Lyapunov exponents, $z_l \equiv \lambda_l + \lambda_{4n-l+1}$ is not constant nor independent of l, as is shown for $E = 1$ by the dashed curve labeled z_l which refers to the expanded scale on the right-hand side of Fig. 4. The same result is also found for a larger friction $\zeta = 1$ in Fig. 5. Thus, the conjugate pairing rule [26, 27] is not obeyed, which would require that $z_l = -\zeta$, independent of l. This is a consequence of the term containing $\mathbf{A}_i$ on the right-hand side of Eq. (25), which significantly reduces the net heat flow to the bath. Thus, the rule

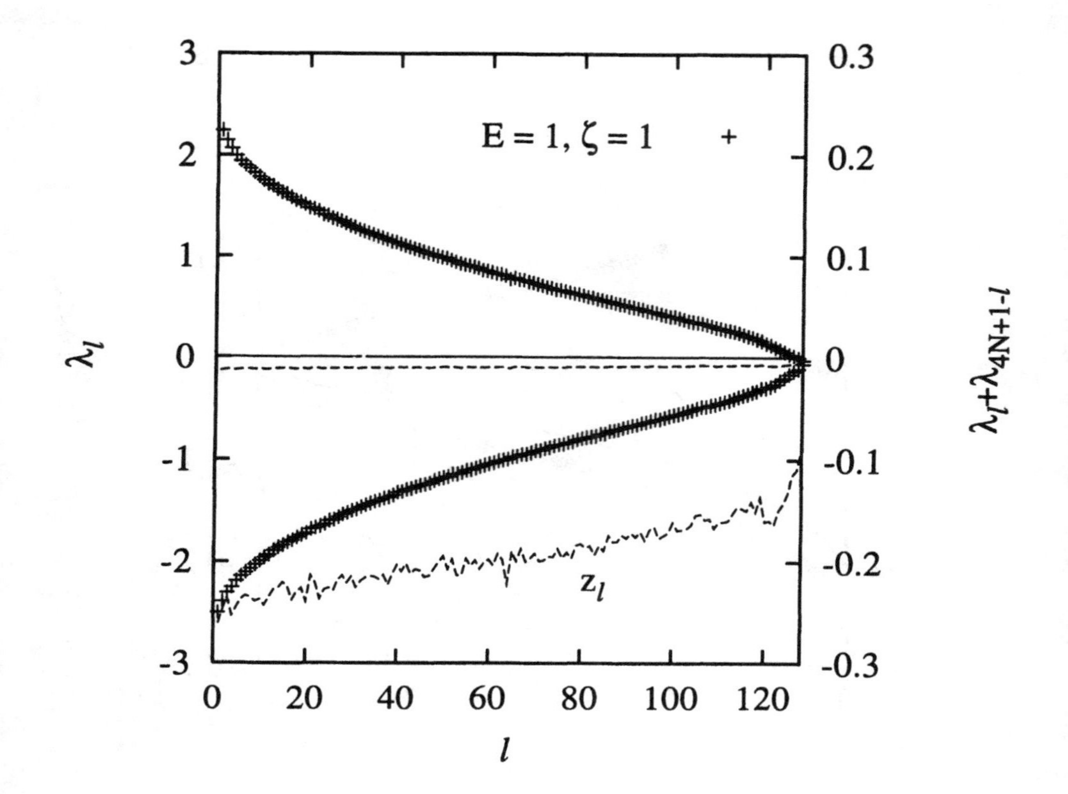

Figure 5. Lyapunov spectrum for 64 Brownian soft disks of unit mass m and a particle density $\rho = 0.2\sigma^{-2}$ immersed in a bath with friction coefficient $\zeta = 1$ and field $E = 1$. The arithmetic mean of conjugate Lyapunov exponents for $E = 1$ is shown by the dashed line parallel to the abscissa. Twice that value, $z_l \equiv \lambda_l + \lambda_{4N-l+1}$, is also shown on an expanded vertical scale on the right-hand side. The Lyapunov exponents are given in units of $(\epsilon/(m\sigma^2))^{1/2}$, and fields in units of $\epsilon/(\sigma c)$. The Lyapunov spectra are only defined for integer values of the index l

needs to be generalized to be applicable to this case.

We list in Table 3 also the color conductivity $\kappa = (\langle \mathbf{J} \rangle \cdot \mathbf{E})/NE^2$ and the time-averaged magnitude of the color current, $\langle J \rangle$. The latter is found to obey the relation $\langle J \rangle = \langle \dot{Q} \rangle / E$, which follows from Eqs. (23–25). Just as with the deterministic thermostats discussed in Section 2, the negative sum of Lyapunov exponents is an indication of the presence of a multifractal strange attractor in the phase space of our system for stationary nonequilibrium states. From the Kaplan-Yorke formula [17] we obtain the information dimension D_1 and the reduction in dimensionality $\Delta D_1 = 4N - D_1$, which is also listed in Table 3. ΔD_1 does not increase with the square of the field as is the case for the deterministic thermostats, but for a trivial reason: the temperature of the Brownian particles is not constant.

The Brownian particle system discussed in this Section is related to a model investigated recently by one of us [33]. This model corresponds to the limit that the bath is at a very low temperature $T_0 = 0$, and that there is only a friction force $-\zeta \mathbf{p}_i$ but no stochastic force $\bar{\mathbf{A}}_i$ acting on a particle. As a consequence, the system freezes in equilibrium. In the nonequilibrium steady state and for not too-large densities, the system phase separates into two counterpropagating jets of particles moving without collision and with constant speed $p_i/m = c_i E/\zeta$. Our system studied in Section 4 removes some of this artificial behavior through the introduction of the stochastic force.

5. Conclusions

The multifractal nature of the phase-space distribution and the reduction of the information dimension for stationary nonequilibrium systems with time-reversible dynamic thermostats has been established numerically for many nonequilibrium flows [1, 34, 35, 10], and has been confirmed theoretically [36, 37]. We show here that this concept also applies to a qualitatively different class of models, for which the heat flow through the system is controlled by a stochastic map. It may still be related to the exponential rate of shrinkage of small phase volumes comoving with the flow. It is important to realize that in the models studied here, the stochastic map affects the reference trajectory and a perturbed trajectory alike by the same sequence of random numbers.

All our considerations apply to purely classical systems, and the concept of multifractal attractors in phase space with ever smaller scales and a diverging entropy seems to be unreconcilable with quantum mechanics [12]. Quantum mechanics, as formulated by Schrödinger or Feynman, is just not capable of following actual trajectories. It can only generate a linear combination of them which has no connection to the single-trajectory approach followed here. Attempts of introducing thermostats have been made [38], and random matrix theory seems to be a useful method for studying finite-temperature dynamics in complex quantum systems [39]. We do not know if any of these methods will be successful to study stationary nonequilibrium quantum systems. However, we believe that the classical approach, which so successfully links the appearance of multifractal structures with the validity of the Second Law of thermodynamics [1, 4] and which provides a connection between the Lyapunov instability of such systems with the respective transport properties, has a counterpart also for quantum mechanical systems. It remains for future research to uncover these relationships.

Acknowledgements

We thank Bob Dorfman, Tamas Tél, Henk van Beijeren, and Nico van Kampen for interesting discussions, and in particular Christoph Dellago for generating the data in Fig. 1 and Table 1. Support by the Austrian Fonds zur Förderung der wissenschaftlichen Forschung, grants P09677-PHY and P11428-PHY, is gratefully acknowledged. Work at the Lawrence Livermore National Laboratory was performed under the auspices of the University of California, through Department of Energy contract W-7405-eng-48, and was further supported by grants from the Advanced Scientific Computing Initiative, the Accelerated Strategic Computing Initiative, and the Department of Mechanical Engineering at LLNL.

References

1. B. L. Holian, W. G. Hoover, and H. A. Posch, Phys. Rev. Lett.**59**, 10 (1987).
2. W.G. Hoover, H.A. Posch, B.L. Holian, M.J. Gillan, M. Mareschal, and C. Massobrio, Molecular Simulations **1**, 79 (1987).
3. H.A. Posch, and W.G. Hoover, Phys. Rev. A **38**, 473 (1988).
4. H.A. Posch, Ch. Dellago, W.G. Hoover, and O. Kum, in *Pioneering Ideas for the Physical and Chemical Sciences: Josef Loschmidt's Contributions and Modern Developments in Structural Organic Chemistry, Atomistics, and Statistical Mechanics* W. Fleischhacker and T. Schönfeld, eds., (Plenum, New York, 1997).
5. Wm.G. Hoover, D.J. Evans, H.A. Posch, B.L. Holian, and G.P Morriss, Phys. Rev. Lett., **80**, 4103 (1998).
6. D.J. Evans, D.J. Searles, Wm.G. Hoover, C. G. Hoover, B.L. Holian, H.A. Posch, and G. P. Morriss, J. Chem. Phys., **108**, 4351 (1998).
7. D. J. Evans and G. P. Morriss, *Statistical Mechanics of Nonequilibrium Liquids*, (Academic Press, London, 1990).
8. W. G. Hoover, *Computational Statistical Mechanics* (Elsevier, Amsterdam, 1991).
9. H.A. Posch and W.G. Hoover, in *Molecular Liquids: New Perspectives in Physics and Chemistry*, José J.C.Teixeira-Dias, ed., NATO-ASI Series C: Mathematical & Physical Sciences, (Kluwer Academic Publishers, Dordrecht, 1992), p. 527.
10. Wm. G. Hoover, *Time Reversibility, Computer Simulation and Chaos*, (World Scientific, Singapore, 1999).
11. See the discussions in the Proceedings of the NATO Advanced Science Institute, *Microscopic Simulations of Complex Hydrodynamic Phenomena*, Physics B **292**, M. Mareschal and B.L.Holian, eds., (Plenum, New York, 1992).
12. Nico G. van Kampen, in *Proceedings of the XXI Solvay Conference on Dynamical Systems and Irreversibility*, Keihanna, November 1 - 5, 1998.
13. S. Goldstein, C. Kipnis, and N. Ianiro, J. Stat. Phys. **41**, 915 (1985).
14. See the discussions in the Proceedings of the NATO Advanced Science Institute, *Microscopic Simulations of Complex Hydrodynamic Phenomena*, Physics B **292**, M. Mareschal and B.L.Holian, eds., (Plenum, New York, 1992).
15. Wm. G. Hoover, and H. A. Posch, Phys. Lett. A **246**, 247 (1998).
16. H. A. Posch and Wm. G. Hoover, Phys. Rev. E, **58**, 4344 (1998).
17. E. Ott, *Chaos in Dynamical Systems*, (Cambridge University Press, Cambridge,

1993).
18. V. I. Oseledec, Trans. Moscow Math. Soc. **19**, 197 (1968).
19. J.-P. Eckmann and D. Ruelle, Rev. of Modern Phys. **57**, 617 (1985).
20. G. Benettin, L. Galgani, A. Giorgilli, and J.M. Strelcyn, Meccanica **15**, 9 (1980).
21. A. Wolf, J. B. Swift, H. L. Swinney, and J. A. Vastano, Physica D **16**, 285 (1985).
22. D.J. Evans and D. MacGowan, Phys. Rev. A **36**, 948 (1987).
23. Wm. G. Hoover, K. Boercker, and H. A. Posch, Phys. Rev. E **57**, 3911 (1998).
24. Ch. Dellago, H.A. Posch, and W.G.Hoover, Phys. Rev. E **53**, 1485 (1996).
25. Ch. Dellago and H.A. Posch, Physica A, **240**, 68 (1997).
26. D. J. Evans, E. G. D. Cohen, and G. P Morriss, Phys. Rev. A **42**, 5990 (1990).
27. S. Sarman, D. J. Evans, and G. P. Morriss, Phys. Rev. A **45**, 2233 (1992).
28. Ya. B. Pesin, Sov. Math. Dokl. **17**, 196 (1976).
29. C.W. Oseen, *Hydrodynamik* (Akademischer Verlag, Leipzig, 1927).
30. J. Riseman and J.G. Kirkwood, in *Rheology Theory and Applications* (Academic, New York,1956), Chap. 13, p. 495.
31. J.M. Deutch and I. Oppenheim, J. Chem. Phys. **54**, 3547 (1971).
32. S. Chandrasekhar, Rev. Mod. Phys. **15**, 1 (1943); reprinted in *Selected Papers on Noise and Stochastic Processes*, N. Wax, ed., (Dover Publications, New York, 1954).
33. Wm. G. Hoover, Phys. Lett. A **255**, 37 (1999).
34. W. G. Hoover, and B. Moran, Phys. Rev. A **40**, 5319 (1989).
35. H.A. Posch and W.G. Hoover, Phys. Rev. A **39**, 2175, (1989).
36. N. I. Chernov, G. L. Eyink, J. L. Lebowitz, and Ya. G. Sinai, Comm. Math. Phys. **154**, 569 (1993).
37. W. N. Vance, Phys. Rev. Lett. **69**, 1356 (1992).
38. J. Schnack, Physica A **259**, 49 (1998).
39. D. Kusnezov, Czechoslovak J. of Phys. **49**, 35 (1999).

A MODEL OF NON-EQUILIBRIUM STATISTICAL MECHANICS

J. PIASECKI
Institute of Theoretical Physics, University of Warsaw
Hoża 69, 00-681 Warsaw, Poland

AND

YA.G. SINAI
Department of Mathematics, Princeton University, Princeton, NJ 08544-1000, USA, and L.D. Landau Institute of Theoretical Physics, Moscow, Russia

1. Introduction

We consider a three-dimensional system composed of point particles of mass m enclosed in a cylinder of length $2L$. The cross section of the cylinder is a circle of area A. We choose the coordinate system whose X-axis coincides with the symmetry axis of the cylinder. The cylinder extends from $X = -L$ to $X = L$.

At time $t = 0$, the total volume $2LA$ is divided into two equal subvolumes LA by a circular piston with vanishing width, situated at the origin $X = 0$. The mass m of the piston is equal to the mass of the point particles filling the cylinder. The number of the particles to the left and to the right of the piston are N^- and N^+, respectively. The piston is assumed to move along the X-axis without friction. Its collisions with the gas particles are perfectly elastic, and consist in instantaneous exchanges of velocities. As the piston has no internal degrees of freedom it represents a mobile adiabatic wall.

Our object here is to study the dynamical evolution of the piston supposing that initially the particles to the left and to the right of it are at thermal equilibrium with inverse temperatures β^- and β^+, respectively. The collisions of the particles with the surface of the cylinder are assumed to be specular reflections. The present study is a contribution to the discussion of the adiabatic piston problem in the particular case where the

J. Karkheck (ed.),
Dynamics: Models and Kinetic Methods for Non-equilibrium Many Body Systems, 191–199.

mass of the piston is equal to the mass of the fluid particles (see [1] and references given therein).

Clearly, this problem is essentially one-dimensional. Indeed, only the projections of the particle velocities to the X-axis, perpendicular to the surface of the piston and to the lateral circular boundaries at $X = -L$ and $X = L$, are relevant. As the piston has one degree of freedom and can move only along the symmetry axis of the cylinder, the components of the velocities parallel to its surface do not influence its motion. We can thus take into account only the X-components of the velocities, and consider a purely one-dimensional motion.

In order to get an equivalent problem for a one-dimensional gas one has to keep in mind that the number densities $\rho^{\pm} = (N^{\pm}/AL)$ of the three-dimensional system when projected to the X-axis become $N^{\pm}/L$, and thus tend to infinity with N in the thermodynamic limit. Taking this remark into account it is then sufficient to assume that the binary collisions between the point particles moving on the line are perfectly elastic. As a consequence the particles just exchange their velocities at encounters and, from the point of view of the motion of the piston, one can as well suppose that there is no interaction between them at all. We recover in this way precisely the three-dimensional situation for point particles where the collisions with the piston and with the surface of the cylinder are separated by periods of free motion. This is why we concentrate in the sequel on the one-dimensional version of the problem. Its solution provides at the same time the solution for the three-dimensional case.

2. Adiabatic Piston in One Dimension

We consider here a one-dimensional ensemble of statistical mechanics consisting of classical point particles of equal mass m, situated within the interval $[-L, L]$. Initially, at time $t = 0$, the interval $[-L, 0)$ contains N^- particles, the interval $(0, L]$ contains N^+ particles, and a distinguished particle is at the origin $X = 0$. This distinguished particle will be called a piston.

The initial coordinates of the particles are $X_i(0)$, $-N^- \leq i \leq N^+$. They are numbered so that

$$-L \leq X_{-N^-}(0) \leq X_{-N^-+1}(0) \leq \ \dots \ \leq X_0(0) = 0 \leq X_1(0) \leq$$

$$\dots \leq X_{N^+}(0) \leq L.$$

In the sequel we shall also use the rescaled coordinates

$$x_i = X_i L^{-1}, \quad -N^- \leq i \leq N^+ \tag{1}$$

and also the rescaled time

$$\tau = tL^{-1}. \tag{2}$$

Notice that the free motion is invariant under this scaling.

The initial velocities of the particles are denoted by $v_i(0), \quad -N^- \leq i \leq N^+$. The phase spaces of the ensembles of particles in $[-L, 0)$ and in $(0, L]$, are M^- and M^+, respectively, where M^- consists of $Z^- = (\{X_i, v_i\}, \ -N^- \leq i < 0)$, while M^+ consists of $Z^+ = (\{X_i, v_i\}, \ 0 < i \leq N^+)$. We define $M = M^- \times M^+$.

The dynamics of the system is the free motion combined with elastic binary collisions between the particles, and elastic reflections from the walls at $\pm L$. The corresponding group of time translations is denoted by $\{T^t\}$.

We introduce the probability distribution

$$P = P^- \times P^+$$

defined on the Borel σ – algebra of subsets of M, where P^- and P^+ are probability distributions on the the Borel σ – algebras of M^+ and M^-, respectively. Here P^- is such that the conditional distribution of the velocities $\{v_i, \ -N^- \leq i < 0\}$ is the direct product of Gaussian distributions with the density

$$\phi(v; \beta^-(L)) = \sqrt{\frac{m\beta^-(L)}{2\pi}} \exp\left\{-\frac{m\beta^-(L)}{2} v^2\right\}, \tag{3}$$

i.e. it is the Maxwell distribution with the inverse temperature $\beta^-(L)$.

The distribution of the coordinates is the usual distribution of a classical ideal gas, i.e. it is the image of the direct product of uniform distributions on $(-L, 0)$ under the action of the group of permutations. We could equivalently consider the Poisson distribution with the parameter

$$\rho^-(L) = \frac{N^-}{L}. \tag{4}$$

In the same way one defines the distribution P^+ with parameters $\beta^+(L)$ and $\rho^+(L) = N^+/L$.

From the physical point of view the initial state described above is thus an equilibrium state of two volumes of the ideal gas separated by a piston fixed at $X = 0$. The temperatures and the densities in the two halves of the volume $2L$ are in principle arbitrary. The case of equal pressures

$$p^- = \frac{\rho^-}{\beta^-} = \frac{\rho^+}{\beta^+} = p^+ \tag{5}$$

is of particular interest, as it corresponds to the mutual compensation of macroscopic forces acting on the initially fixed piston. After the piston is released, the dynamics will thus reflect in this case the pure effect of fluctuations.

We shall study in the next section the evolution of the state of the system in the thermodynamic limit

$$L \to \infty, \qquad \frac{N^-}{L} = \rho^-(L) \to \rho^-, \quad \frac{N^+}{L} = \rho^+(L) \to \rho^+ \tag{6}$$
$$\beta^-(L) \to \beta^-, \quad \beta^+(L) \to \beta^+$$

taking exactly into account the effect of collisions of the particles with the boundaries at $\pm L$.

3. Asymptotic Equilibrium State

In this section we prove the following

THEOREM:

One can find a function $y(\tau)$, $0 \leq \tau < \infty$, such that

$$\lim_{\tau\to\infty} y(\tau) = y_\infty,$$

where y_∞ is the equilibrium position of the piston determined by the equation

$$\frac{\rho^-}{1+y_\infty} = \frac{\rho^+}{1-y_\infty}, \tag{7}$$

so that for any finite interval $[0, \mathcal{T}]$, and any $\epsilon > 0$

$$\lim_{L\to\infty} P\{\max_{0\leq\tau\leq\mathcal{T}} |x_0(\tau) - y_\tau| \leq \epsilon\} = 1. \tag{8}$$

This theorem shows that the rescaled trajectory of the piston approaches a deterministic function $y(\tau)$ of the rescaled time τ, and thus converges asymptotically to the position corresponding to a uniform distribution of particles in the interval $[-L, +L]$.

PROOF

The proof is based on a very useful notion of the trajectory of a given velocity which has been used in developing a rigorous description of the hard rod dynamics [2],[3]. Suppose that we have a particle starting from $(X_i(0), v_i(0))$. For some time it moves freely with velocity v_i. When it collides with another particle the instantaneous exchange of velocities takes place. However, if we consider the coordinate $X_{v_i}(t)$ such that at $X_{v_i}(t)$ there is a point mass whose velocity is v_i, then $X_{v_i}(t)$ moves uniformly over the whole interval $(-L, +L)$. If $v_i > 0$, it reaches L, gets reflected, and moves with velocity $-v_i$. In other words, the trajectory of each velocity is periodic with period $4L/|v_i|$.

Assume that the initial velocities v_i, $-N^- \leq i \leq N^+$ are such that $v_i \neq \pm v_j$ for any pair (i,j), $i \neq j$. Then the trajectory of each v_i is a periodic function of time with period $4L/|v_i|$. When $v_i > 0$, it covers $[X_i, L]$ with positive velocity v_i, then covers $[-L, +L]$ with the negative velocity $-v_i$, and then returns back to the initial point. The analogous behavior is true for negative velocities.

Fix some x, $-1 < x < 1$, and $\tau > 0$, and introduce the random variable $\xi(x,\tau)$ equal to the number of coordinates $X_i(\tau L)$ of the particles satisfying the condition

$$X_i(\tau L) \leq xL. \tag{9}$$

We shall calculate

$$m(x,\tau) = \lim_{L\to\infty} \frac{1}{L}\mathrm{E}[\xi(x,\tau)] \tag{10}$$

and the variance

$$\mathcal{D} = \lim_{\tau\to\infty} \mathrm{E}[\frac{1}{L}\xi(x,\tau) - m(x,\tau)]^2. \tag{11}$$

To calculate the expectation $m(x,\tau)$ assume that the coordinates $X_i(0)$ are given, and find the set of velocities $\Delta_\tau(X_i(0))$ such that if v belongs to $\Delta_\tau(X_i(0))$ then the trajectory of the velocity v at time $t = \tau L$ belongs to $[-L, xL]$. Since the trajectory of each v is a periodic function of time it can make several revolutions between the end points $-L$ and $+L$ before landing at some point within $[-L, xL]$. Denote by r the number of these revolutions. Then

$$\Delta_\tau(X_i(0)) = \bigcup_{r\geq 0} \Delta_\tau^r(X_i(0)),$$

where $\Delta_\tau^r(X_i(0))$ is the interval of velocities making r revolutions.

For positive v the boundaries of this interval V_r^-, V_r^+ for $r \geq 1$ can be found from the equations

$$\begin{aligned} -L - X_i(0) + 4rL - L - xL &= V_r^- t \\ -L - X_i(0) + 4rL - L - xL + 2(L + xL) &= V_r^+ t \end{aligned} \tag{12}$$

or

$$\begin{aligned} \frac{-2 - x_i(0) + 4r - x}{\tau} &= V_r^- \\ \frac{-x_i(0) + 4r + x}{\tau} &= V_r^+. \end{aligned} \tag{13}$$

It is readily checked that for $v < 0$ similar intervals are given again by equation (12), but now with $r \leq -1$. Finally, (V_0^-, V_0^+) are the extremities of the interval corresponding to the trajectory shorter than $4L$. From the formulae (13) we see that the length of $\Delta_\tau^r(X_i(0))$ equals

$$l[\Delta_\tau^r(X_i(0))] = \frac{2(1+x)}{\tau}, \tag{14}$$

and does not depend either on r or on the initial point $X_i(0)$. The same is true for the distance between the right end-point of $\Delta_\tau^r(X_i(0))$ and the left end-point of $\Delta_\tau^{r+1}(X_i(0))$: it is equal to $2(1-x)/\tau$.

Therefore, using equation (13) we find

$$\begin{aligned} m(x,\tau) &= \lim_{L\to\infty} \frac{1}{L} E[\xi(x,\tau)] \qquad (15) \\ &= \rho^- \int_{-1}^{0} dz \sum_{-\infty<r<+\infty} \int_{\Delta_\tau^r(zL)} dv \sqrt{\frac{m\beta^-}{2\pi}} \exp\left\{-\frac{m\beta^-}{2}v^2\right\} \\ &+ \rho^+ \int_{0}^{1} dz \sum_{-\infty<r<+\infty} \int_{\Delta_\tau^r(zL)} dv \sqrt{\frac{m\beta^+}{2\pi}} \exp\left\{-\frac{m\beta^+}{2}v^2\right\}, \end{aligned}$$

where

$$\int_{\Delta_\tau^r(zL)} dv \ldots \equiv \int_{(-2-z+4r-x)/\tau}^{(-z+4r+x)/\tau} dv \ldots \tag{16}$$

Clearly, the right hand side of (15) is a monotonically increasing function of x, in accordance with the definition of the stochastic variable ξ. In fact, $m(x,\tau)$ grows from $m(-1,\tau) = 0$ to the maximal value $m(1,\tau) = \rho^- + \rho^+$. It follows that there exists a point $y(\tau)$ such that

$$m(y(\tau),\tau) = \rho^-. \tag{17}$$

It can be readily checked that $y(0) = 0$. Indeed, the implicit definition (17) of $y(\tau)$ can be written in the form

$$\begin{aligned} \rho^- &= \rho^- \int_{-1}^{0} dz \sum_{-\infty<r<+\infty} \int_{(-2-z+4r-y)}^{(-z+4r+y)} dv \frac{1}{\tau}\phi\left(\frac{v}{\tau};\beta^-\right) \qquad (18) \\ &+ \rho^+ \int_{0}^{1} dz \sum_{-\infty<r<+\infty} \int_{(-2-z+4r-y)}^{(-z+4r+y)} dv \frac{1}{\tau}\phi\left(\frac{v}{\tau};\beta^+\right), \end{aligned}$$

where $\phi(v;\beta)$ denotes the Gaussian distribution (see (3)). As

$$\lim_{\tau\to 0} \frac{1}{\tau}\phi\left(\frac{v}{\tau};\beta^\pm\right) = \delta(v), \tag{19}$$

only the term with $r = 0$ survives in (18). So, the asymptotic form of (18) for $\tau \to 0$ reads

$$\begin{aligned} \rho^- &= \rho^- \int_{-1}^{0} dz \int_{(-2-z-y)}^{(-z+y)} dv \frac{1}{\tau} \phi\left(\frac{v}{\tau}; \beta^-\right) \\ &+ \rho^+ \int_{0}^{1} dz \int_{(-2-z-y)}^{(-z+y)} dv \frac{1}{\tau} \phi\left(\frac{v}{\tau}; \beta^+\right). \end{aligned} \tag{20}$$

When $\tau \to 0$, the terms on the right hand side of (20) contribute provided

$$z < y, \quad -(z+2) < y.$$

Taking this into account we find that equation (20) at $\tau = 0$ takes the form

$$\rho^- = [1 + y(0)\theta(-y(0))]\rho^- + y(0)\theta(y(0))\rho^+, \tag{21}$$

where θ is the unit step function. The unique solution of (21) is $y(0) = 0$, in accordance with the assumed initial state.

For $\tau > 0$ the point $y(\tau)$ continues to characterize the position of the piston. Indeed, at the initial moment the number of particles to the left of the piston is equal to N^-. This number is conserved by the dynamics. Therefore, the point at which $m(x, \tau)$ attains the value $\rho^- = \lim_{L\to\infty, N^-\to\infty} N^-/L$ does correspond to the position of the piston.

As $\tau \to \infty$, the length $2(1+x)/\tau$ of the intervals $\Delta_\tau^r(X_i(0))$ tends to zero. Also the length of the intervals separating Δ_τ^r from Δ_τ^{r+1} approaches zero as $2(1-x)/\tau$. Therefore, when $\tau \to \infty$

$$\sum_{-\infty<r<+\infty} \int_{\Delta_\tau^r(xL)} dv \sqrt{\frac{m\beta}{2\pi}} \exp\left\{-\frac{m\beta}{2} v^2\right\} \to \frac{1+x}{2}, \tag{22}$$

and the whole expression (15) converges to

$$(\rho^- + \rho^+)\frac{1+x}{2}. \tag{23}$$

Hence, $\lim_{\tau\to\infty} y(\tau) = y_\infty$ satisfies the equation

$$(\rho^- + \rho^+)\frac{1+y_\infty}{2} = \rho^- \tag{24}$$

or

$$\frac{\rho^-}{1+y_\infty} = \frac{\rho^+}{1-y_\infty}. \tag{25}$$

This gives the asymptotics of $y(\tau)$ as $\tau \to \infty$. The condition (25) reflects the fact that the final equilibrium state has a uniform distribution of particles with density $(\rho^- + \rho^+)/2$. The mechanical equilibrium (constant pressure) implies then also the uniformity of temperature, where by temperature we understand the mean kinetic energy. Clearly, the dynamics of the system conserves the fraction of particles moving with velocities $\pm v$. So, in the asymptotic uniform stationary state the velocity distribution becomes simply a weighted combination of the initial Maxwell distributions

$$\frac{\rho^-}{\rho^- + \rho^+}\phi(v;\beta^-) + \frac{\rho^+}{\rho^- + \rho^+}\phi(v;\beta^+).$$

Assuming that the stochastic variable $\xi(x,\tau)$ has the Poisson distribution with parameter $Lm(x,\tau)$ we find that the variance $\mathrm{E}[\xi(x,\tau) - Lm(x,\tau)]^2$ grows like $Lm(x,\tau)$. From this it follows by standard methods that, when $L \to \infty$,

$$P\left[\left|\frac{X_0(\tau L)}{L} - y(\tau)\right| \geq \epsilon\right] \to 0, \tag{26}$$

for any given τ.

4. Discussion

The asymptotic uniformity of the density, of the temperature and of the pressure is attained for any initial values of parameters $\rho^-, \rho^+, \beta^-, \beta^+$. In particular, the theorem proved in the previous section applies to the case where the initial pressures on both sides of the fixed piston are equal. So, before the piston is released the momentum fluxes on both sides compensate each other being equal to

$$p = \frac{\rho^-}{\beta-} = \rho^- \int_0^\infty dv\, v2mv\phi(v;\beta^-) = \rho^+ \int_{-\infty}^0 dv\, v2mv\phi(v;\beta^+) = \frac{\rho^+}{\beta+}. \tag{27}$$

According to the proved theorem, if at some moment the piston is released, the dynamical evolution will put it asymptotically in the position

$$y_\infty = \frac{\rho^- - \rho^+}{\rho^- + \rho^+} = \frac{\beta^- - \beta^+}{\beta^- + \beta^+}. \tag{28}$$

The piston will be thus displaced in the direction of higher temperature, despite the initial absence of macroscopic forces (a similar effect has been found for an infinite system within the linear Boltzmann theory ([4]). From the physical point of view this is due to the asymmetry of fluctuations which disappears only after the entirely uniform equlibrium state is attained.

It is interesting to investigate under what non-equilibrium conditions the time derivative of $y(\tau)$ vanishes at $\tau = 0$. To perform the calculation one can use again the asymptotic formula (20). Taking the time derivative of both sides of (20) and evaluating the limit $\tau \to 0$ yields the formula

$$y'(0) = \frac{dy}{d\tau}(\tau = 0) = \left\{ \frac{\rho^-}{m\beta^-}\phi[y'(0);\beta^-] - \frac{\rho^+}{m\beta^+}\phi[y'(0);\beta^+] \right\} D^{-1}, \quad (29)$$

where

$$D = \rho^- \int_{-\infty}^{0} dv\phi[v - y'(0);\beta^-] + \rho^+ \int_{0}^{\infty} dv\phi[v - y'(0);\beta^+].$$

Requiring the equality $y'(0) = 0$, we find the relation

$$\frac{\rho^-}{\sqrt{\beta^-}} = \frac{\rho^+}{\sqrt{\beta^+}}. \quad (30)$$

Equation (30) expresses the compensation of the initial particle fluxes arriving from the left- and the right-hand side of the piston

$$\int_{0}^{\infty} dv\, \rho^- v\phi(v;\beta^-) = \int_{-\infty}^{0} dv\, \rho^+ v\phi(v;\beta^+). \quad (31)$$

However, if the condition (30) is satisfied but the system is not uniform ($\rho^- \neq \rho^+$), the position of the piston will change in the course of time until it reaches the equlibrium value y_∞ given by (28). Only then the densities on both sides of the piston acquire the common value $(\rho^- + \rho^+)/2$, and all the derivatives of $y(\tau)$ vanish.

Acknowledgement

Discussions with J.L. Lebowitz are gratefully acknowledged.

References

1. Ch. Gruber, Europ. J. Phys. **20**, 259 (1999).
2. J.L. Lebowitz and J.K. Percus, Phys. Rev. **155**, 122 (1967).
3. D.W. Jepsen, J. Math. Phys. **6**, 405 (1965).
4. J. Piasecki and Ch. Gruber, Physica A **265**, 463 (1999).

STATISTICAL PROPERTIES OF CHAOTIC SYSTEMS IN HIGH DIMENSIONS

N. CHERNOV
Department of Mathematics
University of Alabama at Birmingham
Birmingham, AL 35294, USA

Abstract. Certain classical models, such as dispersing billiards, exhibit strong chaotic behavior but are highly nonlinear and contain singularities. It was a long standing conjecture that, due to singularities, the rate of the decay of correlations in such models is subexponential. Recently, L.-S. Young disproved this conjecture – she established an exponential decay of correlations for a periodic Lorentz gas with finite horizon. We extend this result to all the major classes of dispersing billiards. We also discuss many-particle interacting systems with exponential decay of correlations.

1. Introduction

Strong statistical properties – exponential decay of correlations (EDC) and central limit theorem (CLT) – for smooth uniformly hyperbolic dynamical systems, namely Anosov and Axiom A diffeomorphisms, have been proven by Ya. Sinai, D. Ruelle and R. Bowen in the seventies [23, 20, 2]. For piecewise smooth or nonuniformly hyperbolic systems, however, statistical properties are often weaker – the correlations decay slowly or the central limit theorem fails, and, in any case, these properties are very hard to prove.

We discuss systems with uniform hyperbolicity, i.e. such that one step expansion and contraction factors are bounded away from unity, but we do not require smoothness everywhere, i.e. allow singularities. Well studied and physically important systems of this kind are dispersing billiards, in particular the periodic Lorentz gas (a particle moving freely or in a small external field and bouncing off a periodic array of convex molecules, called scatterers). Another popular class of systems with singularities is that of attractors, e.g. Lorenz, Lozi and Belykh attractors [1].

J. Karkheck (ed.),
Dynamics: Models and Kinetic Methods for Non-equilibrium Many Body Systems, 201–214.

In billiards and other conservative systems there is a natural Liouville invariant measure, which is equivalent to the volume in the phase space. In many other systems such a measure is lacking. In dissipative systems, the nonequilibrium steady state is a singular measure, which has continuous distributions on unstable manifolds [14]. We call it Sinai-Ruelle-Bowen (SRB) measures.

Definition. A Sinai-Ruelle-Bowen (SRB) measure is an invariant measure for a hyperbolic dynamical system whose conditional distributions on unstable manifolds are absolutely continuous.

SRB measures are the only physically observable invariant measures for smooth or piecewise smooth hyperbolic dynamical systems. In the conservative case, the Liouville measure is an SRB measure automatically. In other cases, SRB measures are weak Cesaro limits of iterations of smooth measures on M. Furthermore, for any ergodic SRB measure μ there is a positive volume set consisting of μ-generic points, i.e. points $x \in M$ such that $\frac{1}{n}\sum_{i=0}^{n-1} f \circ T^i(x) \to \int f\,d\mu$ for all continuous functions $f : M \to \mathbb{R}$, see, e.g., [24]. (This property is sometimes taken as the definition of SRB measures.) We will only consider hyperbolic systems with their SRB measures.

L. Bunimovich and Sinai [3, 4] developed techniques to study the statistical properties of planar Lorentz gases with finite horizon (uniformly bounded free path). They constructed Markov partitions, which were necessarily countable. The latter made it impossible to effectively use the regular Perron-Frobenius operator, due to its non-compactness. Bunimovich and Sinai instead approximated the dynamics by a countable Markov chain and proved an analogue of the Doeblin condition of probability theory. In this way they obtained a CLT (with the weak invariance principle) and a subexponential (stretched exponential) bound on correlations.

These techniques and results were improved and extended to other planar dispersing billiards in 1991, see [7]. There, the authors abandoned Markov partitions due to their ineffectiveness, and instead used a family of finite Markov chains that approximated the dynamics. In this way the authors improved a bound on correlation functions, but it was still subexponential. Later the author obtained similar statistical properties for multidimensional periodic Lorentz gases with finite horizon in 1994, see [9]. Again, the bound on correlation function was stretched exponential rather than exponential.

There was much discussion in the physics and mathematics communities in the eighties about the actual rate of the decay of correlations in dispersing billiards, whether it was truly exponential or slower. Numerical experiments produced inconclusive or contradictory estimates, see a recent discussion and further references in [16]. It seems now that in physical models, there are two rates of the decay of correlations: one, initial, lasts a few dozen of iterations and is slow, and the real rate of decay is seen after 20-50 iterations, and is much faster. Early numerical experiments done on slower computers could only reveal the initial decay, while later experiments on faster computers showed the two-rate picture.

In the early nineties, the 'exponential' point of view got the upper hand in the theoretical area, too. In 1992, the EDC was established for piecewise linear hyperbolic 2-D toral automorphisms by the author [8]. In 1994, Liverani [17] established the EDC for 2-D piecewise smooth area-preserving uniformly hyperbolic systems. The singularities in the above papers were artificially very 'nice' – they consisted of a finite number of smooth curves on which the dynamics was discontinuous but had one-sided derivatives (in particular, the derivatives were uniformly bounded). In addition, the singularity curves were transversal to unstable vectors (this assumption was recently relaxed by H. van den Bedem, though.) Those classes did not cover billiards, where derivatives are always unbounded. Still the above results showed in principle that singularities did not necessarily slow down the decay of correlations. While Chernov's and Liverani's techniques were completely different, both arrived at the same exponential asymptotics for correlations. At this point, attempts to cover billiard systems were unsuccessful.

A breakthrough occurred in 1996 when Young [24] proved the EDC for quite generic hyperbolic systems. Young used a tower construction: she fixed a hyperbolic product structure (basically, a rectangle in Bowen's sense) and considered its iterations until parts of that rectangle returned to itself in a regular (Markov) way. She proved that if the probability of long returns decayed exponentially then the correlations also decayed exponentially. Her proof of exponential decay of correlations was based on the Perron-Frobenius operator, after she managed to compactify it by introduction of a special norm in the function space. Young also verified the exponential tail bound on return times for two classes of systems. One was that of 2-D piecewise smooth uniformly hyperbolic maps of Liverani type, without assuming the preservation of areas. The other consisted of a planar periodic Lorentz gas with finite horizon. Thus, Young established, for the first time ever, an exponential bound on correlations for a billiard model.

The author recently extended Young's result to other classes of hyperbolic systems with singularities, including billiards. These results and other possible extension to chaotic models of statistical mechanics are discussed in this note.

2. Statement of the General Result

Let M be an open subset in a d-dimensional C^∞ Riemannian manifold, such that $\bar{M}$ is compact (the sets M and $\bar{M}$ are not necessarily connected), and let $\Gamma \subset \bar{M}$ be a closed subset. We consider a map $T : M \setminus \Gamma \to M$, which is a C^2 diffeomorphism of $M \setminus \Gamma$ onto its image.

The set Γ will be referred to as the singularity set for T. For $n \geq 1$ denote by

$$\Gamma^{(n)} = \Gamma \cup T^{-1}\Gamma \cup \cdots \cup T^{-n+1}\Gamma \tag{1}$$

the singularity set for T^n. For any $\delta > 0$ denote by $\mathcal{U}_\delta$ the δ-neighborhood of the closed set $\Gamma \cup \partial M$.

Notation. We denote by ρ the Riemannian metric in M and by m the Lebesgue measure (volume) in M. For any submanifold $W \subset M$ we denote by ρ_W the metric on W induced by the Riemannian metric in M, by m_W the Lebesgue measure on W generated by ρ_W, and by $\mathrm{diam} W$ the diameter of W in the ρ_W metric.

Below we formulate all our assumptions.

Hyperbolicity. We assume that T is fully and uniformly hyperbolic, i.e. there exist two families of cones C^u_x and C^s_x in the tangent spaces $\mathcal{T}_x M$, $x \in \bar{M}$, such that $DT(C^u_x) \subset C^u_{Tx}$ and $DT(C^s_x) \supset C^s_{Tx}$ whenever DT exists, and

$$|DT(v)| \geq \Lambda |v| \quad \forall v \in C^u_x$$
$$|DT^{-1}(v)| \geq \Lambda |v| \quad \forall v \in C^s_x$$

with some constant $\Lambda > 1$. These families of cones are continuous on $\bar{M}$, their axes have the same dimensions across the entire $\bar{M}$, and the angles between C^u_x and C^s_x are bounded away from zero. Denote by d_u and d_s the dimensions of the axes of C^u_x and C^s_x, respectively. The full hyperbolicity here means that $d_u + d_s = \mathrm{dim} M$.

For any $x \in M$ we set

$$E^s_x = \cap_{n \geq 0} DT^{-n}(C^s_{T^n x}), \qquad E^u_y = \cap_{n \geq 0} DT^n(C^u_{T^{-n} y})$$

respectively. It is standard, see, e.g., [19], that the subspaces E^s_x, E^u_x are DT-invariant, depend on x continuously, $\mathrm{dim} E^{u,s}_x = d_{u,s}$, and $E^s_x \oplus E^u_x = \mathcal{T}_x M$ for $x \in M$.

As a consequence, there can be no zero Lyapunov exponents on M. The space E^u_x is spanned by all vectors with positive Lyapunov exponents, and E^s_x by those with negative Lyapunov exponents.

We call a submanifold $W^u \subset M$ a local unstable manifold (LUM), if T^{-n} is defined and smooth on W^u for all $n \geq 0$, and $\forall x, y \in W^u$ we have $\rho(T^{-n}x, T^{-n}y) \to 0$ as $n \to \infty$ exponentially fast. Similarly, local stable manifolds (LSM), W^s, are defined. Obviously, $\mathrm{dim} W^{u,s} = d_{u,s}$. We denote by $W^u(x)$, $W^s(x)$ local unstable and stable manifolds containing x, respectively.

Since we study SRB measures, we will primarily work with LUM's, and for brevity we will denote them by just W, suppressing the superscript u. Denote by $J^u(x) = |\det(DT|E^u_x)|$ the jacobian of the map T restricted to $W(x)$ at x, i.e. the factor of the volume expansion on the LUM $W(x)$ at the point x.

We assume the following standard properties of unstable manifolds:

Bounded curvature. The sectional curvature of any LUM W is uniformly bounded by a constant $B \geq 0$.

Distorsion bounds. Let x, y be in one connected component of $W \setminus \Gamma^{(n-1)}$, denote it by V. Then

$$\log \prod_{i=0}^{n-1} \frac{J^u(T^i x)}{J^u(T^i y)} \leq \varphi\left(\rho_{T^n V}(T^n x, T^n y)\right), \tag{2}$$

where $\varphi(\cdot)$ is some function, independent of W, such that $\varphi(s) \to 0$ as $s \to 0$.

Absolute continuity. Let W_1, W_2 be two sufficiently small LUM's, such that any LSM W^s intersects each of W_1 and W_2 in at most one point. Let $W_1' = \{x \in W_1 : W^s(x) \cap W_2 \neq \emptyset\}$. Then we define a map $h : W_1' \to W_2$ by sliding along stable manifolds. This map is often called a holonomy map. We assume that it is absolutely continuous with respect to the Lebesgue measures m_{W_1} and m_{W_2}, and its jacobian (at any density point of W_1') is bounded, i.e.

$$1/C' \leq \frac{m_{W_2}(h(W_1'))}{m_{W_1}(W_1')} \leq C' \tag{3}$$

with some $C' = C'(T) > 0$.

Non-branching of unstable manifolds. LUM's are locally unique, i.e. for any two LUM's $W^1(x), W^2(x)$ we have $W^1(x) \cap B_\varepsilon(x) = W^2(x) \cap B_\varepsilon(x)$ for some $\varepsilon > 0$. Here $B_\varepsilon(x)$ is the ε-ball centered at x. Furthermore, let $\{W_n^1\}$ and $\{W_n^2\}$ be two sequences of LUM's that have a common limit point $x \in \bar{M}$, i.e. $\rho(x, W_n^i) \to 0$ as $n \to \infty$ for $i = 1, 2$. Assume also that $\exists \varepsilon > 0$ such that $\rho(x, \partial W_n^i) > \varepsilon$ for all $n \geq 1$ and $i = 1, 2$. Then $\rho_H(W_n^1 \cap B_\varepsilon(x), W_n^2 \cap B_\varepsilon(x)) \to 0$ as $n \to \infty$, where

$$\rho_H(A, B) = \max\{\sup_{x \in A} \rho(x, B), \sup_{y \in B} \rho(y, A)\}$$

is the Hausdorff distance between sets.

u-SRB measures. A unique probability measure ν_W, absolutely continuous with respect to the Lebesgue measure m_W, is defined on any LUM W

by the following equation:

$$\frac{\rho_W(x)}{\rho_W(y)} = \lim_{n\to\infty} \prod_{i=1}^{n} \frac{J^u(T^{-i}y)}{J^u(T^{-i}x)} \qquad \forall x, y \in W, \tag{4}$$

where $\rho_W(x) = d\nu_W/dm_W(x)$ is the density of ν_W with respect to m_W. The existence of the limit in (4) is guaranteed by (2). We call ν_W the u-SRB measure on W. Observe that u-SRB measures are conditionally invariant under T, i.e. for any submanifold $W_1 \subset TW$, the measure $T_*\nu_W|W_1$ (the image of ν_W under T conditioned on W_1) coincides with ν_{W_1}.

SRB measure. We assume that the map T preserves an ergodic Sinai-Bowen-Ruelle (SRB) measure μ. That is, there is an ergodic probability measure μ on M such that for μ-a.e. $x \in M$ a LUM $W(x)$ exists, and the conditional measure on $W(x)$ induced by μ is absolutely continuous with respect to $m_{W(x)}$. In fact, that conditional measure coincides with the u-SRB measure $\nu_{W(x)}$. See a discussion in the next section on the necessity of this assumption.

δ_0-LUM's. Let $\delta_0 > 0$. We call W a δ_0-LUM if it is a LUM and $\operatorname{diam} W \leq \delta_0$. For an open subset $V \subset W$ and $x \in V$ denote by $V(x)$ the connected component of V containing the point x. Let $n \geq 0$. We call an open subset $V \subset W$ a (δ_0, n)-subset if $V \cap \Gamma^{(n)} = \emptyset$ (i.e., the map T^n is defined on V) and $\operatorname{diam} T^n V(x) \leq \delta_0$ for every $x \in V$. Note that $T^n V$ is then a union of δ_0-LUM's. Define a function $r_{V,n}$ on V by

$$r_{V,n}(x) = \rho_{T^n V(x)}(T^n x, \partial T^n V(x)). \tag{5}$$

Note that $r_{V,n}(x)$ is the radius of the largest open ball in $T^n V(x)$ centered at $T^n x$. In particular, $r_{W,0}(x) = \rho_W(x, \partial W)$.

We now turn to the key assumptions on the growth of unstable manifolds that will ensure a fast decay of correlations.

One-step growth of unstable manifolds. We assume that there are constants $\alpha_0 \in (0, 1)$ and $\beta_0, D_0, \kappa, \sigma, \zeta > 0$ with the following property. For any sufficiently small $\delta_0, \delta > 0$ and any δ_0-LUM W there is an open $(\delta_0, 0)$-subset $V_\delta^0 \subset W \cap \mathcal{U}_\delta$ and an open $(\delta_0, 1)$-subset $V_\delta^1 \subset W \setminus \mathcal{U}_\delta$ (one of these may be empty) such that $m_W(W \setminus (V_\delta^0 \cup V_\delta^1)) = 0$ and $\forall \varepsilon > 0$

$$m_W(r_{V_\delta^1,1} < \varepsilon) \leq \alpha_0 \Lambda \cdot m_W(r_{W,0} < \varepsilon/\Lambda) + \varepsilon \beta_0 \delta_0^{-1} m_W(W) \tag{6}$$

$$m_W(r_{V_\delta^0,0} < \varepsilon) \leq D_0 \delta^{-\kappa}\, m_W(r_{W,0} < \varepsilon) \tag{7}$$

and

$$m_W(V_\delta^0) \leq D_0\, m_W(r_{W,0} < \zeta\delta^\sigma). \tag{8}$$

We now state a general result, followed by the necessary definitions.

Theorem 2.1 *Let T satisfy the above assumptions. If the system (T^n, μ) is ergodic for all $n \geq 1$, then the map T has exponential decay of correlations (EDC) and satisfies the central limit theorem (CLT) for Hölder continuous functions on M.*

Let $\mathcal{H}_\eta$ be the class of Hölder continuous functions on M with exponent $\eta > 0$:

$$\mathcal{H}_\eta = \{f : M \to \mathbb{R} \,|\, \exists C > 0 : |f(x) - f(y)| \leq C\rho(x,y)^\eta, \ \forall x, y \in M\}.$$

Exponential decay of correlations. We say that (T, μ) has exponential decay of correlations for Hölder continuous functions if $\forall \eta > 0 \ \exists \gamma = \gamma(\eta) \in (0,1)$ such that $\forall f, g \in \mathcal{H}_\eta \ \exists C = C(f,g) > 0$ such that

$$\left| \int_M (f \circ T^n) g \, d\mu - \int_M f \, d\mu \int_M g \, d\mu \right| \leq C\gamma^{|n|} \qquad \forall n \in \mathbb{Z}.$$

Central limit theorem. We say that (T, μ) satisfies central limit theorem (CLT) for Hölder continuous functions if $\forall \eta > 0, f \in \mathcal{H}_\eta$, with $\int f \, d\mu = 0$, $\exists \sigma_f \geq 0$ such that

$$\frac{1}{\sqrt{n}} \sum_{i=0}^{n-1} f \circ T^i \xrightarrow{\text{distr}} \mathcal{N}(0, \sigma_f^2).$$

Furthermore, $\sigma_f = 0$ iff $f = g \circ T - g$ for some $g \in L^2(\mu)$

3. Outline of the Proof

The key point in the study of hyperbolic systems with singularities is to ensure that small unstable manifolds grow exponentially fast. Precisely, if W is an unstable manifold of 'size' $\varepsilon > 0$, then its image $T^n W$ has 'size' of order one for $n \approx \text{const} \cdot \ln \varepsilon^{-1}$. For uniformly hyperbolic systems without singularities this fact is an immediate consequence of the uniform expansion of W in all directions under T.

For hyperbolic systems with singularities, $T^n W$ is, in general, broken into a finite or countable number of unstable manifolds, some of them might be arbitrarily small. Hence, the 'size' of $T^n W$ must be measured 'on the average', and this is the one of the technical problems. It was successfully solved (for two-dimensional systems) by Bunimovich and Sinai [5], see also in [7, 24]. Note that in this case unstable manifolds are one-dimensional (curves) and the size of each can be simply identified with its length. There is another problem. To prevent the creation of arbitrarily many small pieces of $T^n W$, one needs to assume that $\exists K \geq 1$ and $\varepsilon_0 > 0$ such that no unstable fiber of length $< \varepsilon_0$ can be broken into more than K pieces by T. Furthermore, one has to assume that $K < \Lambda$, where $\Lambda > 1$ is the minimum expansion factor. (This often can be achieved by taking a higher iteration of T.) One then obtains that the measure of points $x \in W$ whose images $T^i x$, $1 \le i \le n$, are all in short components of $T^i W$ (of length $< \varepsilon_0$) is bounded by const$\cdot(K/\Lambda)^n$, i.e. it decreases exponentially with n, since $K/\Lambda < 1$. In other words, we can sort out large and small components of $T^n W$ and bound the total measure of the small ones.

The situation dramatically changes when $\dim W > 1$. In this case, even a single singularity manifold can cut some small unstable manifolds into infinitely many pieces. The above approach falls apart. In addition, it is not clear at all how to measure the 'size' of unstable manifolds, since they are not round. These troubles were so great that no successful approaches to the problem have been developed, until recently. We describe the approach used in [10, 11].

Z-function. This is the main technical tool for the bookkeeping of unstable manifolds in any dimension. Let W be a δ_0-LUM, $n \ge 0$, and $V \subset W$ an open (δ_0, n)-subset of W. We define the Z-function introduced in [10] by

$$Z[W,V,n] = \sup_{\varepsilon>0} \frac{m_W(x \in V : r_{V,n}(x) < \varepsilon)}{\varepsilon \cdot m_W(W)}. \tag{9}$$

The supremum here is not necessarily finite. It will be finite if the boundary $\partial T^n V$ is 'regular enough'. In particular, if $\partial T^n V$ is piecewise smooth (i.e., consists of a finite number of smooth compact submanifolds of dimension $\le d_u - 1$), then $Z[W,V,n] < \infty$, see e.g. [13]. In the case $m_W(W \setminus V) = 0$, the value of $Z[W,V,n]$ characterizes, in a certain way, the 'average size' of the components of $T^n V$ – the larger they are the smaller $Z[W,V,n]$. In particular, the value $Z[W,W,0]$ characterizes the size of W in the following way.

Examples. Let W be a ball of radius r, then $Z[W,W,0] \sim r^{-1}$. Let W be a cylinder whose base is a ball of radius r and height $h \gg r$, then again $Z[W,W,0] \sim r^{-1}$. Let W be a rectangular box with dimensions $l_1 \times l_2 \times \cdots \times l_{d_u}$, then $Z[W,W,0] \sim 1/\min\{l_1, \ldots, l_{d_u}\}$.

Notation. Let $\delta_{\max} > 0$ be so small that $\alpha := \alpha_0 e^{6\varphi(\delta_{\max})} < 1$. Denote also $\beta := \beta_0 e^{6\varphi(\delta_{\max})}$ and $D := D_0 e^{6\varphi(\delta_{\max})}$. We will always assume that $\delta_0 < \delta_{\max}$, so that $\alpha < 1$. Next, put

$$\bar\beta = 2\beta/(1-\alpha)$$

and

$$a = -(\ln \alpha)^{-1} \quad \text{and} \quad b = \max\{0, -\ln(\delta_0(1-\alpha)/\beta)/\ln \alpha\}.$$

We also put

$$\delta_1 = \delta_0/(2\bar\beta). \tag{10}$$

δ-Filtration. Let $\delta_0, \delta > 0$ and W be a δ_0-LUM. Two sequences of open subsets $W = W_0^1 \supset W_1^1 \supset W_2^1 \supset \cdots$ and $W_n^0 \subset W_n^1 \setminus W_{n+1}^1$, $n \geq 0$, are said to make a *δ-filtration* of W, denoted by $\{W_n^1, W_n^0\}$ if[1] $\forall n \geq 0$
(a) the sets W_n^1 and W_n^0 are (δ_0, n)-subsets of W;
(b) $m_W(W_n^1 \setminus (W_{n+1}^1 \cup W_n^0)) = 0$.
(c) $T^n W_{n+1}^1 \cap \mathcal{U}_{\delta\Lambda^{-n}} = \emptyset$ and $T^n W_n^0 \subset \mathcal{U}_{\delta\Lambda^{-n}}$.

We put $W_\infty^1 = \cap_{n\geq 0} W_n^1$. This set consists of points whose forward images stay far away from singularities. It is quite standard that for any $x \in W_\infty^1$ a stable manifold $W^s(x)$ exists and $\rho_{W^s(x)}(x, \partial W^s(x)) \geq \delta$, i.e. the stable manifold $W^s(x)$ contains a disk of radius δ centered at x. For a point $x \in W \setminus W_\infty^1$ we cannot secure a large enough stable manifold, because its forward trajectory comes too close to the singularity at least once, when $T^n x \in \mathcal{U}_{\delta\Lambda^{-n}}$. We say that $T^n x$ falls in the gap $\mathcal{U}_{\delta\Lambda^{-n}}$ created by the singularity.

Put also $w_n^1 = m_W(W_n^1)/m_W(W)$ and $w_n^0 = m_W(W_n^0)/m_W(W)$. Observe that $w_n^1 = 1 - w_0^0 - \cdots - w_{n-1}^0$ and $w_n^1 \searrow w_\infty^1 := m_W(W_\infty^1)/m_W(W)$ as $n \to \infty$.

Theorem 3.1 (n-step growth of unstable manifolds) *Let W be a δ_0-LUM and $\delta > 0$. Then there is a δ-filtration $(\{W_n^1\}, \{W_n^0\})$ of W such that*
(i) *$\forall n \geq 1$ and $\forall \varepsilon > 0$ we have*

$$m_W(r_{W_n^1,n} < \varepsilon) \leq (\alpha\Lambda)^n \cdot m_W(r_{W,0} < \varepsilon/\Lambda^n) + \tag{11}$$

$$\varepsilon\beta\delta_0^{-1}(1 + \alpha + \cdots + \alpha^{n-1})\, m_W(W).$$

Furthermore, $\forall n \geq 0$ and $\forall \varepsilon > 0$

$$m_W(r_{W_n^0,n} < \varepsilon) \leq D\delta^{-\kappa}\Lambda^{\kappa n}\, m_W(r_{W_n^1,n} < \varepsilon) \tag{12}$$

and

$$m_W(W_n^0) \leq D\, m_W(r_{W_n^1,n} < \zeta\delta^\sigma\Lambda^{-\sigma n}); \tag{13}$$

(ii) *we have $\forall n \geq 1$*

$$Z[W, W_n^1, n] \leq \alpha^n Z[W, W, 0] + \beta\delta_0^{-1}(1 + \alpha + \cdots + \alpha^{n-1}); \tag{14}$$

(iii) *$Z[W, W_n^1, n] \leq \bar{\beta}/\delta_0 = (2\delta_1)^{-1}$ for all $n \geq a \ln Z[W, W, 0] + b$;*
(iv) *for any $n \geq 0$ we have $Z[W, W_n^0, n] \leq D\delta^{-\kappa}\Lambda^{\kappa n} \cdot Z[W, W_n^1, n]$;*
(v) *for any $n \geq 0$ we have $w_n^0 \leq D\zeta\delta^\sigma\Lambda^{-\sigma n} \cdot Z[W, W_n^1, n]$.*

Proof of (11)-(13) goes by induction on n. The base of induction, $n = 1$ in (11) and $n = 0$ in (12)-(13) follow from our assumptions (6)-(8). The inductive transition from n to $n + 1$ is a direct calculation, with the help

[1] In [10], it was called a refined u-filtration.

of the distorsion bound (2). Lastly, the parts (ii)-(v) follow directly from (11)-(13), respectively, upon dividing by $m_W(W)$ and using (9).

Remark. Effectively, the theorem asserts that if an unstable manifold W is small or thin, so that $Z[W,W,0]$ is very large, then the connected components of $T^n W_n^1$ grow larger, on the average, so that $Z[W,W_n^1,n]$ decreases exponentially in n, until it becomes small enough, $\leq (2\delta_1)^{-1}$. This is our exact version of the basic principle that 'small unstable manifolds grow exponentially in size' in the context of high dimension. The parts (iv) and (v) of the theorem simply estimate the relative measure of the parts of $T^n W$ falling through the gaps around the singularities.

The rest of the proof of Theorem 2.1 goes more or less along standard lines of [24, 10, 11]. One deploys a finite number of rectangles in M such that any unstable manifold containing a disk of radius δ_0 crosses at least one of the rectangles. Then Young's tower is defined like this. One starts with a rectangle and iterates it until the unstable manifolds in its image get large (have inner radius $> \delta_0$). At this time there is a proper intersection with some other rectangle. The points landing in the intersection make successful return. The rest of the rectangle keeps going. After the removal of the returned part the remaining components get smaller, of course, and it takes time to grow them again. The alternating process of growth and removal of returned parts continues, so that almost every point of the original rectangle returns. The exponential tail bound on the return times is a result of calculations based on Theorem 3.1.

Remarks and discussion. Since we actually prove, under the conditions of Section 2, Young's exponential tail bound mentioned above, our conditions are more restrictive than Young's. On the other hand, our conditions are a little easier to check than Young's: in particular, they involve only one iteration of the map rather than all its positive iterations. We take full advantage of this simplification when applying our general theorem to billiards. The verification of our conditions for billiards is then not so hard a job, compared to the sophisticated analysis of billiard dynamics done in early papers [7, 9].

Theorem 2.1 obviously holds for functions that are only Hölder continuous on the connected components of the set $M \setminus \Gamma^{(m)}$ for some $m \geq 1$. Moreover, it can be naturally extended to a wider class of the so called piecewise Hölder continuous functions, as defined in [7, 9].

In applications, it is often enough to prove Theorem 2.1 for any power, T^m, of the map T:

Proposition 3.2 *Let $m \geq 2$. Assume that the map T^l is Hölder continuous (with some exponent $\eta_l > 0$) on every connected component of $M \setminus \Gamma^{(l)}$ for each $l = 1, \dots, m$. If T^m enjoys exponential decay of correlations, then so does T.*

Proof. Let $n \geq 1$, and $n = km + l$ with some $0 \leq l \leq m - 1$. Let $f, g \in \mathcal{H}_\eta$. Then

$$\begin{aligned}\int_M (f \circ T^n) g \, d\mu &= \int_M \left(f \circ T^n - f \circ T^{km}\right) g \, d\mu + \int_M (f \circ T^{km}) g \, d\mu \\ &= \int_M (h_l \circ T^{km}) g \, d\mu + \int_M (f \circ T^{km}) g \, d\mu, \qquad (15)\end{aligned}$$

where $h_l = f \circ T^l - f$. The function h_l is Hölder continuous (with exponent $\eta_l \eta > 0$) on each connected component of $M \setminus \Gamma^{(l)}$. Since l takes a finite number of values, both integrals in (15) are exponentially small in k. □

The assumption on the existence of an ergodic SRB measure μ does not seem to be necessary. Indeed, it can be often proved under various general assumptions similar to ours, see [19, 21, 24], and the proof is normally easier than that of statistical properties of μ. We intentionally left out this problem altogether, in order to focus on the EDC and CLT. Note, however, that the other assumptions in Section 2 do not logically imply the existence of SRB measures, as the following example shows.

Example. Let $R = \{(x, y) : 0 < x < 1, y > 1\}$ be an open strip in $\mathbb{R}^2$, and let $M' = \{(s, t) : 0 \leq s \leq 1, 0 \leq t \leq 1\}$ with the identification of $s = 0$ and $s = 1$ be a closed cylinder. Let $T_1 : R \to R$ be given by $(x, y) \to (x/3 + 1/3, 2y - 1)$ and $T_2 : R \to M'$ be defined by $s = y \pmod 1$ and $t = e^{-y} + x(e^{-y-1} - e^{-y})$. Then $M = T_2(R)$ is an open subset of M', and the map $T = T_2 \circ T_1 \circ T_2^{-1}$ takes M to M. It satisfies all the assumptions of Section 2 (other than the existence of an SRB measure), with $\Gamma = \emptyset$, but has no SRB measure.

4. Applications

Our main theorem can be applied to dispersing billiards and some other chaotic models of statistical mechanics.

Dispersing billiards. Let Q be a bounded connected domain in $\mathbb{R}^2$ or on a 2-D torus $\mathbf{T}^2$ with a piecewise C^3 smooth boundary ∂Q. If the boundary ∂Q is strictly concave inward at every point of smoothness, then the billiard in Q is said to be dispersing, or a Sinai billiard. A particularly important class of dispersing billiards is that with entirely smooth boundary (no corner points), i.e. such that ∂Q is a finite disjoint union of C^3 smooth simple closed curves. Such tables only exist on the 2-torus $\mathbf{T}^2$, and they are known in physics as periodic Lorentz gases.

Let $M = \partial Q \times [-\pi/2, \pi/2]$ be the standard cross-section of the billiard flow, $T : M \to M$ the first return map (billiard ball map), and $\tau(x) > 0$ the return time (the length of the free path till the next collision), see details in [7]. The coordinates on M are denoted by (r, φ), where $r \in \partial Q$ is the arc length parameter and $\varphi \in [-\pi/2, \pi/2]$ is the angle of reflection. The map T preserves the smooth measure $d\mu = c_\mu \cos\varphi \, dr \, d\varphi$, where c_μ is the normalizing constant. It is known that μ is an SRB measure, the system (T, μ) is ergodic, mixing, K-mixing and Bernoulli [22, 15].

Theorem 4.1 (see [11]) *For periodic Lorentz gases, the billiard ball map (T, μ) enjoys exponential decay of correlations.*

This theorem, under the additional assumption of finite horizon (bounded free path) was proved by Young [24], where she directly verified the exponential tail bound on return times. In [10], the author provided a shorter proof by verifying the assumptions of Section 2, and covered the gases without horizon as well.

Next, consider dispersing billiard tables $Q \subset \mathbb{R}^2$. They necessarily have corner points, i.e. intersections of smooth curves of ∂Q. We assume, as usual [6, 7], that all such intersections are transversal, i.e. the angle made by the sides of Q at each corner point is positive. This is widely believed to be a necessary assumption for exponential decay of correlation, because otherwise the decay seems to be polynomial [18].

In billiards with corner points, the uniform hyperbolicity in the sense of Sect. 2 is lost: during a series of reflections in the vicinity of a corner point the one-step expansion factors of unstable manifolds are not bounded away from unity. Fortunately, when all the intersections at corner points are transversal, the lengths of such series are uniformly bounded. Let m_0 denote the maximal number of reflections in a 'corner series'. Then the map T^{m_0} is uniformly hyperbolic.

We need one more assumption. For $m \geq 1$ denote by K_m the maximal number of smooth components of the singularity set for the map T^m that intersect or terminate at any one point of M. We assume that K_m does not grow too fast with m. Specifically, there is a large enough m such that

$$K_m < C \cdot \Lambda^{m/m_0} - 1, \tag{16}$$

where the constant $\Lambda > 1$ is the minimum expansion factor for the map T^{m_0} and $C = C(Q) > 0$ another constant specified in [11]. This is quite a standard assumption. Similar bounds have been assumed in the literature [6, 7, 17, 24]. The bound (16) is widely believed to hold for generic billiard tables [6], even though this is not proven yet.

Theorem 4.2 (see [11]) *For dispersing billiard tables with corner points, the billiard ball map (T, μ) enjoys exponential decay of correlations.*

The rest of this section is devoted to tentative applications. The work on some of them is currently under way.

Multidimensional Lorentz gas. Let B be a finite union of disjoint convex closed subsets of the torus $\mathbf{T}^d$, $d \geq 3$, with C^3 smooth boundary. The billiard in the region $Q = \mathbf{T}^d \setminus B$ is a dispersing one, it is known in physics as d-dimensional periodic Lorentz gas. We assume a finite horizon (bounded free path), not only for technical convenience, but also because otherwise the uniform hyperbolicity fails, see [9].

A stretched exponential bound on correlations was proved in [9]. A CLT and Donsker's invariance principle were proved as well. It is also conjectured in [9] that the actual rate of decay of correlations is exponential. It seems now possible to prove this based on the present results.

Lorentz gases in small external fields. Let the billiard particle obey the equations of motion

$$\dot{q} = p, \qquad \dot{p} = F, \tag{17}$$

where $F = F(q,p)$ is a small force. Upon reaching the boundary of the billiard table, the particle reflects elastically, just as in ordinary billiards.

If the dynamics in the original billiard table is uniformly hyperbolic, as in the Lorentz gas with finite horizon, it will be so under small perturbations. One has to ensure, in addition, that a smooth function is preserved by the dynamics (e.g., the speed of the particle) to prevent the creation of an additional dimension in the phase space. Various physically meaningful models allow this, e.g. thermostatted Lorentz gases.

It was proved in [12] for a thermostatted Lorentz gas in a small constant electrical field that an SRB measure exists and has strong statistical properties. Important physical applications then followed – Ohm's law, the Einsten relation and the Green-Kubo formula. It was all proved by using quite sophisticated Markov approximations, and the bound on the correlation function was subexponential.

It seems now possible to apply the general theorem 2.1. This will simplify mathematical arguments, produce a better bound on correlations (an exponential one), and extend these results to Lorentz gases in more general external fields, with possible interesting physical implications.

Many-particle systems. The above model is still low-dimensional – just one moving particle. Now consider N independent particles moving freely or in an external field and colliding with periodic obstacles. This is a direct product of N copies of the Lorentz gas dynamics. Thus, it is hyperbolic and has strong statistical properties. Now, allow small interaction between particles. Thus we make a perturbation, which presumably preserves hyperbolicity. It will be interesting to prove the existence of SRB measure and its strong statistical properties. This is a truly high-dimensional model, with many possible implications for statistical mechanics.

References

1. V.S. Afraimovich, N.I. Chernov, and E.A.Sataev, Chaos **5**, 238 (1995).
2. R. Bowen, Lect. Notes Math., **470**, (Springer-Verlag, Berlin, 1975).
3. L.A. Bunimovich and Ya.G. Sinai, Comm. Math. Phys. **73**, 247 (1980).
4. L.A. Bunimovich and Ya.G. Sinai, Comm. Math. Phys. **78**, 479 (1981).
5. L.A. Bunimovich and Ya.G. Sinai, Comm. Math. Phys. **107**, 357 (1986).
6. L.A. Bunimovich, Ya.G. Sinai, and N.I. Chernov, Russ. Math. Surv. **45**, 105 (1990).
7. L.A. Bunimovich, Ya.G. Sinai, and N.I. Chernov, Russ. Math. Surv. **46**, 47 (1991).
8. N.I. Chernov, J. Statist. Phys. **69**, 111 (1992).
9. N.I. Chernov, J. Statist. Phys. **74**, 11 (1994).
10. N.I. Chernov, Discr. Cont. Dynam. Syst. **5**, 425 (1999).
11. N.I. Chernov, J. Statist. Phys. **94**, 513 (1999).
12. N.I. Chernov, G.L. Eyink, J.L. Lebowitz, and Ya.G. Sinai, Commun. Math. Phys. **154**, 569 (1993).
13. H. Federer, Trans. Amer. Math. Soc. **93**, 418 (1959).
14. G. Gallavotti and E.G.D. Cohen, J. Statist. Physics **80**, 931 (1995).
15. G. Gallavotti and D. Ornstein, Comm. Math. Phys. **38**, 83 (1974).
16. P. Garrido and G. Gallavotti, J. Statist. Phys. **76**, 549 (1994).
17. C. Liverani, Annals of Math. **142**, 239 (1995).
18. J. Machta, J. Statist. Phys. **32**, 555 (1983).
19. Ya. Pesin, Ergod. Th. Dyn. Sys. **12**, 123 (1992).
20. D. Ruelle, Amer. J. Math. **98**, 619 (1976).
21. E.A. Sataev, Russ. Math. Surv. **47**, 191 (1992).
22. Ya.G. Sinai, Russ. Math. Surv. **25**, 137 (1970).
23. Ya.G. Sinai, Russ. Math. Surv. **27**, 21 (1972).
24. L.-S. Young, Ann. Math. **147**, 585 (1998).

DYNAMICAL SYSTEMS AND STATISTICAL MECHANICS: LYAPUNOV EXPONENTS AND TRANSPORT COEFFICIENTS

E. G. D. COHEN

Center for the Studies in Physics and Biology
The Rockefeller University
1230 York Avenue, New York, NY 10021, USA

Abstract. A brief survey is given of the derivation of an expression for the shear viscosity coefficient of a fluid in a nonequilibrium stationary state in terms of its maximal Lyapunov exponents. This is done by considering the fluid as a dynamical system in its multidimensional phase space. A numerical test of this expression, in which the Conjugate Pairing Rule is used, is presented.

1. Introduction

The main content of this lecture was published previously in reference [1]. In this paper I will sketch the main points of that paper and mention a few results obtained after this paper was written. For details, I refer to the original paper [2] and [1].

Perhaps one of the most striking connections between dynamical systems theory on the one hand, i.e., considering, e.g., a fluid as a large dynamical system consisting of $N >> 1$ particles and the statistical mechanics of such macroscopic systems, on the other hand, is a connection between the sum of the (dynamical) Lyapunov exponents of such a system in its $2dN$-dimensional phase space in a nonequilibrium stationary state (NESS) and its physical three-dimensional transport coefficients. I will restrict myself here to the shear viscosity η, arguably the most important transport coefficient.

J. Karkheck (ed.),
Dynamics: Models and Kinetic Methods for Non-equilibrium Many Body Systems, 215–220.

2. The Shear Viscosity

To make this connection precise is the main purpose of this paper. Let me begin with the definition of the viscosity, η, by the relation

$$F_{xy} = \eta \frac{\partial u_x(y)}{\partial y} = \eta\gamma. \tag{1}$$

Here F_{xy} is the shear stress between two adjacent layers, i.e., the force between these two layers per unit (layer) area. Or, alternatively, the average flow, per unit time and per unit area, of x-momentum in the y-direction in a fluid subject to a constant gradient in the y-direction of the flow velocity in the x-direction, u_x: $\frac{\partial u_x}{\partial y} = \gamma$ =constant, where γ is the shear rate (see Fig. 1). The η can depend on γ, i.e. exhibit a rheological behavior. The hydrodynamical Newtonian viscosity η_N is independent of γ and applies in the linear regime, where $F_{xy} \sim \gamma$ and $\eta = \eta_N$.

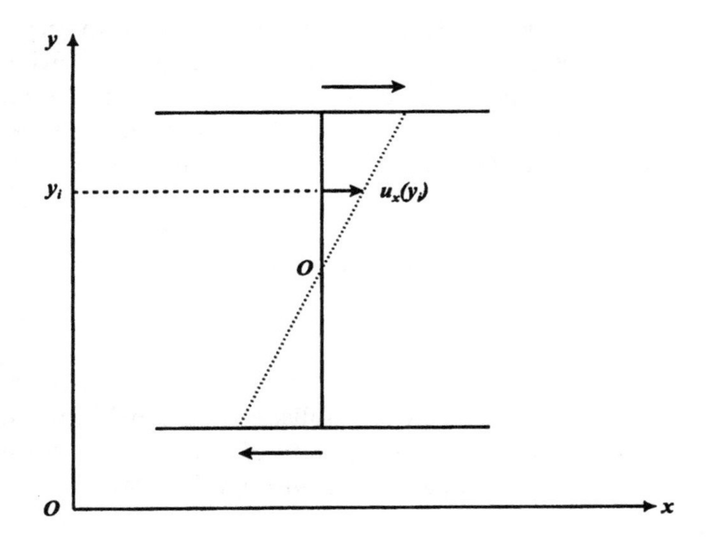

Figure 1. Linear velocity profile (dotted line) of the flow of a viscous fluid between two parallel plates, which move with equal but opposite velocities. The local fluid velocity at the position (x_i, y_i) of particle i is $u_x(y_i)$. The flux of x-momentum in the y-direction through unit area of a plane perpendicular to the y-axis is the shear stress F_{xy} between two fluid layers adjacent to the plane.

3. Dynamical Representation of a Shearing Fluid

Considering the shearing fluid as a dynamical system of N particles, one can describe this system dynamically by the so-called SLLOD equations of motion ($i = 1, ..., N$) [3]:

$$\begin{aligned} \dot{\mathbf{q}}_i &= \mathbf{p}_i + \mathbf{i}\gamma y_i \\ \dot{\mathbf{p}}_i &= \mathbf{F}_i - \mathbf{i}\gamma p_{y_i} - \alpha(q,p)\mathbf{p}_i. \end{aligned} \tag{2}$$

Here $\mathbf{i}$ is a unit vector in the x-direction, $y_i \equiv q_{y_i}$, $\mathbf{p}_i$ is the peculiar momentum of particle i [1,2], $\mathbf{F}_i$ the force on particle i due to the other $(N-1)$ particles, and $-\alpha(q,p)\mathbf{p}_i$ a "thermostatting" term which allows these equations of motion to describe a fluid in a NESS. This is done by determining $\alpha(q,p)$ (where q, p stand for the collection of all $\mathbf{q}_i, \mathbf{p}_i$ from $i = 1, ..., N$) in such a way that either the kinetic energy (isokinetic (IK) constraint) or the kinetic and potential energy (isoenergetic (IE) constraint) of the system remain constant during its time evolution. In particular, for an IE constraint, $\alpha_{IE}(q,p)$ can be determined using Ganss' principle of minimum constraint [3] to be

$$\alpha_{IE}(q,p) = -\frac{P_{xy}(q,p)V\gamma}{dNk_BT_{ss}}. \tag{3}$$

Here $P_{xy}(q,p)$ is the microscopic pressure tensor defined by

$$P_{xy}(q,p) = \sum_{i=1}^{N} [\frac{p_{ix}p_{iy}}{m} + F_{ix}y_i], \tag{4}$$

and dNk_BT_{ss} results from a denominator $\sum_{i=1}^{N} \frac{P_i^2}{m}$, twice the kinetic energy of the system, which for a macroscopic system, with $N >> 1$, can be set equal to dNk_BT_{ss}, where d is the dimensionality of (ordinary) space, k_B Boltzmann's constant, T_{ss} the kinetic temperature of the fluid in the NESS and V the volume of the system.

The equations of motion (2) and (3) have to be complemented by appropriate boundary conditions: the Lees-Edwards boundary conditions, to maintain a stationary viscous flow state in a finite volume [1, 3].

For the numerical simulations carried out for the fluid, described by the eqs. (2)–(4), a Weeks-Chandler- Andersen (WCA) interparticle potential was used, a purely repulsive finite range potential, obtained from a Lennard-Jones potential $\phi(r)$ by shifting it upwards over its maximum depth ϵ, so that its minimum touches the r-axis and cutting it off right there, i.e. at $r = 2^{1/6}\sigma$. Here σ is the molecular "diameter" associated with the Lennard-Jones potential ($\phi(r = \sigma) = 0$) and r the interparticle distance between two particles. The simulations were carried out in two dimensions, so that then $d = 2$ in eq. (3). Associated with the eqs. (2)–(4) are a set of $2dN$ Lyapunov exponents $\{\lambda_i\}$, with $i = 1, ..., 2dN$, of the dynamical system, i.e. of the fluid [3]. They characterize the exponential separation ($\lambda_i > 0$) or approach ($\lambda_i < 0$), in time, of two initially very nearby phase points.

In particular, in the NESS, one has for a macroscopic system the simple relation [1]

$$dN < \alpha_{IE}(q,p) >_{ss} = -\sum_{i=1}^{2dN} \lambda_i^{ss}, \tag{5}$$

where the average $< ... >_{ss}$ is a time average over the NESS and λ_i^{ss} is the i-th Lyapunov exponent in the NESS.

4. The Viscosity in Terms of the Lyapunov Exponents

The viscosity η of a fluid in a NESS can be expressed in terms of the Lyapunov exponents of the fluid in the NESS by using its definition in terms of the macroscopic, i.e. time averaged, pressure tensor

$$< P_{xy}(q,p) >_{ss} = -F_{xy} = -\eta(\gamma)\gamma, \tag{6}$$

where eq. (1) has been used.

Using then the eqs. (3), (5) and (6), one obtains for the viscosity the expression [1]

$$\eta_L(\gamma) = -\frac{k_B T_{ss}}{V\gamma^2} \sum_{i=1}^{2dN} \lambda_i^{ss}(\gamma). \tag{7}$$

Here the subscript L on the left hand side (l.h.s.) refers to an expression for η in terms of Lyapunov exponents.

5. Conjugate Pairing Rule

In practice it is very difficult to obtain the viscosity $\eta(\gamma)$ from the sum of all Lyapunov exponents on the right hand side (r.h.s.) of eq. (7) for a large system.

A great simplification is obtained by using a conjugate pairing rule (CPR) for the Lyapunov exponents, λ_i^{ss}, which reads

$$\lambda_i^{ss}(\gamma) + \lambda_{i'}^{ss}(\gamma) = - < \alpha(q,p) >_{ss}, \tag{8}$$

where i and i' are conjugate Lyapunov exponents:

$$\lambda_{max}^{ss}(\gamma) + \lambda_{min}^{ss}(\gamma) \equiv \lambda_1^{ss}(\gamma) + \lambda_{2dN}^{ss}(\gamma) = \lambda_2^{ss}(\gamma) + \lambda_{2dN-1}^{ss} = ...$$

Here the $\{\lambda_i^{ss}\}$ for $i = 1,...2dN$ determine the Lyapunov spectrum in the NESS, which, in general, consists of an (approximately) equal number of positive $(\lambda_1(\gamma), \lambda_2(\gamma), ...)$ and negative $(\lambda_{2dN}(\gamma), \lambda_{2dN-1}(\gamma), ...)$ Lyapunov exponents (see Fig. 13 in ref. [1]). The CPR has been proven for thermostatted Hamiltonian systems by Dettmann and Morriss for the IK constraint for all N [4], but appears to hold for the IE constraint for such systems only in the large system ($N >> 1$) limit [5]. Unfortunately, the thermostatted SLLOD equations (2) are not those for a thermostatted Hamiltonian system, since the eqs. (2) are not Hamiltonian for $\alpha = 0$. It seems that the CPR is obeyed in general with an accuracy of about 5% for the numerical simulations for $\eta(\gamma)$, to be described below. The most recent simulations point to a possible violation of the SLLOD equations (plus the Lees-Edwards boundary conditions) of about 7%, except for the sum of the two maximal Lyapunov exponents, which seem to satisfy (8) to less than 1% [6].

Using the CPR (8) in eq. (7), this equation reduces, for $d = 3$, to the simple form

$$\eta_L(\gamma) = -\frac{3nk_BT_{ss}}{\gamma^2}[\lambda_{max}^{ss}(\gamma) + \lambda_{min}^{ss}(\gamma)]. \tag{9}$$

Here the pair of maximal Lyapunov exponents, which is easiest to determine numerically, has been chosen to represent the sum on the r.h.s. of eq. (7) and n is the number density of the fluid.

In the linear regime, when $\gamma \to 0+$, the eq. (9) gives the Newtonian viscosity

$$\eta_N = \lim_{\gamma \to 0+} \eta_L(\gamma) = -\frac{3nk_BT_{ss}}{2} \lim_{\gamma \to 0+} \frac{\partial^2}{\partial \gamma^2}[\lambda_{max}^{ss}(\gamma) + \lambda_{min}^{ss}(\gamma)]. \tag{10}$$

In (10) one has used that on the r.h.s. of eq. (9), for $\gamma \to 0+$, the $\lambda^{ss}(\gamma)$ can be expanded in powers of γ and that the constant terms cancel. For, although not Hamiltonian, the phase-space volume of the adiabatic SLLOD equations ($\alpha = 0$) does not change [3]. In addition, the linear terms in γ must also cancel, since the viscosity must be invariant under a sign change of γ.

For the numerical determination of the λ_{max}^{ss} and λ_{min}^{ss} I refer to the literature [1,2].

6. Results

In order to test the Lyapunov expression (9) for $\eta_L(\gamma), \eta(\gamma)$ was also determined numerically directly from its definition (6) by computing numerically the ratio $- < P_{xy}(q,p) >_{ss} /\gamma$, using IE Non-Equilibrium Molecular Dynamics (NEMD) [2, 3]. The $\eta(\gamma)$ computed from eqs. (6) and (9) are compared in Fig. 2. The agreement of the two expressions for $\eta(\gamma)$ is within about 2% here.

As pointed out before, for large IE thermostatted Hamiltonian systems, CPR can be expected to apply only with $0(\frac{1}{N})$ corrections, if N is the number of particles in the system.

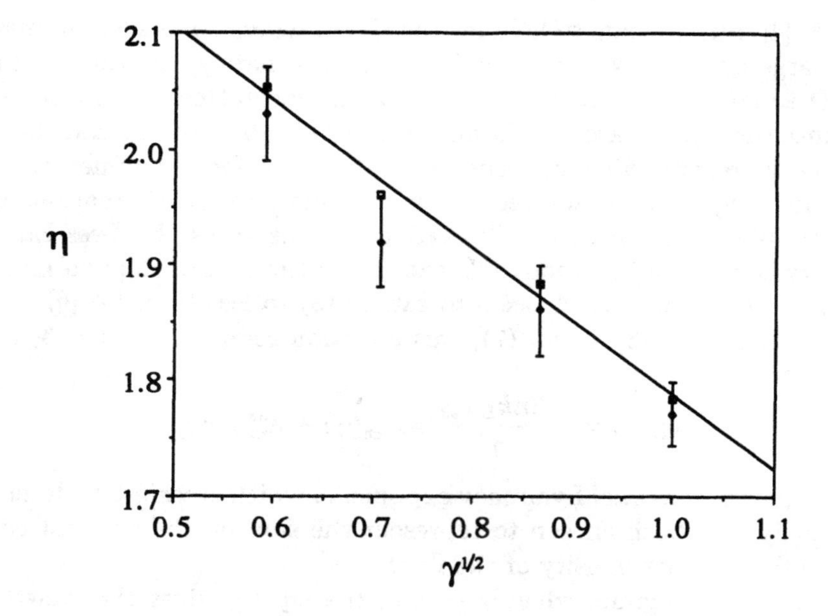

Figure 2. Comparison of Lyapunov $\eta_L(\gamma)$ (diamonds) and numerical $\eta(\gamma)$ (boxes) shear viscosities for a dense fluid, with reduced energy per particle $e = E/N\epsilon = 1.93019$, density $n = 0.8442$ and $N = 108$. Both sets of data are consistent with each other and with a $\gamma^{1/2}$ dependence, indicated by the straight line. A WCA interparticle potential is employed.

Acknowledgement

The author is indebted to D. J. Evans and D. J. Searles for providing up-to-date numerical data for the conjugate pairing rule for the viscosity and to H. Van Beijeren and J. R. Dorfman for a reading of the manuscript. He also gratefully acknowledges financial support by the Engineering Research Program of the Office of Basic Energy Sciences at the Department of Energy under Grant number DE-FG-02-88-13857.

References

1. E. G. D. Cohen, Physica A **213**, 293 (1995).
2. D. J. Evans, E. G. D. Cohen, and G. P. Morriss, Phys. Rev. A **42**, 5990 (1990).
3. D. J. Evans and G. P. Morriss, *Statistical Mechanics of Nonequilibrium Liquids*, (Academic Press, London, 1990).
4. C. Dettmann and G. P. Morriss, Phys. Rev. E **53**, 5545 (1996).
5. F. Bonetto, E. G. D. Cohen, and C. Pugh, J. Stat. Phys. **92**, 587 (1998).
6. D. J. Searles, D. J. Evans, and D. J. Isbister, Chaos **8**, 337 (1998).

SECOND INTERLUDE

BOLTZMANN AND STATISTICAL MECHANICS†

E. G. D. COHEN
Center for the Studies in Physics and Biology
The Rockefeller University
1230 York Avenue, New York, NY 10021, USA

"O! immodest mortal! Your destiny is the joy of watching the evershifting battle!"

Ludwig Boltzmann

1. Introduction

I received two invitations to this meeting: in the first unofficial one I was asked to speak "on the transport properties of dense gases". This involves the generalization of the Boltzmann equation to higher densities, a topic on which I have worked for more than 35 years. Later, I also received an official invitation, in which I was asked to give a lecture of "a generalized character". Although the first topic would be a natural and relatively easy one, since I have spoken on it often and thought about it a lot, the second one seemed much more difficult but irresistibly challenging, in allowing me to view Boltzmann's work in the last century from the perspective of the end of this century. This seems at first sight to be a precarious undertaking for a research scientist, but, as I hope to make clear to you, there may be advantages to this. While the historian of science is able to place the work of a scientist of the past in the context of that of his contemporaries, the research scientist can place the work of that scientist in the context of present day research and, up to a point, identify with his difficulties and achievements in the past on the basis of his own experience in the present day. I embark then on my perilous self-imposed task in the hope of providing some new perspectives on Boltzmann and his work, which are,

†The author and editor are very grateful to the Accademia dei Lincei in Rome for giving their permission to republish this lecture, which was first published in issue 131 of the Atti Dei Convegni Lincei. That issue contains this and all other lectures presented at an International Conference "Boltzmann's Legacy - 150 Years After His Birth", organized by the Accademia dei Lincei, 25-28 May, 1994, in Rome.

J. Karkheck (ed.),
Dynamics: Models and Kinetic Methods for Non-equilibrium Many Body Systems, 223–238.

I hope, historically not too inaccurate as far as the past is concerned, and stimulating, if not provocative, as far as the future is concerned.

2. Boltzmann, Mechanics and Statistics.

At the time Boltzmann began his career, Mechanics was the queen of theoretical physics, by far the most completely developed part of theoretical physics, the example as well as the ultimate goal for all other branches of theoretical physics. In the second half of the 19th century two major obstacles were to present themselves to this: the Second Law of Thermodynamics and Electromagnetism. Boltzmann's first attempt to "mechanize" the Second Law can be found already in his second paper, published at age 22 in 1866, entitled "On the Mechanical Meaning of the Second Law of the Theory of Heat"[1]. It is good to keep in mind that the Second Law consists of two parts; 1. a reversible and 2. an irreversible one. Part 1 introduces the existence of an integrating factor, the inverse absolute temperature $1/T$, for the heat dQ reversibly supplied to or removed from a system, such that $dQ/T = dS$, the (total) differential of the entropy S; part 2 states that the entropy of an isolated (adiabatic) system can never decrease. In section IV of Boltzmann's 1866 paper, "Proof of the Second Law of the Mechanical Theory of Heat", he mainly addresses the first aspect and is very cavalier about the second, the more difficult or, perhaps better, intractable one. He first deals with the case that heat is supplied to a system under the condition of equality of the inner and outer pressure of the system, and he shows that in that case $\oint dQ/T = \oint dS = 0$. He then argues that if this equality of pressures does not obtain, dQ must be smaller, so that in that case $\oint dQ/T < 0$. This is clearly at best a physical argument not a mechanical proof!

Two years later in a paper called: "Studies on the Equilibrium of the Kinetic Energy Between Moving Material Points" [2], he follows Maxwell in introducing probability concepts into his mechanical considerations and discusses a generalization of Maxwell's distribution function for point particles in free space to the very general case that "a number of material points move under the influence of forces for which a potential function exists. One has to find the probability that each one of them moves through a given volume with a given velocity and velocity direction". This is the first of many papers in which Boltzmann discusses and generalizes Maxwell's velocity distribution for point particles in free space to the case that external forces are present [2, 3] and to (polyatomic) molecules [4], leading to the Boltzmann factor and the Maxwell-Boltzmann distribution function [5].

Yet in 1872, when Boltzmann derived in his paper: "Further Studies on Thermal Equilibrium Between Gas Molecules" what we now call the

Boltzmann equation for the single particle position and velocity distribution function in a dilute gas [6] and used, following Clausius and Maxwell, what the Ehrenfests called the Stosszahl Ansatz, he does not seem to have fully realized the statistical nature of this assumption and therefore also of the ensuing H-theorem, for the approach to equilibrium. He says [7]: "One has therefore rigorously proved that, whatever the distribution of the kinetic energy at the initial time might have been, it will, after a very long time, always necessarily approach that found by Maxwell".

The beauty of the H-theorem was that it derived in one swoop both aspects of the Second Law: first the (irreversible) approach to thermal equilibrium and then, from the value of the H-function *in* equilibrium, the connection between the H-function and Clausius' entropy S: H = –const. S + const. Only later forced by Loschmidt's Reversibility Paradox [8] and Zermelo's Recurrence Paradox [9], as the Ehrenfests were to call them [10], did Boltzmann clearly state the probabilistic nature of the Stosszahl Ansatz, viz. that the Stosszahl Ansatz and the H-theorem only held for disordered states of the gas and that these states were much more probable than the ordered ones, since the number of the first far exceeded that of the second.

In the paper itself, however, this is never mentioned; it is as if the Ansatz was self evident. Therefore, Boltzmann did *not* derive here the Second Law purely from mechanics alone either and till the present day, no mechanical derivation of the Boltzmann equation exists, although the Stosszahl Ansatz must ultimately be derivable from the mechanics of a very large number N of particles, i.e., from "large N-dynamics".

I must admit that I find it difficult to assess Boltzmann's precise attitude towards the mixture of mechanics and statistics that a description of the behavior of macroscopic systems – gases mainly for him – necessitates. Uhlenbeck, a student of Ehrenfest's, who was himself a student of Boltzmann's, told me several times: "Boltzmann was sometimes confusing in his writings on the statistical aspects of his work and this, in part, prompted the Ehrenfests to write their clarifying and in a way definitive article to answer his opponents [10]".

The depth of ill-feelings generated by Boltzmann's exhausting discussions with his German colleagues, especially the Energeticists [11], and the resistance to his ideas, in particular with regard to the H-theorem that surrounded him, still resonated for me when Uhlenbeck said to me one day in some mixture of anger and indignation: "that damned Zermelo, a student of Planck's, nota bene"[12], an echo after two generations of past injustice and pain inflicted on Boltzmann by his hostile environment. Let me quote Boltzmann himself in his introduction to his response to Zermelo in 1896, for another aspect of his isolation and near desperation. After having explained that he has repeatedly and as clearly as possible emphasized in his publications that the Maxwell distribution function as well as the Second Law are of a statistical nature, he says [9a]: "Although the treatise by Mr. Zermelo "On a Dynamical Theorem and the Mechanical Theory of Heat" admittedly shows that my above mentioned papers have still not been understood, nevertheless I have to be pleased with his article as being the first proof that these papers have been noticed in Germany at all".

I want to cite a second indication of this solitude. Boltzmann wrote a letter to H. A. Lorentz in 1891, in response to a letter Lorentz sent to him, which pointed out for the second time an error in one of his papers [13]: "Already from the postmark and the handwriting I knew that the letter came from you and it pleased me. Of course, each letter from you implies that I have made an error; but I learn then always so much, that I would almost wish to make still more errors to receive even more letters from you". This quotation must be seen in the above mentioned context of Boltzmann's isolation in the German speaking countries, since in an earlier letter to Lorentz in 1886, in response to the above mentioned earlier error he made, he says [14]: "I am very pleased that I have found in you someone who works on the extension of my ideas about the theory of gases. In Germany, there is almost no one who understands this properly".

Probably motivated by the opposition of his contemporaries to his mechanical or kinetic method to prove the Second Law on the basis of the Boltzmann equation via the H-theorem, Boltzmann switched completely in 1877, when he introduced his statistical method, as the Ehrenfests called it, with no mechanical component in it at all, leading to the famous relation between entropy and probability: $S = k \log W$. In this 1877 paper [15], "On the Relation Between the Second Law of Thermodynamics and Probability Theory with Respect to the Laws of Thermal Equilibrium" he begins by saying [16]: "A relation between the Second Law and probability theory showed, as I proved [4c], that an analytical proof of it [the Second Law] is possible on no other basis than one taken from probability theory".

Boltzmann emphasizes this necessity of probability concepts for understanding the Second Law throughout the 1890's, when he mainly argued with his opponents over the interpretation of the H-theorem; his creative period had lasted about twenty years and one could ask to what extent this had been influenced by his difficulties with his contemporaries. To be sure, Boltzmann's explanations of the crucial points concerning the interplay between mechanics and probability, which were at the heart of Loschmidt's and Zermelo's objections to the H-theorem, although basically correct, did not capture all the subtleties of the necessary arguments and, together with the hostile Zeitgeist, made his efforts largely unsuccessful, certainly for himself. One must admit, though, that even today it is not easy to explain the paradoxes clearly, even to a sympathetic audience!

In particular, in his rebuttal of Zermelo's Recurrence Paradox [9] in 1896 and 1897 Boltzmann argued as clearly as he could, from a great variety of points of view, just as he had done twenty years earlier against Loschmidt's Reversibility Paradox [8]. It seems like a last vigorous attempt to show once and for all that there was really no conflict between mechanics and his kinetic theory. It was to no avail and this must have greatly depressed him. When Einstein in 1905 proved the existence of atoms by Brownian motion [17], it was far too late: Einstein was still unknown in 1905 and Boltzmann had probably given up long before then. I do not know whether Boltzmann ever read or heard of Einstein's paper, but if he did, although other causes undoubtedly played a role, it might – considering the state he must have been in – have contributed to, rather than prevented, his suicide in 1906. A systematic, critical and very structured account of Boltzmann's arguments was finally presented in 1909 - 1911 by the Ehrenfests' "Apologia" in their above mentioned Encyclopedia article [10].

It is ironic perhaps to note that Boltzmann's second approach, the statistical method, introducing what we now call Boltzmann statistics, has been in retrospect much more influential than the first, the kinetic method. This is in part because it has turned out that a meaningful generalization of Boltzmann's equation to higher densities, as well as obtaining concrete results from such an equation, have proved very difficult. Nevertheless, many new deeper insights into the behavior of dense nonequilibrium fluids have resulted from this work [18]. Boltzmann himself clearly preferred the kinetic method over the statistical method, because it was based on the dynamics, i.e., the collisions between the molecules of which the gas consists, and therefore allowed a direct connection with the motion of the particles. That statistics also came in was finally due to the presence of very many particles, but no substitute for the basic mechanical nature of the behavior of gases. As if he had a premonition of this future development, Boltzmann's summarizing "Lectures on Gas Theory" [19] are almost exclusively devoted to the kinetic method and hardly mention the statistical method at all.

Boltzmann never lost his predilection for mechanics. In his 1891-1893 lectures on Maxwell's theory of electromagnetism [20], he used, wherever he could, elaborate mechanical analogies, by endowing the ether with all kinds of intricate mechanical properties. This had been started by Maxwell himself in 1856 [21] but later Maxwell abandoned this approach in his presentations. Boltzmann was very well aware of his "old-fashioned" mechanical predilection. I quote what he said in a lecture "On Recent Developments of the Methods of Theoretical Physics" at a Naturforschung meeting in Munich in 1899 [22]. After having described the situation in theoretical physics as it existed at the beginning of his studies when [23] "the task of physics seemed to reduce itself, for the entire future, to determining the force between any two atoms and then integrating the equations that follow from all these interactions for the relevant initial conditions," he continues: "How everything has changed since then! Indeed, when I look back at all these developments and revolutions, I see myself as an old man in scientific experience. Yes, I could say, I am the only one left of those who still embrace the old wholeheartedly, at least I am the only one who still fights for it as much as he can ... I present myself therefore to you as a reactionary, a straggler, who adores the old, classical rather than the newer things...".

Although it was Boltzmann himself who introduced the idea of an ensemble of systems (he called it "Inbegriff" or "collection"), this (probabilistic) "trick", as he later called it, did not deter him from a deep mechanical point of view as well. The ensemble is first mentioned in the beginning of a paper in 1871 entitled [4b] "Some General Theorems on Thermal Equilibrium": "One has a very large number of systems of material points (similar to a gas which consists of very many molecules, each of which itself is again a system of material points). Let the state of any one of these point systems at any time t be determined by n variables $s_1 ... s_n$; .... we have only to assume that between the material points of the various point systems no interaction ever occurs. What one calls in the theory of gases the collisions of the molecules, will be excluded in the present investigation......The number of variables s that determines the state as well as the differential equations [of motion] should be the same for all systems. The initial values of the variables s and consequently the states at an arbitrary time t on the other hand should be different for the various point systems"[24]. It was a neat trick to compute the macroscopic observables of a gas as an average over many samples of the gas, each with different initial coordinates and momenta of the gas particles, assuming tacitly equal a priori probability of all the possible microstates of the gas with the same total energy. It was much easier than following the motion of all the particles in a given system in time and then taking a time average. Boltzmann used it to determine the equilibrium distribution function for a gas in thermal equilibrium. Curiously enough, Maxwell used the same idea independently in an 1879 paper entitled [25]: "On Boltzmann's Theorem on the Average Distribution of Energy in a System of Material Points". He says [26]: "I have found it convenient, instead of considering one system of material particles, to consider a large number of systems similar to each other in all respects, except in the initial circumstances of the motion, which are supposed to vary from system to system, the total energy being the same in all. In the statistical investigation of the motion, we confine our attention to the *number* of those systems which at a given time are in a phase such that the variables which define it lie within given limits". Maxwell does not mention Boltzmann here because he probably stopped reading Boltzmann after 1868, due to the -

for him - excruciating amount of detail in the latter's papers [27].

That Boltzmann had indeed not abandoned his hopes of giving a mechanical interpretation of the Second Law is borne out by his work on Helmholtz's monocycles, first published in 1884, in his paper [28]: "On the Properties of Monocyclic and Other Related Systems", which allowed a formal analogy between appropriate changes of these simple mechanical systems characterized by a single frequency [29] and those appearing in the First and the Second Law for reversible thermodynamic changes. He begins this paper as follows [30]: "The most complete mechanical proof of the Second Law would clearly consist in showing that for each arbitrary mechanical process equations exist that are analogous to those of the Theory of Heat. Since, however, on the one hand the Law does not *seem* [my italics] to be correct in this generality and on the other hand, because of our ignorance of the nature of the *so-called* [my italics] atoms, the mechanical conditions under which the heat motion proceeds cannot be precisely specified, the problem arises to investigate in which cases and to what extent the equations of mechanics are analogous to those of the Theory of Heat". He continues: "One will not be concerned here with the construction of mechanical systems, which are completely identical with warm bodies, but rather with identifying all systems, that exhibit behavior more or less analogous to that of warm bodies". He further elaborates on this analogy in two more papers in the following years [31]. I quote the beginning of the third paper [31b] "New Proof of a Theorem Formulated by Helmholtz Concerning the Properties of Monocyclic Systems", to illustrate the importance Boltzmann attached to this work: "The great importance which the introduction of the notion of monocyclic systems and the development of their most important properties has for all investigations concerning the Second Law should make the following considerations appear as not completely superfluous...".

As an aside, I remark that Boltzmann's writings on the existence of atoms seem to be ambivalent and a mixture of, on the one hand, actually using atomism all through his works and elaborately discussing the many arguments in favor of it [32a], while, on the other hand, stating in writing the possibility of other equally valid descriptions of nature, as provided by the Energeticists or phenomenologists. Thus in his 1899 lecture mentioned above, he says [32b]: "From this follows that it cannot be our task to find an absolutely correct theory, but rather the simplest possible picture which represents experiment as best as possible. One could even think of the possibility of two entirely different theories, which are both equally simple and agree with the phenomena equally well, which therefore, although completely different, are both equally correct". Although true, I find it hard to escape the impression that this statement was meant more as an attempt to assuage his Energeticist opponents than as an account of his actual position. The difficulty, of course, was that, at the time, evidence for the existence of atoms was ultimately only circumstantial and that no direct experimental demonstration in any fashion had yet been given.

In the same 1884 paper mentioned above, Boltzmann introduces the notion of Ergoden [33]. It was used by Boltzmann as an equilibrium ensemble of systems (Gibbs' micro-canonical ensemble). According to the Ehrenfests [10], Boltzmann defines an ergodic system as one whose unperturbed motion goes, when indefinitely continued, finally *"through each phase point"* that is consistent with its given total energy. In this way Boltzmann suggested how to understand on the basis of the dynamics of the gas molecules, i.e., from mechanics, that "ergodic (ensemble) averages"[34] in phase space could replace the time averages through which the macroscopic properties of the gas were defined. In their article, the Ehrenfests argued that Boltzmann's requirement for an ergodic system was too strong and they replaced it by introducing a quasi- ergodic system "which approaches each point of the energy surface arbitrarily close". The idea is that under such conditions the time and ensemble averages would still be the same. The classical ergodic theory culminated in Birkhoff's ergodic theorems [35] proving the existence of the time average and then for metrically transitive systems the equality of time and phase space averages. Later, it was the application of dynamical systems theory and the introduction of the concept of measure which allowed the more precise unification of mechanics and statistics that Boltzmann had in mind.

As in electromagnetic theory, so in the theory of gases, the mechanical aspects have been obliterated. In the theory of gases this occurred not only through Boltzmann's statistical method but mainly through Gibbs' 1902 book "Elementary Principles in Statistical Mechanics"[36]. It was also there that the term "Statistical Mechanics" was first publicly introduced [36]. Here, on purpose, all reference to the molecular "constitution of matter" was as much as possible avoided to achieve a generality similar to that of thermodynamics. The book was mainly devoted to a study of thermal equilibrium: the canonical ensemble, already introduced by Boltzmann [4b, 4c] for a system in thermal equilibrium at a given temperature, as well as the micro-canonical ensemble were given their names and extensively studied. Their connection with thermodynamics was made via a thermodynamic analogy [37], as had been done earlier by Boltzmann [4c, 15, 28]. While thermodynamics never changed and was essentially unaffected by the advent of quantum mechanics - except for the addition of the third law, which is not as absolute and general as the first two - Gibbs' ensembles had only to be modified slightly to accommodate quantum mechanics.

The Ehrenfests were rather critical of Gibbs' contributions to statistical mechanics and they gave a very subjective presentation of his accomplishments [38]. The main criticism was that Gibbs had only devoted essentially one chapter - and a purely descriptive one at that - to the problem of the approach to equilibrium (ch.12), a problem that was central in the considerations of Boltzmann and the Ehrenfests. In this chapter Gibbs describes a generalization of Boltzmann's H-theorem in μ-space, the phase space of one molecule, to Γ-space, the phase space of the entire gas. This consideration, as well as the rest of the book, was entirely based on probability notions, and the whole basic molecular mechanism of collisions - viz. that of binary collisions for the dilute gas which Boltzmann had considered - was completely absent. I think the Ehrenfests' critical attitude in this has turned out to be unjustified since Gibbs' statistical mechanics has been far more influential than they surmised at the time. In fact, it has dominated the entire twentieth century and only now, with a renewed interest in nonequilibrium phenomena, is a revival of Boltzmann's mechanistic approach reemerging. Here one should keep in mind two things. First, even simple nonequilibrium phenomena are often far more difficult to treat than many rather complicated equilibrium phenomena. Second, starting with L. S. Ornstein's Ph.D. thesis "Application of the Statistical Mechanics of Gibbs to Molecular-Theoretical Questions" written under Lorentz's direction in 1908 [39], the calculation of the thermodynamic properties of a gas in thermal equilibrium via the canonical and related distribution functions turned out to be far simpler than those based on the microcanonical ensemble. Gibbs' statistical mechanics has not only led to enormous advances in equilibrium statistical mechanics but also, by virtue of, for instance, the introduction of the Renormalization Group in the theory of critical phenomena, to entirely new ways of thinking about what is relevant in nonequilibrium statistical mechanics as well. And yet ...

3. Return of Mechanics - Boltzmann's Heritage

The revival of the role of mechanics in statistical mechanics is due to important new developments in the 1950's in mechanics itself or as it is now called the theory of dynamical systems, especially by the Russian school of mathematics, emanating from A. N. Kolmogorov and Ya. G. Sinai. This has led in recent years to a beginning of a dynamical formulation of statistical mechanical problems, especially for nonequilibrium systems, near or far from equilibrium. Far from equilibrium means here a system with large gradients and therefore large deviations from Maxwell's (local) equilibrium velocity distribution function, where hydrodynamics cannot be applied, not a turbulent system, where a hydrodynamic description is still (believed to be) applicable. I will quote three examples.

1. Boltzmann's notion of ergodicity for Hamiltonian systems in equilibrium on the energy surface has, in a way, been generalized by Sinai, Ruelle and Bowen (SRB) [40] to dissipative systems in a nonequilibrium stationary state. In that case the attractor, corresponding to the nonequilibrium stationary state, plays the role of the energy surface in equilibrium and the SRB measure on the attractor in terms of expanding, i.e., positive Lyapunov exponents (which determine the rate of exponential separation of two initially very close trajectories) replaces the Liouville measure on the energy surface for systems in equilibrium. This allows the assignment of purely dynamical weights to a macroscopic system even far from equilibrium, since the molecules, whose dynamical properties one uses, do not know how far the system is from equilibrium, as equilibrium is a macroscopic non-molecular concept, or, to paraphrase Maxwell [41]: "When one gets to the molecules the distinction between heat and work disappears, because both are [ultimately molecular] energy". This approach seems to differ in principle from the conventional ones, based on extending Gibbs' equilibrium ensembles to nonequilibrium, e.g. via a Chapman-Enskog-like solution of the Liouville equation for the entire system (instead of for the Boltzmann equation, for which it was originally designed). To be sure, these Gibbsian nonequilibrium ensembles also contain dynamics - no nonequilibrium description is possible without it - but not as unadulterated as in the SRB-measure. The farthest one has gone with these nonequilibrium Gibbs ensembles is the Kawasaki distribution function [42] and it is not excluded that there is an intimate connection between this distribution function and the SRB measure.

Recently the SRB measure has been checked for the first time far from equilibrium for a many (56) particle system by a computer experiment [43]. Here one studies very large temporary fluctuations of a shearing fluid in stationary states with very large shear rates. The ratio of the fluctuations of the stress tensor to have, during a finite time, a given value parallel or opposite to the applied shear stress, i.e. consistent with or "in violation of" the Second Law, respectively, was measured and found to be given correctly on the basis of the SRB measure. In fact, one has recently been able to indicate how this, in a way typical nonequilibrium statistical mechanical system, can perhaps be discussed on the basis of the SRB measure: at least a scenario has been formulated, where one could hope, at least in principle, to derive the just mentioned result rigorously from the SRB measure with "large N-dynamics" [44].

2. Relations between the transport coefficients and the Lyapunov exponents of a fluid in a nonequilibrium stationary state have been uncovered. While the former refer to hydrodynamic, i.e., nonequilibrium properties in ordinary three dimensional space, the latter refer to the dynamical behavior

on the attractor in the multidimensional phase-space of the entire system. For the above mentioned many-particle shearing fluid in a nonequilibrium stationary state, one has obtained an explicit expression for the viscosity coefficient of this fluid in terms of its two maximal - i.e., its largest and its smallest - Lyapunov exponents [45]. This expression gives a value for the viscosity which agrees numerically with that found directly from computer simulations for a 108 and 864 particle shearing fluid, not only in the linear (hydrodynamic) regime near equilibrium, where the viscosity is independent of the imposed shear rate, but even in the non-linear (rheological) regime, far from equilibrium, where the viscosity coefficient itself depends on the shear rate.

3. A somewhat analogous expression has been derived for the linear diffusion coefficient of a point particle in a regular triangular array of hard disks [46]. In addition, the diffusion coefficient [47, 48a] as well as the pressure [48b] for this system have been computed using a cycle-expansion, i.e., an expansion in terms of the periods and Lyapunov exponents of the unstable periodic orbits of the particle in the hard disk system. Furthermore a number of rigorous results have been proved for the linear transport behavior of this system based on the SRB measure [49].

In all these cases the re-emergence of dynamics appears in a global Gibbsian-sense, in that it involves global Lyapunov exponents of the entire system, not detailed collision dynamics between small groups of particles as in kinetic theory.

It appears therefore that Boltzmann's attachment to mechanics, as exemplified by his unceasing attempts at a mechanical interpretation of the macroscopic behavior of many-particle systems, in particular of the Second Law of thermodynamics, has re-emerged after a hundred years, thanks to new developments in mechanics. I note that these new developments, apart from the above mentioned statistical-mechanical problems, have so far been applied mainly to simple (one-dimensional) maps and few-particle (i.e., few degrees of freedom) systems. For the connection with statistical mechanics it seems important to introduce methods into these dynamical considerations which use explicitly the very large number of degrees of freedom typical for statistical mechanical, i.e., for macroscopic systems. Such "large N-dynamics" could lead to a "statistical" dynamics that might be crucial to bridge the gap between the prevailing rigorous treatments of dynamical systems of a few degrees of freedom on the one hand and statistical mechanics on the other hand. For example, important distinctions in "small N-dynamics" as to the exact number of conservation laws (or of Axiom-A systems) [50] might be less relevant in the "large N-dynamics" for macroscopic systems.

A case in point would be to establish a connection between dynamical system theory and Boltzmann's kinetic theory of dilute gases. One could then ask: what is the connection between the linear viscosity of a dilute gas as given by the Boltzmann equation in terms of binary collision dynamics (e.g. as related to an eigenvalue of the linear Boltzmann collision operator [51]) and the above mentioned expression for the viscosity in terms of its two maximal Lyapunov exponents?

Before I end this mostly scientific presentation of Boltzmann's work in statistical mechanics and its legacy, I would like to remark that I have not just illustrated Boltzmann's work from a scientific but also, from time to time, from a human or psychological point of view. I believe that the latter aspects are too often missing in the discussion of scientists, especially those of the recent past. The "psychological" remarks are usually confined to anecdotes and occasional non-scientific comments. This is in contrast to what happens in the arts, where musicians, painters and especially writers are critically "psychoanalyzed" as to their behavior, their motivations and the connection between their personality and their work. In fact, this does exist to some degree for some scientists of the remote past, e.g. for Newton [52]. Of all nineteenth century scientists, Boltzmann seems to be one of the most openly human and deeply tragic, i.e., an obvious candidate for such an endeavor. I hope that this often missing human dimension in the discussion of scientists and their work will be developed, although it is admittedly difficult to find writers who have both the scientific and the human perception and depth to do this [53].

I would like to conclude with two quotations, one about the work and one about the man. The first is from the obituary lecture by Lorentz, given one year after Boltzmann's death [54]. This quotation is perhaps even more applicable now than it was then and expresses beautifully Boltzmann's message for the future when Lorentz says: "The old of which Boltzmann speaks [see above] has in our days, thanks especially also to his own efforts, flowered to new, strong life, and even though its appearance has changed and will certainly often change in the course of time, we may yet hope that it will never get lost for science".

The other is from Boltzmann's 1899 lecture "On Recent Developments of the Methods of Theoretical Physics", mentioned before. It demonstrates that Boltzmann's deep love for science transcended all his suffering in practicing it. Here, after discussing the many achievements of atomism and the molecular theory of matter, which cannot at all be obtained by just using macroscopic equations alone without any further microscopic foundations - as is done in phenomenology or energetics - he asks [55]: "Will the old mechanics with the old forces ... in its essence remain, or live on one day only in history ... superseded by entirely different notions? Will the essence of the present molecular theory, in spite of all amplifications and modifications, yet remain, or will one day an atomism totally different from the present prevail, or will even, in spite of my proof [to the contrary][56], the notion of an absolute continuum prove to be the best picture?" He concludes: "Indeed interesting questions! One almost regrets to have to die long before they are settled. O! immodest mortal! Your destiny is the joy of watching the ever-shifting battle!"[57].

Acknowledgement

I have very much profited from various papers of M. J. Klein and from the books *The Kind of Motion We Call Heat* by S. G. Brush. I am also indebted to my colleagues J. R. Dorfman, G. Gallavotti, A. J. Kox, A. Pais, L. Spruch, and especially M. J. Klein for helpful remarks. I thank my secretary, S. Rhyne for preparing the manuscript and her as well as my daughter, A. M. Cohen, for their help in translating Boltzmann's German and to the latter also for her help with the text.

References

1. L. Boltzmann, "Über die mechanische Bedentung des Zweiten Hauptsatzes der Wärmetheorie", Wien. Ber. **53**, 195-220 (1866); Wissenschaftliche Abhandlungen (W.A.), F. Hasenöhrl, ed., (Chelsea Publ. Co., New York, 1968) Band I, pp. 9-33.
2. L. Boltzmann, "Studien über das Gleichgewicht der lebendigen Kraft zwischen bewegten Materiellen Punkten", Wien. Ber. **58**, 517-560 (1868); W.A. Band I, pp. 49-96; id., "Lösung eines mechanisches Problems", Wien. Ber. **58**, 1035-1044 (1868); W. A. Band I, p. 97.
3. See also, L. Boltzmann, Über das Wärmegleichgewicht van Gasen auf welche äuszere Kräfte wirken", Wien. Ber. **72**, 427-457 (1875); W.A. Band II, pp. 1-30.
4. L. Boltzmann, (a) "Über das Wärmegleichgewicht zwischen mehratomigen Gasmolekulen", Wien. Ber. **63**, 397-418 (1871); W.A. Band I, pp. 237-258; (b) id., "Einige allgemeine Sätze über Wärmegleichgewicht", Wien. Ber. **63**, 679-711 (1871); W.A. Band I, pp. 259-287; (c) id., "Analytischer Beweis des zweiten Hauptsatzes der mechanischen Wärmetheorie aus den Sätzen über das gleichgewicht der lebendigen Kraft", Wien. Ber. **63**, 712-732 (1871); W.A. Band I, pp. 288-308; (d) id., "Neuer Beweis zweier Sätze über das Wärmegleichgewicht unter mehratomigen Gasmolekülen", Wien. Ber. **95**, 153-164 (1887); W.A. Band III, pp. 272-282.
5. Of Boltzmann's about 140 scientific papers around 18 deal with the Second Law and 16 deal with Maxwell's equilibrium distribution function or both.
6. L. Boltzmann, "Weitere Studien über das Wärmegleichgewicht unter Gasmolekülen", Wien. Ber. **66**, 275-370 (1872); W.A. Band I, pp. 316-402.
7. See ref. 6, W.A., Band I. p. 345.
8. L. Boltzmann, "Bemerkungen über einige Probleme der mechanischen Wärmetheorie", Wien. Ber. **74**, 62-100 (1877), section II; W.A. Band II, pp. 112- 148, section II.
9. (a) L. Boltzmann, "Entgegnung auf die Wärmetheoretischen Betrachtungen des Hrn. E. Zermelo", Wied. Ann. **57**, 778-784 (1896); W.A. Band III, pp. 567-578; (b) id. "Zu Hrn. Zermelos Abhandlung 'Über die mechanische Erklärung irreversibler Vorgänge'", Wied. Ann. **60**, 392-398 (1897), W. A. Band III, pp. 579-586; (c) id. "Über einen mechanischen Satz Poincaré's", Wien. Ber. **106**, 12-20 (1897); W. A.

Band III, pp. 587-595.
10. (a) P. and T. Ehrenfest, "Begriffliche Grundlagen der statistischen Auffassung in der Mechanik", Enzycl. d. Mathem. Wiss. IV, 2, II, Heft 6, 3-90 (1912), pp. 30-32 or in P. Ehrenfest, *Collected Papers* 213-309 (North-Holland, Amsterdam, 1959), pp. 240-242; (b) P. and T. Ehrenfest, *The Conceptual Foundations of the Statistical Approach in Mechanics*, M.J. Moravcsik, tr., (Cornell University Press, Ithaca, NY 1959), pp. 20-22.
11. The Energeticists wanted to explain all natural phenomena on the basis of energy alone, the most general "substance" present in the world. All phenomena were then continuous transformations of energy.
12. "note well".
13. A. J. Kox, "H. A. Lorentz's Contributions to Kinetic Gas Theory", Ann. Sci. **47**, 591-606 (1990), p. 602.
14. A. J. Kox, l.c. p. 598.
15. L. Boltzmann, "Über die Beziehung swischen dem zweiten Hauptsatz der mechanischen Wärmetheorie und der Wahrscheinlichkeitsrechnung respektive den Sätzen über des Wärmegleichgewicht", Wien. Ber. **76**, 373-435 (1877); W. A. Band II, pp. 164-223.
16. Ref. 15, p. 164.
17. Cf. A. Einstein, "Autobiographical Notes" in *Albert Einstein: Philosopher - Scientist*, P. A. Schilp, ed., (The Library of Living Philosophers, Vol. VII, Evanston, IL, 1949), pp. 47-49. See also: B. Hoffman, *Albert Einstein, Creator and Rebel*, (New Amer. Libr., New York 1972) pp. 58-59.
18. See, e.g., E. G. D. Cohen, "Fifty Years of Kinetic Theory", Physica A **194**, 229-257 (1993).
19. L. Boltzmann, *Vorlesungen über Gastheorie*, 2 vols (Barth, Leipzig, 1896, 1898); engl. transl. by S. G. Brush as *Lectures on Gas Theory*, (Univ. of Calif. Press, Berkeley, 1964).
20. L. Boltzmann, "Vorlesunger über Maxwell's Elektrizitätstheorie" (Aus den Mitteilungen des naturwissenschaftlichen Vereins in Graz. August 1873.) in *Populäre Schriften*, 2nd ed., (Barth, Leipzig, 1919) pp. 11-24.
21. J. C. Maxwell, "On Faraday's Lines of Force", Trans. Cambr. Phil. Soc. **10**, 27-83 (1856); *Scientific Papers* 1, (Cambridge, U. K., 1890), p. 155.
22. L. Boltzmann, "Über die Entwicklung der Methoden der theoretischen Physik in neuerer Zeit", in *Populäre Schriften*, 2nd ed., (Barth, Leipzig, 1919) pp. 198-227.
23. Ref. 22, pp. 204-205.
24. Ref. 4b, pp. 259-260.
25. J. C. Maxwell, "On Boltzmann's theorem on the average distribution of energy in a system of material points", Cambr. Phil. Soc. Trans. **12**, 547-575 (1879); *Scientific Papers* 2, (Cambridge, U. K., 1890), pp. 713-741. See also ref. 28, p. 123.
26. Ref. 25, p. 715.
27. See, e.g. M.J. Klein, "The Maxwell-Boltzmann Relationship", in A.I.P. Conference Proceedings, *Transport Phenomena*, J. Kestin, ed., (Amer. Inst. Phys., New York, 1973) pp. 300-307.
28. L. Boltzmann, "Über die Eigenschaften monozyklischen und andere damit verwandter Systeme", Zeitschr. f. R. u. Angew. Math (Crelles Journal) **98**, 68-94 (1884); W. A. Band III, pp. 122 - 152. See also: G. Gallavotti, "Ergodicity, Ensembles and Irreversibility", J. Stat. Phys. **78**, 1571-1589 (1995).
29. Ref. 28, p. 123, footnote 1.
30. Ref. 28, p. 122.
31. (a) L. Boltzmann, "Über einige Fälle, wo die lebendige Kraft nicht integrierender Nenner des Differentials der zugeführten Energie ist", Wien. Ber. **92**, 853-875 (1885); W.A. Band III, pp. 153-175; (b) id., "Neuer Beweis eines von Helmholtz aufgestellten Theorems betreffende die Eigenschaften monozyclischen Systeme", Gött. Nachr.

209-213 (1886); W.A. Band III, pp. 176-181.

32. (a) See e.g., L. Boltzmann, "Über die Unentbehrlichkeit der Atomistik in der Naturwissenschaften" in *Populäre Schriften*, 2nd ed., (Barth, Leipzig, 1919) pp. 141-157; (b) ref. 22, p. 216.
33. See ref. 28, p. 134.
34. See ref. 10a, p. 30, footnote 83; ref.10b, p. 89, note 88.
35. G. D. Birkhoff, (a) "Proof of the Ergodic Theorem", Proc. Nat. Acad. USA **17**, 656-660 (1931); (b) id., "Probability and Physical Systems", Bull. Amer. Math. Soc., 361-379 (1932); (c) G. D. Birkhoff and B. 0. Koopman, "Recent Contributions to the Ergodic Theory", Proc. Nat. Acad. USA **18**, 279-287 (1932) and (d) G. D. Birkhoff and P. A. Smith, "Structure Analysis of Surface Transformations", J. Math. Pures Appl. **7**, 345-379 (1928). See also: A. I. Khinchin, *Mathematical Foundations of Statistical Mechanics*, (Dover, New York, 1949) Ch. II, pp. 19-32; E. Hopf, *Ergodentheoric*, (Chelsea Publ. Co., New York, 1948).
36. J. W. Gibbs, *Elementary Principles in Statistical Mechanics*, (Yale University Press, New Haven, 1902; also Dover Publications, New York, 1960).
37. Ref 36, pp. 42-45.
38. Ref 10a, pp. 53-70; 10b, pp. 44-63.
39. L. S. Ornstein, "Toepassing der Statistische Mechanica van Gibbs op Molekulair-Theoretische Vraagstukken", Leiden (1908).
40. See e.g. J. P. Eckmann and D. Ruelle, "Ergodic Theory of Chaos and Strange Attractors", Rev. Mod. Phys. **57**, 617-656 (1985), p. 639.
41. J. C. Maxwell, "Tait's Thermodynamics", Nature **17**, 257-259 (1878); *The Scientific Papers of James Clerk Maxwell* Vol. 2, (Dover Publ., New York, 1952), p. 669.
42. D. J. Evans and G. P. Morriss, *Statistical Mechanics of Nonequilibrium Liquids*, (Academic Press, New York, 1990) p. 171.
43. D. J. Evans, E. G. D. Cohen, and G. P. Morriss, "Probability of Second Law Violations in Shearing Steady States", Phys. Rev. Lett. **71**, 2401-2404 (1993); id. **71**, 3616 (1993).
44. G. Gallavotti and E. G. D. Cohen, "Dynamical Ensembles in Nonequilibrium Statistical Mechanics", Phys. Rev. Lett. **74**, 2694-2697 (1995); ibid., "Dynamical Ensembles in Stationary States", J. Stat. Phys. **80**, 931-970 (1995).
45. D. J. Evans, E. G. D. Cohen, and G. P. Morriss, "Viscosity of a simple fluid from its maximal Lyapunov exponents", Phys. Rev. A **42**, 5990-5997 (1990); See also: H. A. Posch and W. G. Hoover, "Lyapunov Instability of Dense Lennard-Jones Fluids", Phys. Rev. A **38**, 473-482 (1988); id., "Equilibrium and Nonequilibrium Lyapunov Spectra for Dense Fluids and Solids", Phys. Rev. A **39**, 2175-2188 (1989).
46. P. Gaspard and G. Nicolis, "Transport Properties, Lyapunov Exponents and Entropy Per Unit Time", Phys. Rev. Lett. **65**, 1693-1696 (1990).
47. (a) P. Cvitanovič, P. Gaspard, and T. Schreiber, "Investigation of the Lorentz Gas in Terms of Periodic Orbits", Chaos **2**, 85-90 (1992); (b) W. W. Vance, "Unstable Periodic Orbits and Transport Properties of Nonequilibrium Steady States", Phys. Rev. Lett. **69**, 1356-1359 (1992).
48. (a) G. Morriss and L. Rondoni, "Periodic Orbit Expansions for the Lorentz Gas", J. Stat. Phys. **75**, 553-584 (1994); (b) G. P. Morriss, L. Rondini, and E. G. D. Cohen, "A Dynamical Partition Function for the Lorentz Gas", J. Stat. Phys. **80**, 35-43 (1994).
49. N. J. Chernov, G. L. Eyink, J. L. Lebowitz, and Ya. G. Sinai, "Derivation of Ohm's Law in a Deterministic Mechanical Model", Phys. Rev. Lett. **70**, 2209-2212 (1993); id. "Steady State Electrical Conduction in the Periodic Lorentz Gas", Comm. Math. Phys. **154**, 569-601 (1993).
50. Ref. 40, p. 636.
51. I am indebted for this suggestion to Prof. B. Knight of The Rockefeller University.
52. See, e.g., F. E. Manuel, *A Portrait of Isaac Newton*, (Harvard Univ. Press, Cam-

bridge, MA 1968).
53. Exceptions are, e.g., M. J. Klein, *Paul Ehrenfest*, Vol. I, (North-Holland, Amsterdam, 1970); C. W. F. Everitt, "Maxwell's Scientific Creativity," in *Springs of Scientific Creativity*, R. Aris, H. T. Davis, and R. H. Stuewer, eds., (Univ. Minnesota Press, Minneapolis, MN 1983) Ch. 4, p. 71-141; M. Dresden, *H. A. Kramers: Between Tradition and Revolution*, (Springer Verlag, New York, 1987) and W. Moore, *Schrödinger*, (Cambridge University Press, Cambridge, 1989).
54. H. A. Lorentz, "Ludwig Boltzmann", Commemoration oration at the meeting of the German Physical Society, 17 May 1907; Verhandl. Deutsch. Physik. Gesells. **12**, 206-238 (1907); *Collected Works*, Vol. IX, p. 389.
55. Ref. 22, pp. 226, 227.
56. See e.g. L. Boltzmann, "Statistische Mechanik" in ref. 22, p. 358.
57. Ref. 22, p. 227: "O unbescheidener Sterbliche! Dein Los ist die Freude am Anblicke des wogenden Kampfes!".

KINETIC THEORY OF GRANULAR FLUIDS: HARD AND SOFT INELASTIC SPHERES

M.H. ERNST
Institute for Theoretical Physics, University of Utrecht
Princetonplein 5, 3584 CC Utrecht, The Netherlands

1. Introduction

The basic concept that rapid granular flows can be considered as a collection of particles with short range interactions, moving ballistically and suffering instantaneous and inelastic binary collisions, is formulated in Haff's seminal paper *Grain flow as a fluid mechanical phenomenon* [1].

In the present proceedings of the NATO Advanced Study Institute on kinetic methods and modeling a wide area of research in granular fluids has been covered. Experimental studies in this field were presented by J.M. Huntley and H.L. Swinney. More theoretically oriented contributions ranged from the high frequency and wavenumber inelastic collapse phenomenon by S. McNamara, to stationary states in systems with dissipative interactions by J. Piasecki, to model kinetic equations by J.J. Brey, to the low frequency and wavenumber phenomena of granular fluid dynamics by J.T. Jenkins on shear flows in inelastic spheres, A. Hansen on flows driven by gravity, and the present article on undriven granular fluids.

An essential feature of inelastic fluids is of course that collisions are inelastic and kinetic energy is not conserved in collisions, but total momentum is. The latter property guarantees that the flow field $\mathbf{u}(\mathbf{r}, t)$ is a slowly changing macroscopic variable. Therefore granular matter in rapid flow qualifies as a fluid. The former property makes the temperature of the fluid drop, when not driven. To maintain a steady state, energy has to be put into the system through external driving fields, shear flow or heat sources.

Haff's paper has laid the foundations for using the methods of kinetic theory in this field, which have been further developed in several classic papers in the eighties [2]-[4], and a continuous flow of publications, relevant to this field, thereafter [5]-[47]. On the one hand, the randomizing dynamics

J. Karkheck (ed.),
Dynamics: Models and Kinetic Methods for Non-equilibrium Many Body Systems, 239–266.

of short-range strongly repulsive binary collisions will help the system to explore large regions of phase space, which is a prerequisite for applying the methods of non-equilibrium statistical mechanics and kinetic theory. On the other hand, due to the dissipative dynamics the flow in phase space does not preserve volume, but is in general contracting. Therefore it is not a priori clear that macroscopic quantities can be calculated as averages over some initial non-equilibrium ensemble. The basic *assumption* in this article is that for *weakly inelastic* granular fluids macroscopic properties can indeed be calculated from averages over distribution functions.

Granular fluids can be considered as dense fluids with very short range and strongly repulsive interactions, modeled as inelastic soft and hard spheres, in which the particles follow free particle trajectories until a collision occurs. In the first model of *soft* inelastic or visco-elastic spheres the particles interact with a rather stiff elastic repulsion, and lose energy during contact through a frictional force, which is proportional to their relative velocities [37]-[47]. In the second model of inelastic *hard* spheres [5]-[36] there is an instantaneous loss of kinetic energy on contact, also proportional to the relative velocity of the colliding pair.

Kinematics and dynamics, as well as elementary kinetic methods, for *soft* inelastic spheres have been widely discussed in the literature [24, 25, 37, 39, 40], but more sophisticated kinetic equations for the single-particle distribution function seem to be rare [41]. For *hard* spheres, on the other hand, kinetic theory is well developed [1]-[4], [26]-[33], and based on the Enskog-Boltzmann equation. Inherent to this description is the molecular chaos assumption of uncorrelated binary collisions. However, there exist in inelastic fluids [12, 16, 18] long-range spatial correlations in density and flow fields, which cannot be understood on the basis of a mean-field type kinetic equation, like the Boltzmann or Enskog-Boltzmann equation. Therefore, kinetic theory has to be extended to account for dynamic correlations, neglected through the molecular chaos assumption. For *elastic* hard sphere fluids this has been done through ring kinetic theory, which will be generalized to inelastic particles in the present article. The phenomenological counterpart of ring kinetic theory, valid on large spatial and temporal scales, is mode coupling theory and nonlinear fluctuating hydrodynamic equations [48]-[51].

In the last 35 years many-body theories have been developed to account for these dynamic correlations in systems of microscopic particles obeying the standard conservation laws. The fundamental concept to describe these dynamic correlations are 'ring collisions', *i.e.* sequences of correlated binary collisions, which lead to the so called *ring kinetic theory* [50]-[54]. This ring kinetic theory for systems of smooth elastic hard spheres has been the basis of all major developments in nonequilibrium statistical mechanics over the last three decades: it explains the breakdown of the virial expansion for transport coefficients and their logarithmic density dependence [52]; the algebraic long time tails of the velocity autocorrelation function and similar current-current correlation functions [50]-[56]; the non-analytic dispersion relations for sound propagation and for the relaxation rates of hydrodynamic excitations [54]-[56], as well as the breakdown of the Navier-Stokes equations in two-dimensional fluids at very long times [53]-[56], the non-existence of linear transport coefficients in two dimensions and of Burnett coefficients [54]-[56] in three dimensions; moreover, it explains the existence of long range spatial correlations in nonequilibrium stationary states [57], driven by reservoirs which impose shear rates or temperature gradients, or in driven diffusive systems. Such systems violate the conditions of detailed balance and the stationary states are non-Gibbsian states [57].

The goal of this article is twofold. In the first part we use elementary kinetic methods to explain some of the more interesting observations in rapid granular flows. This is intended to inform the reader about the complex phenomena occurring in granular fluids, and to introduce intuitive concepts and basic understandings. The second part is more formal, and shows how kinetic equations can be derived for different models. We start from the Liouville equation, *i.e.* the time evolution equation for the N-particle density in phase space for fluids with inelastic soft and hard sphere interactions. Once the Liouville equation has been formulated, the methods of kinetic theory can be applied to derive kinetic equations.

2. Generic Features of Inelastic Fluids

2.1. DECAY OF ENERGY

2.1.1. *Hard Spheres*

Standard fluids, when out of equilibrium, will rapidly decay to local equilibrium within a few mean free times. The subsequent decay of spatial inhomogeneities towards global equilibrium is controlled by the slow hydrodynamic time evolution. A prototypical model for such fluids is a system of N smooth elastic hard spheres. Compare this system with a system of smooth inelastic hard spheres, where a fraction of the kinetic energy is lost in each collision. The lack of energy conservation makes this fluid, whether driven by gravity or shear stresses, or freely evolving, behave very differently from a standard fluid [6].

The system of inelastic hard spheres represents an idealized model for *rapid granular flows*, where the dynamics of individual particles is described by binary collisions, separated by free propagation over a typical mean free path. To study this in more detail one considers a system of N inelastic hard spheres of diameter σ and mass m, contained in a volume $V = L^d$

with periodic boundary conditions, and one performs molecular dynamics simulations in two or three dimensions [5]-[22]. At the initial time, the system starts off in a spatially homogeneous equilibrium state of elastic hard spheres, and stays for many collision times in a so-called *homogeneous cooling state.* The corresponding single-particle distribution function is essentially a Maxwellian with a time dependent kinetic energy per particle $E(t)$ [6, 26, 30, 35]. As long as the system is in this state, the energy per particle and the temperature are simply related as $E(t) = \frac{1}{2}dT(t)$.

A simple inelastic collision law is $g_n^* = -e_n g_n$, where g_n and g_n^* are the components of the pre- and postcollision relative velocity at contact, normal to the surface, and e_n is the coefficient of normal restitution with $e_n \in (0,1)$, which depends in general on the impact velocity, but is assumed here to be a constant. The case $e_n = 1$ represents elastic hard spheres. On average a particle loses per collision a fraction of its energy, proportional to $1 - e_n^2$, which leads to cooling. Let $2\gamma_0$ be the fraction of energy lost on average in a collision, and let ω be the average collision frequency, then the rate of cooling in an inelastic hard-sphere fluid is given by

$$\frac{\mathrm{d}T}{\mathrm{d}t} = -\Gamma = -2\gamma_0 \omega T. \tag{1}$$

The quantity γ_0, referred to as the *coefficient of inelasticity*, can be calculated from kinetic theory [29, 30] and turns out to be $\gamma_0 = (1 - e_n^2)/(2d)$. In the present discussion the collision frequency $\omega(T) = \omega_E(T)$ is assumed to be given by the Enskog theory $\omega_E(T(t)) \sim n\sigma^{d-1}\chi v_0(t)$ (see Ref. [12] for details), and depends explicitly on time through the r.m.s. velocity or 'thermal' velocity $v_0(t) = \sqrt{2T(t)/m}$ and $\chi = g(\sigma)$ is the radial distribution function of hard spheres at contact. Using the relation $\omega \sim \sqrt{T}$, the temperature equation can readily be integrated to give Haff's *homogeneous cooling law* [1],

$$T(t) = T_0/(1 + \gamma_0 t/t_0)^2 = T_0 e^{-2\gamma_0 \tau}, \tag{2}$$

as plotted in Fig. 1a,b. Here $t_0 = 1/\omega(T_0)$ is the mean free time in the initial equilibrium state at temperature T_0, and $t_e = t_0/\gamma_0$ is the characteristic time for homogeneous cooling. It predicts an algebraic decay in the external time t. From a physical point of view the evolution and aging of the system is determined by the number of collisions $C(t)$ that has occurred in the system during the time t, or more conveniently, by the 'internal' or 'proper' time, $\tau(t) = 2C(t)/N$, which counts the mean number of collisions suffered per particle within a time t. The factor 2 accounts for the fact that every collision involves two particles. According to kinetic theory, 'internal' and 'external' time are related by the differential relation, $\mathrm{d}\tau = \omega(T(t))\mathrm{d}t$. This relation in combination with Eq. (1) can be trivially integrated to give the second equality in Eq. (2). We note that the exponential decay in

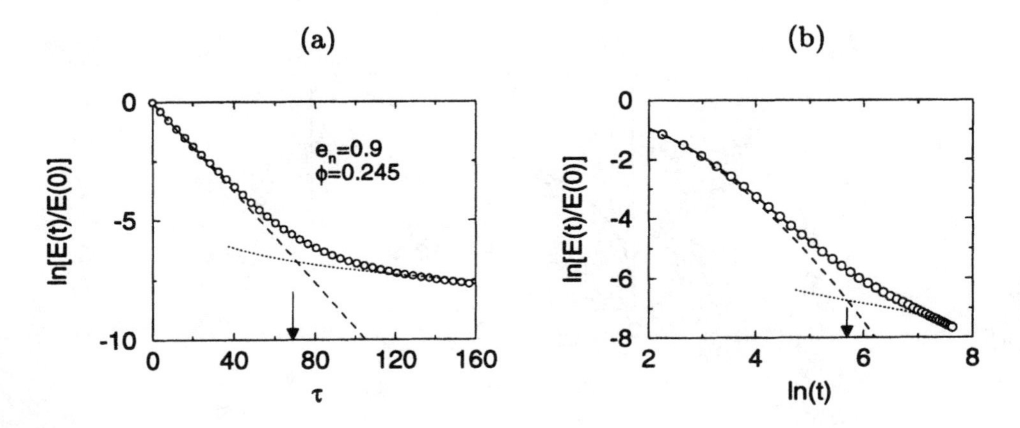

Figure 1. Energy decay (a) versus number of collisions τ and (b) versus external time t at packing fraction $\phi = 0.245$ and low inelasticity $e_n = 0.9$. MD results [19] for $N = 50000$ hard disks compared with theoretical predictions (dashed lines) for short times (Haff's law (2)) and for long times (Eq. (7)). The intersection defines the crossover time $\tau_{cr} \simeq 69$ and $t_{cr} \simeq 300$ (see down arrows in Fig. 1 and 2).

$T(t) = T_0 \exp(-2\gamma_0\tau)$ has a rather *universal* form. It is independent of the density in the inelastic hard-sphere fluid, and no a priori knowledge about the t-dependence of $\omega(T(t))$ is required to obtain this result.

Equation (2) is valid as long as the collisional dissipation Γ is correctly given by the factorized expression on the right hand side of Eq. (1). This is a typical mean-field approximation, which is based on the assumption of *molecular chaos*, that neglects the dynamic correlations between the particles. Of course, the differential relation combined with $\omega \sim \sqrt{T}$ and Eq. (2) can be integrated explicitly to give the relation, in the homogeneous cooling state of the inelastic hard-sphere fluid,

$$e^{\gamma_0\tau} = 1 + \gamma_0 t/t_0, \tag{3}$$

as plotted in Fig. 2. In the elastic limit ($\gamma_0 \to 0$) the internal time becomes $\tau = t/t_0$. As the cooling proceeds, the mean free time $1/\omega(T(t))$ between collisions increases, and the internal clock of the particles runs more slowly than the laboratory clock. However, the mean free path, $l_0 = v_0(t)/\omega(T(t))$, remains constant.

2.1.2. *Soft Spheres*

Next we consider a modification of the inelastic hard-sphere model to account for visco-elasticity. It is suggested by a very simple model, introduced by Hwang and Hutter [40]. During free flight the particles have only kinetic energy. In a (central) collision the relative kinetic energy E_K is converted into vibrational energy of an elastic wave (damped because of visco-elastic

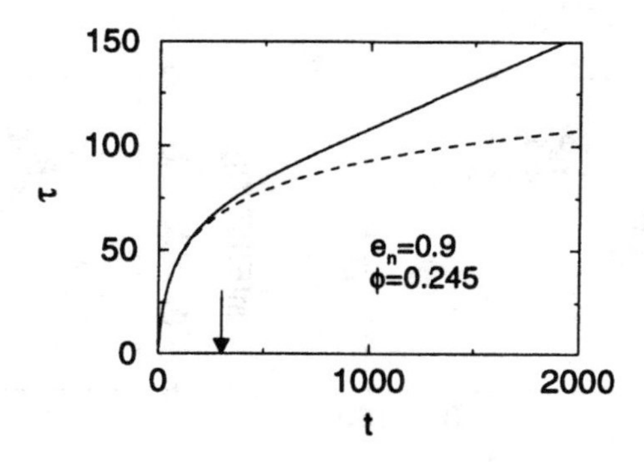

Figure 2. **Number of collisions per particle (internal time) τ versus external time t. Comparison of simulations (solid line) and short-time (dashed line) prediction (3). Same parameters as in Fig. 1.**

effects), which is reflected and refracted by the boundaries of the sphere, and converted back into kinetic energy, $E_K^* < E_K$. The collisional loss, $\Delta E_K \simeq -\gamma E_K$, is proportional to the friction coefficient γ of the viscoelastic medium [58].

Let τ_E be the mean time between successive encounters, let t_c be the mean contact time in inelastic collisions, and t_E the mean free flight time. Then one estimates $\tau_E \simeq t_E + t_c$, where $t_E = 1/\omega_E$ is estimated here to be Enskog's mean free time, and the contact time is estimated as $t_c \simeq a_c\sigma/c$, where the constant $a_c \simeq 2$, and $c = \sqrt{\mathcal{E}/m}$ is the speed of the elastic wave propagating inside the granule of mass m, diameter σ and $\mathcal{E}/\sigma^d$ is the elastic modulus. Consequently the collision frequency of the visco-elastic medium is modified to

$$\omega = 1/(t_E + t_c) = \omega_E/(1 + t_c\omega_E). \tag{4}$$

Of course this estimate will break down when the stiffness and/or the packing fraction increase, so that t_c/t_E becomes large, and multiple contacts will dominate the collision frequency. Inserting this relation into Eq. (1) and introducing the internal time variable τ yields exactly the same universal relation as before, $T(t) = T_0 \exp(-2\gamma_0\tau)$, independent of the t–dependence of T and τ. The relations (2) and (3) for the homogeneous cooling state are modified to

$$\begin{aligned} \sqrt{T_0/T} + \tfrac{1}{2}\,\delta \ln(T_0/T) &= 1 + \gamma_0 t/t_0 \\ e^{\gamma_0\tau} + \delta\gamma_0\tau &= 1 + \gamma_0 t/t_0. \end{aligned} \tag{5}$$

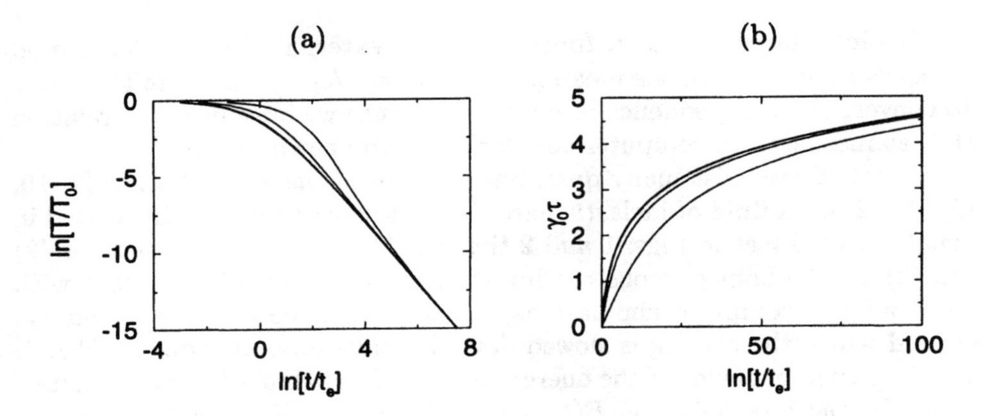

Figure 3. **Generalization of (a) Haff's law and (b) $t-$dependence of τ from Eq.(5) for soft inelastic spheres, with t measured in units of homogeneous cooling times $t_e = t_0/\gamma_0$. The stiffness $\mathcal{E}$ (see Eq. (6)) decreases from top to bottom.**

Here T_0 is the initial granular temperature, and the parameter δ is

$$\delta = t_c/t_0 = t_c \omega_E(T_0) \sim n\sigma^d \chi \sqrt{T_0/\mathcal{E}}, \tag{6}$$

which decreases as $\sim 1/\sqrt{\mathcal{E}}$ with increasing stiffness. The case of elastic hard spheres is recovered for $\mathcal{E} \to \infty$ or $t_c \to 0$. The functions $\tau(t)$ and $T(t)$ are shown in Fig. 3a,b for several values of δ. For large t or small δ one recovers Eqs. (2) and (3).

The time-dependent state with the slowly decaying energy in Eq. (3), evolving from a homogeneous initial state, is the homogeneous cooling state. In *lowest approximation* it can be interpreted as an adiabatic equilibrium state with a homogeneous density n, a vanishing flow field $\mathbf{u} = 0$, and a temperature $T(t)$, that changes adiabatically slowly according to Haff's law. Strong additional support for this statement comes from studies on the velocity distribution in the homogeneous cooling state, both from computer simulations [6, 8], from kinetic theory [26], [29]-[33], as well as from Direct Simulation Monte Carlo (DSMC) methods to simulate the solutions of the nonlinear Boltzmann equation [34]-[36].

2.1.3. *Hydrodynamic Decay*

As long as the system stays spatially homogeneous the temperature $T(t) = (1/dN)\sum_i m(\mathbf{v}_i - \mathbf{u}(\mathbf{r}_i))^2$ and the energy per particle, $E(t) = (1/2N)\sum_i m\mathbf{v}_i^2$, are simply proportional. In Fig.1 we have plotted the energy $E(t)$, as measured in MD simulations [17], as a function of the internal and external time.

To plot the energy as a function of the external time we have used Enskog's prediction for the mean free time $t_0 = 1/\omega(T_0)$ in the initial state. To convert the τ-dependence into a t-dependence we have used the relation $\tau(t)$, as measured in computer simulations, and shown in Fig. 2.

In fact, there exist many quantitative verifications of Haff's law [7, 10, 12, 25, 22] for a fluid of inelastic hard disks at short times, as illustrated in Fig. 1a,b. One sees in Figs. 1 and 2 that the kinetic theory predictions (2) and (3) for the homogeneous cooling state are in excellent agreement with the results of computer simulations, at least until the crossover time τ_{cr}, beyond which the cooling is slowed down considerably, as shown in Fig. 1. The *long* time behavior of the energy deviates from Haff's law and is determined by the flow field, *i.e.* $E(t) \sim (1/2N) \int d\mathbf{r}\rho u^2$. For small inelasticity γ_0 only shear modes contribute, and one obtains the long time behavior for thermodynamically large systems in a straightforward manner [17],

$$\frac{E(t)}{E_0} \simeq \frac{d-1}{dn} \left(\frac{\omega}{8\pi\nu\tau} \right)^{d/2}, \tag{7}$$

where ν is the kinematic viscosity. The agreement between the theoretical prediction and MD simulations is excellent. It holds for $\tau > \tau_{cr}$, far into the clustering regime. Note that the time $\tau = 160$ in Fig. 1 corresponds to the snapshot of the density field in Fig. 4. Moreover the asymptotic formula applies to soft inelastic spheres as well. An accurate estimate of the crossover time τ_{cr} is obtained by equating the right hand sides of Eqs. (2) and (7).

One observes in MD simulations [17, 19] that the internal time τ becomes again *linear* in the external time t, say, $\tau \simeq \tau_o + \omega_{cl} t$, for $\tau \gg \tau_{cl}$, with a 'generic' slope, $\omega_{cl} = d\tau/dt$, as shown in Fig. 2. This slope is independent of the realized configurations, and much smaller than the initial slope.

Combining the phenomenological input with the analytic result (7), one obtains for very long times $E(t) \sim t^{-d/2}$. This t^{-1} - tail, for d=2, has indeed been observed in large scale MD simulations with 10^6 inelastic disks by Herrmann and collaborators [34, 22]. The $t^{-3/2}$ - tail has been observed by Chen *et al.* [47] in MD simulations with 10^6 soft inelastic spheres. We also deduce from Figs. 1 and 2 that the time interval over which the asymptotic relation (7) holds is in fact quite a bit larger than the $t-$ and $\tau-$interval, over which Haff's law is valid.

As hydrodynamic fluctuations are outside the scope of this article, we will not pursue this topic any further (see [17, 19]).

2.2. HYDRODYNAMIC EQUATIONS

Let us consider the macroscopic equations for the inelastic hard-sphere fluid, where $n(\mathbf{r}, t)$ is the local density, $\mathbf{g}(\mathbf{r}, t) = mn(\mathbf{r}, t)\mathbf{u}(\mathbf{r}, t)$ the local momentum density, and $e(\mathbf{r},t) = \frac{1}{2}mn(\mathbf{r},t)\mathbf{u}^2(\mathbf{r},t) + \frac{d}{2}n(\mathbf{r},t)T(\mathbf{r},t)$ the local energy density. For weakly dissipative granular fluids the hydrodynamic fields obey the standard hydrodynamic equations, except that the temperature balance equation contains an additional energy sink term, given in mean field approximation by $\Gamma = -2\gamma_0\omega T$ [1]-[4]. They read

$$\begin{aligned} \partial_t n + \boldsymbol{\nabla}\cdot(n\mathbf{u}) &= 0 \\ \partial_t \mathbf{u} + \mathbf{u}\cdot\boldsymbol{\nabla}\mathbf{u} &= -\frac{1}{\rho}\boldsymbol{\nabla}\cdot\boldsymbol{\Pi} \\ \partial_t T + \mathbf{u}\cdot\boldsymbol{\nabla}T &= -\frac{2}{dn}(\boldsymbol{\nabla}\cdot\mathbf{J} + \boldsymbol{\Pi}:\boldsymbol{\nabla}\mathbf{u}) - \Gamma. \end{aligned} \tag{8}$$

Here $\rho = mn$, $\mathbf{u}$ the flow velocity, and $\frac{1}{2}dnT$ the kinetic energy density in the local rest frame of the inelastic hard-sphere fluid. The pressure tensor $\Pi_{\alpha\beta} = p\delta_{\alpha\beta} + \delta\Pi_{\alpha\beta}$ contains the local pressure p and the dissipative momentum flux $\delta\Pi_{\alpha\beta}$, which is proportional to $\nabla_\alpha u_\beta$ and contains the kinematic and longitudinal viscosities ν and ν_l [12]. The constitutive relation for the heat flux, $\mathbf{J} = -\kappa\boldsymbol{\nabla}T$, defines the heat conductivity κ. The pressure is assumed to be given by its value for *elastic* hard spheres. Similarly, the transport coefficients ν, ν_l, and κ.are assumed to be given by the Enskog theory for a dense gas of *elastic* hard spheres or disks [62]. These equations also apply to a fluid of visco-elastic spheres, where the excess pressure $p^{exc} = p - nT$, the collisional damping Γ and the transport coefficients ν, ν_l, and κ are all proportional to the modified collision frequency in Eq.(4).

2.3. VORTEX STRUCTURES (PENEPLANATION)

How do the hydrodynamic fields behave? Initially the system is started in a spatially homogeneous state with a constant temperature and a vanishing flow field. This situation lasts for many collisions per particle. Gradually, the flow field starts to show local alignment of the particle velocities, while the density field remains homogeneous. This alignment induces an energy decay, which is slower than the decay in the homogeneous cooling state, as given by Haff's law. The local alignment leads on intermediate time scales to the formation of large eddies or vortex structures, which slowly grow in size and become more pronounced. At later times the inhomogeneities in the density start to become visible. This is illustrated in Fig. 4.

We first note a striking feature of the flow field, which is the large degree of local alignment or 'parallelization', called 'coherent motion' by Goldhirsch *et al.* [6], called 'noise reduction in the velocity distribution' by Brito and the author [17], and 'ordered flow' by Luding *et al.* [22]. How can these observations be understood? When two inelastic hard spheres collide, they lose a fraction of their *relative* velocity $\mathbf{v}_{12}$. As a result of

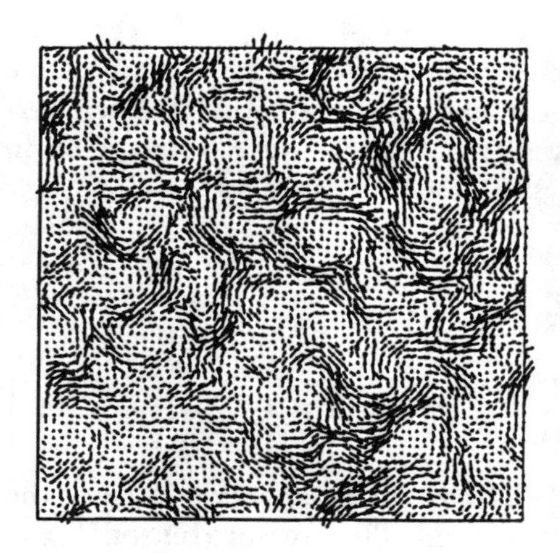

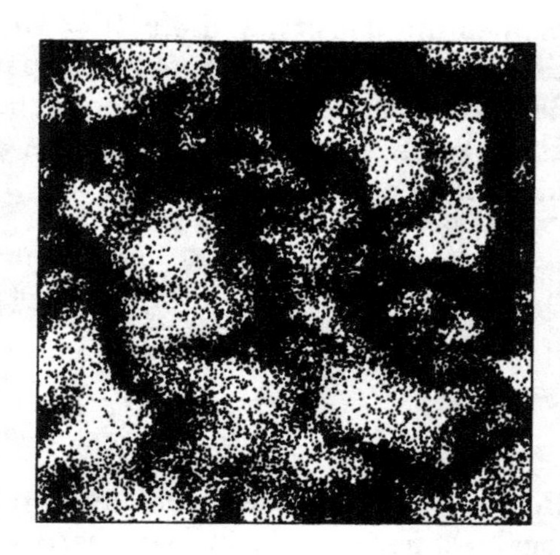

Figure 4. Snapshots of momentum field (left) and density field (right) for $\phi = 0.245$, $e_n = 0.9$ at $\tau = 80$ (left) and $\tau = 160$ (right) for $N = 50000$ disks.

such repeated collisions, the spheres tend to move more and more parallel. The larger part of a particle's velocity, $\mathbf{v} = \mathbf{u} + \mathbf{V}$, is given by the mean flow $\mathbf{u}$; the smaller part is the random fluctuating ('thermal') velocity $\mathbf{V}$, whose typical size is given by the r.m.s. velocity, $\sqrt{\langle \mathbf{V}^2 \rangle} \sim \sqrt{T}$, which is proportional to the speed of sound. Consequently, free granular flows are typically 'supersonic'[1], with $|\mathbf{V}| \sim \sqrt{T} \ll |\mathbf{u}|$, whereas flows in elastic fluids are typically 'subsonic' with $|\mathbf{V}| \sim \sqrt{T} \gg |\mathbf{u}|$. Therefore, the total kinetic energy of granular fluids is essentially the kinetic energy of the flow field. Figures 5a,b show a blow up of part of an actual configuration in a late stage of evolution showing the instantaneous positions and velocities of the hard disks in the laboratory frame (a) and in the local rest frame (b) of a typical particle (shaded). The local flow is supersonic with quite a large Mach number.

It is of interest to briefly discuss the mechanisms of pattern formation leading to the large eddies and density clusters, as observed in Fig. 4. To quantify the development of patterns we consider the structure factors

$$\begin{aligned} S_{nn}(k,t) &= V^{-1}\langle\, |\delta n(\mathbf{k},t)|^2 \rangle \\ S_{\perp}(k,t) &= V^{-1}\langle\, |u_{\perp}(\mathbf{k},t)|^2 \rangle, \end{aligned} \tag{9}$$

where $\delta n(\mathbf{k},t)$ and $u_{\perp}(\mathbf{k},t)$ are Fourier modes of the fluctuations in density and transverse flow field, and Fig. 6a,b shows some typical results for the structure factors. An elastic fluid in thermal equilibrium does not show any structure on *hydrodynamic* length scales, where $k \lesssim \min\{1/\ell_0, 1/\sigma\}$ with

Figure 5. Supersonic flow: blow-up of a local $\{\mathbf{r}_i, \mathbf{v}_i\}$–configuration of inelastic hard disks (a) in the laboratory frame and (b) in the local rest frame of the shaded particle. The length of the arrows in (b), when compared to (a), is magnified by a factor M, which is on the order of the Mach number of the flow. The parameters are $N = 50000, \phi = 0.05, e_n = 0.85$, and the crossover time $\tau_{cr} \simeq 59$.

ℓ_0 the mean free path and σ the particle diameter[1]. This means that the hydrodynamic structure factors $S_{nn}(k)$ and $S_{\perp}(k)$ are totally flat, independent of k. The corresponding hydrodynamic correlation functions are short ranged on these length scales. Development of structure on length scales larger than the microscopic scales $\{l_0, \sigma\}$ will manifest itself in the appearance of one or more maxima or peaks in the structure factors. A *linear instability* will manifest itself in a structure factor that grows exponentially in time. With these concepts in mind, we analyze the structure factors for the inelastic hard-sphere fluid, as we want to determine which physical excitations are responsible for the features observed in the MD simulations and in the numerical solutions.

The physical implications of the growing vortices in Fig. 6a are quite interesting. This figure shows the phenomenon of *noise reduction* [17] at small wavelengths. With increasing time the noise strengths $S_{\perp}(k,t)$ of the most dominant fluctuations in the transverse flow field decrease at larger k values and remain bounded for all t by their initial equipartition value $S_{\perp}(k,0) = T_0/mn$, which is independent of k. This can be rephrased by stating that the flow field exhibits only a 'relative' instability. Nothing increases, but the surroundings of $\mathbf{k} = 0$ decrease.

[1]Note that in a dense fluid of hard spheres $l_0 < \sigma$ for large packing fractions ϕ. For instance, hard disks at $\phi = 0.2$ and 0.4 have respectively $l_0/\sigma \simeq 1.1$ and 0.34.

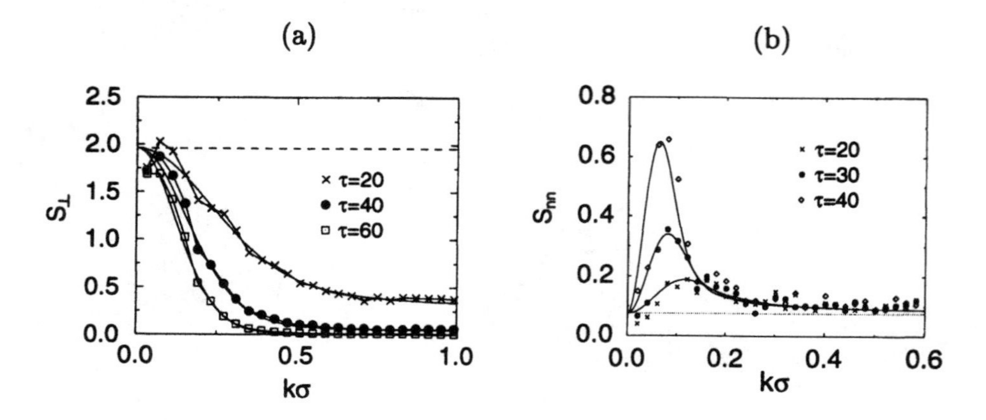

Figure 6. Structure factors (a) $S_\perp(k,t)$ and (b) $S_{nn}(k,t)$ versus k at different τ values, showing structure formation of (a) large vortices and (b) density clusters through respectively (a) peneplanation and (b) a spinodal decomposition-type instability. Simulation data [19] compared with theory (smooth solid lines) [13]. Parameters are $N = 50000, \phi = 0.4, e_n = 0.9$.

The noise reduction is a direct consequence of the microscopic inelastic collisions, which force the particles to move more and more parallel in successive collisions. It is this 'physical coarse graining' process that selectively suppresses the shorter wavelength fluctuations in the flow field in an ever-increasing range of wavelengths.

So, noise reduction is the pattern selection mechanism responsible for the large scale eddies observed in Fig. 4. An interesting analogue of the peak, formed at $\mathbf{k} = \mathbf{0}$ in reciprocal space, is peneplanation, a mechanism for pattern formation in structural geology caused by selective erosion [61], of which Ayer's Rock (Mount Uluru) in the center of Australia is an outstanding example.

2.4. DENSITY CLUSTERS

The most spectacular feature in undriven granular flows is the clustering instability, *i.e.* the spontaneous formation of high density clusters, a phenomenon discovered by Goldhirsch and Zanetti [6]. Local fluctuations in the density around the homogeneous cooling state do not decay on average, as expected on the basis of Onsager's regression hypothesis [63], but slowly grow to macroscopic size (clusters), as illustrated in Fig. 6b.

The basic explanation of the density instability comes from the inelasticity of the collisions. Suppose we start from a homogeneous initial state. A positive density fluctuation δn at position $\mathbf{r}$ leads to an increased collision rate, which in turn leads to cooling ($\delta T < 0$) and to a pressure drop

($\delta p < 0$). Then particles from the surrounding (hotter) regions flow into the region around $\mathbf{r}$, which in turn increases the local density again, and the process keeps repeating itself. This instability is driven by a pressure gradient, which yields a flow field. The process leads to a 'phase separation' into regions with cold dense clusters and hot dilute regions, turning gradually into voids.

A more quantitative and more macroscopic explanation is suggested by comparing this pattern forming mechanism with spinodal decomposition [60], where the observed phenomena are similar in several respects. In spinodal decomposition a mixture is prepared in a spatially homogeneous state (analogous to the homogeneous cooling state) by a deep temperature quench into the unstable immiscible region. Here long wavelength composition fluctuations are unstable (analogous to density fluctuations in granular fluids), and lead to phase separation and pattern formation (analogous to the formation of density clusters and vortex structures in Fig. 4). This instability and the concomitant pattern formation are explained by the Cahn-Hilliard theory [60]. A similar theory for undriven granular flows has been proposed in Refs. [10, 12, 13]. This theory enables one to calculate the structure factors, and the growing size $L_{cl}(t) \simeq \sqrt{\tau/\gamma_0}$ of the high density clusters, at least in the initial stages, $\tau < \tau_{cr}$.

So far we have described and partially explained some phenomena, occurring in undriven granular fluids. In the second part of this article we turn to a more mathematical derivation of the kinetic equations for granular fluids.

3. Liouville and Kinetic Equations

3.1. CONSERVATIVE INTERACTIONS

To give a didactic presentation we start from the Liouville equation for conservative interactions, and then systematically extend it to include nonconservative forces such as friction.

In nonequilibrium statistical mechanics one can calculate the mean value of a dynamical variable $A(\mathbf{x})$ from either of the following expressions

$$\int d\mathbf{x}\, \rho(\mathbf{x}, 0) A(\mathbf{x}(t)) = \int d\mathbf{x}\, \rho(\mathbf{x}, t) A(\mathbf{x}), \tag{10}$$

where $\mathbf{x} = \{x_1, x_2, \ldots x_N\}$ with $x_i = \{\mathbf{r}_i, \mathbf{v}_i\}$ is a point in the N-particle phase space. On the left hand side the time dependence is assigned to the dynamical variable $A(\mathbf{x}(t)) \equiv S_t(\mathbf{x}) A(\mathbf{x})$ with $S_t(\mathbf{x})$ the time evolution or streaming operator, and on the right hand side to the N-particle distribution function $\rho(\mathbf{x}, t)$. This can be done by considering expression (10) as an inner product. The time dependence of the N-particle distribution function is then given by

$$\rho(\mathbf{x},t) = S_t^\dagger \rho(\mathbf{x},0), \tag{11}$$

where $S_t^\dagger(\mathbf{x})$ is the adjoint of $S_t(\mathbf{x})$.

If the interactions are conservative and additive, the force between the pair (ij) is $\mathbf{F}_{ij} = -\partial V(r_{ij})/\partial \mathbf{r}_{ij}$, where $V(r)$ is the pair potential and the streaming operator is given by

$$S_t(\mathbf{x}) = \exp[tL(\mathbf{x})] = \exp\left[t\sum_i L_i^0 - t\sum_{i<j}\theta(ij)\right], \tag{12}$$

where $L(\mathbf{x})\ldots = \{H(\mathbf{x}),\ldots\}$ is the Poisson bracket with the Hamiltonian. This yields

$$\begin{aligned} L_i^0 &= \mathbf{v}_i \cdot \frac{\partial}{\partial \mathbf{r}_i} \\ \theta(ij) &= \frac{1}{m}\frac{\partial V(r_{ij})}{\partial \mathbf{r}_{ij}} \cdot \left(\frac{\partial}{\partial \mathbf{v}_i} - \frac{\partial}{\partial \mathbf{v}_j}\right). \end{aligned} \tag{13}$$

In this case $S_t(\mathbf{x})$ is a unitary operator, $S_t^\dagger(\mathbf{x}) = S_{-t}(\mathbf{x})$, and $L^\dagger = -L$. The free streaming operator $S_t^0(\mathbf{x}) = \exp[t\sum_i L_i^0]$ generates the free particle trajectories,

$$S_t^0(\mathbf{x})A(\mathbf{r}_i,\mathbf{v}_i) = A(\mathbf{r}_i + \mathbf{v}_i t, \mathbf{v}_i). \tag{14}$$

The Liouville equation, which is the evolution equation for the phase-space density in conservative systems, follows from (11), and reads

$$\left(\partial_t + \sum_i L_i^0\right)\rho(\mathbf{x},t) = \sum_{i<j}\theta(ij)\rho(\mathbf{x},t), \tag{15}$$

which is an expression of the incompressibility of the flow in phase space.

An equivalent representation of the time evolution of the system can be given in terms of reduced s-particle distribution functions ($s = 1,2,\ldots$), defined as

$$f_{12\ldots s}(t) \equiv f^{(s)}(x_1, x_2, \ldots x_s, t) = \frac{N!}{(N-s)!}\int \mathrm{d}x_{s+1}\ldots\mathrm{d}x_N \rho(\mathbf{x},t), \tag{16}$$

where $\rho(\mathbf{x},t)$ is normalized to unity. Integrating (15) over $x_{s+1}\ldots x_N$ yields the BBGKY hierarchy for the reduced distribution functions, named after Bogolyubov, Born, Green, Kirkwood and Yvon. We only quote the first two equations of the hierarchy,

$$\begin{aligned} (\partial_t + L_1^0)f_1 &= \int \mathrm{d}x_2 \theta(12) f_{12} \\ \left[\partial_t + L_1^0 + L_2^0 - \theta(12)\right] f_{12} &= \int \mathrm{d}x_3 [\theta(13) + \theta(23)] f_{123}. \end{aligned} \tag{17}$$

This set of equations is an open hierarchy, which expresses the time evolution of the s-particle distribution function in terms of the $(s+1)$-*th* function.

It is a minor generalization to include external conservative force fields. In that case the infinitesimal generator L_i^0 of the free-particle motion should be replaced by

$$L_i^0 = \mathbf{v}_i \cdot \frac{\partial}{\partial \mathbf{r}_i} + \mathbf{a}_i \cdot \frac{\partial}{\partial \mathbf{v}_i}, \tag{18}$$

where $\mathbf{a}_i$ is an external conservative force per unit mass, acting on the i-*th* particle, for instance gravity.

To obtain the Boltzmann equation for dilute gases with short-range repulsive and non-dissipative interparticle forces, one may follow Boltzmann's intuitive derivation, based on the assumption of molecular chaos [62]. It yields

$$\left(\partial_t + \mathbf{v}_1 \cdot \frac{\partial}{\partial \mathbf{r}}\right) f(r, \mathbf{v}_1, t) = I(f, f)$$

$$\equiv \int d\mathbf{v}_2 \int d\mathbf{b} v_{12} \left[f(\mathbf{r}, \mathbf{v}_1' t) f(\mathbf{r}, \mathbf{v}_2', t) - f(\mathbf{r}, \mathbf{v}_1, t) f(\mathbf{r}, \mathbf{v}_2, t)\right], \tag{19}$$

where $\mathbf{v}_{12} = \mathbf{v}_1 - \mathbf{v}_2$ is the relative velocity, $\mathbf{b} = \mathbf{r}_{12} \times \hat{\mathbf{v}}_{12}$ is the impact parameter, $d\mathbf{b} = db$ with $|b| \le \sigma$ in two dimensions, and $d\mathbf{b} = b\, db\, d\varphi$ with $b \le \sigma$ in three dimensions and an azimuthal angle $\varphi \in (0, 2\pi)$.

3.2. FRICTIONAL FORCES

Our next goal is to generalize this result to include frictional forces, exerted by external sources or by interparticle interactions in complex fluids. Suppose the N-particle system is described by the equations of motion

$$\frac{d\mathbf{r}_i}{dt} = \mathbf{v}_i \qquad \frac{d\mathbf{v}_i}{dt} = \mathbf{a}_i, \tag{20}$$

where $\mathbf{a}_i$ is the force per unit mass. As the total force is vanishing, *i.e.* $\sum_i \mathbf{a}_i = \mathbf{0}$, the average momentum density satisfies a local conservation law, implying that the local flow velocity $\mathbf{u}(\mathbf{r}, t)$ is a slowly varying macroscopic field. The systematic forces may be conservative or frictional. They may be caused by external sources, or by interparticle interactions, as used to model complex fluids.

To derive the equation of motion for the N-particle distribution $\rho(\mathbf{x}, t)$ — referred to as Liouville equation in the present context — we write $\rho(\mathbf{x}, t) = \langle \delta(\mathbf{x} - \hat{\mathbf{x}}(t))\rangle$, where $\hat{\mathbf{x}}(t)$ denotes the trajectory started at the phase point $\hat{\mathbf{x}} = \hat{\mathbf{x}}(0)$, and $\langle \ldots \rangle$ denotes an average over an initial distribution $\rho(\hat{\mathbf{x}}, 0)$. Then

$$\partial_t \rho \quad = \quad \sum_i \frac{\partial}{\partial x_i} \left\langle \delta(\mathbf{x} - \hat{\mathbf{x}}(t)) \frac{d\hat{x}_i(t)}{dt} \right\rangle$$

$$= - \sum_i \left[\frac{\partial}{\partial \mathbf{r}_i} \cdot \mathbf{v}_i + \frac{\partial}{\partial \mathbf{v}_i} \cdot \mathbf{a}_i \right] \rho, \tag{21}$$

where $\mathbf{a}_i$ represents a conservative or dissipative force, caused by external fields or by interparticle forces.

The Liouville equation (21) does *not* describe an incompressible flow in phase space, as its Hamiltonian counterpart (15) does. If the forces are conservative, then $\mathbf{a}_i$ depends only on positions, and the Liouville equation (21) reduces to its standard form (15), with L_i^0 given by (18) in case the system is subject to an external conservative force field. If the forces are impulsive, the method of 'pseudo-pair forces' with $T(ij)$ and $\overline{T}(ij)$ operators can be used, as will be discussed in the next section. If the forces are frictional, $\mathbf{a}_i$ may depend on the velocities of the particle, and cannot be interchanged with $\partial/\partial\mathbf{v}$.

Assume that there are no external driving forces, so that $m\mathbf{a}_i = \sum_j \mathbf{F}_{ij}$ originates from interparticle forces. Then the Liouville equation (21) takes the form

$$\Big(\partial_t + \sum_i L_i^0\Big)\rho(\mathbf{x},t) = - \sum_{i<j} \theta^\dagger(ij)\rho(\mathbf{x},t) \tag{22}$$

with

$$\theta^\dagger(ij) = \frac{1}{m}\left(\frac{\partial}{\partial \mathbf{v}_i} - \frac{\partial}{\partial \mathbf{v}_j}\right) \cdot \mathbf{F}_{ij}. \tag{23}$$

The corresponding BBGKY hierarchy for the reduced distribution functions takes the same form as in (17) with $\theta(ij)$ replaced by $-\theta^\dagger(ij)$. They can be used as a formal starting point to derive the Boltzmann equation for models with dissipative interactions, such as visco-elastic spheres.

The above equation can be applied directly to the model of soft inelastic spheres, which is frequently used to model granular fluids [42]-[47]. In this model the interparticle forces $m\mathbf{a}_i = \sum_{j(\neq i)} \mathbf{F}_{ij}$ with $\mathbf{F}_{ij} = \mathbf{F}_{ij}^C + \mathbf{F}_{ij}^D$ contain a rather stiff elastic repulsion, $\mathbf{F}_{ij}^C$, and a fluid-type frictional force, $\mathbf{F}_{ij}^D$, proportional to the relative velocity $\mathbf{v}_{ij}$, *i.e.*

$$\begin{aligned} \mathbf{F}_{ij}^C &= \mathcal{E}(\sigma - r_{ij})\hat{\mathbf{r}}_{ij} w(r_{ij}) \\ \mathbf{F}_{ij}^D &= -\gamma_n \mathbf{v}_{ij} \cdot \hat{\mathbf{r}}_{ij}\hat{\mathbf{r}}_{ij} w(r_{ij}), \end{aligned} \tag{24}$$

where $w(r)$ is the overlap function, defined as $w(r) = 1$ for $r < \sigma$, and $w(r) = 0$ for $r > \sigma$. Here σ is the diameter of the soft sphere, $\hat{\mathbf{a}}$ is a unit vector along $\mathbf{a}$, the relative distance is $\mathbf{r}_{ij} = \mathbf{r}_i - \mathbf{r}_j$ with length r_{ij}. Moreover, the coefficient $\mathcal{E}$ is the stiffness or Young modulus and γ_n the coefficient of (normal) friction. The frictional force above is directed normal to the surface. One may also add a tangential friction,

$$\mathbf{F}_{ij}^{D\prime} = -\gamma_t(\mathbf{v}_{ij} - \mathbf{v}_{ij} \cdot \hat{\mathbf{r}}_{ij}\hat{\mathbf{r}}_{ij}) w(r_{ij}). \tag{25}$$

The forces in (24) and (25) are only nonvanishing when the particles are in contact, *i.e.* $r_{ij} < \sigma$.

If the Young modulus and friction coefficients are small, the particles are only weakly scattered in binary collisions, and a weak coupling kinetic equation may be derived, as has been done by Marsh *et al.* [64] for the somewhat related model of 'dissipative particle dynamics'. However, in applications to granular fluids the hard-sphere limit $\mathcal{E} \longrightarrow \infty$ is more appropriate, and the kinetic equation of Ref.[64] is not applicable. In the hard-sphere limit, the forces become impulsive, and the method of the next section is more suitable. However, the relationship between the friction coefficients γ_n and γ_t, and the restitution coefficient, e_n and e_t of the next section, is only approximately known from quasi-static energy considerations [25].

4. Impulsive Forces and Binary Collision Operators

If the interactions have a hard core, as for inelastic hard spheres, the derivatives of the potential in Eq. (13) are ill-defined, but the trajectories in phase space are well-defined. They consist of free propagation and instantaneous collisions with well-defined collision laws.

For an explicit description we need to fully specify the *inelastic hard sphere* model. The interparticle interactions are modeled by instantaneous collisions as in the case of elastic hard spheres. During a collision momentum will be transferred instantaneously along the line joining the centers of the two colliding particles, indicated by the vector $\boldsymbol{\sigma}$ pointing from the center of particle 2 to that of particle 1. The collision rules for the dissipative reflection laws yield postcollision velocities,

$$\begin{aligned} v^*_{12,n} &= -e_n v_{12,n} \\ \mathbf{v}^*_{12,t} &= e_t \mathbf{v}_{12,t}, \end{aligned} \tag{26}$$

where $0 < e_n \leq 1$ and $-1 \leq e_t \leq 1$. The coefficients e_n and e_t are the coefficients of normal and tangential restitution, which are assumed to be independent of the impact velocities. The case $e_n = e_t = 1$ corresponds to elastic hard spheres. The normal (n) component and the $(d-1)$ tangential (t) components are respectively parallel and perpendicular to vector $\boldsymbol{\sigma} = \sigma\hat{\boldsymbol{\sigma}}$, connecting the centers of colliding spheres at contact. Combination of these reflection laws with momentum conservation $\mathbf{v}^*_1 + \mathbf{v}^*_2 = \mathbf{v}_1 + \mathbf{v}_2$, yields explicit expressions for the postcollision velocities $\mathbf{v}^*_i (i = 1, 2)$ in terms of precollision ones, *i.e.*

$$\begin{aligned} \mathbf{v}^*_1 &= \tfrac{1}{2}(\mathbf{v}_1 + \mathbf{v}_2) + \tfrac{1}{2}e_t\mathbf{v}_{12} - \tfrac{1}{2}(e_n + e_t)(\mathbf{v}_{12} \cdot \hat{\boldsymbol{\sigma}})\,\hat{\boldsymbol{\sigma}} \\ \mathbf{v}^*_2 &= \tfrac{1}{2}(\mathbf{v}_1 + \mathbf{v}_2) - \tfrac{1}{2}e_t\mathbf{v}_{12} + \tfrac{1}{2}(e_n + e_t)(\mathbf{v}_{12} \cdot \hat{\boldsymbol{\sigma}})\,\hat{\boldsymbol{\sigma}}. \end{aligned} \tag{27}$$

The restituting (precollision) velocities $(\mathbf{v}_1^{**}, \mathbf{v}_2^{**})$ leading to $(\mathbf{v}_1, \mathbf{v}_2)$, are found by inverting collision rule (27) and given by

$$\begin{aligned}
\mathbf{v}_1^{**} &= \tfrac{1}{2}(\mathbf{v}_1 + \mathbf{v}_2) + \tfrac{1}{2e_t}\mathbf{v}_{12} - \tfrac{1}{2}\left(\tfrac{1}{e_n} + \tfrac{1}{e_t}\right)(\mathbf{v}_{12} \cdot \hat{\boldsymbol{\sigma}})\,\hat{\boldsymbol{\sigma}} \\
\mathbf{v}_2^{**} &= \tfrac{1}{2}(\mathbf{v}_1 + \mathbf{v}_2) - \tfrac{1}{2e_t}\mathbf{v}_{12} + \tfrac{1}{2}\left(\tfrac{1}{e_n} + \tfrac{1}{e_t}\right)(\mathbf{v}_{12} \cdot \hat{\boldsymbol{\sigma}})\,\hat{\boldsymbol{\sigma}}.
\end{aligned} \tag{28}$$

Note that this inversion is not possible for $e_n = 0$ or $e_t = 0$. We note that the components of $\mathbf{v}_{12}$ in Eq. (26) decrease upon collision. In a binary collision between particles 1 and 2, an amount ΔE of the total energy is lost, *i.e.*

$$\begin{aligned}
\Delta E &= \tfrac{1}{2}m(\mathbf{v}_1^{*2} + \mathbf{v}_2^{*2} - \mathbf{v}_1^2 - \mathbf{v}_2^2) \\
&= \tfrac{1}{4}m\left[(1 - e_n^2)v_{12,n}^2 + (1 - e_t^2)\mathbf{v}_{12,t}^2\right],
\end{aligned} \tag{29}$$

where the deviations of e_n and e_t from unity are a measure of the inelasticity of the collision. As mentioned already, the total momentum is conserved. However, total (orbital) angular momentum, $\mathcal{L} = m\sum_i \mathbf{r}_i \times \mathbf{v}_i$, is not conserved, but an amount, $\Delta\mathcal{L} = \frac{1}{2}m(1 - e_t)\,\boldsymbol{\sigma} \times \mathbf{v}_{12}$, is lost in every (12)-collision because the tangential forces are non-central. We note in addition that periodic boundary conditions do not conserve angular momentum.

The inelastic hard-sphere model incorporates the most fundamental feature of the dissipative dynamics of rapid granular flows, namely the irreversible loss of kinetic energy in collisions. Moreover, it is a many-body system with well-defined and relatively simple dynamics, to which the many-body methods of statistical mechanics can be applied.

Next, we construct the streaming operators S_t for inelastic hard-sphere fluids. The results described here have been derived in [29, 30]. For particles with hard-core interactions the dynamics is undefined for physically inaccessible configurations, where the particles are overlapping. Such configurations have a vanishing weight in Eq. (10), since $S_t(\mathbf{x})$ only appears in the combination $\rho(\mathbf{x}, 0)S_t(\mathbf{x})$ which vanishes for overlapping initial configurations. Therefore, it suffices to consider $W_N(\mathbf{x})S_t(\mathbf{x})$ with $W_N(\mathbf{x}) = \prod_{i<j} W(r_{ij})$, where the overlap function is $W(r) = 1$ for $r > \sigma$ and $W(r) = 0$ for $r < \sigma$. However, the methods of many-body theory require formal perturbation expansions and subsequent resummations. To do so, the time evolution operator $S_t(\mathbf{x})$ needs to be defined for *all* configurations, including the unphysical overlapping configurations. A standard representation, defined for all points in phase space, has been developed for elastic hard spheres in Ref. [54], and is based on the binary collision expansion of $S_t(\mathbf{x})$ in terms of binary collision operators.

The binary collision operator, $T(ij)$, is defined in terms of two-body dynamics through the time displacement operator $S_t(12)$ as

$$S_t(12) = S_t^0(12) + \int_0^t \mathrm{d}\tau S_\tau^0(12)T(12)S_{t-\tau}^0(12). \tag{30}$$

Following the argument of Ref. [54] for the case of elastic hard spheres step by step, the binary collision operator $T(12)$ for inelastic hard spheres is constructed as [54, 29, 30]

$$T(12) = \sigma^{d-1} \int^{(-)} \mathrm{d}\hat{\boldsymbol{\sigma}} |\mathbf{v}_{12} \cdot \hat{\boldsymbol{\sigma}}| \delta(\mathbf{r}_{12} - \boldsymbol{\sigma})(b_\sigma^* - 1). \tag{31}$$

Here $\int^{(\pm)}$ indicates that the $\hat{\sigma}$-integration is restricted to the precollision $(-)$ or postcollision $(+)$ hemisphere, and b_σ^* is an operator that replaces all (precollision) velocities $\mathbf{v}_i$ $(i = 1, 2)$ appearing to its right by postcollision velocities $\mathbf{v}_i^*$.

The operator $T(12)$ is defined for overlapping and nonoverlapping configurations of two hard spheres. It extends the definition of $S_t(12)$ to all points in phase space. In the ensemble average considered in Eq. (10), the overlap function $W_N(\mathbf{x})$ contains a factor $W(r_{12})$ which vanishes whenever $r_{12} < \sigma$. The generator $S_t(12)$ for two-particle dynamics is only defined for *positive* times. The conservation laws imply $T(12)(a(\mathbf{v}_1) + a(\mathbf{v}_2)) = 0$, where $a(\mathbf{v})$ is a collisional invariant with $a(\mathbf{v}) = \{1, \mathbf{v}, v^2\}$ for elastic hard spheres, and $a(\mathbf{v}) = \{1, \mathbf{v}\}$ for inelastic hard spheres.

Combinations of T operators and free streaming operators S_t^0, preceded by appropriate combinations of overlap functions, can be used to construct the time displacement operators S_t for dynamical variables of the many-body problem. To discuss the Liouville equation and describe the time evolution of the reduced distribution functions we need in addition to consider the adjoint time displacement operators, $S_t^\dagger$, which is defined through the inner product (10). This yields $S_t^{0\dagger} = S_{-t}^0$, and the adjoint $\overline{T}$ of T is constructed as

$$\overline{T}(12) = \sigma^{d-1} \int^{(+)} \mathrm{d}\hat{\boldsymbol{\sigma}} (\mathbf{v}_{12} \cdot \hat{\boldsymbol{\sigma}}) \left(\frac{1}{e_n^2 e_t^{d-1}} \delta(\mathbf{r}_{12} - \boldsymbol{\sigma}) b_\sigma^{**} - \delta(\mathbf{r}_{12} + \boldsymbol{\sigma}) \right). \tag{32}$$

Here b_σ^{**} acts on the velocities $\mathbf{v}_i$ $(i = 1, 2)$ to its right and replaces them by the restituting velocities, $\mathbf{v}_i^{**}$, as defined in collision rule (28).

The time displacement operators $S_t(12)$ can be put in a more convenient form by using the property, $T(12)S_t^0(12)T(12) = 0$, valid for any $t > 0$. It also holds with T replaced by $\overline{T}$. This relation expresses the fact that two hard spheres cannot collide more than once with only free propagation in between. Using this property, the time displacement operator can be written as

$$W(12)S_t(12) = W(12)\exp\Big[tL^0(12) + tT(12)\Big]$$
$$S_t^\dagger(12)W(12) = \exp\Big[-tL^0(12) + t\overline{T}(12)\Big]W(12), \qquad (33)$$

where $L^0(12) = L_1^0 + L_2^0$. This can readily be generalized to the full N-particle operators, and be represented in the compact form of pseudo-streaming operators,

$$W_N(\mathbf{x})S_t(\mathbf{x}) = W_N(\mathbf{x})\exp\Big[tL^0(\mathbf{x}) + t\sum_{i<j}T(ij)\Big]$$
$$S_t^\dagger(\mathbf{x})W_N(\mathbf{x}) = \exp\Big[-tL^0(\mathbf{x}) + t\sum_{i<j}\overline{T}(ij)\Big]W_N(\mathbf{x}), \qquad (34)$$

with $L^0(\mathbf{x}) = \sum_i L_i^0$ the free particle streaming operator. The time evolution operators are defined everywhere in phase space, and the overlap function gives a vanishing weight to unphysical configurations, provided that $W_N(\mathbf{x})$ appears to the left of T operators, or to the right of $\overline{T}$ operators.

The time evolution of the N-particle distribution function $\rho(\mathbf{x},t)$ is given by the pseudo-Liouville equation. For conservative Hamiltonian systems this equation is the Liouville equation. According to Eqs. (11) and (34) the time evolution of the distribution function for inelastic hard-sphere fluids is given by the pseudo-Liouville equation

$$\Big[\partial_t + L^0(\mathbf{x})\Big]\rho(\mathbf{x},t) = \sum_{i<j}\overline{T}(ij)\rho(\mathbf{x},t). \qquad (35)$$

The binary collision operators $\{T(ij), \overline{T}(ij)\}$ for elastic or inelastic hard-sphere fluids account for the impulsive hard-sphere interactions, and may be considered as 'pseudo-pair forces'. Hence, the name pseudo-Liouville equation.

An equivalent representation of the time evolution of the system can be given in terms of reduced s-particle distribution functions, defined in (16). Integration of Eq. (35) yields the BBGKY hierarchy for the reduced distribution functions. We only quote the first two hierarchy equations for elastic and inelastic hard spheres, in the absence of external driving forces ($L_i^0 = \mathbf{v}_i \cdot \partial/\partial\mathbf{r}_i$),

$$\Big(\partial_t + L_1^0\Big)f_1 = \int \mathrm{d}x_2\overline{T}(12)f_{12}$$
$$\Big[\partial_t + L_1^0 + L_2^0 - \overline{T}(12)\Big]f_{12} = \int \mathrm{d}x_3\Big[\overline{T}(13) + \overline{T}(23)\Big]f_{123}. \qquad (36)$$

The pseudo-Liouville equation and related BBGKY hierarchy are the *exact* evolution equations for elastic ($e_n = 1$, $e_t = 1$) and inelastic hard spheres, which are *not driven* by external forces. The operators T and $\overline{T}$ have also been constructed for rough hard spheres and needles [31].

5. Kinetic Equations for Inelastic Hard Spheres

5.1. ENSKOG-BOLTZMANN EQUATION

In this section we discuss the Boltzmann equation, the Enskog-Boltzmann equation, and the ring kinetic equations for inelastic hard-sphere systems. In the literature on kinetic theory of inelastic hard-spheres, the first equation of the BBGKY hierarchy has been derived intuitively [26, 27] and used as a starting point to obtain the Enskog-Boltzmann equation for the single-particle distribution function. Using the explicit expression (32) for $\overline{T}(12)$ the first hierarchy equation can be written in full detail as

$$\Big(\partial_t + \mathbf{v}_1 \cdot \frac{\partial}{\partial \mathbf{r}_1}\Big) f(\mathbf{r}_1, \mathbf{v}_1, t) = \sigma^{d-1} \int d\mathbf{v}_2 \int^{(+)} d\hat{\boldsymbol{\sigma}}\, (\mathbf{v}_{12} \cdot \hat{\boldsymbol{\sigma}}) \times$$
$$\left\{ \frac{1}{e_n^2 e_t^{d-1}} f^{(2)}(\mathbf{r}_1, \mathbf{v}_1^{**}, \mathbf{r}_1 - \boldsymbol{\sigma}, \mathbf{v}_2^{**}, t) - f^{(2)}(\mathbf{r}_1, \mathbf{v}_1, \mathbf{r}_1 + \boldsymbol{\sigma}, \mathbf{v}_2, t) \right\}. \quad (37)$$

This equation contains the pair distribution function $f^{(2)}$ of two spheres just *before* touching, *i.e.* at $r_{12} = \sigma + 0$.

In order to derive a closed equation for the single-particle distribution function f, a closure relation is required to express f_{12} in terms of f. The basic ansatz to do so is Boltzmann's molecular chaos assumption, requiring that the velocities of two particles, just before collision, are *uncorrelated*, *i.e.* the pair distribution function factorizes into

$$f_{12} = f(\mathbf{r}_1, \mathbf{v}_1, t) f(\mathbf{r}_2, \mathbf{v}_2, t). \quad (38)$$

This assumption implies that the time evolution of the single-particle distribution function is only determined by sequences of *uncorrelated* binary collisions, whereas sequences of *correlated* binary collisions, *e.g.* (12) (13) (23), can be neglected. To state this even more emphatically: as far as the time evolution of $f(x,t)$ is concerned, it is assumed that no particle collides more than once with any other particle in its whole time evolution. It is clear that this assumption can only be correct for low densities. Here we assume that Boltzmann's assumption of molecular chaos also holds for a dilute gas of *inelastic* hard spheres, at least for *small* inelasticity.

When the density increases the molecular chaos assumption in elastic hard-sphere systems is *violated* [52], due to the increasing importance of ring collisions, which create dynamic correlations between the velocities of particles. Ring kinetic theory [52]-[54] takes these correlated collisions into account. The same scenario applies to inelastic hard spheres.

Here we derive the Boltzmann equation and the ring kinetic equation for the inelastic hard-sphere gases. The Boltzmann equation is obtained from the first hierarchy equation by keeping only terms to dominant order in the density and using Eq. (37).

Furthermore, in the low density limit the spatial separation between the colliding particles can be neglected, and the binary collision operator $\overline{T}(12)$, entering in the BBGKY hierarchy, reduces to

$$\overline{T}(12) = \delta(\mathbf{r}_{12})\overline{T}_0(12) = \delta(\mathbf{r}_{12})\,\sigma^{d-1}\int^{(+)} d\hat{\boldsymbol{\sigma}}\,(\mathbf{v}_{12}\cdot\hat{\boldsymbol{\sigma}})\left(\frac{b_\sigma^{**}}{e_n^2 e_t^{d-1}} - 1\right). \quad (39)$$

Then the nonlinear Boltzmann equation for the single-particle distribution function $f(\mathbf{r}, \mathbf{v}_1, t)$ in a dilute gas of inelastic hard spheres becomes

$$\left(\partial_t + \mathbf{v}_1 \cdot \frac{\partial}{\partial \mathbf{r}}\right) f(\mathbf{r}, \mathbf{v}_1) = I(f, f) \equiv \sigma^{d-1}\int d\mathbf{v}_2 \,\times$$

$$\int^{(+)} d\hat{\boldsymbol{\sigma}}\,(\mathbf{v}_{12}\cdot\hat{\boldsymbol{\sigma}})\left\{\frac{1}{e_n^2 e_t^{d-1}} f(\mathbf{r}, \mathbf{v}_1^{**}) f(\mathbf{r}, \mathbf{v}_2^{**}) - f(\mathbf{r}, \mathbf{v}_1) f(\mathbf{r}, \mathbf{v}_2)\right\}. \quad (40)$$

where the t–dependence of f has been suppressed. There are several significant differences with the Boltzmann equation for the elastic case: *(i)* the occurrence of a factor $1/[e_n^2 e_t^{d-1}]$ in the gain term on the right hand side of (40); a factor $1/J$ with $J = e_n e_t^{d-1}$ comes from the Jacobian $d\mathbf{v}_1^{**} d\mathbf{v}_2^{**} = (1/J)\, d\mathbf{v}_1 d\mathbf{v}_2$ and the other one from the reflection law $\mathbf{v}_{12}^{**}\cdot\hat{\boldsymbol{\sigma}} = -(1/e_n)\mathbf{v}_{12}\cdot\hat{\boldsymbol{\sigma}}$. In the more common models with only normal restitution, one should set $e_t = 1$. *(ii)* In the inelastic case, the restituting precollision velocities, which yield $(\mathbf{v}_1, \mathbf{v}_2)$ as postcollision velocities, are different from the postcollision velocities $(\mathbf{v}_1^*, \mathbf{v}_2^*)$, which result from the direct precollision velocities $(\mathbf{v}_1, \mathbf{v}_2)$. In the elastic case $(e_n = e_t = 1)$ the relation $\mathbf{v}_i^* = \mathbf{v}_i^{**}$ holds. *(iii)* Because energy is not conserved in binary collisions, the terms inside the curly brackets on the right hand side of (40) do not cancel for a Maxwellian velocity distribution. *(iv)* The Boltzmann equation (40) does not satisfy an H–theorem, and does not have a non-trivial stationary solution.

For the elastic hard-sphere fluid Enskog has given a semi-phenomenological derivation of the Boltzmann equation valid for liquid densities. This equation is obtained by replacing f_{12} in the first hierarchy equation of (36) by the 'Stosszahlansatz' $f_{12} = \chi(\mathbf{r}_1, \mathbf{r}_2) f_1 f_2$, where $\chi(\mathbf{r}_1, \mathbf{r}_2)$ is the radial distribution function of elastic hard spheres at contact in a spatially nonuniform equilibrium state. This version of the molecular chaos assumption still neglects the velocity correlations, built up by sequences of correlated binary collisions, but does account for static short-range correlations, caused by excluded volume effects. The same Stosszahlansatz for the inelastic hard-sphere fluid yields the Enskog equation for the single-particle distribution function $f(\mathbf{r}, \mathbf{v}_1, t)$ of the inelastic hard-sphere fluid,

$$\left(\partial_t + \mathbf{v}_1 \cdot \frac{\partial}{\partial \mathbf{r}}\right) f(\mathbf{r}, \mathbf{v}_1) = I_E(f,f) \equiv \sigma^{d-1} \int d\mathbf{v}_2 \int^{(+)} d\hat{\boldsymbol{\sigma}}\, (\mathbf{v}_{12} \cdot \hat{\boldsymbol{\sigma}}) \times$$
$$\left\{ \frac{1}{e_n^2 e_t^{d-1}} \chi^{(-)}(\mathbf{r}) f(\mathbf{r}, \mathbf{v}_1^{**}) f(\mathbf{r} - \boldsymbol{\sigma}, v_2^{**}) - \chi^{(+)}(\mathbf{r}) f(\mathbf{r}, \mathbf{v}_1) f(\mathbf{r} + \boldsymbol{\sigma}, \mathbf{v}_2) \right\}, \quad (41)$$

where $\chi^{(\pm)}(\mathbf{r}) = \chi(\mathbf{r}, \mathbf{r} \pm \boldsymbol{\sigma})$. In Enskog's original derivation χ is the local equilibrium radial distribution function, $g(\sigma | n(\mathbf{R}^{(\pm)}, t))$, where the local density $n(\mathbf{R}^{(\pm)}, t)$ is taken at the point of contact, $\mathbf{R}^{(\pm)} = \mathbf{r} \pm \frac{1}{2}\boldsymbol{\sigma}$, of the two colliding spheres. The Enskog equation for elastic hard spheres has been derived in Ref. [65] as an exact short-time limit of the pseudo-Liouville equation (35), which yields a Markovian approximation to the single-particle kinetic equation. In this theory $\chi^{(\pm)}(\mathbf{r})$ is found as the radial distribution function in a non-uniform equilibrium state. This version of the Enskog theory, called *Revised Enskog Theory* (RET), is frequently used to analyze dynamic structure functions and transport properties of liquid argon and liquid sodium in neutron scattering experiments [66].

5.2. RING EQUATION

As the density increases, the contributions of correlated collision sequences to the collision term on the right hand side of (37) become more and more important. The most simple sequences of correlated collisions are the so called *ring* collisions; for example (12)(13)(14)...(23′)(24′)...(12), ending with a recollision of the pair (12), which was involved in the first collision. In the intermediate time particle 1 collides, say, s times with s different particles (3,4,...), and particle 2 collides s' times with *another* set of s' different particles (3′,4′,...). When particles 1 and 2 are about to recollide, they are dynamically correlated through their collision history, and the molecular chaos assumption (38) is no longer valid, *i.e.* $g_{12} \equiv f_{12} - f_1 f_2 \neq 0$.

A simple way to take these correlations into account at moderate densities has been given in Ref. [54]. The method is based on a cluster expansion of the s-particle distribution functions, defined recursively as

$$\begin{aligned} f_{12} &= f_1 f_2 + g_{12} \\ f_{123} &= f_1 f_2 f_3 + f_1 g_{23} + f_2 g_{13} + f_3 g_{12} + g_{123}, \end{aligned} \quad (42)$$

etc. Here g_{12} accounts for pair correlations, g_{123} for triplet correlations, etc. The molecular chaos assumption implies $g_{12} = 0$, which is equivalent to (38). The basic assumption to obtain the ring kinetic equation is that the pair correlations are dominant and higher order terms in (42) can be neglected, *i.e.* $g_{123} = g_{1234} = \cdots = 0$.

Substitution of (42) into (36) and elimination of $\partial f_i/\partial t$ $(i = 1, 2)$ from the second hierarchy equation using the first one, yields the *ring kinetic equations* for inelastic hard spheres with the appropriate $\overline{T}$–operators from Eq.(32),

$$
\begin{aligned}
(\partial_t + L_1^0) f_1 &= \int \mathrm{d}x_2 \overline{T}(12)(f_1 f_2 + g_{12}) \\
\Big[\partial_t + L_1^0 + L_2^0 &- \overline{T}(12) - (1 + \mathcal{P}_{12}) \int \mathrm{d}x_3 \overline{T}(13)(1 + \mathcal{P}_{13}) f_3\Big] g_{12} \\
&= \overline{T}(12) f_1 f_2.
\end{aligned}
\tag{43}
$$

Here $\mathcal{P}_{ij}$ is a permutation operator that interchanges the particle labels i and j. The second equation is the so called *repeated ring* equation for the pair correlation function g_{12}. If the operator $\overline{T}(12)$ on the left hand side of the second equation is dropped, one obtains the simple *ring* equation. Formally solving this equation for g_{12} yields an expression in terms of the single-particle distribution functions f_i $(i = 1, 2, 3)$, and subsequent substitution into the first hierarchy equation above yields the generalized Boltzmann equation in ring approximation. For a more detailed discussion of the collision sequences taken into account by Eqs. (43) we refer to the original literature [52, 54].

What is our motivation for studying the ring kinetic equations for inelastic hard-sphere fluids? We recall from the introduction that all major developments in the theory of elastic fluids of the last 35 years are based on ring kinetic equations, or on their phenomenological counterparts: mode coupling theory and fluctuating nonlinear hydrodynamic equations. The analogous equations for complex fluids open up the possibility to investigate all these phenomena for inelastic granular fluids, such as long-range spatial and temporal correlations.

In fact, one of the few applications up to now of the ring kinetic theory to inelastic hard-sphere fluids has been given in [30], where the ring equations have been analyzed to establish a more fundamental justification of the basic assumption upon which the phenomenological theory of fluctuating hydrodynamics of Ref. [12] has been built.

6. Conclusions

In the first part of these lectures we have taken undriven inelastic hard and soft sphere fluids as examples to describe some interesting phenomena observed in rapid granular flows, and explained them on the basis of the microscopic inelastic collision processes using elementary kinetic theory concepts.

In the second part we have presented the fundamental and more sophisticated approach of Boltzmann-type kinetic equations, and we have outlined how such equations can be derived starting from the many-body 'Liouville' equation for conservative, impulsive or dissipative forces, using the methods of non-equilibrium statistical mechanics and kinetic theory. Many of the results in the first part can be explained in a more quantitative manner using the kinetic equations of the second part.

The macroscopic equations for granular fluid dynamics, given in section 2.2, have been derived from the Boltzmann or Enskog equation [2]-[4]. Grad's moment method or multi-time scale techniques, like the Chapman-Enskog method [62], have been used to solve these kinetic equations, derive the proper constitutive relations, and to calculate not only the transport coefficients of viscosity and heat conductivity, but also new ones [27, 29] that are absent in elastic fluids.

The homogeneous cooling state in section 2.1 can be investigated in much greater detail by analyzing the Boltzmann or Enskog equation for the single-particle distribution function. The homogeneous state is described by a scaling solution, $f(\mathbf{v},t) = (n/v_0^d(t))\tilde{f}(v/v_0(t))$, where $v_0(t)$ is the r.m.s. velocity. If the scaling form would be a Maxwellian, then this state would indeed be the adiabatically changing equilibrium state, discussed in section 2.1. The solution is indeed found to be close to a Maxwellian. Its higher cumulants are small, but non-vanishing [26, 30, 33, 35, 36], and the high energy tail of the distribution shows overpopulation [9, 30].

An interesting new development is the application of Bird's method or Direct Simulation Monte Carlo (DSMC) method to simulate the solution of the nonlinear Boltzmann or Enskog equation [45, 35, 36]. These methods are very powerful tools to test approximate analytic solutions, and assess their range of validity. Of course these methods are also able to bypass the analysis of kinetic equations, and compare the numerical results of DSMC methods directly with the results of molecular dynamics simulations of the many-body problem [22]. So, one can establish a priori the range of densities and inelasticities for which the Boltzmann or Enskog equations can give reliable predictions for the behavior of elastic and inelastic fluids, as well as for more complex fluids, such as granular flows.

This article is intended as an introduction into the kinetic theory of granular fluids. It is presented in an upside-down fashion, by first explaining the observed phenomena by elementary kinetic theory methods, and then deriving the fundamental kinetic equations through which it can all be done better.

Acknowledgements

It is a pleasure to thank R. Brito, T.P.C. van Noije, J.A.G. Orza, I. Goldhirsch, J.T Jenkins, S. Luding, H.J. Herrmann, T. Pöschel, N.V. Brilliantov and A. Goldshtein for interesting discussions and correspondence, and for providing important references.

References

1. P.K. Haff, J. Fluid Mech. **134**, 401 (1983).
2. J.T. Jenkins and S.B. Savage, J. Fluid Mech. **130**, 187 (1983).
3. C.K.K. Lun, S.B. Savage, D.J. Jeffrey, and N. Chepurniy, J. Fluid Mech. **140**, 223 (1984).
4. J.T. Jenkins and M.W. Richman, Arch. Rat. Fluid Mech. **192**, 313 (1988).
5. M.A. Hopkins and M.Y. Louge, Phys. Fluids A **3**, 47 (1991).
6. I. Goldhirsch and G. Zanetti, Phys. Rev. Lett. **70**, 1619 (1993); I. Goldhirsch, M-L. Tan, and G. Zanetti, J. Scient. Comp. **8**, 1 (1993).
7. S. McNamara, Phys. Fluids A **5**, 3056 (1993).
8. S. McNamara and W.R. Young, Phys. Rev. E **50**, R28 (1994); S. McNamara and W.R. Young, Phys. Rev. E **53**, 5089 (1996).
9. S.E. Esipov and T. Pöschel, J. Stat. Phys. **86**, 1385 (1997).
10. P. Deltour and J.-L. Barrat, J. Phys. I (France) **7**, 137 (1997).
11. D.R. Williams and F.C. MacKintosh, Phys. Rev. E **54**, R9 (1996).
12. T.P.C. van Noije, M.H. Ernst, R. Brito, and J.A.G. Orza, Phys. Rev. Lett. **79**, 411 (1997); T.P.C. van Noije, R. Brito, and M.H. Ernst, Phys. Rev. E **57**, R4891 (1998); J.A.G. Orza, R. Brito, T.P.C. van Noije, and M.H. Ernst, Int. J. Mod. Phys. C **8**, 953 (1997).
13. T.P.C. van Noije and M.H. Ernst, Phys. Rev. E submitted, June 1999; T.P.C. van Noije, *Ph.D. thesis*, Universiteit Utrecht, September 1999.
14. A. Puglisi, V. Loreto, U. Marini Bettolo Marconi, A. Petri, and A. Vulpiani, Phys. Rev. Lett. **81**, 3848 (1998).
15. M.R. Swift, M. Boamfă, S.J. Cornell, and A. Maritan, Phys. Rev. Lett. **80**, 4410 (1998).
16. G. Peng and T. Ohta, Phys. Rev. E **58**, 4737 (1998).
17. R. Brito and M.H. Ernst, Europhys. Lett. **43**, 497 (1998); Int. J. Mod. Phys. C **9**, 1339 (1998).
18. T.P.C. van Noije, M.H. Ernst, E. Trizac, and I. Pagonabarraga, Phys. Rev. E **59**, 4326 (1999).
19. J.A Orza, R. Brito, and M.H. Ernst, preprint August 1999.
20. C. Bizon, M.D. Shattuck, J.B. Swift, and H.L. Swinney, this volume, p. 361.
21. S. Luding and S. McNamara, Granular Matter, to appear.
22. S. Luding and H.J. Herrmann, Chaos, **9**, 25 (1999).
23. H.M. Jaeger, S.R. Nagel, and R.P. Behringer, Rev. Mod. Phys. **68**, 1259 (1996).
24. T. Pöschel, *Dynamik Granularer Systeme*, Habilitationsschrift, Humboldt Universität, Berlin, November 1998.
25. S. Luding, *Die Physik kohäsionsloser granularer Medien*, Habilitationsschrift, Universität Stuttgart (Logos Verlag, Berlin, 1998).
26. A. Goldshtein and M. Shapiro, J. Fluid Mech. **282**, 75 (1995).
27. N. Sela, I. Goldhirsch, and S.H. Noskowicz, Phys. Fluids **8**, 2337 (1996), I. Goldhirsch and M-L. Tan, Phys. Fluids **8**, 1752 (1996); N. Sela and I. Goldhirsch, J. Fluid. Mech. **361**, 41 (1998).
28. M.H. Ernst, J.R. Dorfman, W.R. Hoegy, and J.M.J. van Leeuwen, Physica **45**, 127 (1969).

29. J.J. Brey, J.W. Dufty, and A. Santos, J. Stat. Phys. **87**, 1051 (1997); J.J. Brey, J.W. Dufty, C.S. Kim, and A. Santos, Phys. Rev. E **58**, 4638 (1998).
30. T.P.C. van Noije, M.H. Ernst, and R. Brito, Physica A **251**, 266 (1998); T.P.C. van Noije and M.H. Ernst, Granular Matter **1**, 57 (1998).
31. M. Huthmann and A. Zippelius, Phys. Rev. E **56**, R6275 (1997). M. Huthmann, T. Aspelmeier, and A. Zippelius, Phys. Rev. E **60**, 654 (1999); T. Aspelmeier, M. Huthmann, and A. Zippelius, in *Proceedings of 215th WE-Heraeus Seminar on Granular Gases*, Bad Honnef (Germany), March 8-12 1999.
32. N.V. Brilliantov and T. Pöschel, e-print cond-mat/9906404.
33. M. Huthmann, J.A.G. Orza, and R. Brito, preprint (1999).
34. M. Müller and H.J. Herrmann, in *Physics of Dry Granular Media*, H.J. Herrmann, J.-P. Hovi, and S. Luding. eds., (Kluwer Academic Publishers, Dordrecht, 1998).
35. J.J. Brey, M.J. Ruiz-Montero, and D. Cubero, Phys. Rev. E **54**, 3664 (1996); J.J. Brey, D. Cubero, and M.J. Ruiz-Montero, Phys. Rev. E **59**, 1256 (1999).
36. J.M. Montanero and A. Santos, submitted to Granular Matter, July 1999.
37. N.V. Brilliantov, F. Spahn, J.M. Hertzsch, and T. Pöschel, Phys. Rev. E **53**, 5382 (1996); T. Schwager and T. Pöschel, Phys. Rev. E **57**, 650 (1996).
38. G. Kuwabara and K. Kono, Jpn. J. Appl. Phys. **26**, 1230 (1987).
39. W.A.M. Morgado and I. Oppenheim, Phys. Rev. E **55**, 1940 (1998).
40. H. Hwang and K. Hutter, Continuum Mech. and Thermodynamics **7**, 357 (1995).
41. J. Trulsen, Astrophys. Space Sci. **12**, 329 (1971); K.A. Hameen-Anttila, Astrophys. Space Sci. **58**, 477 (1978); P. Goldreich and S. Tremaine, *ICARUS* **34**, 227 (1978).
42. C.K.K. Lun and S.B. Savage, Acta Mechanica **63**, 15 (1986).
43. O.R. Walton and R.L. Braun, J. Rheol. **30**, 949 (1986); Acta Mech. **63**, 73 (1986).
44. C.S. Campbell, Annu. Rev. Fluid Mech. **22**, 57 (1990), C.S. Campbell and C.E. Brennen, J. Fluid Mech. **151**, 167 (1985).
45. H.J. Herrmann, Physica A **191**, 263 (1992).
46. S. Luding, E. Clement, A. Blumen, J. Rajchenbach, and J. Duran, Phys. Rev. E **50**, 4113 (1994).
47. S. Chen, Y. Deng, X. Nie, and Y. Tu, unpublished.
48. K. Kawasaki, in *Phase Transitions and Critical Phenomena*, Vol. 5A, C. Domb and M.S. Green, eds., (Academic Press, London, 1976).
49. T. Keyes, in *Modern Theoretical Chemistry: Statistical Mechanics*, B.J. Berne, ed., (Academic Press, New York, 1977).
50. B.J. Alder and W.E. Alley, Physics Today **37**, 56 (1984); E.G.D. Cohen, Physics Today **37**, 64 (1984).
51. J.-P. Hansen and I.R. McDonald, *Theory of Simple Liquids* (Academic Press, London, 1986).
52. J.R. Dorfman and H. van Beijeren, The Kinetic Theory of Gases, in *Statistical Mechanics, Part B: Time-Dependent Processes*, B.J. Berne, ed., (Plenum Press, New York, 1977), Chap. 3.
53. J.R. Dorfman and E.G.D. Cohen, Phys. Rev. Lett. **25**, 1257 (1970).
54. M.H. Ernst and J.R. Dorfman, Physica **61**, 157 (1972).
55. M.H. Ernst, E.H. Hauge, and J.M.J. van Leeuwen, Phys. Rev. A **4**, 2055 (1971); M.H. Ernst and J.R. Dorfman, J. Stat. Phys. **15**, 311 (1976).
56. Y. Pomeau and P. Resibois, Phys. Rep. C **19**, 63 (1975).
57. J.R. Dorfman, T.R. Kirkpatrick, and J. Sengers, Ann. Rev. Phys. Chem. **45**, 213 (1994).
58. L. Landau and E.M. Lifshitz, *Theory of Elasticity* (Pergamon Press, London, 1959).
59. L. Landau and E.M. Lifshitz, *Fluid Mechanics* (Pergamon Press, London, 1959), ch. 3, 17.
60. J.S. Langer, in: *Solids Far from Equilibrium*, C. Godrèche, ed., (Cambridge University Press, 1992), p. 297; J.K. Dhont, this volume, p. 73.
61. *Holmes' Principles of Physical Geology*, D. Duff, ed., (Chapman and Hall, Boca

Raton, U.S.A., 1994), p. 368.

62. S. Chapman and T.G. Cowling, *The Mathematical Theory of Non-uniform Gases* (Cambridge University Press, London, 1970).
63. S.R. de Groot and P. Mazur, *Non-equilibrium Thermodynamics* (North Holland Publ. Co, Amsterdam, 1963) p.100.
64. C.A. Marsh, G. Backx, and M.H. Ernst, Phys. Rev. E **56**, 1676 (1997).
65. H. van Beijeren and M.H. Ernst, Phys. Lett. A **43**, 367 (1973); Physica **68**, 437 (1973); Physica **70**, 225 (1973).
66. For a review, see E.G.D. Cohen, Physica A **194**, 241 (1993).

INELASTIC COLLAPSE

S. MCNAMARA
Levich Institute, City College of New York
New York, NY 10031, USA

1. Introduction

Inelastic (energy-dissipating) particles can collide an infinite number of times in finite time. This phenomenon, called inelastic collapse, is an extreme example of correlated collisions. Inelastic collapse is present in the "inelastic hard-sphere" model of granular materials, where granular particles are modeled as identical spheres which interact only via instantaneous collisions. Unlike the classical hard-sphere fluid, collisions are inelastic, i.e. they dissipate energy. If two particles with relative velocity $\mathbf{v}$ collide, the component of the velocity along the line of centers is reduced by a factor r:

$$\mathbf{v}' \cdot \hat{\mathbf{n}} = -r\mathbf{v} \cdot \hat{\mathbf{n}}, \tag{1}$$

where $\hat{\mathbf{n}}$ is a unit vector pointing along the line of centers, r is the restitution coefficient, and $\mathbf{v}'$ is the relative velocity after collision. Energy is dissipated for $0 \leq r < 1$, and conserved for $r = 1$. Energy is added for $r > 1$, and the particles interpenetrate for $r < 0$, so r is confined between 0 and 1. It is possible to use more complicated collision rules to mimic the behavior of a specific granular material, but we confine ourselves here to Eq. (1). There are also various schemes for evading inelastic collapse, but the focus of this article is the phenomenon itself.

2. Inelastic Collapse

Inelastic collapse was first identified while studying the one-dimensional granular medium[1, 2, 3]. The simplest example of inelastic collapse requires two very inelastic ($r \ll 1$) particles thrown against an inelastic wall in one dimension. Space-time trajectories of the collapsing particles are shown in the left panel of Fig.1. We will trace the evolution of the system, neglecting corrections of size $O(r)$. Let V be the order of magnitude of the velocities at

J. Karkheck (ed.),
Dynamics: Models and Kinetic Methods for Non-equilibrium Many Body Systems, 267–277.

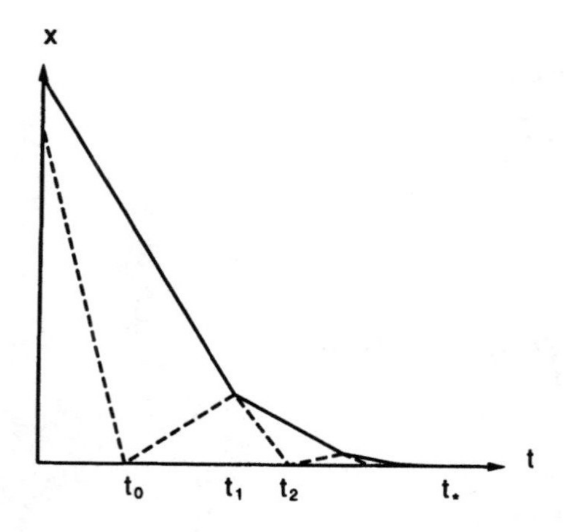

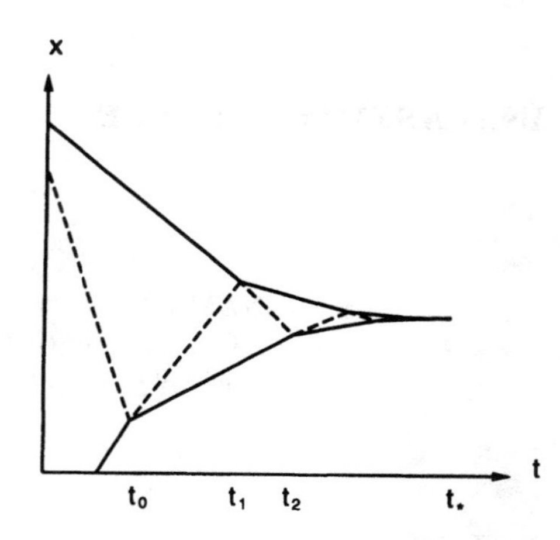

Figure 1. Sketches of the two simplest examples of inelastic collapse. *Left:* Two particles collapsing against a wall (requires $r < 3 - 2\sqrt{2} \approx 0.171573$), *Right:* Three particles in free space (requires $r < 7 - 4\sqrt{3} \approx 0.071797$).

time t_0, and H be the height of the upper particle. At time t_0, the bottom particle strikes the wall and rebounds with a speed of $rV \ll V$. The upper particle continues its descent, and strikes the lower particle after a time $t_1 = t_0 + H/[V(1+r)] \approx t_0 + H/V$. At this time, the lower particle has a height of $rV(t_1 - t_0) \approx rH$. After the collision, the particles separate at the speed of rV, but are moving at $V/2$ with respect to the wall. At time $t_2 = t_1 + 2rH/V$, the lower particle is just about to strike the bottom wall again. By this time, the particles have a separation of $rV(t_2 - t_1) = 2r^2H$. In every respect, the system is exactly as it was at t_0, except that the distances and velocities have been rescaled. Note that the distance scales contract much more quickly than the velocity scales: $H(t_2) = 2r^2H(t_0) \ll H(t_0)$, but $V(t_2) = V(t_0)/2$. The cycle then repeats on every smaller distance and time scales. One can show that $(t_4 - t_2) = 4r^2(t_2 - t_0)$, and that an infinite number of cycles take place in finite time: $t_\infty - t_0 = (t_2 - t_0)/(1 - 4r^2)$.

The above example of inelastic collapse requires that $r \ll 1$. As r increases, eventually we will reach a value where collapse no longer occurs. We denote this critical value of r as r_*^{wall}. For any number of particles N thrown against a wall, collapse occurs for $r < r_*^{\mathrm{wall}}(N)$, and the particles disperse for $r > r_*^{\mathrm{wall}}(N)$. As N increases, so does $r_*^{\mathrm{wall}}(N)$; in fact $r_*^{\mathrm{wall}}(N) \to 1$ as $N \to \infty$.

Inelastic collapse can also occur in free space (right panel of Fig. 1), therefore we define $r_*^{\mathrm{free}}(N)$ to be the critical value of r for a cluster of N particles in free space. It has usually been assumed that $r_*^{\mathrm{free}}(2N) = r_*^{\mathrm{wall}}(N)$, because an exactly symmetric cluster of $2N$ particles mimics a cluster of N particles colliding with a wall.

3. Calculations of $r_*(N)$

3.1. EXACT RESULTS

When the sequence of collisions is known, it is possible to calculate r_* exactly. We now consider two particles colliding with a wall. Let v_1 be the velocity of the lower particle, and v_2 be that of the upper. Positive velocities are directed upwards, so that the particles disperse if $v_2 > 0$. The vector $\mathbf{V} = (v_1, v_2)$ contains all the information needed to determine the sequence of collisions and all future velocities. The effect of two collisions can be written as a matrix $\mathcal{C}$, so that $\mathbf{V}(t_2) = \mathcal{C}\mathbf{V}(t_0)$, and the velocities at any later time can be calculated by repeatedly applying $\mathcal{C}$. By examining the eigenvalues of $\mathcal{C}$, one can determine the fate of the particles.[1] When $r < r_* = 3 - 2\sqrt{2} \approx 0.171573$, $\mathcal{C}$ has two real eigenvalues between 0 and 1. This corresponds to inelastic collapse: a vector $\mathbf{V}$ with $v_2 < 0$ will be reduced in magnitude by $\mathcal{C}$, but never rotated into the part of the (v_1, v_2) plane where $v_2 > 0$. Using this same method, it is also possible to show that $r_*^{\text{free}}(3) = r < 7 - 4\sqrt{3} \approx 0.071797$.

This calculation furnishes an additional prediction: when r is slightly greater than r_*, the eigenvalues have an imaginary part proportional to $\sqrt{r - r_*}$. Therefore, each application of $\mathcal{C}$ rotates $\mathbf{V}$ through some angle proportional to $\sqrt{r - r_*}$. Eventually, $\mathbf{V}$ reaches the upper half plane, and the particles disperse. Thus, the number of cycles needed to disperse the cluster scales as $(r - r_*)^{-1/2}$.

Benedetto and Caglioti[4] have proved that $r_*^{\text{free}}(N) < 1 - (2\ln 2)/N$. They also constructed initial conditions which collapse for $r < r_*^{\text{free}}(N) = 2r_*^{\text{wall}}(N)$, where $r_*^{\text{wall}}(N)$ is given below in Eq. (2). For $N \leq 4$, they were able to show that there is a set of non-zero measure around these initial conditions which collapses also. Sec. 3.3 confirms their results for $N \leq 4$ and suggests that these collapsing solutions for $N > 4$ are unstable.

3.2. APPROXIMATE RESULTS

We now discuss two approximate methods for estimating $r_*(N)$. A central concept to both methods is the *collision wave*, illustrated in Fig. 2. These waves are more easily seen than explained. In the left panel of Fig. 2, nine particles are at rest just above a wall at $x = 0$. A particle moving at high velocity crashes into this "cushion" of particles, shocking the top particle.

[1] One must also consider the initial condition, but as long as $v_2 < 0$ and $-(1-r)/(r+r^2) < v_1/v_2 < (1-r)/(1+r)$, the eigenvalues will determine if the particles collapse. These conditions are not restrictive when $r \ll 1$.

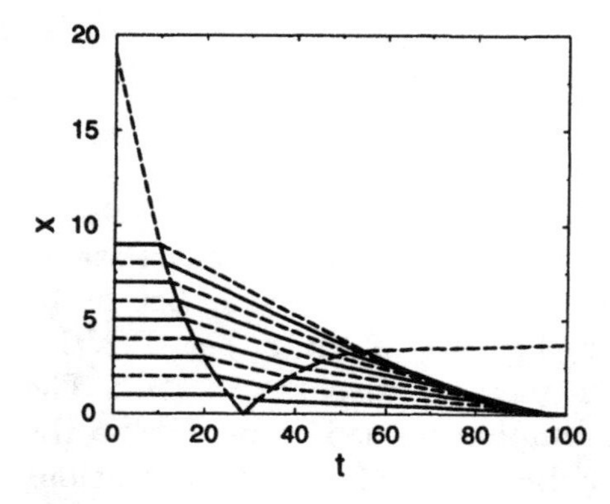

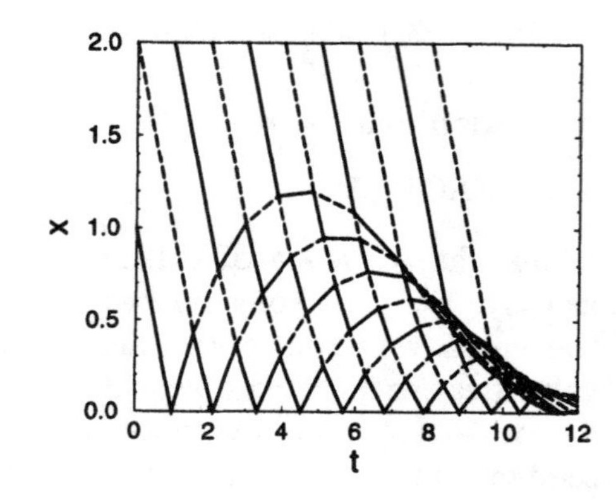

Figure 2. Trajectories of 10 particles at $r = 0.75$ showing examples of collision waves. The trajectories of the particles are shown alternately in solid and dashed lines to show the identities of the particles. There is a wall at $x = 0$. *Left:* a collision wave starts on the outermost particle, then travels through the "cushion" of particles near the wall, reflects off the wall, and then travels back through the cushion. *Right:* A group of equally spaced particles traveling at the same speed incident on a wall.

Since all collisions in one dimension are head-on, the particles very nearly exchange velocities. The top particle then hits the next particle down in the cushion, which then hits the next one, etc. The impulse or "collision wave" travels all the way down to the wall, where it reflects, and then travels back out towards the first particle, which is 'waiting' for it at the top of the cushion. If we attempt to describe this figure from the point of view of the particles, it is very complicated. Each particle in the cushion is hit twice, once from above, and then from below. On the other hand, discussing it in terms of collision waves is simple: the wave starts on the outermost particle, travels through the cushion, reflects off the wall, and travels back out. Both estimates of $r_*(N)$ given below use this concept.

3.2.1. *The Independent Collision Wave (ICW) model*

Imagine N particles thrown against a wall (Fig. 2-right). The lowermost particle hits the wall and is reflected ($t = 1$ in Fig. 2-right). The resulting collision wave then travels upwards through the descending particles. The bottom particle hits the wall a second time at $t \approx 2$, and a second collision wave propagates upwards. Each time the lower particle hits the bottom, a collision wave begins its journey upwards. The effects on the particle velocities induced by the passage of one collision wave can be represented by a matrix which operates on the vector of particle velocities [1]. As in Sec. 3.1, the presence of a real eigenvalue less than 1 indicates that the cycle can go on forever. This leads to

$$r_*^{\text{wall}}(N) = \tan^2\left[\frac{\pi}{4}\left(1 - \frac{1}{N}\right)\right]. \tag{2}$$

The assumption underlying this method is that the collision waves are independent, i.e. a collision wave propagates through *all* the particles before it meets another collision wave coming up from the bottom. In Fig. 2, we see that this is not exactly true. At $t \approx 7.5$, the first and second collision wave originating at the bottom interact before they have traveled through all the particles. It is hoped that the inaccuracies introduced will not be significant.

Note that Eq. (2) reproduces the exact result for $N = 2$, and for $N = 3$ in free space if one assumes $r_*^{\text{wall}}(N) = r_*^{\text{free}}(2N)$.

3.2.2. *The cushion model*

Another approach is to consider the initial condition shown in the left panel of Fig. 2: $N - 1$ particles are at rest above the wall (the "cushion"), and the outermost particle strikes the cushion ($t \approx 10$ in Fig. 2-left). It generates a collision wave which propagates to the wall, and reflects off it ($t \approx 28$). During its passage towards the wall, the wave has deposited some of its momentum in the "cushion", and, as a result, these particles are moving slowly towards the wall. The collision wave must now fight its way against all these particles to get back to the outermost particle. When it gets there ($t \approx 55$), the remaining particles have slow velocities towards the wall. These particles will eventually collide with the wall, suffer very many collisions, and dissipate most of their energy, reforming the cushion ($t \approx 100$). If the outermost particle is still moving towards the wall after the collision wave reaches it, it will eventually collide again with the cushion, and the process will repeat itself indefinitely. On the other hand, if it is moving away from the wall, it will never collide again, and the cluster will disperse. (The $N - 1$ particles in the cushion will eventually disperse; if N particles disperse, then $N - 1$ will disperse also.) Therefore, everything depends on the velocity of the outermost particle after the collision wave has returned to it [2]. This velocity can be calculated as a function of N and r, assuming the "cushion" is initially motionless. Setting this velocity to 0 allows determination of the number of particles required for collapse, $N_*^{\text{wall}}(r)$, when the restitution coefficient is r,

$$N_*^{\text{wall}}(r) = \ln(2/q)/(2q), \tag{3}$$

where $q \equiv (1 - r)/2$, and $q \ll 1$ has been used to write Eq. (3) in an elegant form. The approximation underlying the cushion model is that the outermost particle possesses most of the kinetic energy, and is therefore the most important particle.

3.3. RESULTS OF SIMULATIONS

We will now compare these results to simulations. The simulation algorithm must be designed carefully because the length and time scales contract geometrically. For example, suppose a cluster of particles are collapsing at $x = 1$. In the beginning, the position of two particles might be, say, $x_1 = 1.1$ and and $x_2 = 1.2$. But after several cycles, the positions may become $x_1 = 1.0000001$ and $x_2 = 1.0000002$. When the particle separation $x_2 - x_1$ is calculated, the results will be inaccurate. A similar problem occurs with the velocities and with the times between collisions. These problems can be eliminated by working only with particle spacings and relative velocities: i.e., instead of storing the position x_i of each particle, we store $x_i - x_{i-1}$, its distance from its neighbor. To eliminate inaccuracies in time, the time origin can be frequently redefined .

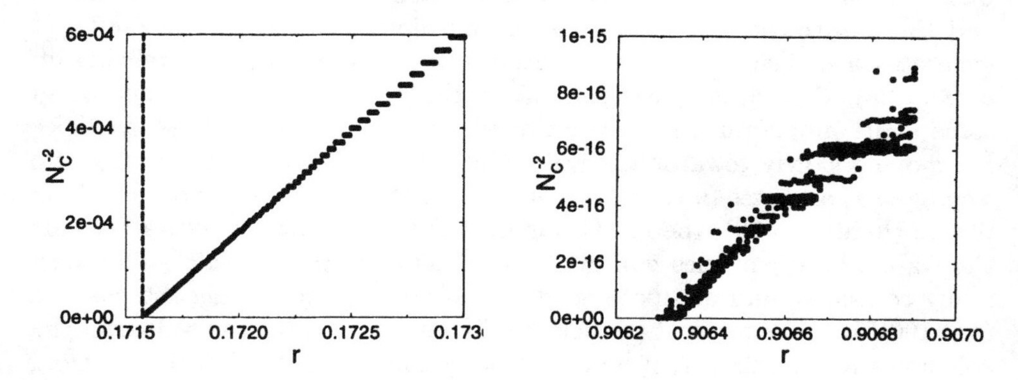

Figure 3. $1/N_c^2$ (N_c is the number of collisions required to disperse a group of particles thrown against a wall) plotted against r. A straight line drawn through the points intersects the horizontal axis at r_*. *Left:* $N = 2$ particles. The vertical dashed line is the theoretical prediction of $r_* = 3-2\sqrt{2}$. *Right:* $N = 40$ particles; the theoretical predictions here are $r_* = 0.90618$ for the cushion model and $r_* = 0.9244$ for the ICW model.

The first theoretical prediction we test is that $N_c \sim (r - r_*)^{-1/2}$ for r just a bit larger than r_*, where N_c is the number of collisions required to disperse a cluster. In Fig. 3, we plot N_c^{-2} against r. If the theory is correct, the points should fall on a straight line, intercepting the x-axis at r_*. This is indeed the case, not only for small numbers of particles ($N = 2$), but also for large numbers ($N = 40$). In both cases, N_c^{-2} appears to become quantized as r becomes larger than r_*. For $N = 2$, this is because there can only be an integer number of collisions. For $N = 40$, the reason is different:

there is a cycle, like the one assumed in the cushion model, which takes approximately the same number of collisions each time. A cluster disperses after an integer number of cycles, so that the total number of collisions will be approximately an integer times the number of collisions per cycle. The data of Fig. 3 suggest that each cycle contains roughly 8×10^6 collisions.

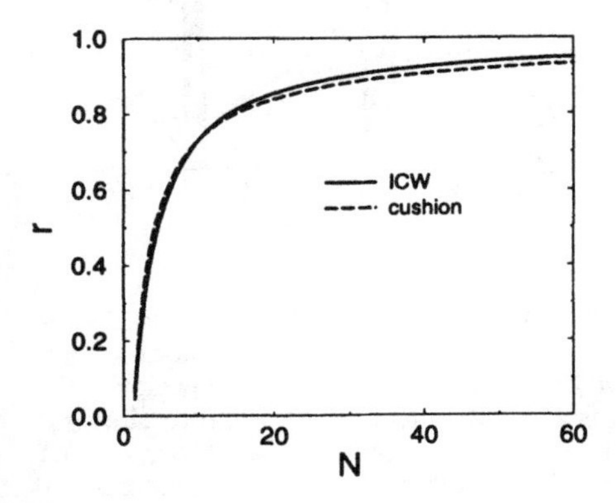

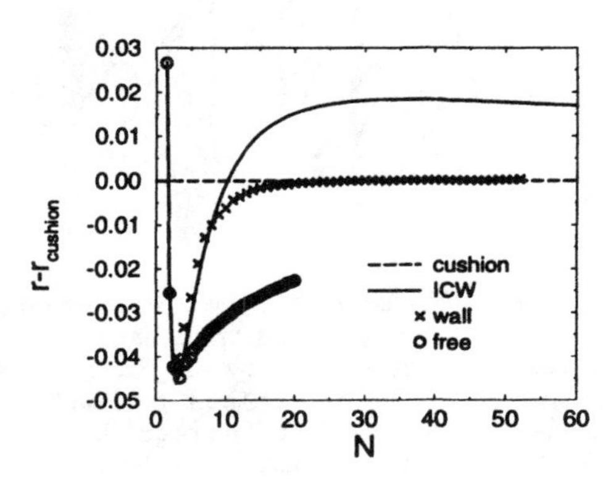

Figure 4. *Left:* Theoretical predictions of $r_*^{\text{wall}}(N)$. *Right:* Comparison with simulations. 'Wall' and 'free' indicate the results of simulations. To check the assumption $r_*^{\text{free}}(2N) = r_*^{\text{wall}}(N)$, the 'free' simulations actually include twice as many particles as indicated on the graph. Thus, the circle at $N = 20$ indicates the result of a simulation of 40 particles. To make visible the difference between the theories, r_* predicted by the cushion model is subtracted from all values of r. Hence a point directly on the horizontal line indicates agreement with the cushion model, and a point on the curve labeled 'ICW' indicates agreement with the ICW model.

In Fig. 4, we compare the results of simulations with the cushion and ICW models. We see that the ICW model is more accurate for $N < 15$, but the cushion model is better when $N > 15$. The reason for this can be seen by examining Fig. 5, which shows the motion of the outermost eight of forty particles colliding with a wall. As assumed by the cushion model, the outermost particle does indeed possess most of the energy, and all collisions occur in short bursts triggered by the outermost particle's collisions with the "cushion". A simulation of eight particles colliding with a wall show no signs of this structure; here the ICW model is accurate.

Note that the assumption $r_*^{\text{free}}(2N) = r_*^{\text{wall}}(N)$ is exact at $N = 4$, (At $N = 2$, the 'wall' and 'free' point coincide), but is invalid for large N. This suggests large clusters disperse asymmetrically.

4. Inelastic Collapse in Two Dimensions

Inelastic collapse has also been documented in two dimensions [5]. If a group of two-dimensional particles are precisely lined up, they will recreate ex-

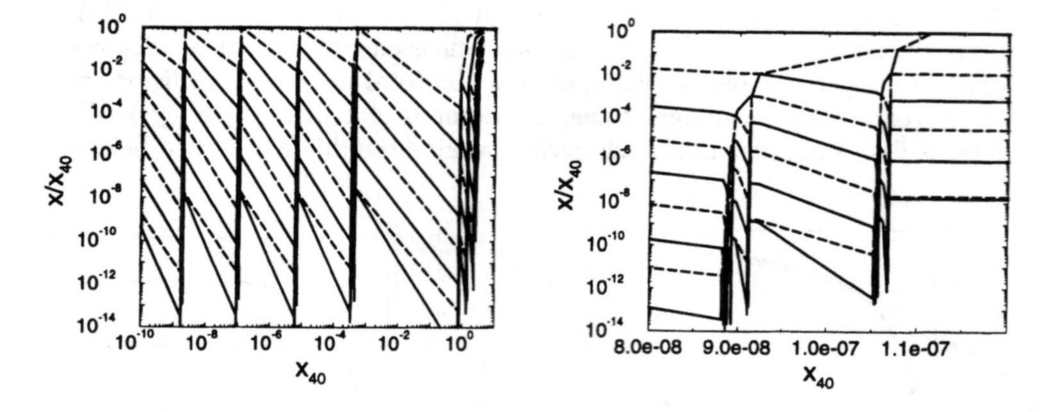

Figure 5. Trajectories of the outermost nine particles in a collapsing $N = 40$, $r = 0.9063$ simulation. The particles are collapsing against a wall at $x = 0$. The positions of the particles are normalized by x_{40}, the position of the outermost particle. Therefore, the solid line at $x/x_{40} = 1$ at the top of each graph is the outermost particle. Note that x_{40} is also the abscissa. *Left:* The beginning of the collapse, with several cycles shown. *Right:* An expanded view of the single event near $x_{40} = 10^{-7}$. Each cycle is actually composed of four separate bursts of collisions.

actly the one-dimensional sequence of collisions, thus reproducing inelastic collapse. Inelastic collapse can also occur in two dimensions when the particles are not exactly lined up. In this case, the criterion for collapse contains three parameters: r and N (as in one dimension), and some measure of the disalignment of the particles.

Detecting inelastic collapse in two-dimensional simulations is more delicate than in one dimension. There is no comparable simple algorithm which works with the particle separations instead of the positions. Since we must work with the positions, once separations between particles decrease below 10^{-15} of a particle diameter, they can no longer be calculated accurately with double precision numbers, and the computed sequence of collisions will diverge from the predictions of the model. Usually, this divergence is not catastrophic, and the collapsing line of particles disperses. Thus, it is possible to have inelastic collapse in two dimensions without noticing it. In the following, we define inelastic collapse to have occurred if, at the time of collision between two particles, a third particle can be found at a distance of less than 10^{-13} of a particle diameter from either colliding particle.

4.1. SMALL NUMBERS OF PARTICLES

The only analytic work concerning inelastic collapse in two dimensions [6] considers three collapsing particles. The deviation of the particles from a straight line is quantified by the angle θ (see Fig. 6). Assuming the particles

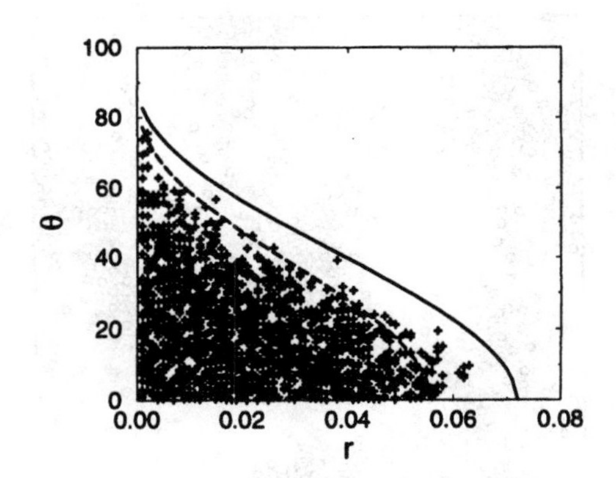

Figure 6. *Left:* Definition sketch of the angle θ. *Right:* A comparison of Eqs.(4) [solid line] and (5) [dashed line] against simulations. One hundred simulations were done at each value of r. If collapse was detected within 90 collisions, θ is measured, and a point is put on the graph. If no collapse occurred, no point is put on the graph.

to be at rest and considering only the change in velocities induced by the collisions, the requirement for collapse is

$$\cos\theta \geq 4\sqrt{r}/(1+r). \tag{4}$$

Accounting for the motion of the particles in a simple way gives a stricter condition:

$$\cos\theta \geq 2r^{1/3}(1+r^{1/3})/(1+r). \tag{5}$$

This equation implies that $r < 9 - 4\sqrt{5} \approx 0.055728$ is required for collapse in two dimensions. In Fig. 6, we test these criteria. The majority of the simulations obey Eq. (5), and all of them obey Eq. (4).

4.2. LARGE NUMBERS OF PARTICLES

In Fig. 7, we show two examples of inelastic collapse with many particles. The initial condition of these pictures were generated by running a simulation at $r = 1$ for several hundred collisions per particle. During this time, the density becomes uniform, and the velocities are distributed according to the Maxwellian velocity distribution. Then, at $t = 0$, the r is reduced to a value less than 1, and the system evolves without input of energy. The kinetic energy decreases monotonically (since every collision dissipates energy), and the system is said to "cool". Clusters spontaneously appear: they are entirely distinct from inelastic collapse, and are related to the stability of the homogeneous cooling state (see related chapters in this volume). The clusters organize the particles into lines. If these lines contain enough

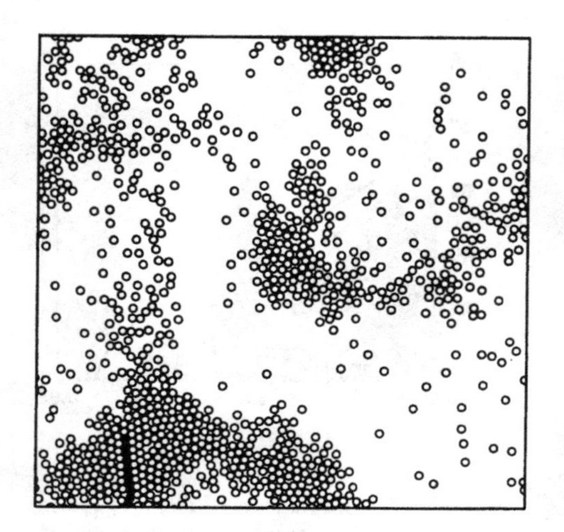

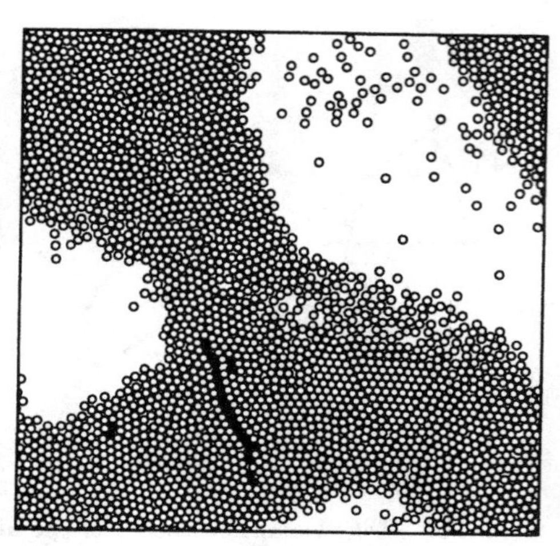

Figure 7. Inelastic collapse with many particles in 2-D. *Left:* $N = 1000$ particles with area fraction $\nu = 0.2$, $r = 0.55$. Collapse (particle separations less than 10^{-13} of a particle radius) occurs after 170.8 collisions per particle. The shaded particles were involved in the last 100 collisions before collapse was detected. *Right:* same as at the left, except $\nu = 0.5$, $N = 2500$, $r = 0.66$, and collapse occurs after 362.0 collisions. Particles involved in the last 200 collisions are shaded.

particles and are straight enough, collapse can occur. An approximate criterion for collapse in these "cooling" simulations is $r < r_*^{\rm wall}(\nu L/d)$ where $\nu = \pi N d^2/(4L^2)$ is the fraction of the total area covered by disks, d is the diameter of the disks, and L is the length of the simulation domain [7].

5. The Physical Relevance of Inelastic Collapse

In physical systems, grains never suffer an infinite number of collisions in finite time for two reasons. First of all, the restitution coefficient increases as the impact velocity $v_{\rm imp}$ decreases. Assuming that the particles are viscoelastic gives $1 - r \sim v_{\rm imp}^{1/5}$ [8, 9]. Experiments also show that $r \to 1$ as $v_{\rm imp} \to 0$ [10]. The restitution coefficient is nearly constant at large velocities, but begins to increase towards 1 as $v_{\rm imp}$ approaches 0. If a group of particles begins to undergo collapse, their velocities will decrease and their restitution coefficient will eventually increase above r_* so the cluster will disperse [11]. A variable restitution coefficient is easily incorporated into computer simulations, and would be sufficient to remove inelastic collapse on a computer with infinite precision. In practice, the particle separations can still decrease below machine precision because the length scales decrease much more rapidly than the velocity scales.

Secondly, colliding particles remain in contact for a finite time t_c. When the inelastic hard-sphere model predicts a time between collisions Δt, less than t_c, there is really a multiparticle interaction. These multiparticle events dissipate less energy than an equivalent series of inelastic collisions [12, 13]. The contact time can be included in numerical simulations using the "TC model" [14]: each particle carries a clock which records the time since its last collision. When two particles collide, their clocks are examined, and if either particle's clock shows a time less than t_c, the collision dissipates no energy. Adding this rule to event-driven simulations removes the inelastic collapse.

Note that both of these mechanisms will always interrupt inelastic collapse: the first because v_{imp} is always decreasing, and the second because Δt is always diminishing. In general, Δt decreases much more rapidly than v_{imp}, so the author guesses that the second mechanism will usually be more important. This guess is supported by the fact that computer simulations incorporating a realistic variable-restitution coefficient still can show unphysically small particle separations, whereas inelastic collapse can be removed by the TC model.

References

1. B. Bernu and R. Mazighi, J. Phys. A: Math. Gen. **23**, 5745 (1990).
2. S. McNamara and W.R. Young, Phys. Fluids A **4**, 496 (1992).
3. P. Constantin, E. L. Grossman and M. Mungan, Physica D **83**, 409 (1995).
4. D. Benedetto and E. Caglioti, Physica D **132**, 457 (1999).
5. S. McNamara and W.R. Young, Phys. Rev. E **50**, R28 (1994).
6. T. Zhou and L. P. Kadanoff, Phys. Rev. E **54**, 623 (1996).
7. S. McNamara and W.R. Young, Phys. Rev. E **53**, 5089 (1996).
8. G. Kuwabara and K. Kono, Jap. J. of Appl. Phys. **26**, 1230 (1987).
9. T. Schwager and T. Pöschel, Phys. Rev. E **57**, 650 (1998).
10. F.G. Bridges, A. Hatzes and D.N.C. Lin, Nature **309**, 333 (1984).
11. D. Goldman, M.D. Shattuck, C. Bizon, W.D. McCormick, J.B. Swift and H.L. Swinney, Phys. Rev. E **57**, 4831 (1998).
12. S. Luding, E. Clément, A. Blumen, J. Rajchenbach and J. Duran, Phys. Rev. E **50**, 4113 (1994).
13. E. Falcon, C. Laroche, S. Fauve and C. Coste, Eur. Phys. J. B **5**, 111 (1998).
14. S. Luding and S. McNamara, Granular Matter **1**, 111 (1998).

STATIONARY STATES IN SYSTEMS WITH DISSIPATIVE INTERACTIONS

J.PIASECKI
Institute of Theoretical Physics, University of Warsaw
Hoża 69, PL-00 681 Warsaw, Poland

1. Introduction

Recent developments in the physics of granular matter [1, 2] motivate the analysis of the effects of dissipative interactions on the structure of stationary states. The series of theoretical models described below is sufficiently simple to allow rigorous analytic analysis to be performed, but it contains at the same time some features relevant to the theory of granular flows. Thus, a description of coupling to a vibrating base via stochastic boundary conditions, or dissipative binary collisions are ingredients of the models considered.

In the search for understanding the new features induced by dissipation it is interesting to compare the situations with and without dissipative coupling. The simplest model system discussed in section 2 permits to pursue this idea. It consists of a single mass falling in the gravitational field and colliding with a thermalizing base, characterized by an appropriate rebound velocity distribution (the case of a sinusoidally oscillating base has been studied experimentally and theoretically [3]). When the flight between collisions is free, interaction with the base produces a Maxwell-Boltzmann equilibrium state. We consider the most common mechanism of dissipation observed at the macroscopic level - the friction. It seems very interesting to study the nature of changes in the stationary state when the flying particle suffers friction forces from the surrounding medium. The rigorous analysis in section 2 reveals the character of the stationary velocity distribution both in the case of a solid and a fluid friction. The simplicity of the model stems from assuming a macroscopic description of friction, so that the particle trajectory between collisions is uniquely defined for a given initial velocity gained when rebounding from the base.

In section 3, the dynamics of inelastic binary collisions is studied for a one-dimensional column of N particles, $N \geq 2$, falling in vacuum in the

J. Karkheck (ed.),
Dynamics: Models and Kinetic Methods for Non-equilibrium Many Body Systems, 279–295.

gravitational field and colliding with the base. Such a system with a base represented by a vibrating plate has been the object of intensive studies [7]. In contradistinction to the one-particle case, the source of dissipation is not the coupling to the environment, but rather the internal dynamics involving inelastic binary collisions within the N-particle system. Remarkably organized stratified periodic states are observed in the case of a singular rebound velocity distribution, represented by a Dirac-δ. The conditions for their occurence turn out to be rather restrictive, apart from the case of $N = 2$.

A more microscopic theory of dissipative coupling to the medium can be based on the assumption of inelastic collisions between the accelerated particle and the scatterers filling the space it is flying through. This case will be discussed in section 4 in the framework of the Lorentz model [4] where, from the point of view of the propagating particle, the external scatterers are supposed to be infinitely massive, and thus represent a zero-temperature static environment. The linear Boltzmann equation, suitably adapted to inelastic collisions, is then the basis of the analysis. It has been proved that there was no stationary state in the absence of dissipation [5, 6]. So, the inelasticity of collisions is essential in forming the stationary velocity distribution. Rigorous results described here show the net effect of dissipative interactions.

In the closing section 5, the stationary state of a single particle coupled to a thermostat via inelastic collisions is discussed. A remarkable prediction of the linear Boltzmann theory is the building up of a Maxwellian distribution with an effective temperature. The notion of granular temperature receives here a precise meaning.

2. Dissipation through Friction

Consider a particle of mass m acted upon by a constant and uniform gravitational field oriented along the z-axis. The particle falls with acceleration $(-g)$ until it touches the base situated at $z = 0$. It is then instantaneously sent back into the space with initial velocity w, distributed according to some given probability density $\phi(w)$. The shape of ϕ characterizes the base. It is supposed that during the flight through the surrounding medium the particle trajectory is uniquely defined by the initial condition $(z = 0, v = w)$.

The kinetic equation satisfied by the probability density $f(z, v; t)$ for finding the particle at time t at point z with velocity v reads

$$\left(\frac{\partial}{\partial t} + L\right) f(z, v; t) = R(z, v; t) \tag{1}$$

Figure 1. Schematic representation of the system. $\phi(w)$ is the rebound velocity distribution.

where

$$R(z,v;t) = \delta(z)[\phi(v)\int dw|w|\theta(-w)f(0,w;t) + v\theta(-v)f(z,v;t)]. \quad (2)$$

Here L is the phase-space volume preserving generator of collisionless motion, and $R(z,v;t)$ denotes the collision term. As the post-collisional velocity is independent of the velocity of approach, the gain term in $R(z,v;t)$ contains the product of the total collision frequency

$$\nu(t) = \int_{-\infty}^{0} dw|w|f(0,w;t) \quad (3)$$

and the rebound probability density $\phi(v)$.

In order to write the integral form of equation (1) we introduce the jacobian $J(z,v;-t)$ of the transformation

$$\begin{aligned} z &\rightarrow z(-t) \\ v &\rightarrow v(-t) \end{aligned} \quad (4)$$

where $[z(-t), v(-t)]$ is the state reached from (z,v) by a collisionless motion during time t, backward in time. Whereas $J(z,v;-t) = 1$ for a canonical transformation, $J(z,v;-t)$ differs from 1 when dissipation contracts the volume in the velocity space. Hence, the kinetic equation (1) is in general equivalent to

$$f(z,v;t) = J(z,v;-t)f[z(-t),v(-t);0] \quad (5)$$

$$+\int_0^t d\tau\, J(z,v;-\tau)\delta[z(-\tau)]\,\phi[v(-\tau)]\,\nu(t-\tau).$$

Notice that the loss term in (2) does not contribute to the right hand side of (5). This is because starting with negative velocity from $z(-\tau) = 0$ it is not possible to reach point $(z > 0, v)$ through collisionless motion (see the conditions imposed by the $\delta(z)$-distribution and the step function $\theta(-v)$).

Taking now the $t \to \infty$ limit we obtain the equation satisfied by the stationary distribution $F(z, v) = \lim_{t\to\infty} f(z, v; t)$

$$F(z,v) = \nu(\infty) \int_0^\infty d\tau\, J(z,v;-\tau)\, \delta[z(-\tau)]\, \phi[v(-\tau)]. \tag{6}$$

2.1. MOTION IN VACUUM: THERMALIZING REBOUND DISTRIBUTION

When falling with acceleration $(-g)$ takes place in vacuum, the generator of collisionless motion reads

$$L = v\frac{\partial}{\partial z} - g\frac{\partial}{\partial v} \tag{7}$$

and we have

$$J(z,v;-\tau) = 1, \quad z(-\tau) = z - v\tau - \frac{1}{2}g\tau^2 \;, \quad v(-\tau) = v + g\tau. \tag{8}$$

A straightforward calculation based on (6) yields then the formula

$$F(z,v) = \nu(\infty)\frac{\phi[w(z,v)]}{w(z,v)}, \tag{9}$$

where $w(z, v) > 0$ is the rebound velocity which makes the particle attain velocity v at altitude $z > 0$. The energy conservation implies

$$m[w(z,v)]^2/2 = E(z,v), \quad E(z,v) = mgz + mv^2/2.$$

We thus find

$$F(z,v) = \nu(\infty)\frac{\phi(\sqrt{v^2+2zg})}{\sqrt{v^2+2zg}}. \tag{10}$$

The normalization condition

$$\int dz \int dv F(z,v) = 1$$

determines the value of the stationary collision frequency

$$\nu(\infty) = \frac{g}{\int dw|w|\phi(|w|)}. \tag{11}$$

The fundamental conclusion from equation (10) concerns the form of the rebound distribution ϕ which induces the equilibrium Maxwell-Boltzmann state

$$F_T(z,v) = \frac{gm}{k_BT}\sqrt{\frac{m}{2\pi k_BT}}\exp\left(-\frac{m}{2k_BT}[v^2+2zg]\right). \tag{12}$$

We infer from (10) that the thermalizing base is characterized by

$$\phi(v) = \phi_T(v) \equiv \theta(v)\frac{mv}{k_BT}\exp\left(-\frac{mv^2}{2k_BT}\right). \tag{13}$$

It turns out that the rebound distribution (13) can be realized experimentally by using a sinusoidally oscillating base in the regime where the frequency of oscillation is very high compared to the collision frequency [3].

2.2. EFFECT OF SOLID FRICTION

The solid friction adds to gravitational acceleration $(-g)$ an additional term $[\theta(-v)-\theta(v)]\alpha$, always opposed to the velocity, and thus depending on its orientation. The upward and the downward motions have the generator of the form (7) with g replaced by $(g+\alpha)$ and by $(g-\alpha)$, respectively. So, equation (9) can be still applied, but the rebound velocity $w(z,v)>0$ depends now on the orientation of v. For $v>0$ we simply get

$$w(z,v) = \sqrt{v^2+2z(g+\alpha)}, \quad v>0. \tag{14}$$

When $v<0$, the particle first arrives at the maximum altitude z_{max} with acceleration $-(g+\alpha)$, and then falls down to the level z with acceleration $-(g-\alpha)$. The energy conservation equations here read

$$w^2 = 2(g+\alpha)z_{max}, \quad v^2 = 2(g-\alpha)(z_{max}-z),$$

yielding the formula

$$w(z,v) = \sqrt{\frac{g+\alpha}{g-\alpha}v^2+2z(g+\alpha)}, \quad v<0. \tag{15}$$

Combining these results and adopting the thermalizing rebound distribution (13) we find the stationary state of the form

$$F(z,v;\alpha) = 2n(z;\alpha)\left[P^{\uparrow}\theta(v)\phi^m(v;T)+P^{\downarrow}\theta(-v)\sqrt{\frac{g+\alpha}{g-\alpha}}\phi^m(v;T_\alpha)\right], \tag{16}$$

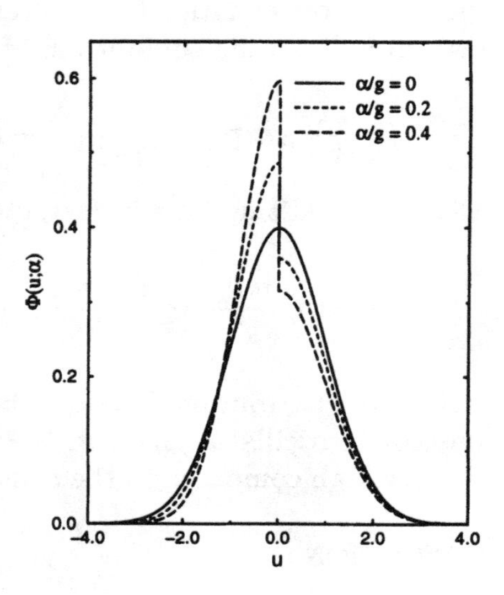

Figure 2. Velocity distribution $\Phi(v;\alpha)$ for various values of the friction coefficient α, according to equation (20). Variable $u = v(m/k_BT)^{1/2}$ is the dimensionless velocity.

where $n(z;\alpha)$ is the normalized Boltzmann spatial density

$$n(z;\alpha) = \frac{m}{k_BT}(g+\alpha)\exp\left(-\frac{m}{k_BT}(g+\alpha)z\right), \tag{17}$$

ϕ^m denotes the Maxwell velocity distribution

$$\phi^m(v;T) = \sqrt{\frac{m}{2\pi k_BT}}\exp\left(-\frac{mv^2}{2k_BT}\right), \tag{18}$$

and the α-dependent temperature T_α is given by

$$T_\alpha = \left(\frac{g-\alpha}{g+\alpha}\right)T. \tag{19}$$

The probabilities $P^\uparrow$ and $P^\downarrow$ of the upward and downward motion, appearing in (16), can be determined by requiring the vanishing of the particle current $< v >= 0$ (we denote by $< ... >$ the mean value with respect to the density F). The final formula for the stationary state reads

$$F(z,v;\alpha) = n(z;\alpha)\,\Phi(v;\alpha), \tag{20}$$

where

$$\Phi(v;\alpha) = \frac{2\left[\theta(v)\sqrt{g-\alpha}\phi^m(v;T) + \theta(-v)\sqrt{g+\alpha}\phi^m(v;T_\alpha)\right]}{\sqrt{g+\alpha}+\sqrt{g-\alpha}}.$$

The factorized structure of distribution $F(z,v;\alpha)$ reveals the absence of correlations between the position and the velocity variables. Moreover, much as at equilibrium (12), the spatial density remains represented by the Boltzmann factor (17) with however an effective mass $m^{eff} = m(1+\alpha/g)$. An important change occurs for the velocity distribution. From the equilibrium Maxwellian forced by the thermalizing base there remains only one half corresponding to ascending motion ($v > 0$). The probability density for negative velocities, although also gaussian, has a lower temperature $T_\alpha < T$ (see (19)). This asymmetry precludes the description of the velocity distribution in terms of some effective temperature. The mean kinetic energy, proposed to define granular temperature T_{gr} through the equality $< mv^2 >= k_B T_{gr}$, would introduce here the geometric mean $T_{gr} = \sqrt{TT_\alpha}$. When α is close to g (strong asymmetry) T_{gr} does not provide useful information about the velocity distribution.

2.3. ACTION OF FLUID FRICTION

The fluid friction exerts a force $(-\alpha v)$, which leads to exponential contraction of the volume in the velocity space. Indeed, in this case the transformation (4) takes the form

$$\begin{aligned} z(-t) &= z + \frac{mg}{\alpha}t - \frac{m}{\alpha}\left(v + \frac{mg}{\alpha}\right)\left[\exp\left(\frac{\alpha}{m}t\right) - 1\right] \\ v(-t) &= \left(v + \frac{mg}{\alpha}\right)\exp\left(\frac{\alpha}{m}t\right) - \frac{mg}{\alpha}. \end{aligned} \tag{21}$$

The corresponding jacobian equals

$$J(z,v;-t) = \frac{\partial[z(-t),v(-t)]}{\partial[z,v]} = \exp\left(\frac{\alpha t}{m}\right) \tag{22}$$

(forward in time evolution would get the contracting factor $\exp(-\alpha t/m)$). The generator L in the kinetic equation (1) is correspondingly given by the formula

$$L = -\frac{\alpha}{m} + v\frac{\partial}{\partial z} - \left(g + \frac{\alpha}{m}v\right)\frac{\partial}{\partial v}. \tag{23}$$

The term $(-\alpha/m)$ adds up here to the generator of the particle trajectory to balance the exponential factor (22) and conserve the normalization of the evolving probability density. Consequently, the equation (6) satisfied by the stationary distribution $F(z,v;\alpha)$ reads

$$F(z,v;\alpha) = \nu(\infty;\alpha)\int_0^\infty d\tau \exp\left(\frac{\alpha}{m}\tau\right)\delta[z(-\tau)]\,\phi_T[v(-\tau)]. \tag{24}$$

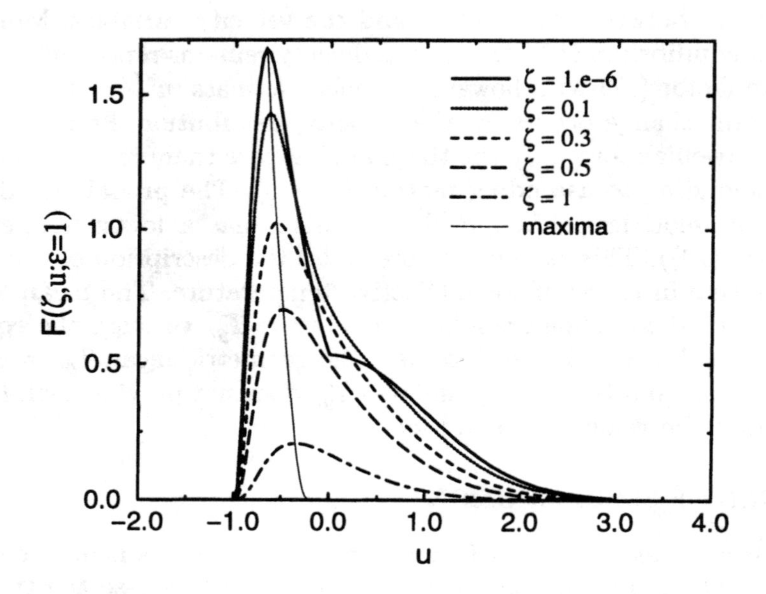

Figure 3. Velocity distribution at different altitudes for $\epsilon = v_T/|v_{min}| = 1$. Here $v_T = (k_B T/m)^{1/2}$ is the thermal velocity, and $v_{min} = -mg/\alpha$. Dimensionless position and velocity are defined by: $\zeta = mgz/k_B T$, $u = v/v_T$.

Inserting into (24) the trajectory (21) we arrive at an explicit formula

$$F(z, v; \alpha) = \int_0^\infty d\tau \delta \left\{ z + \frac{mg}{\alpha}\tau - \frac{m}{\alpha}\left(v + \frac{mg}{\alpha}\right)\left[\exp\left(\frac{\alpha}{m}\tau\right) - 1\right]\right\} \quad (25)$$

$$\times \exp\left(\frac{\alpha}{m}\tau\right) \phi_T \left\{ \left(v + \frac{mg}{\alpha}\right) \exp\left(\frac{\alpha}{m}\tau\right) - \frac{mg}{\alpha} \right\} \nu(\infty; \alpha).$$

As before, the collision frequency $\nu(\infty; \alpha)$ is to be determined from the normalization of F. Equation (25) implies the vanishing of $F(z, v; \alpha)$ for $v < -mg/\alpha$ (the argument of the δ-distribution is strictly positive therein). There is thus a cut-off in the spectrum of possible negative velocities, corresponding to the balance between the gravitational force $(-mg)$ and friction $(-\alpha v)$. This shows an important difference with respect to the equilibrium Maxwell state. A straightforward calculation based on formula (25) yields the form of the velocity distribution for the ascending motion $(v > 0)$

$$\theta(v)\Phi(v; \alpha) = \frac{m\nu(\infty; \alpha)}{\alpha v + mg} \exp(-mv^2/2k_B T). \quad (26)$$

The nature of the deviation from the gaussian shape is here explicitly shown.

Another important change with respect to the equilibrium state (12) is the appearance of correlations between positions and velocities: the stationary state (25) is not a product of a spatial density profile and a velocity distribution. In particular, at very low altitudes positive velocities are not too perturbed by friction, and a wing of the equilibrium gaussian distribution appears (see fig.3). It is to be noted that a discontinuity in the derivative of the velocity distribution develops at the connection point $v = 0$. A thorough analysis of formula (25) also shows that with increasing friction the probability weight in the velocity space gets shifted more and more toward the negative interval $[-mg/\alpha, 0]$ [8]. The fluid friction introduces thus important qualitative changes, inducing an asymmetric, correlated stationary state.

3. Dissipative Collisions Balanced by a Single Rebound Velocity

Our object now will be to study the effects of inelastic binary collisions taking place within the N-particle system, moving in vacuum (no friction), and forming a column of identical masses falling with acceleration $(-g)$ in an external field. The energy losses due to binary collisions between the particles are balanced by encounters with the base, playing the role of the energy source, and the question is to study the resulting stationary state. We are here close to typical questions concerning the dynamics of fluidized granular matter.

To begin with, we choose the simplest possible system with $N = 2$. The states of the two mases will be denoted by (z_1, v_1) and (z_2, v_2), respectively, particle 1 being closer to the base: $0 \leq z_1 \leq z_2$ (linear ordering is preserved by the dynamics).

We shall consider here the simplest case where the rebound velocity at the base has a singular distribution

$$\phi(w) = \delta(w - w_0) \tag{27}$$

centered on a fixed characteristic value $w_0 > 0$. At a binary collision between masses 1 and 2 , their velocities of approach (w_1, w_2) instantaneously take post-collisional values (w_1', w_2') given by

$$w_1' = w_1 - \frac{1+\alpha}{2} w_{12}, \quad w_2' = w_2 + \frac{1+\alpha}{2} w_{12}, \tag{28}$$

where $w_{12} = w_1 - w_2$. When $\alpha = 1$, the particles just exchange their velocities, and the total kinetic energy is conserved. For $\alpha < 1$, the dissipation occurs. In order to perform the quantitative analysis let us denote by $E_1^{\uparrow}$

the energy of particle 1 in its ascending motion after collision with the base. In accordance with (27)

$$E_1^{\uparrow} = mw_0^2/2. \tag{29}$$

Suppose that particle 2 is falling then with energy $E_2^{\downarrow}$. The collision law (28) implies that after the binary encounter particle 1 is sent back to the base with energy

$$E_1^{\downarrow} = \frac{1-\alpha}{2}E_1^{\uparrow} + \frac{1+\alpha}{2}E_2^{\downarrow} - \frac{1-\alpha^2}{8}mw_{12}^2, \tag{30}$$

whereas particle 2 acquires the energy

$$E_2^{\uparrow} = \frac{1-\alpha}{2}E_2^{\downarrow} + \frac{1+\alpha}{2}E_1^{\uparrow} - \frac{1-\alpha^2}{8}mw_{12}^2. \tag{31}$$

Adding up equations (30) and (31) we obtain the value of the dissipated kinetic energy

$$[E_1^{\downarrow} + E_2^{\uparrow}] - [E_1^{\uparrow} + E_2^{\downarrow}] = -\frac{1-\alpha^2}{4}mw_{12}^2. \tag{32}$$

The computer visualization of the motion of this system revealed a remarkable synchronization in the stationary state: particles 1 and 2 perform then periodic motions colliding always at the same altitude z_{coll}. This information can be used to construct the stationary distribution analytically. First of all, in such a system the energy of particle 2 stays constant, the collisions with 1 just reversing its velocity. From (27), (29) and (32) we get thus two relations

$$E_1^{\downarrow} = mw_0^2/2 - (1-\alpha^2)mw_{12}^2/4, \quad -w_2 = w_2 + (1+\alpha)w_{12}/2. \tag{33}$$

Moreover, the periods of motion must be the same for both particles. The time $\tau^{\uparrow}$ needed by particle 1 to attain the precollisional velocity w_1 starting from the base equals

$$\tau^{\uparrow} = (w_0 - w_1)/g.$$

To reach back to the base, the particle uses time $\tau^{\downarrow}$ during which its postcollisional velocity $[w_1 - (1+\alpha)w_{12}/2)]$ attains the value $-\sqrt{2E_1^{\downarrow}/m}$. So,

$$-\sqrt{\frac{2E_1^{\downarrow}}{m}} = \left[w_1 - \frac{1+\alpha}{2}w_{12}\right] - g\tau^{\downarrow}.$$

Now, the period of motion τ of particle 2 is simply the time needed to perform a free flight starting from altitude z_{coll}, and coming back to it

$$\tau = -2w_2/g.$$

For consistency, the equation

$$g\tau = -2w_2 = g(\tau^{\uparrow} + \tau^{\downarrow}) = w_0 - \frac{1+\alpha}{2} w_{12} + \sqrt{\frac{2E_1^{\downarrow}}{m}} \tag{34}$$

must hold.

Equations (33), (34) suffice to evaluate the parameters characterizing the stationary state F. We find

$$E_1^{\uparrow} = \frac{mw_0^2}{2}, \quad E_1^{\downarrow} = E_1^{\uparrow}\left(\frac{1+3\alpha}{3+\alpha}\right)^2, \quad w_{12} = \frac{4}{3+\alpha} w_0 \tag{35}$$

$$E_2 = E_1^{\uparrow}\left(\frac{1+14\alpha+\alpha^2}{(3+\alpha)^2}\right).$$

Taking into account the normalization condition and the fact that the relative velocity after collision gets multiplied by $(-\alpha)$, we can write the explicit formula for the stationary state $(z_1 > 0, z_{21} > 0)$

$$\begin{aligned} F(z_1, v_1; z_2, v_2) = {} & \frac{mg^2}{w_0}\left(\frac{3+\alpha}{1+\alpha}\right)\delta\left[E(z_2, v_2) - \frac{(1+14\alpha+\alpha^2)}{(3+\alpha)^2}\frac{mw_0^2}{2}\right] \\ & \times\left\{\delta\left[v_{12} - \frac{4}{3+\alpha}w_0\right]\delta\left[E(z_1, v_1) - \frac{mw_0^2}{2}\right]\right. \\ & \left. +\delta\left[v_{12} + \frac{4\alpha}{3+\alpha}w_0\right]\delta\left[E(z_1, v_1) - \left(\frac{1+3\alpha}{3+\alpha}\right)^2\frac{mw_0^2}{2}\right]\right\}, \end{aligned} \tag{36}$$

where the notation $E(z, v) \equiv [mgz + mv^2/2]$ for the energy in the gravitational field has been used.

The probability density (36) involves products of delta distributions confining the stationary state to a one-dimensional manifold in the four-dimensional phase space. For a two-particle system, the state (36) will be approached in the long-time limit from any initial state.

In principle, analogous periodic states could exist for a column of $N > 2$ masses. The particles would remain confined to adjacent volumes in the position space, suffering inelastic collisions with their nearest neighbors always at the same altitudes. The visualization of this highly synchronized motion is quite impressive. However, when $N > 2$ one cannot expect to observe these stratified structures for any values of coefficient α. In fact, one can derive a rigorous necessary condition for the existence of N-particle single-period stationary states [8] . It has the form

$$(1-\alpha)^2 < 6\gamma - 2\gamma^2, \quad \text{with } \gamma = (1-\alpha)N. \tag{37}$$

The relevance of parameter $\gamma = (1-\alpha)N$ appearing in (37) has been recognized in the study of a clustering transition in a column of beads [7]. Inequality (37) does not represent any restriction only for $N = 2$. However, already for $N = 3$ not all values of α are admissible. Clearly, for large values of N only an extremely small inelasticity $(1-\alpha) \to 0$ allows the simple periodic motion to appear.

4. Dissipation through Inelastic Scattering: the Lorentz Gas

The friction was described in previous sections on the macroscopic level. It entered via the friction coefficient into the deterministic equations describing propagation through the surrounding medium (see (21)). In order to analyze this propagation on a more microscopic level we shall consider now the coupling to the environment through inelastic binary collisions.

The system under consideration is again a single particle (or a gas of non-interacting particles) moving with a constant acceleration g. It is supposed to collide on its way with scattering centers, randomly distributed in space with some number density n. Continuing to study the one-dimensional propagation we suppose that the velocity v of the particle at encounters gets instantaneously changed according to the law

$$v \quad \to \quad v' = -\alpha v. \tag{38}$$

When $\alpha = 1$, the particle is scattered elastically as if it encountered an immobile infinite mass. The system under consideration represents then a one-dimensional version of the Lorentz gas. It has been proved that in this case no stationary state could exist [5, 6, 9]. In fact, for $\alpha = 1$, the only way to lose energy is to move against the field, owing to the elasticity of the scattering. But this mechanism does not suffice to balance the energy flow from the accelerating external field. The system gets heated beyond any bounds, its mean kinetic energy diverging asymptotically as $t^{2/3}$ in the long time limit $t \to \infty$. When $0 \leq \alpha < 1$, the kinetic energy is dissipated at inelastic collisions according to the transformation

$$E = mv^2/2 \quad \to E' = E - (1-\alpha^2)mv^2/2. \tag{39}$$

This creates the possibility for the appearance of a stationary state.

In order to investigate this question let us write down the one-dimensional version of the linear Boltzmann equation corresponding to the collision law (38)

$$\left(\frac{\partial}{\partial t}+g\frac{\partial}{\partial v}\right)f(v;t)=\frac{|v|}{\lambda\pi}\left\{\alpha^{-2}f(-\alpha^{-1}v;t)-f(v;t)\right\}. \tag{40}$$

Here $\lambda\sim n^{-1}$ denotes the mean free path. The factor α^{-2} in the gain term compensates for the contraction of the velocity space at collisions (38), and assures the correct value $(-v/\alpha)$ of the velocity of approach before collision (the inverse transformation to (38) is obtained by replacing α by α^{-1}).

Passing to the dimensionless velocity variable

$$u=v/\sqrt{g\lambda} \tag{41}$$

and denoting by $F(u)$ the stationary dimensionless distribution we find from (40) the equation

$$\frac{d}{du}F(u)=|u|\left[\alpha^{-2}F(-\alpha^{-1}u)-F(u)\right]. \tag{42}$$

The amplitude g of the external acceleration disappeared from equation (42). The field dependence of the moments of F follows thus directly from the dimensional velocity scaling (41). In particular we find

$$<v>=\sqrt{g\lambda}<u>\sim\sqrt{g},\qquad <v^2>\sim g. \tag{43}$$

So, if the stationary state F existed, the particle current would be proportional to the square root of the external field (no linear response). In fact, the only energy scale in the model is related to acceleration g, because the medium composed of immobile scatterers does not introduce a non-zero temperature. The predictions (43) could be thus deduced by dimensional arguments.

In order to prove the existence of a stationary state it is convenient to use an auxiliary function G defined by

$$F(u)\equiv G(u|u|/2). \tag{44}$$

It satisfies the equation

$$G'(s)+G(s)=\alpha^{-2}G(-\alpha^{-1}s), \tag{45}$$

where G' denotes the derivative of G. The Fourier transform

$$\hat{G}(k)=\int ds\exp(-iks)G(s)$$

can be readily calculated from (45). One finds

$$\hat{G}(k) = (1+ik)^{-1}\hat{G}(-\alpha^2 k) = \hat{G}(0)\prod_{r=0}^{\infty}[1+i(-\alpha^2)^r k]^{-1}. \tag{46}$$

For $0 \leq \alpha < 1$ the infinite product converges showing the existence of the stationary state. The dissipative dynamics (38) introduces thus a qualitative change with respect to the case of elastic collisions. It turns out that an arbitrary degree of inelasticity suffices to balance the energy flow from the external field and to generate a stationary state.

The case of perfectly inelastic collisions ($\alpha = 0$) is particularly simple. Equation (46) reduces to

$$\hat{G}(k) = \frac{\hat{G}(0)}{(1+ik)}. \tag{47}$$

Inverting the Fourier transform and using (44) we find

$$F(u) = \theta(u)\sqrt{\frac{2}{\pi}}\exp(-u^2/2). \tag{48}$$

In one dimension each collision with $\alpha = 0$ dissipates the whole energy absorbed from the field (the particle gets stopped). It follows that the velocities oriented against the field are not possible in a stationary flow (the θ factor in (48)). Equation (48) predicts a gaussian probability weight in the half space of possible velocities. The dynamics of the approach to the stationary distribution (48) has been studied in [10].

We turn now to the study of the limit of low inelasticity $\epsilon = (1-\alpha) \to 0$. For $\epsilon \ll 1$ one can expect a very high kinetic energy of the propagating particle. The collision frequency becomes then also very big, which must make the velocity distribution almost symmetric. It is thus natural to rewrite equation (45) in terms of the symmetric part $G_+(s) = [G(s) + G(-s)]/2$ and the antisymmetric part $G_-(s) = [G(s) - G(-s)]/2$ of G. We find the system of two coupled equations

$$\begin{aligned} G'_+(s) + G_-(s) &= -\alpha^{-2}G_-(s/\alpha^2) \\ G'_-(s) + G_+(s) &= \alpha^{-2}G_+(s/\alpha^2). \end{aligned} \tag{49}$$

The integral form of the second relation in (49) reads

$$G_-(s) = \int_s^{s/(1-\epsilon)^2} G_+. \tag{50}$$

Substitution of (50) into (49) yields a closed equation for the symmetric part

$$G'_+(s) = -\int_s^{s/(1-\epsilon)^2} G_+ \;-\; \frac{1}{(1-\epsilon)^2}\int_{s/(1-\epsilon)^2}^{s/(1-\epsilon)^4} G_+. \qquad (51)$$

When $\epsilon \to 0$, (51) reduces asymptotically to the differential relation $G'_+(s) = -4\epsilon s G_+(s)$, and we find

$$G_+(s) = C\epsilon^{1/4}\exp(-2\epsilon s^2). \qquad (52)$$

Using then relations (50) and (44) we determine the asymptotic form of the velocity distribution F in the limit of low inelasticity

$$F(u) = \frac{C\epsilon^{1/4}}{\sqrt{2}}(1+\epsilon u|u|)\exp(-\epsilon u^4/2), \quad \epsilon \ll 1 \qquad (53)$$

where

$$C^{-1} = \int dw \exp(-2w^4)$$

(normalization condition).

As expected, the dominant term in (53) has the spherical symmetry. It depends on the velocity via the scaled variable (ϵu^4). This implies the divergence $\sim \epsilon^{-1/2}$ in the kinetic energy $< u^2 >$ for $\epsilon \to 0$, reflecting the infinite heating of the system with elastic collisions.

As the high collision frequency makes the asymptotic distribution (53) almost symmetric, a very weak mean velocity $< u >$ characterizes the stationary particle flow

$$< u >= C\,\epsilon^{1/4}/\sqrt{2}. \qquad (54)$$

Our analysis shows that the stationary velocity distribution changes from completely asymmetric half-gaussian (48) at the strongest dissipation, to an almost symmetric scaled distribution $\sim \exp(-\epsilon u^4)$ close to the elastic limit. The studies of the classical Lorentz model showed that the qualitative behavior of the system was independent of the spatial dimension [5]. We can thus expect both the existence of the stationary state in three dimensions and the same scaling structure (53) in the elastic limit [11].

5. Coupling to a Thermostat via Dissipative Collisions

The unsatisfactory feature of the Lorentz gas from the point of view of the theory of granular matter is that particles propagate there through a zero-temperture medium. It would be very interesting to extend the theory giving a positive temperature $T > 0$ to the scatterers. We shall consider here the first step in this direction by describing the stationary velocity distribution in the absence of the external field. We thus assume that a particle of mass m moves freely between collisions through a medium at temperature T, filled with scattering centers of mass M at uniform density n. The state of the thermostat as seen by the particle is thus

$$n\phi^M(w;T) = n\sqrt{\frac{M}{2\pi k_B T}}\exp\left(-\frac{Mw^2}{2k_B T}\right). \tag{55}$$

Keeping to the one-dimensional model we shall assume here that at a binary encounter the velocities v and w of the particle and the scatterer take post-collisional values

$$v' = v - (1+\alpha)\mu(v-w), \quad \text{and} \quad w' = w + (1+\alpha)(1-\mu)(v-w), \tag{56}$$

respectively, with

$$\mu = M/(m+M).$$

The inverse transformation is obtained by replacing α by α^{-1}. The linear Boltzmann equation for the stationary velocity distribution $F(v)$ of the particle has the form

$$\int dw\,|w|\left\{-F(v)\phi^M(w;T) + \right. \tag{57}$$

$$\left. \alpha^{-2}F[v-(1+\alpha^{-1})\mu(v-w)]\phi^M[w+(1+\alpha^{-1})(1-\mu)(v-w);T]\right\} = 0.$$

It is quite remarkable that the solution to this equation is a Maxwell distribution

$$F(v) = \phi^m[v;T^*(\alpha, M/m)] \tag{58}$$

with an effective temperature

$$T^*(\alpha, M/m) = \left[\frac{1+\alpha}{2+(1-\alpha)M/m}\right]T. \tag{59}$$

The validity of equations (58), (59) can be verified by straightforward calculation. The reported result remains valid also in three dimensions [12] with the same effective temperature (59), which is quite remarkable.

We have here a beautiful example where the granular temperature can be given a precise meaning. The dissipative inelastic collisions have the effect of preventing the thermostat from transmitting its own temperature T to the particle. Nevertheless, thermalization takes place in the sense that a Maxwell distribution results from the particle-thermostat coupling for any value of the restitution parameter α. Clearly, the effective temperature T^* is lower than T, attaining the minimal value $[mT/(2m+M)]$ for $\alpha = 0$.

References

1. H.M. Jaeger and S.R. Nagel, Science **255**, 1523 (1992).
2. H.M. Jaeger and S.R. Nagel, Rev.Mod.Phys. **68**, 1259 (1996).
3. S. Warr, W. Cooke, R.C. Ball, and J.M. Huntley, Physica A **231**, 551 (1996).
4. H.A. Lorentz, Arch.Néerl. **10**, 336 (1905)
5. J. Piasecki and E. Wajnryb, J.Stat.Phys. **21**, 549 (1979).
6. J. Piasecki, Am.J.Phys. **61**, 718 (1993).
7. B. Bernu, F. Delyon, and R. Mazighi, Phys. Rev. E **50**, 4551 (1994).
8. T. Biben and J.Piasecki, Phys. Rev. E **59**, 2192 (1999).
9. K. Olaussen and P. Hemmer, J.Phys.A: Math.Gen. **15**, 3255 (1982).
10. J. Piasecki, J.Stat.Phys. **30**, 185 (1983).
11. Ph.A. Martin and J. Piasecki, Physica A **265**, 19 (1999).
12. Ph.A. Martin and J. Piasecki, Euro.Phys.Lett. **46**, 613 (1999).

MOTION ALONG A ROUGH INCLINED SURFACE

ALEX HANSEN
Institutt for Fysikk,
Norges Tekniske-Naturvitenskapelige Universitet,
N–7491 Trondheim, Norway

1. Introduction

Somebody has remarked in connection with the development of quantum field theory that in classical mechanics, the three-body problem is too difficult. With the advent of quantum mechanics, the two-body problem has become too difficult, while quantum field theory has made the zero-body problem too difficult.

This is of course a caricature. Since the early work of giants such as Boltzmann, physics has learned to deal with systems containing huge numbers of particles, provided the right questions are to be answered.

Granular media, such as a heap of sand, contain huge numbers of particles - sand grains - and it would not have been unreasonable to assume that statistical physics should have been capable of describing their behavior. However, it has turned out to be extraordinarily difficult to get a handle on these systems. Granular media are typically very dissipative systems - just think how long the sugar in a box keeps moving after one has stopped shaking the box. The growing experimental body on granular media shows that they exhibit an extraordinarily wide range of very different behaviors depending on the circumstances.

But, what about the behavior of one or two or just a few grains? It is the aim of this small review to demonstrate that the remark made at the very beginning of this article more or less applies to granular media as well. We will discuss the motion of a single spherical object rolling on a rough inclined plane. As will be evident, there are many open questions remaining in this seemingly simple problem.

In the next section, we discuss why it is of interest to study the motion of a single grain in a static environment when the problems encountered in granular media usually entail huge numbers of moving and interacting

J. Karkheck (ed.),
Dynamics: Models and Kinetic Methods for Non-equilibrium Many Body Systems, 297–312.

grains. In Section 3, we describe some experimental results in connection with the study of a single spherical grain rolling on a rough surface consisting of fixed randomly placed smaller spherical grains. Section 4 is devoted to a simple model of a particle jumping along a rough inclined line. This model shows clearly the origin of elastic collapse. In Section 5, we discuss the intricacies of the restitution coefficients that are at the heart of the inelastic collapse singularity. Section 6 demonstrates how any velocity-dependency of the normal restitution coefficient will remove the inelastic collapse problem in the simple particle-jumping-on-a-line model. We conclude with some brief comments in Section 7.

2. Algorithmic vs. Microscopic Approach to Avalanches

Avalanches in granular media have attracted much attention in the physics community since they were used as a prime example in the early work on self-organized criticality by Bak et. al. [1]. In other parts of society, they have long been of great interest for practical reasons: avalanches in snow-covered mountain valleys can have devastating consequences for those caught up in them. The same goes for the potentially even more serious cases of mud and soil slides that, e.g., may be triggered by earth quakes.

The approach to complex collective phenomena, such as avalanches, that illustrate self-organized criticality, is to search for a model which typically is considerably simpler than the physical phenomenon it is to describe. The basic idea is that even though the model seems oversimplified on the small scale, the large scale collective behavior of the model is the same as that of the real system with respect to certain aspects. In connection with avalanches, such an aspect would for example be their size statistics. We may refer to such modeling as an *algorithmic approach.*

However, other questions need a detailed knowledge of the small-scale aspects of the system at hand. For example, in connection with snow avalanches, it is of great importance to know what kind of avalanches may result from the structure of the snow flakes that consitute the snow: What kind of snow is "safe" and what kind of snow is dangerous?

Such questions need a very different approach than the algorithmic one. These questions are as complex, however, as those posed when dealing with algorithmic modeling. Thus, simplifications are necessary, but these must be of a very different kind than those employed in algorithmic modeling: They must preserve the relevant properties of the microscopic constituents of the system. Thus, we may refer to this as *microscopic modeling.*

There are thus two difficult aspects to the avalanche problem: There are a lot of grains participating in an avalanche and the grains have structure that determines the collective behavior of the avalanche. In the algorithmic approach we simplify the grains and the details of their interactions, in the microscopic approach we simplify the avalanche itself.

In this review, we describe a microscopic approach to avalanches. Imagine *freezing* all grains except one. That is, think of a three-dimensional snapshot of the avalanche; each grain has been frozen in the position in space it occupied at the time of the snapshot. Then, choose one of the grains, and let this one keep on moving. It will of course collide with the other grains that have been frozen. If the frozen grains are to stay frozen during such a collision, one has to assume that they have become infinitely massive in order to conserve momentum.

Such a system is simple to model on the computer, but it is of course hopeless to try to devise an experimental setup of this kind. As the aim of the study is to eventually understand the influence of the microscopic properties of the grains on the avalanche, experiments *are* necessary. Thus, further simplifications are called for.

We move all the frozen grains onto a plane. In practice, simply glue the grains to the surface. Then, pick one grain and let it roll on the plane, which is held at some inclination, and study the motion of this grain.

Of course, the dynamics of a ball rolling on a rough surface, into which we have transformed and simplified the initial avalanche problem, is of great interest in itself. We describe some of the experimental results on this problem in the next section.

3. Experimental Techniques and Results

In the experimental setup of Riguidel [2, 3], the inclined plane consists of a 1 cm thick rectangular glass plate mounted on a support that can be tilted along the short axis. The glass plate measures 100 cm × 70 cm. Contact paper is placed on top of the glass plate with the glue-covered side pointing upwards. Glass spheres with an average diameter ranging from 0.5 mm to 2 mm are poured onto the contact paper and spread out as evenly as possible over the entire surface using a large ruler. This ensures that a homogeneous, locally disordered monolayer of glass spheres will cover the glass plate. The packing fraction obtained this way is of the order of 0.7.

The "grains" rolling on the rough inclined plane are steel spheres with diameters ranging from 0.75 mm to 5.15 mm.

In a later series of experiments by Henrique et. al. [4], sifted sand with average diameters between 0.2 mm and 0.25 mm was used to provide the roughness of the plane, thus obtaining a packing fraction of 0.8. The sand grains were much more irregularly shaped than the glass spheres of Riguidel. Furthermore, the glue used by Henrique et. al. was stiffer than the one used by Riguidel, thus changing the properties of the collisions between the rolling sphere and the fixed grains. The steel spheres of Henrique et. al. ranged in size from 1.6 mm to 10.3 mm.

If d_1 is the diameter of the fixed grains and d_2 is that of the rolling sphere, we define the ratio $\Phi = d_2/d_1$. The angle of inclination of the plane is θ. In the (Φ, θ)-plane three regimes may be identified: (1) a *retardation regime* where the sphere decelerates until it stops, (2) a *rolling regime* where the sphere attains a well-defined average speed while seemingly staying in contact with the surface at all times, and (3) a *jumping regime* where the sphere makes jumps between each contact with the surface. Whether this last regime may be subdivided into two further regimes, one where a well-defined average speed is attained, and one where the sphere will keep on accelerating, making larger and larger jumps, was not possible to determine as this requires a very long surface.

Jan et. al. [5] have done experiments on a one-dimensional version of this problem. The sphere then rolls along a channel whose bottom is covered with a one-dimensional string of grains of varying diameter.

All those experiments have mostly concentrated on the rolling regime. Among several interesting observations, we concentrate first on the relation between the average velocity component of the sphere along the direction of the incline as a function of the inclination angle θ. The gravitational force, $m\vec{g}$, acting on the sphere having a mass m points downward. Furthermore, there is an average force $\langle \vec{f} \rangle$ pointing upward along the incline stemming from the collisions between the sphere and the fixed grains. In order for the sphere to attain a fixed average velocity, the forces on the sphere must balance. Thus,

$$\langle f \rangle = mg \sin \theta \ . \tag{1}$$

Clearly, $\langle f \rangle$ must depend both on m and the average speed of the sphere, $\langle v \rangle$. We guess a very general dependency,

$$\langle f \rangle \propto m^{\alpha} \langle v \rangle^{\beta} \ , \tag{2}$$

where α and β are exponents to be determined through the experiment. Combining (1) and (2), we find

$$\langle v \rangle \propto m^{(1-\alpha)/\beta} \sin^{1/\beta} \theta \ . \tag{3}$$

Before revealing the experimental result, let us analyze the situation to see what to expect.

The magnitude of the average force $\langle f \rangle$ is determined by two sources: (1) the number of collisions per unit time, ν, and (2) the momentum transfer per collision, $\langle \Delta p \rangle$,

$$\langle f \rangle \propto \nu \langle \Delta p \rangle \ . \tag{4}$$

As the sphere is essentially in contact with the surface at all times, we expect the frequency of collisions to be proportional to the average velocity

$$\nu \propto \langle v \rangle \,. \tag{5}$$

Furthermore, we expect the average momentum transfer, $\langle \Delta p \rangle$, to be proportional to the average momentum of the sphere,

$$\langle \Delta p \rangle \propto m \langle v \rangle \,. \tag{6}$$

Combining Equations (5) and (6), we find

$$\langle f \rangle \propto m \langle v \rangle^2 \tag{7}$$

thus resulting in $\alpha = 1$ and $\beta = 2$. Inserting these into (3) leads to

$$\langle v \rangle \propto \sin^{1/2} \theta \,. \tag{8}$$

This simple argument is equivalent to the one proposed by Bagnold [6] for the stress distribution in granular shear flow in the so-called grain-inertia regime.

The one-dimensional experiments of Jan et. al. [5] do indeed follow Eq. (8). However, the two-dimensional experiments [2, 3, 4] do not. By one and two dimensions we refer to the dimensionality of the rough surface. One finds in the two-dimensional case

$$\langle v \rangle \propto \sqrt{m} \sin \theta \,, \tag{9}$$

corresponding to $\alpha = 1/2$ and $\beta = 1$, so that

$$\langle f \rangle \propto \sqrt{m} \langle v \rangle \,. \tag{10}$$

This is a surprising result given the straight-forwardness of the argument presented above that leads to Eq. (7).[1]

We present one more experimental result obtained in two dimensions [2, 3, 4]. In the rolling regime, it also happens that the sphere may abruptly stop. The speed of the sphere fluctuates around some mean, and from time to time a fluctuation large enough to stop the sphere occurs. One may measure the average distance, L_*, the sphere moves before stopping as a function of the inclination of the plane and the sphere's mass. One finds

$$L_* \propto e^{am \sin^2 \theta} \,. \tag{11}$$

[1] It is, however, interesting to note that when these results are presented to physicists with a background in transport theory, they react oppositely to that of those with a background in granular flow: It is the *one-dimensional* result, Equation (7), which is the strange one, and not the two-dimensional one, (10). The reason for this is the superficial similarity with linear response.

Through a not completely straight-forward argument, one may show that (11) is consistent with Eq. (10). The argument is based on extreme statistics [7].

Let us assume that the force, f, arising from the collisions of the rolling sphere is gaussian around the average value $\langle f \rangle$ with a width σ_f. For f very different from $\langle f \rangle$, the cumulative Gaussian distribution is

$$P(f) = 1 - e^{-(f-\langle f \rangle)^2/2\sigma_f^2} . \tag{12}$$

Each time there is a collision, a new force f is "drawn" from the distribution (12). Suppose there are N collisions all together. The expected largest force, $f_{\max}$, encountered during those N events may be estimated through [7]

$$P(f_{\max}) = 1 - \frac{1}{N} . \tag{13}$$

We are looking for an $f_{\max}$ that is actually able to stop the sphere. This happens if

$$f_{\max} - \langle f \rangle = \langle f \rangle , \tag{14}$$

leading to $f_{\max} = 2\langle f \rangle$. When the sphere stops after having moved a distance L_*, it has suffered $N = L_*/d_1$ collisions. Thus, combining Eqs. (12), (13) and (14), we find

$$L_* = d_1 e^{2\langle f \rangle^2/\sigma_f^2} . \tag{15}$$

Using Eq. (10) for $\langle f \rangle$ in Eq. (15), we find

$$L_* = d_1 e^{(2a/\sigma_f^2) m \sin^2 \theta} . \tag{16}$$

Thus, we reproduce the experimental result, Eq. (11), when assuming that σ_f is independent of m and $\langle v \rangle$. However, we expect this to be so from the Einstein relation, which in this case leads to $\langle f \rangle \propto \sigma_f \langle v \rangle$.

The distribution of stopping lengths, L, is exponential in the rolling regime [2, 3, 4],

$$p(L) = p_0 e^{-L/L_*} , \tag{17}$$

where L_* is given by Eq. (11). However, as the inclination, θ, is reduced, the system shifts from the rolling to the retardation regime. This sets in at some angle θ_T [4], and one finds that near this angle

$$L_* \propto \frac{1}{|\theta - \theta_T|} . \tag{18}$$

For smaller angles so that the system is well into the retardation regime, one finds that

$$L_* \propto \frac{1}{\theta} . \tag{19}$$

There have been a number of numerical studies of this problem, in one dimension [8-10], and in two dimensions [11]. In the latter case, molecular dynamics simulations show a range of inclinations occurs where essentially the behavior indicated by Eq. (9) is reproduced. Still, it is not clear how Eq. (9) emerges. The authors note that the rolling sphere typically collides *several* times with a given sphere that is glued to the surface. An explanation may, perhaps, be found by exploring this remark.

4. Model for A Single Ball Bouncing Along a Rough Line

As mentioned in the last section, it is very difficult to conduct experiments in the bouncing regime, the problem being that the inclined plane has to be very long. However, this case is somewhat simpler to analyse from a theoretical point of view. As we will see, even this case leads to problems.

We will base our discussion on the model of Valance and Bideau [12]. It is a one-dimensional model where the roughness consists of a series of infinitesimal facets whose normals are oriented randomly in some interval $(-\phi_M, \phi_M)$ around the mean normal of the plane. This in turn is inclined at an angle θ with respect to the horizontal. There are no correlations between the orientation of the facets.

The sphere is also taken to be infinitesimally small and rotation is ignored. Between each collision with a facet on the surface the sphere moves by following the laws of free fall. The collisions are instantaneous.

The velocity of the sphere right before it touches the surface for the kth time is $\vec{u}_{k-1}$. We define a normal component and a tangential component of the velocity with respect to the facet at that point having a normal unit vector $\vec{n}$ as

$$u_{k-1}^n = \vec{u}_{k-1} \cdot \vec{n} \; , \tag{20}$$

and

$$u_{k-1}^t = ||\vec{u}_{k-1} - (\vec{u}_{k-1} \cdot \vec{n})\vec{n}|| \; , \tag{21}$$

The velocity right *after* the kth collision we define as $\vec{v}_k$. Defining the normal and tangential *restitution coefficients*, e_n and e_t, in the usual way, we have

$$v_k^n = -e_n u_{k-1}^n \; , \tag{22}$$

and

$$v_k^t = e_t u_{k-1}^t \; . \tag{23}$$

The velocity right after the $(k+1)$th collision is related to the velocity right after the kth collision by the map

$$\vec{v}_{k+1} = e_t \left[\vec{v}_k - \vec{n}_{k+1}(\vec{n}_{k+1} \cdot \vec{v}_k)\right] - e_n \vec{n}_{k+1}(\vec{n}_{k+1} \cdot \vec{v}_k)$$

$$+ \left[e_t \left[\vec{g} - \vec{n}_{k+1}(\vec{n}_{k+1} \cdot \vec{g})\right] - e_n \vec{n}_{k+1}(\vec{n}_{k+1} \cdot \vec{g})\right] \frac{2\vec{v}_k \cdot \vec{N}}{g \cos\theta} , \tag{24}$$

where $\vec{g}$ is the gravitational vector and $\vec{N}$ is the normal to the inclined surface.

Averaging over the disorder in the inclination of the facets leads to [13]

$$\langle \vec{v}_{k+1} \rangle = e_t \left[\langle \vec{v}_k \rangle - \langle \vec{n}\vec{n} \rangle \cdot \langle \vec{v}_k \rangle\right] - e_n \langle \vec{n}\vec{n} \rangle \cdot \langle \vec{v}_k \rangle$$

$$+ \left[e_t \left[\vec{g} - \langle \vec{n}\vec{n} \rangle \cdot \vec{g}\right] - e_n \langle \vec{n}\vec{n} \rangle \cdot \vec{g}\right] \frac{2\langle \vec{v}_k \rangle \cdot \vec{N}}{g \cos\theta} , \tag{25}$$

where

$$\langle n_i n_j \rangle = A\delta_{ij} + B N_i N_j \tag{26}$$

where $A = \phi_M^2/3$ and $B = 1 - 2\phi_M^2/3$ to second order in ϕ_M and where n_i is a component of $\vec{n}$ and N_i is a component of the vector $\vec{N}$. We choose the direction along the inclined plane as the x-direction and the direction of its normal, $\vec{N}$, as the y-direction in the following. Eq. (25) may then be written in matrix form

$$\begin{pmatrix} \langle v_{k+1,x} \rangle \\ \langle v_{k+1,y} \rangle \end{pmatrix} = \begin{pmatrix} \Lambda_1 & \Lambda_3 \\ 0 & \Lambda_2 \end{pmatrix} \cdot \begin{pmatrix} \langle v_{k,x} \rangle \\ \langle v_{k,y} \rangle \end{pmatrix} , \tag{27}$$

where

$$\Lambda_1 = e_t - (e_n + e_t)\, A , \tag{28}$$

$$\Lambda_2 = (e_n + e_t)\,(A + B) - e_t , \tag{29}$$

and

$$\Lambda_3 = 2\Lambda_1 \tan\theta . \tag{30}$$

Analysing the eigenvalues of Eq. (27) shows [13] that the sphere will slow down and stop exponentially if $\Lambda_1 < 1$ and $\Lambda_2 < 1$, it will speed up exponentially if $\Lambda_1 > 1$ and $\Lambda_2 > 1$, and finally it will reach a steady state if $\Lambda_1 = 1$ and $\Lambda_2 < 1$.

These results are curious in that neither Λ_1 nor Λ_2 depend on the inclination of the plane, θ. We also note that if the surface is smooth, that is, $\phi_M = 0$, then $\Lambda_1 = e_t$ and $\Lambda_2 = e_n$. Since $e_n < 1$ and $e_t < 1$, the sphere will always slow down on a smooth surface, no matter how steep it is.

Let us focus on one physical consequence of these equations. The time between collisions k and $k+1$ is

$$\delta t_k = \frac{2\langle v_{k,y} \rangle}{g \cos\theta} = \frac{2 v_{0,y}}{g \cos\theta}\, \Lambda_2^k . \tag{31}$$

When k tends to infinity, the total elapsed time is given by

$$t_* = \sum_{k=0}^{\infty} \delta t_k = \frac{2v_{0,y}}{g\cos\theta} \frac{1}{1-\Lambda_2} . \tag{32}$$

Thus, when $\Lambda_2 < 1$, the sum (32) converges to a finite number. This is an example of *inelastic collapse* [14]. Stated in a dramatic way, time stops in the model: there is a singularity at finite $t = t_*$. Is this singularity real? It is, as we can never get past t_* which is only reached as $k \to \infty$.

Thus, there is a serious problem inherent in the model as it has been defined so far. Something has to be modified. There are at least two obvious candidates for change. (1) Do not assume instantaneous collisions. This is not so desirable from the point of view of building a model. One loses the attractiveness of an *event-driven* algorithm.[2] (2) The restitution coefficients are *not* constants but depend on velocity. This approach would keep the system event driven. We will therefore in the following explore this approach.

5. Coefficients of Restitution

Two spheres both of radius R collide. Their centers are situated at $\vec{r}_1$ and $\vec{r}_2$, respectively. Sphere 1 has velocity $\vec{v}_1$ and rotates with angular velocity $\vec{\omega}_1$, while sphere 2 has velocity $\vec{v}_2$ and rotates with angular velocity $\vec{\omega}_2$. The unit vector pointing from $\vec{r}_1$ to $\vec{r}_2$ is $\vec{n} = (\vec{r}_2 - \vec{r}_1)/||\vec{r}_2 - \vec{r}_1||$, and the relative velocity of the spheres at the point of contact is $\vec{v} = \vec{v}_2 - \vec{v}_1 + \vec{n} \times (R\vec{\omega}_2 - R\vec{\omega}_1)$. We define a normal velocity component $v_n = \vec{v} \cdot \vec{n}$ and a tangential velocity component $v_t = ||\vec{v} - (\vec{v} \cdot \vec{n})\vec{n}||$. Right before the collision, the relative velocity is $\vec{v}^i$, while right after the collision it is $\vec{v}^f$. The normal and tangential restitution coefficients are then defined as

$$e_n = -\frac{v_n^f}{v_n^i} , \qquad e_t = +\frac{v_t^f}{v_t^i} , \tag{33}$$

in agreement with Eqs. (22) and (23).

5.1. NORMAL COEFFICIENT OF RESTITUTION

We study the normal coefficient of restitution first. Our aim is to provide a model for e_n built on a collision that takes time and that deforms the colliding spheres. We will then assume the collision time to be so short that

[2] This is when collision number k fully describes the time in the model. The opposite is a *time-driven* algorithm, where time has to be specified explicitly — as when collisions do take time.

it may be ignored and e_n is determined from the velocities right before and right after the deformation started.

To begin, we study the simplest model possible: the harmonic oscillator model. We define the *deformation* of the spheres as

$$x = \max(0, 2R - ||\vec{r}_2 - \vec{r}_1||) . \tag{34}$$

During the collision process, we assume a damped harmonic force,

$$f(x) = -\kappa x - \gamma \dot{x} , \tag{35}$$

where κ is the spring constant, and γ is the damping. During contact, the motion of the two spheres is governed by the equation

$$m\ddot{x}(t) + \kappa x(t) + \gamma \dot{x}(t) = 0 \tag{36}$$

where m is the reduced mass. The initial conditions are

$$x(0) = 0 , \qquad \dot{x}(0) = v_n^i . \tag{37}$$

The duration of the contact t_n is determined by the equation

$$x(t_n) = 0 . \tag{38}$$

Integrating Eq. (36) and using the definition of e_n, Eq. (33), we find

$$e_n = \frac{\kappa}{m v_n^i} \int_0^{t_n} x(t) dt - 1 \tag{39}$$

Determining $x(t)$ and plugging the result into this equation gives

$$e_n = e^{-(\gamma/2m)t_n} , \tag{40}$$

where

$$t_n = \frac{\pi}{\sqrt{\kappa/m - (\gamma/2m)^2}} . \tag{41}$$

The important point to note here is that e_n is *independent* of v_n^i.

This was an exact calculation of e_n for a simple model. Let us now redo the calculation, but using two approximations that tend in opposite directions, so that the total error is not so substantial. The reason for this exercise is to estimate e_n using more realistic, and therefore more complex, models of the collisional mechanics of the two spheres. The first approximation is to calculate t_n *without* including the damping factor $\gamma\dot{x}$. This underestimates t_n and leads to

$$t_n = \pi\sqrt{\frac{m}{\kappa}} . \tag{42}$$

We then calculate e_n from Eq. (36) *ignoring* the harmonic force $-\kappa x$. This overestimates e_n as a function of t_n, and gives

$$e_n = e^{-(\gamma/m)t_n} , \tag{43}$$

to be compared to Eq. (40). All in all, the result we get is close to the exact one. However, the important point is that no v_n^i-dependency showed up using these two approximations.

The harmonic approximation, however, is far too simplistic. The elastic response of two spheres in contact is *not* linear, but follows the Hertz contact law [15],

$$f(x) = -\kappa x^{3/2} . \tag{44}$$

One may present a crude argument as to why this non-linearity appears. When the two spheres are in contact and have been deformed by an amount x, the radius of their area of contact is $r = \sqrt{R^2 - (R - x/2)^2} \approx \sqrt{Rx}$ when $x \ll R$. The stress due to the deformation of the sphere is essentially due to shear, which is essentially x/r. The force f between the two spheres is then the contact area, which is $\propto r^2 \propto x$ multiplied by the stress, which is proportional to $x/r = \sqrt{x}$. The result is Eq. (44). Replacing the harmonic force in Eq. (35) by the Hertzian force gives

$$f(x) = -\kappa x^{3/2} - \gamma \dot{x} . \tag{45}$$

The coefficient of resitution may be calculated as in Eq. (39), and one finds

$$e_n = \frac{\kappa}{m v_n^i} \int_0^{t_n} x(t)^{3/2} dt - 1 , \tag{46}$$

where $x(t)$ is the solution of the equation

$$m\ddot{x} + \kappa x^{3/2} + \gamma \dot{x} = 0 , \tag{47}$$

with initial conditions Eqs. (37) and (38). We estimate t_n by ignoring the damping, and find

$$t_n = c \left(\frac{m}{\kappa} \right)^{2/5} (v_n^i)^{-1/5} , \tag{48}$$

where $c \approx 3.2\ldots$. We estimate e_n by ignoring the Hertzian force, thus integrating $m\ddot{x} + \gamma\dot{x} = 0$, giving Eq. (43). Combining Eqs. (43) and (48) we find

$$e_n = e^{-c\gamma m^{-3/5} \kappa^{-2/5} (v_n^i)^{-1/5}} . \tag{49}$$

This velocity dependence has been used in numerical studies of granular matter, but as pointed out by Taguchi [16], the velocity dependency of e_n

is clearly wrong: e_n tends to zero as v_n^i tends to zero, and e_n tends to one as v_n^i tends to infinity.

The source of this error is to be found in Eq. (45), where the Hertz contact law has been combined with a linear damping term. Starting with assumed linear elasticity *and* damping of the sphere material and integrating to find the effective force between the two spheres in contact [17, 18] one arrives at the Hertz-Kuwabara-Kono contact law

$$f = -\kappa x^{3/2} - \gamma x^{1/2} \dot{x} . \tag{50}$$

Estimating t_n by ignoring the damping term, one again arrives at Eq. (48). However, estimating e_n from $m\ddot{x} + \gamma x^{1/2}\dot{x} = 0$, thus ignoring the Hertzian force, leads to

$$e_n = e^{-(3/2)c^{3/2}\gamma m^{-2/5}\kappa^{-3/5}(v_n^i)^{1/5}} , \tag{51}$$

which *is* a viable model for e_n. We have that $e_n \to 1$ as $v_n^i \to 0$ and $e_n \to 0$ as $v_n^i \to \infty$, which are sensible.

We have assumed that the colliding particles are spheres. If they are not, however, contact laws other than the Hertzian one may apply. Presumably, these more generalized laws lead to the same type of velocity dependence as e_n in Eq. (51), perhaps not with the power $(-1/5)$ but of similar form.

5.2. TANGENTIAL COEFFICIENT OF RESTITUTION

In experiments, the tangential coefficient of restitution, e_t, defined in Eq. (33), shows much more complex behavior than e_n [19]. Usually, one discusses the behavior of e_t in the *Maw, Barber, and Fawcett representation* [20]. They introduce two tangential velocity parameters

$$\psi^i = \frac{v_t^i}{v_n^i} , \tag{52}$$

and

$$\psi^f = \frac{v_t^f}{v_n^i} . \tag{53}$$

For a head-on collision, $\psi^i = 0$ and for a grazing collision $\psi^i = \infty$. In terms of the two tangential velocity parameters, the tangential restitution coefficient is

$$e_t = \frac{\psi^f}{\psi^i} . \tag{54}$$

Usually, e_t is displayed as the slope of a line from the origin to a point in the (ψ^i, ψ^f) plane (MBF diagram). This point traces a curve that starts out from the origin and typically moves into the quadrant defined by $\psi^i > 0$ and

$\psi^f < 0$. It passes through a minimum and then moves into the quadrant defined by $\psi^i > 0$ and $\psi^f > 0$ where it eventually reaches an asymptotic line of the form $\psi^f = a\psi^i + b$ where $a > 0$ and $b > 0$ are constants. Thus, for collisions that are close to being head on, a *negative* tangential restitution coefficient is found. As the collision becomes more and more oblique, e_t approaches a. The part of the curve from the origin to its minimum value is where the two spheres stick without sliding throught the entire collision process. From the minimum of the curve until it hits $\psi^f = 0$, the spheres stick initially during the collision but then start sliding. The part of the curve where $\psi^f > 0$ describes collisions where the spheres slide throughout the entire collision.

The *Walton model* [21] provides a simplified description of the behavior of e_t based on the MBF diagram. The model contains three parameters: a fixed tangential restitution coefficient from completely sticking collisions, the sliding friction coefficient, and the normal restitution coefficient. Experiments on the tangential restitution coefficient are typically analysed in this framework.

We now turn to the inelastic collapse that was encountered in Eq. (32). Will velocity dependence of the restitution coefficients remove the inelastic collapse singularity?

6. Repairing Inelastic Collapse

Velocity dependence of the restitution coefficients will remove the inelastic collapse singularity. In this section, that will be demonstrated. Our assumption is that the normal coefficient of restitution has the form

$$e_n = 1 - v^\beta \, , \tag{55}$$

for $v \to 0$ and $\beta > 0$. For example, the e_n of Eq. (51) is of this form with $\beta = 1/5$. We assume furthermore that the inclined line, on which the particle is jumping, is smooth, i.e. $\phi_M = 0$. The collapse time is then given by Eq. (32) without averaging over any disorder,

$$t_* = \frac{2}{g \cos\theta} \sum_{k=0}^{\infty} v_{k,y} \, , \tag{56}$$

where $v_{k,y} = e_n(v_{k,y})v_{k-1,y} = e_n(k)v_{k-1,y}$. Thus, we have

$$v_{k,y} = v_{0,y} \prod_{j=1}^{k} e_n(j) \, . \tag{57}$$

We may split the sum of Eq. (56) into two parts,

$$\sum_{k=0}^{\infty} v_{k,y} = v_{0,y} \left[\sum_{k=0}^{M-1} \prod_{j=0}^{k} e_n(j) + \prod_{i=0}^{M} e_n(i) \sum_{k=0}^{\infty} \prod_{j=0}^{k} e_n(M+1+j) \right] . \tag{58}$$

We have that

$$\sum_{k=0}^{\infty} \prod_{j=0}^{k} e_n(M+1+j) \geq \sum_{k=0}^{\infty} [e_n(M+1)]^k = \frac{1}{1-e_n(M+1)} , \tag{59}$$

since $e_n(k) \geq e_n(l)$ when $k \geq l$. We have, furthermore, that

$$\prod_{i=0}^{M} e_n(i) = e^{-\sum_{i=0}^{M} \delta_i} \geq e^{-M\delta_M} , \tag{60}$$

where $\delta_i = 1 - e_n(i) = v_{i,y}^{\beta}$ from Eq. (55) and we assume that δ_i are small. The last inequality comes from $\delta_k \leq \delta_l$ when $k \geq l$. Thus, we have that

$$\prod_{i=0}^{M} e_n(i) \sum_{k=0}^{\infty} \prod_{j=0}^{k} e_n(M+1+j) \geq \frac{e^{-M\delta_M}}{\delta_{M+1}} \geq \frac{e^{-M\delta_M}}{\delta_M} . \tag{61}$$

We now determine δ_M as a function of M. From Eq. (55) and Eq. (57), we have that

$$\begin{cases} v_{0,y} \\ v_{1,y} = (1 - v_{0,y}^{\beta}) v_{0,y} \\ \vdots \\ v_{k,y} = (1 - v_{k-1,y}^{\beta}) v_{k-1,y} \\ \vdots \end{cases} \tag{62}$$

Since $e_n(M) = v_{M,y}/v_{M-1,y}$, we then have

$$e_n(M) = 1 - [e_n(M-1)]^{\beta} [e_n(M-2)]^{\beta} \cdots [e_n(1)]^{\beta} v_{0,y}^{\beta} . \tag{63}$$

Thus, we may write

$$\delta_M = (1-\delta_{M-1})^{\beta} (1-\delta_{M-2})^{\beta} \cdots (1-\delta_1)^{\beta} v_{0,y}^{\beta} \approx \left(e^{-\beta \sum_{i=1}^{M-1} \delta_i} \right) v_{0,y}^{\beta} , \tag{64}$$

and

$$\delta_{M+1} - \delta_M = \left[e^{-\beta \sum_{i=1}^{M} \delta_i} - e^{-\beta \sum_{i=1}^{M-1} \delta_i} \right] = \delta_M \left[e^{-\beta \delta_M} - 1 \right] \approx -\beta \delta_M^2 , \tag{65}$$

again using repeatedly that $\delta_i \ll 1$. Now define $\tau = \epsilon M$, where $\epsilon \ll 1$. Eq. (65) then turns into a differential equation

$$\frac{d}{d\tau}\delta(\tau) = -\frac{\beta}{\epsilon}\delta(\tau)^2 , \tag{66}$$

whose solution is $\delta(\tau) = \epsilon/(\beta\tau)$. In terms of M, this becomes

$$\delta_M = \frac{1}{\beta M} . \tag{67}$$

Combined with Eq. (61), this gives

$$\prod_{i=0}^{M} e_n(i) \sum_{k=0}^{\infty} \prod_{j=0}^{k} e_n(M+1+j) \geq \beta M e^{-1/\beta} , \tag{68}$$

which tends to infinity as $M \to \infty$ for any $\beta > 0$. We may now return to Eqs. (56) and (58), and draw the conclusion that $t_* \to \infty$ for any $\beta > 0$. Thus, the inelastic collapse has been repaired for the Valence and Bideau model [12].

7. Conclusion

We have in this small review attempted to demonstrate that even seemingly simple questions concerning the motion of few - down to one - grains poses deep and still open questions both from a theoretical and experimental point of view. The opening remark, we claim, holds very much for granular media.

Acknowledgement

I thank G.G. Batrouni, D. Bideau, F.X. Riguidel and M.H. Ernst for many interesting discussions on this subject.

References

1. P. Bak, C. Tang, and K. Wiesenfeld, Phys. Rev. Lett. **59**, 381 (1987).
2. F.X. Riguidel, A. Hansen, and D. Bideau, Europhys. Lett. **28**, 12 (1994).
3. F.X. Riguidel, Ph. D. thesis, Université de Rennes 1 (1994).
4. C. Henrique, M.A. Aguirre, A. Calvo, I. Ippolito, S. Dippel, G.G. Batrouni, and D. Bideau, Phys. Rev. E **57**, 4743 (1998).
5. C.D. Jan, H.W. Shen, C.H. Ling, and C.L. Chen, in *Proc. of the 9th Conf. on Eng. Mech.*, L.D. Lutes and J.M. Niedzwecki, eds., (Am. Soc. of Civil Eng., New York, 1992).
6. R.A. Bagnold, Proc. Roy. Soc. London A **225**, 49 (1954).
7. E.J. Gumbel, *Statistics of Extremes,* (Columbia University Press, New York, 1958).
8. G.H. Ristow, F.X. Riguidel, and D. Bideau, J. Phys. 1 (France) **4**, 1161 (1994).

9. G.G. Batrouni, S. Dippel, and L. Samson, Phys. Rev. E **53**, 6496 (1996).
10. S. Dippel, G.G. Batrouni, and D.E. Wolf, Phys. Rev. E **54**, 6845 (1996).
11. S. Dippel, G.G. Batrouni, and D.E. Wolf, Phys. Rev. E **56**, 3645 (1997).
12. A. Valance and D. Bideau, Phys. Rev. E **57**, 1886 (1998).
13. M.H. Ernst, (1998) unpublished.
14. S. McNamara and W.R. Young, Phys. Rev. E **50**, R28 (1994).
15. L. Landau and E.M. Lifshitz, *Theory of Elasticity,* (Pergamon Press, London, 1959).
16. Y.H. Taguchi, Phys. Rev. Lett. **69**, 1367 (1992).
17. G. Kuwabara and K. Kono, Jpn. J. Appl. Phys. **26**, 1230 (1987).
18. N. Brilliantov, F. Spahn, J. Hertzsch, and T. Pöschel, Phys. Rev. E **53**, 5382 (1996).
19. S.F. Foerster, M.Y. Louge, H. Chang, and K. Allia, Phys. Fluids **6**, 1108 (1994).
20. N. Maw, J.R. Barber, and J.N. Fawcett, ASME J. Lub. Technol. **103**, 74 (1981).
21. O.R. Walton, *Granular Flow Project,* Quarterly Report, January–March 1988, UCID–20297-88-1, Lawrence Livermore National Laboratory (1988).

HYDRAULIC THEORY FOR A GRANULAR HEAP ON AN INCLINE

J. T. JENKINS
Department of Theoretical and Applied Mechanics
Cornell University
Ithaca, NY 14853, USA

1. Introduction

We employ an especially simple form of the kinetic theory for identical, frictionless, nearly elastic spheres that applies to relatively dense aggregates in which the fluxes of momentum and energy in a shearing flow are due to interparticle collisions rather than to transport between collisions (e.g.[4]). The theory consists of balance equations for the particle mass, momentum, and energy and constitutive relations that relate the fluxes of momentum and energy to the mean fields of mass density, velocity, and energy and the spatial gradients of velocity and energy.

We also introduce boundary conditions at two types of boundaries, called bumpy and erodible. At a bumpy boundary, the flow slips relative to the boundary and, as a consequence, bumps rigidly attached to the boundary collide with particles of the flow. Momentum is exchanged, part of the energy associated with the working of the slip velocity is converted into energy of velocity fluctuations, and part is dissipated in collisions (e.g.[7]). Near the surface of an erodible boundary, particles identical to those of the flow interact through collisions, but have too high a concentration to participate in the shearing. At such a boundary, the velocity of the flow is equal to the velocity of the boundary and the shear stress is supported by a collisional exchange of momentum in a distorted but otherwise random aggregate of boundary particles. Because the collisions are dissipative and the flow does not slip relative to the boundary, fluctuation energy is lost from the flow to the boundary [2].

The continuum equations of the kinetic theory and the boundary conditions are used to determine the distribution of fluctuation energy and the profile of mean velocity across a steady, fully-developed shearing flow. The

J. Karkheck (ed.),
Dynamics: Models and Kinetic Methods for Non-equilibrium Many Body Systems, 313–323.

flow is maintained in the absence of gravity by an upper erodible boundary moving parallel to a lower bumpy boundary with a fixed velocity. A relation between the dimensionless thickness of the flow and the ratio of shear to normal stress and a relation between the dimensionless velocity of the upper boundary and the stress ratio are obtained as part of the solution of the boundary value problem. These must be satisfied if the flow is to be steady and fully-developed.

We use the second of these relations as input to another problem that takes place at much greater length and time scales. This is the evolution of the height and velocity of a heap of particles moving down an incline under gravity. The heap is assumed to be supported at its base on a collisional shear layer whose thickness is small compared to the height of the heap. Adjustments within the shear layer to changes in the local height and velocity are assumed to take place so quickly that the behavior of the shear layer can be considered to be steady. The weight of the material in the shear layer is assumed to be so small a fraction of the weight of the material supported by it that the influence of gravity on the flow in the shear layer may be neglected. In this case, the results of the analysis of the steady, fully-developed flow in the absence of gravity may be used to specify the shear stress at the base of the heap as a function of the velocity of the heap and its depth. This relation closes the hydraulic equations governing the motion of the heap.

2. Theory

We consider rapid flows of a granular material consisting of identical, nearly elastic, frictionless spheres of mass m and diameter σ. The coefficient of restitution e characterizes the energy lost to the component of velocity normal to the surface of two colliding spheres. The mean fields of interest are the mass density ρ, the product of m and the mean number n of spheres per unit volume; the mean velocity $\mathbf{u}$, about which the actual particle velocities fluctuate; and the granular temperature T that measures the energy per unit mass of the fluctuations in the velocity.

The balance laws for mass, linear momentum, and the fluctuation energy have the familiar local forms:

$$\dot{\rho} + \rho \nabla \cdot \mathbf{u} = \mathbf{0}, \tag{1}$$

where an overdot indicates a time derivative following the mean motion;

$$\rho \dot{\mathbf{u}} = \nabla \cdot \mathbf{t} + n\mathbf{F}, \tag{2}$$

where $\mathbf{t}$ is the symmetric stress tensor and $\mathbf{F}$ is the external force on a sphere; and

$$\frac{3}{2}\rho \dot{T} = -\nabla \cdot \mathbf{Q} + tr\,(\mathbf{t}\nabla \mathbf{u}) - \gamma, \tag{3}$$

where $\mathbf{Q}$ is the flux of translational energy, tr denotes the trace of a tensor, and γ is the rate per unit volume at which the energy of the velocity fluctuations is dissipated in collisions.

We attempt to model the motion of a passive heap of grains that is driven by gravity down an incline while supported on a narrow region of sheared, colliding particles at its base. In general, the free surface of the heap changes in shape as the velocity at its base varies with space and time. We require that the thickness L of the shear layer is a very small fraction of the height H of the heap. We assume that the flow is traveling in the x direction down a slope with an angle of inclination $\phi(x)$, that the velocity at the top of the shear layer is $U(x,t)$, that the y axis is normal to the base and has its origin at the bottom of the shear layer, and that the shape of the free surface of the heap is given by $y = H(x,t)$. We ignore variations of the flow in the z direction and assume that the variations of ϕ, U, and H with x are slow compared to changes with y across the shear layer. In this case, the integrated forms of the balance of mass and momentum governing the evolution of U and H in space and time are (e.g. [8])

$$\frac{\partial H}{\partial t} + \frac{\partial}{\partial x}(UH) = 0, \tag{4}$$

and

$$\frac{\partial U}{\partial t} + U\frac{\partial U}{\partial x} + g\cos\phi(\frac{\partial H}{\partial x} - \tan\phi) = -\frac{S}{\overline{\rho}H}, \tag{5}$$

where $\overline{\rho}$ is the average mass density of the heap, here assumed to be constant, and S is the shear stress at the base of the heap. We note that the pressure p at the top of the shear layer is

$$p = \overline{\rho}gH\cos\phi. \tag{6}$$

In order to close equations (4) and (5), we must relate the shear stress S to U and H or, equivalently, to U and p. To do this, we review an analysis of a shearing flow of colliding grains that is driven by the relative motion of its upper and lower boundaries [2]. We assume that the upper boundary of the shear layer is erodible and the lower boundary of the shear layer is bumpy.

Within the shear layer, we assume that the flow is steady and rectilinear and take the x component u of the velocity, the density ρ, and the fluctuation energy T to depend only on y. The idea is that because the shear layer is so thin, it responds to the relatively slow changes with x and

t in a steady, one-dimensional way. Also, because the layer is so thin, we anticipate that the weight of the particles over a unit area of the base is a small fraction of the pressure at the top of the layer. In such a flow, (1) is satisfied identically, and the x and y components of (2) require that the pressure p and the shear stress S be uniform across the layer.

Then, with $Q \equiv Q_y$, (3) reduces to

$$\frac{dQ}{dy} - S\frac{du}{dy} + \gamma = 0. \tag{7}$$

The mean shear stress working through the gradient of the mean velocity produces fluctuation energy, while the inelastic collisions dissipate it. At any point in the flow at which the rates of production and dissipation of fluctuation energy are not balanced, there is a transport of fluctuation energy to or from neighboring points in the flow.

For the applications that we have in mind, it is reasonable to restrict our attention to flows in which the volume fraction $\nu \equiv n\pi\sigma^3/6$ is relatively large, around 0.5. In this case, collisions between particles, rather than the flight of particles between collisions, are responsible for the fluxes of momentum and energy. In addition, in the expressions for the fluxes derived in the kinetic theory, we retain only those contributions to the collisional fluxes that dominate in the dense limit. Then, with the assumptions that the spheres are nearly elastic, we write the dispersive pressure p and the shear stress S as the high-volume-fraction limits of expressions provided by Chapman & Cowling ([1], Sec. 16.41) for frictionless, elastic spheres:

$$p = 4\rho GT, \tag{8}$$

where

$$G \equiv \frac{\nu(2-\nu)}{2(1-\nu)^3}; \tag{9}$$

and

$$S = \frac{2}{5}J\kappa\frac{du}{dy}, \tag{10}$$

where $J \equiv 1 + \pi/12$ and

$$\kappa \equiv \frac{4}{\pi^{1/2}}\rho\sigma T^{1/2}G. \tag{11}$$

Upon expressing κ in terms of p, we obtain a simple relation between the velocity gradient and the temperature,

$$\frac{du}{dy} = \frac{5\pi^{1/2}}{2J}\frac{T^{1/2}}{\sigma}\frac{S}{p}. \tag{12}$$

We adopt the corresponding expression for the flux of fluctuation energy ([1], Sec. 16.42) that is given in the dense limit by

$$Q = -M\kappa\frac{dT}{dy}, \tag{13}$$

where $M \equiv 1 + 9\pi/32$.

The rate of decrease of fluctuation energy per unit volume is [4]

$$\gamma = \frac{6\kappa T}{\sigma^2}(1 - e). \tag{14}$$

Boundary conditions are derived based upon the balance of collisional exchange of momentum and energy at the boundary. Here we will consider two types of boundaries, a boundary that is a rigid plane to which particles are attached, and a boundary that is a geometric surface that separates colliding particles that are being sheared from colliding particles that are not. The first is called a bumpy boundary, the second is called an erodible boundary. The boundary conditions are applied at the position of the center of a flow particle that touches the point on the boundary that projects furthest into the flow.

The bumpiness of the boundary is characterized by an angle θ that measures the average depth that a flow particle can penetrate between wall particles. A boundary consisting of a plane on which spheres identical to those of the flow have been close-packed has a value of θ near $\pi/6$. Because the particles of the boundary are arranged differently from the particles in the flow, in order to balance the component of linear momentum parallel to the surface of the boundary, the flow must slip with respect to it. We denote the magnitude of the slip velocity by v. At the bottom boundary, the balance of momentum requires that (e.g.[7])

$$v = \left(\frac{\pi}{2}\right)^{1/2} fT^{1/2}\frac{S}{p}, \tag{15}$$

where the slip coefficient f depends only on the bumpiness of the boundary. For small values of θ,

$$f(\theta) = \frac{2}{\theta^2} - \frac{5\pi}{24J} + \frac{25\pi + 300\sqrt{2} - 7J}{360J}\theta^2. \tag{16}$$

In general, collisions between the particles of the flow and the boundary dissipate energy, and the rate D of dissipation per unit area is given in terms of the boundary coefficient of restitution e_w by

$$D = \left(\frac{2}{\pi}\right)^{1/2} (1 - e_w) h T^{1/2} p, \tag{17}$$

where h depends only on the bumpiness,

$$h(\theta) = 1 + \frac{1}{4}\theta^2. \tag{18}$$

The balance of energy at a bumpy boundary requires that the flux of fluctuation energy from the flow plus the rate of working of the shear stress through the slip velocity equals the rate of collisional dissipation. Consequently, at the bottom boundary,

$$-Q + Sv = D, \tag{19}$$

or with (13), (15), and (17),

$$\sigma \frac{dT}{dy} = -2b_0 T, \tag{20}$$

where

$$b_0 \equiv \frac{1}{\sqrt{2}M} \left[\frac{\pi}{2} f \left(\frac{S}{p}\right)^2 - (1 - e_w) h\right]. \tag{21}$$

When b_0 is positive, the boundary provides fluctuation energy to the flow.

At an erodible boundary, particles on both sides of the boundary are agitated and particles from the boundary may be incorporated into the flow. Consequently, at such a boundary, the slip velocity vanishes and we expect the volume fraction to be close to the value of 0.57 for random loose packing [6]. The shear stress exerted on the boundary by the flow is supported by a distortion of the initially isotropic arrangement of the colliding boundary particles. An erodible boundary is purely dissipative [2] and

$$D = \left(\frac{2}{\pi}\right)^{1/2} [6M(1 - e)]^{1/2} T^{1/2} p. \tag{22}$$

Then, when the erodible boundary is at the top of the flow, the balance of energy at the boundary requires that

$$Q = D, \tag{23}$$

or

$$\sigma \frac{dT}{dy} = 2b_1 T, \tag{24}$$

where

$$b_1 \equiv -\frac{1}{\sqrt{M}}\left[3\left(1-e\right)\right]^{1/2}. \tag{25}$$

We employ (12), (13), and (14) in the balance of fluctuation energy, (7), and replace κ by $\sigma p/(\pi T)^{1/2}$ wherever it occurs. The resulting equation for $w \equiv T^{1/2}$ is

$$\sigma^2 \frac{d^2 w}{dy^2} + K^2 w = 0, \tag{26}$$

where

$$K^2 \equiv \frac{1}{M}\left[\frac{5\pi}{4J}\left(\frac{S}{p}\right)^2 - 3(1-e)\right]. \tag{27}$$

The quantity K is a measure of the excess of the rate of production of fluctuation energy by the mean shear over its collisional rate of dissipation. When K is real, the solution of (26) is given in terms of trigonometric functions:

$$w(y) = A \sin\left(\frac{Ky}{\sigma}\right) + B \cos\left(\frac{Ky}{\sigma}\right), \tag{28}$$

where A and B are constants to be determined. The corresponding solution of (12) is

$$u(y) = \frac{5\pi^{1/2}}{2J}\frac{S}{p}\frac{1}{K}\left[-A\cos\left(\frac{Ky}{\sigma}\right) + B\sin\left(\frac{Ky}{\sigma}\right)\right] + C, \tag{29}$$

where C is a constant to be determined. The variation of ν with y in the shear layer may be obtained by inverting (8),

$$\frac{\nu^2(2-\nu)}{(1-\nu)^3} = \frac{\pi}{12}\frac{\sigma^3}{m}\frac{1}{w^2(y)}p. \tag{30}$$

When K is imaginary, we write $K = ik$ and express (28) and (29) in terms of the corresponding hyperbolic functions.

The boundary condition (20) at $y = 0$ requires that

$$B = -\frac{K}{b_0}A. \tag{31}$$

The corresponding boundary condition (24) at $y = L$ does not determine A, but provides a relation between the thickness of the layer and the ratio of shear stress to pressure that must be satisfied in a steady flow:

$$\tan\left(\frac{KL}{\sigma}\right) = \frac{(b_0 + b_1)\,K}{(b_0 b_1 - K^2)}. \tag{32}$$

In order to satisfy this relationship as the stress ratio varies, the thickness L of the shear layer must change by erosion or deposition of particles at its upper boundary. When K is imaginary, $K = ik$, and (32) becomes

$$\tanh\left(\frac{kL}{\sigma}\right) = \frac{(b_0 + b_1)\,k}{(b_0 b_1 + k^2)}. \tag{33}$$

In Figure 1 we show curves of normalized layer thickness, L/σ, versus stress ratio, S/p, for values of the coefficients of restitution thought to be characteristic of glass spheres and for a value of θ equal to $\pi/5$. For $e = 0.85$, three values of e_w were employed. Across the curves, the dissipation of fluctuation energy at the wall increases from left to right. Large layer thicknesses in the steady flow are achieved asymptotically as k approaches b_0. The curves have been terminated at their lower end at a layer thickness of three particle diameters. The results of computer simulations (e.g.[5]) indicate that the layer thickness is predicted with a fair degree of accuracy even for such small thicknesses.

The constant A may be expressed in terms of the value $w_1 \equiv w(L)$ of the fluctuation velocity at the top of the layer,

$$A = -\frac{b_0}{K}\left(\frac{K^2 + b_1^2}{K^2 + b_0^2}\right)^{1/2} w_1, \tag{34}$$

and, through (8), w_1 may be determined in terms of the pressure and the volume fraction ν_1 there,

$$w_1^2 = \frac{\pi}{12}\frac{\sigma^3}{m}\frac{(1-\nu_1)^3}{\nu_1^2(2-\nu_1)}p. \tag{35}$$

As an alternative, we may determine A in terms of the value $w_0 = w(0)$ of the fluctuation velocity at the lower boundary,

$$A = -\frac{b_0}{K}w_0. \tag{36}$$

However, before solving the boundary value problem, we have no knowledge of the volume fraction at the lower boundary, so this serves only as a convenience in the analysis.

We may determine C by evaluating the velocity (29) at the lower boundary, equating it to the slip velocity, and using (15) to express the slip velocity in terms of the stress ratio and w_0,

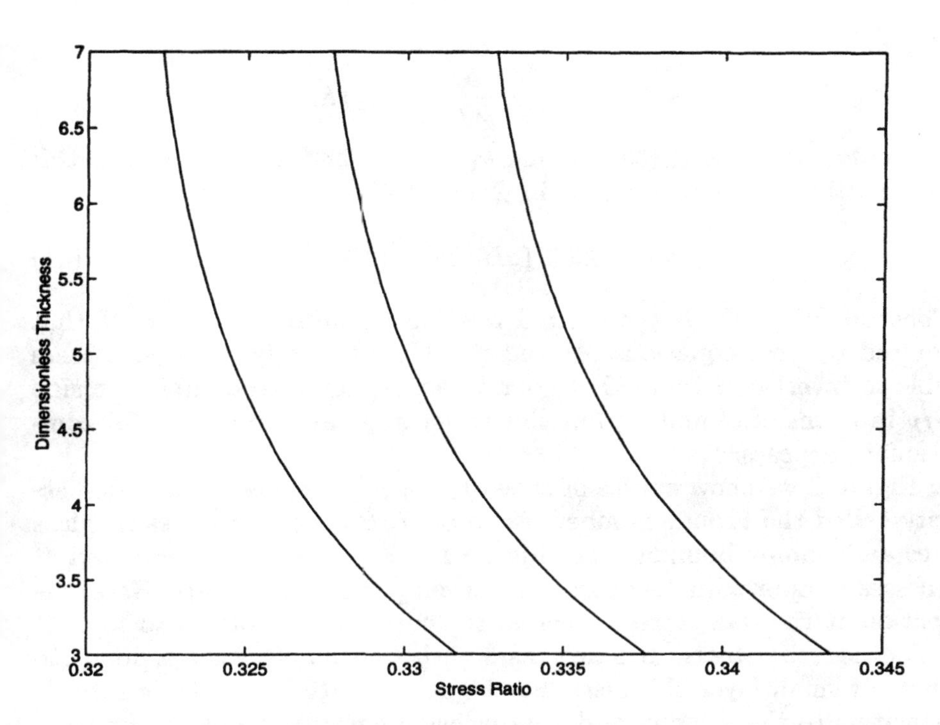

Figure 1. **Dimensionless thickness, L/σ, versus stress ratio, S/p, for $e = 0.85$, $\theta = \pi/5$, and, from left to right, $e_w = 0.90$, 0.85, and 0.80.**

$$C = \left(\frac{\pi}{2}\right)^{1/2} \left(f - \frac{5}{\sqrt{2}J}\frac{b_0}{K^2}\right) \frac{S}{p} w_0. \tag{37}$$

At the top of the shear layer the velocity is U. So, upon evaluating (29) there and using the information that we have already obtained, we may write

$$U = \left(\frac{\pi}{2}\right)^{1/2} \left\{ \frac{5}{\sqrt{2}J}\frac{b_0}{K^2} \left[\cos\left(\frac{KL}{\sigma}\right) + \frac{K}{b_0}\sin\left(\frac{KL}{\sigma}\right)\right] + f - \frac{5}{\sqrt{2}J}\frac{b_0}{K^2} \right\} \frac{S}{p} w_0, \tag{38}$$

or, after eliminating L/σ, simplifying the result, and writing it in terms of w_1, as

$$\frac{U}{w_1} = -\left(\frac{\pi}{2}\right)^{1/2} \frac{1}{K^2} \left[\frac{5}{\sqrt{2}J} b_1 + \frac{\alpha}{M} \left(\frac{K^2 + b_1^2}{K^2 + b_0^2}\right)^{1/2} \right] \frac{S}{p}, \tag{39}$$

where

$$\alpha \equiv 3\,(1-e)\,f - \frac{5}{2J}\,(1-e_w)\,h. \tag{40}$$

Upon employing (6) and (39), taking $\nu_1 = 0.57$, and evaluating $\overline{\rho}$ at this volume fraction, we may write w_1 in terms of H:

$$w_1 = 0.235\,(gH\cos\phi)^{1/2}\,. \tag{41}$$

Consequently, (39) is the desired relation between S, U, and H that is required to close equations (4) and (5). Unfortunately, it doesn't seem possible to invert equation (39) to obtain an explicit analytical expression for S/p in terms of U and H, but this is not of great importance for computational purposes.

In Figure 2 we show curves of stress ratio, S/p, versus a dimensionless velocity called the Froude number, $Fr \equiv U/\sqrt{gH\cos\phi}$, for the same glass spheres and bumpy boundaries of Figure 1. The curves are terminated at thicknesses of approximately three and seven particle diameters. Here, the dissipation at the wall increases across the curves from bottom to top. At low velocities, the stress ratio decreases with velocity; this corresponds to the limit of small layer thickness. For higher velocities, the stress ratio is less sensitive to the velocity and approaches a constant value as the layer thickness becomes larger. However, when operating near this limit, we must be certain to check that the thickness of the shear layer that is predicted does not become too great a fraction of the height of the heap. An hydraulic theory for a heap that is sheared through a significant part of its height is possible, but the influence of gravity must be included and its derivation is somewhat more complicated.

With the relation between S, U, and H provided by (39), the hydraulic equations may be integrated. An example of such an integration for the evolution of the velocity distribution and the shape of the heap is provided by Jenkins and Askari [3] using a simpler model of the shear layer in which the stress ratio is an increasing function of the dimensionless velocity. Here, we will rest content with having shown how the results of the analysis for the shear layer may be used to complete the hydraulic equations. In closing, we note that the methods of the kinetic theory, the solution of a boundary value problem, and the introduction of the depth-averaged equations have permitted a systematic analysis of a problem over length scales that ranged from the typical separation of the particles to the depth of the heap.

Acknowledgement

This research was supported by the U. S. Department of Energy as part of the Granular Flow Advanced Research Objective.

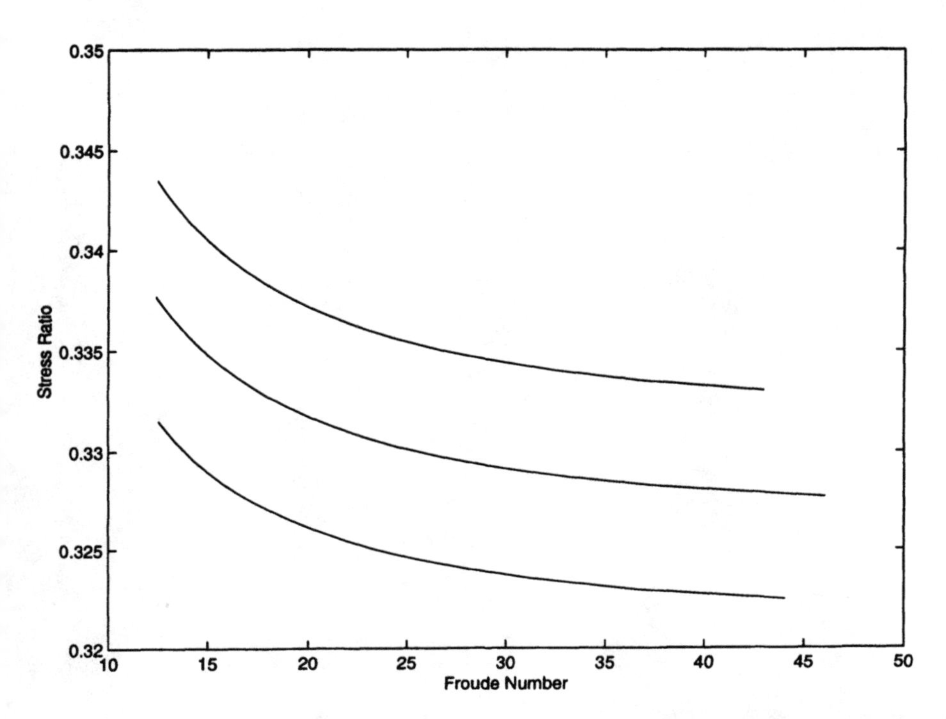

Figure 2. Stress ratio, S/p, versus Froude number, $U/\sqrt{gH\cos\phi}$, for the same values of the parameters as in figure 1 and, from bottom to top, e_w=0.90, 0.85, and 0.80.

References

1. S. Chapman and T.G. Cowling, *The Mathematical Theory of Non-Uniform Gases*, 3rd ed., (Cambridge University Press, Cambridge, 1970).
2. J.T. Jenkins and E. Askari, J. Fl. Mech. **223**, 497 (1991).
3. J.T. Jenkins and E. Askari, Chaos **9**, 654 (1999).
4. J.T. Jenkins and S.B. Savage, J. Fl. Mech. **130**, 187 (1983).
5. M.Y. Louge, J.T. Jenkins, and M.A. Hopkins, Phys. Fl. A **2**, 1042 (1990).
6. G. Onada and E. Liniger, Phys. Rev. Lett. **64**, 2727 (1990).
7. M.W. Richman, Acta Mech. **75**, 227 (1988).
8. G.B. Whitham, *Linear and Nonlinear Waves*, (John Wiley and Sons, New York, 1974).

EXPERIMENTAL STUDIES OF GRANULAR FLOWS

J. M. HUNTLEY AND R. D. WILDMAN
Department of Mechanical Engineering
Loughborough University
Loughborough LE11 3TU, United Kingdom

Abstract. Some of the experimental techniques that have been applied in recent years to the problem of measuring grain position and motion in two- and three-dimensional granular flows are briefly reviewed. The so-called "spin tagging" technique in magnetic resonance imaging provides a direct measure of a single displacement component for a slice through a three-dimensional granular material. Positron emission particle tracking allows the position vector of a single particle to be measured with better time resolution than with MRI, though with lower accuracy. High speed imaging using a video camera provides valuable data in two-dimensional flows, allowing, for example, velocity distributions, self-diffusion coefficients, static structure factors and intermediate scattering functions to be calculated from direct observations of the particle motion. The techniques are illustrated by applications including rotating drum experiments, and convection and fluidization in vertically-vibrated systems.

1. Introduction

In recent years a large body of literature has emerged on the physics of granular materials [1]. Much of this has been concerned with the flow properties resulting from the input of gravitational energy (e.g. chute flow) or kinetic energy (e.g. by acceleration or rotation of the container containing the material). In the case of vertical vibration, for example, a wide range of unusual phenomena is experimentally observed [2], including heaping and convection rolls [3] and size segregation [4-6]. At larger vibration amplitudes, period doubling instabilities lead to both standing waves [7] and travelling waves [8] on the free surface, and eventually the system can become fully fluidized. Numerical methods (e.g. molecular dynamics simula-

J. Karkheck (ed.),
Dynamics: Models and Kinetic Methods for Non-equilibrium Many Body Systems, 325–341.

tions) have become a standard investigative technique for studies of the particle motion on account of their low cost and ability to probe the internal structure of a three-dimensional flow. Experimental verification remains vital, however, and significant developments in applicable techniques have taken place over the past decade or so. In this paper we describe three such techniques: Magnetic Resonance Imaging (MRI), Positron Emission Particle Tracking (PEPT) and high speed photography combined with automated image analysis. These have complementary performances in terms of the positional measurement accuracy, maximum particle velocity, and dimensionality of both the measurement region and of the flow itself. For example, PEPT only provides information at one point, but can be applied to relatively fast particles moving in three dimensions. MRI provides data along a line and can again be used for three dimensional flows, but the allowed particle speeds are lower. High speed photography is restricted to two dimensions, but provides whole field time-resolved data. The principles of the techniques are briefly outlined and are illustrated by at least one application in each case.

2. Magnetic Resonance Imaging (MRI)

MRI is a well-known technique that has been developed primarily for medical applications, and which allows image "slices" to be recorded from three-dimensional objects. Motion within the object can be obtained by comparing two images measured a short time interval apart (e.g. by cross-correlating sub-regions of the images), but a more accurate approach is to label material through a magnetization distribution that varies cyclically with position. The distribution can then be measured a short time later to reveal how far the material within the slice moved in the time interval between the labeling and readout phases [9]. Figures 1 and 2 show the MRI timing pulses and resulting magnetization required to implement the "spin-tagging" technique.

Initially the magnetization of the protons is aligned along the z-axis parallel to the imposed B-field (Figure 2(a)). A radio-frequency pulse causes the magnetization vectors to tilt through an angle θ (Figure 2(b)), all initially with the same phase. A magnetic field with a linear gradient along the z axis is then superimposed for a short time, causing the precession frequencies to have a linear z-dependence. By the end of the field-gradient pulse the magnetization vectors have accumulated a total phase change (measured as the angular coordinate of the projection in the x-y plane) proportional both to z and to the magnitude and duration of the field gradient pulse (Figure 2(c)). A second RF pulse causes the vectors to tilt through an additional angle θ, and since the subsequently measured image intensity is propor-

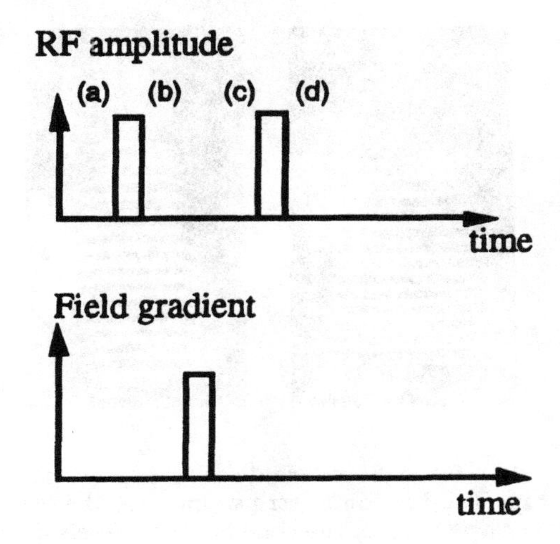

Figure 1. Timing diagram of pulse sequence for MRI spin-tagging (from [9]).

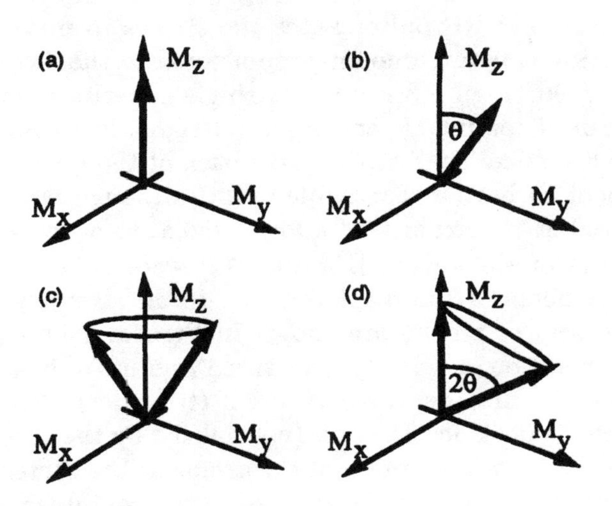

Figure 2. State of magnetization in rotating frame at various times in the sequence of Figure 1, as indicated by (a)-(d) (from [9]).

tional to M_z, (Figure 2(d)), the image intensity is modulated cyclically in the z direction.

Figure 3(left) shows an image from an experiment to measure convection currents in a vertically-vibrated cell (height = 52.5 mm) of poppy seeds of diameter $d = 1$ mm, taken from Reference [10]. The spin-tagging operations result in modulation of the image by a set of horizontal stripes. If the cell is shaken, then movement of material between the field-gradient

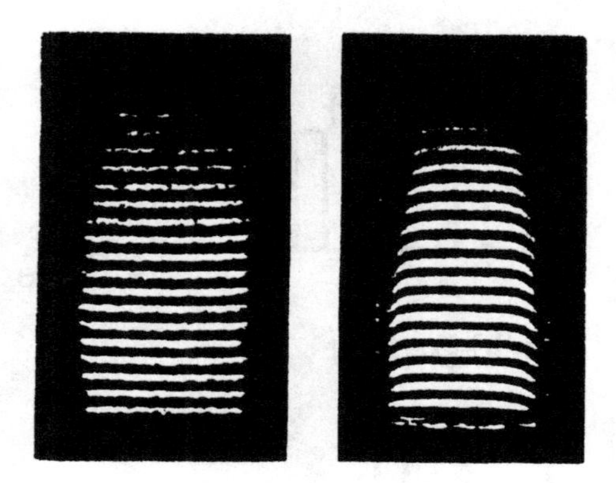

Figure 3. **MRI of slice through cell containing poppy seeds. *Left:* Seeds at rest; *Right:* after a single tap. The horizontal stripes indicate the vertical component of the seed-displacement vector occurring in the time between the field gradient and second RF pulses (from [10]).**

pulse and the second RF pulse causes the stripes to move from their unperturbed positions by an amount proportional to the z-component of the local velocity field (Figure 3(right)). A single experiment measures only a vertical column of the image, and Figure 3(right) is a composite image of a total of 2048 vertical shakes. One drawback of the technique is therefore the requirement to have a repeatable event, although whole-field imaging can be achieved at the expense of a lower signal-to-noise ratio [9].

The authors of References [10] and [11] were able to extract quantitative measurements of the displacements to an accuracy of better than 100 μm. Two sets of results are shown in Figures 4 and 5. In Figure 4, the radial distribution of velocity (measured in units of bead diameters per tap) is plotted at three depths z/d = 9.2 (triangles), 26.5 (squares) and 42.8 (circles) for a peak acceleration (normalized by the acceleration due to gravity) of $\Gamma = 6$. Upward motion of the grains at the center is balanced by downward motion in a narrow layer close to the container walls. In Figure 5, the velocity measured along the cylinder axis is plotted as a function of depth for 5 different acceleration levels. The approximately exponential decay of velocity with depth suggests a constant depth-independent probability for scattering of particles from the downward-moving flow into the upward-moving flow [11].

3. Positron Emission Particle Tracking

PEPT is a second experimental technique which is capable of providing positional data in three-dimensional flows. The system described here has been developed at the University of Birmingham, UK, and the authors

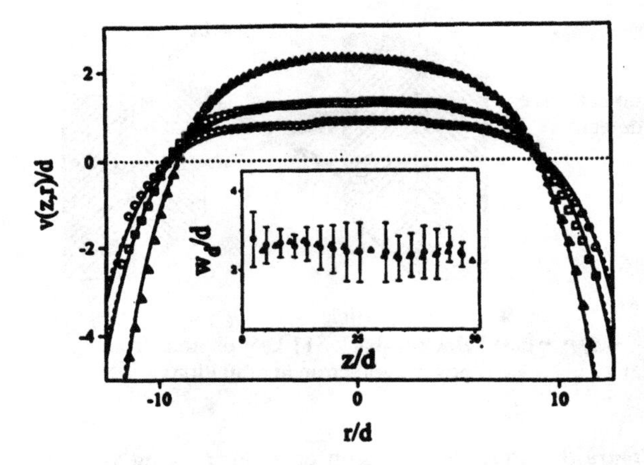

Figure 4. Vertical component of velocity for poppy grains undergoing convection, as a function of radial coordinate, for three depths (see text for details), measured by MRI spin-tagging technique (from [10]).

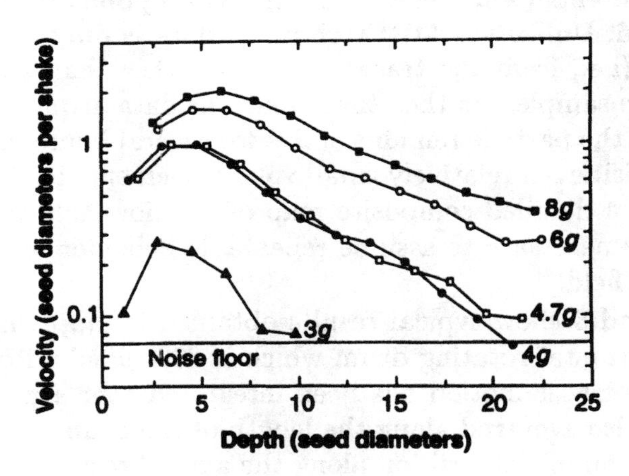

Figure 5. Vertical component of on-axis velocity as a function of depth, for five different peak accelerations, measured by MRI spin-tagging technique (from [11]).

are grateful to Dr David Parker for providing the figures and data presented here. The basic approach is that one follows the trajectory of a single positron-emitting tracer particle. As the positrons annihilate with electrons, pairs of back-to-back 511 keV photons are produced which are then detected in coincidence by a pair of large position-sensitive detectors (see Figure 6). This coincidence defines a line along which the positron annihilated. In theory, just two measured coincidences define lines that cross at the tracer particle position. In practice, many of the detected events are corrupted (e.g. because one of the photons has scattered before detection)

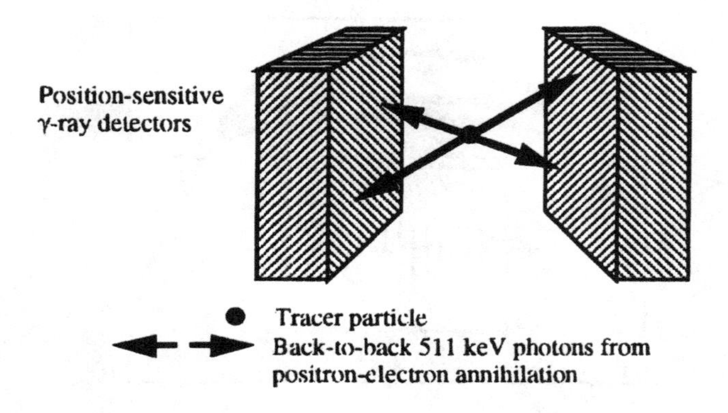

Figure 6. Positron emission particle tracking system.

so one needs to detect a reasonable number (of order 100) in order to distinguish the valid events (which essentially cross at a point) in the presence of this background. Unlike the MRI technique, data is only obtainable from a single point (i.e., from the tracer particle) rather than along an entire line through the sample. On the other hand, the data acquisition process is automated and the particle remains active for several hours. In that time it will normally visit even relatively small volume elements in the flow several times, allowing a detailed composite map of the flow field to be built up. Once again it is necessary to assume repeatable behavior in order to build up the velocity field.

Figures 7 and 8 show typical results obtained by applying the system to a simple horizontal rotating drum which is 30% filled with glass beads. In Figure 7 the tracer motion has been integrated over a period of about one hour, and also averaged along the length of the drum. In Figure 8, the deviation from the initial position along the axial direction is shown at six time intervals. A dispersion coefficient (mean-squared displacement) can be shown to be related to the circulation frequency of the material round the drum [12].

The performance of the current PEPT system in terms of measurement accuracy and maximum particle speed is limited by the photon count rate. The current system uses gas-filled multiwire chambers which have a detection efficiency per photon of around 10%. Both photons emitted by a given positron must be detected to provide an accurate trajectory and therefore the overall detection efficiency is only around 0.5%. In practice this limits the maximum count rate to under 2000/s, allowing a particle moving at 0.1 m/s to be located to within 2 mm 25 times per second, and a particle moving at 1 m/s to be located to within 6 mm 250 times per second. Construction of a new system at Birmingham, based on NaI scintillators,

Figure 7. PEPT measurements on rotating drum with glass beads at three angular velocities. Top row: occupancy (i.e. fraction of time spent at each point). Bottom row: Corresponding average velocity fields. (Courtesy D. Parker).

is under way; the single photon efficiency will be increased to 30% and as a result the expected count rate and positional accuracy will be increased by about one order of magnitude compared with the current system.

4. High Speed Photography

High speed photography is a third experimental technique which, unlike the two previous techniques, can provide whole-field data (a two-dimensional image) with time resolution of 1 ms or better. Its main drawback is its inability to probe the internal structure of three-dimensional flows, but model two-dimensional systems can still generate much useful data.

The system used in the authors' group is a Kodak Ektapro 1000 high speed video camera. The resulting images are stored digitally on processor boards within the camera. When images are stored to the full size of 239 x 192 pixels, a maximum framing rate of 1000 frames per second can be achieved giving 1.6 s of recording time. The framing rate can be increased by decreasing the image size with an upper limit of 12,000 frames per second. Figure 9 shows a typical image from a size segregation experiment.

Images are transferred by GPIB interface to a Sun IPX SPARCstation where the in-plane positions of particles are located using digital image processing software. There are three main steps to the image analysis: (a) the edges of the particles are detected by means of a Sobel filter; (b) the

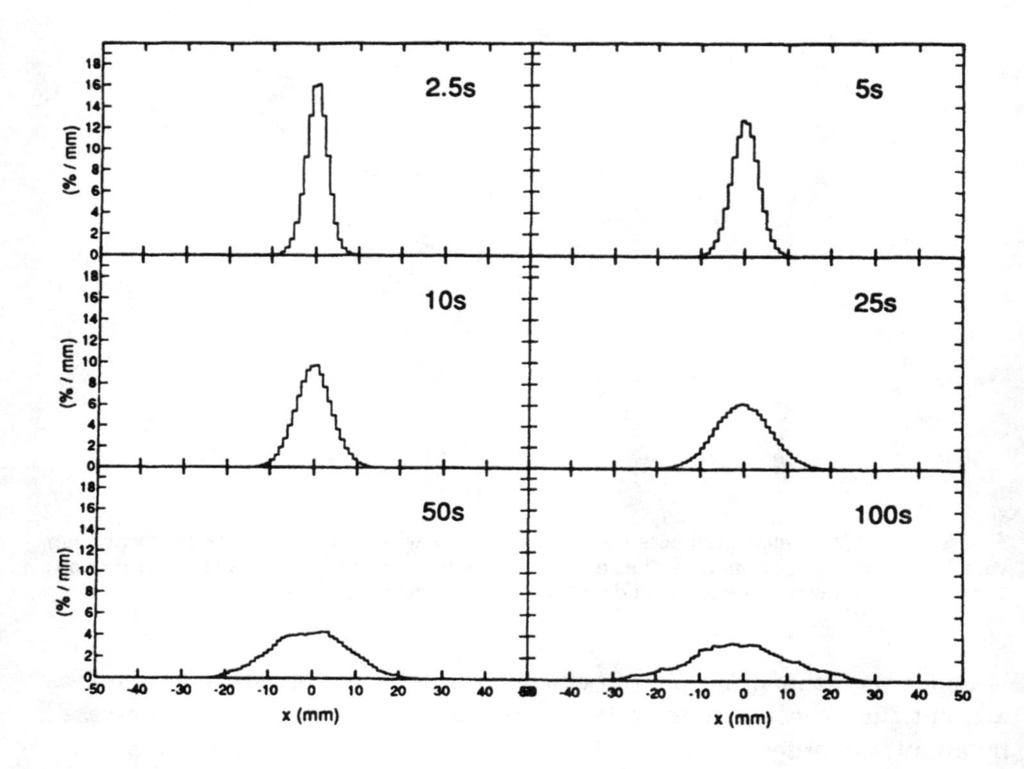

Figure 8. PEPT measurements on a rotating drum containing glass beads: axial dispersion. (Courtesy D. Parker).

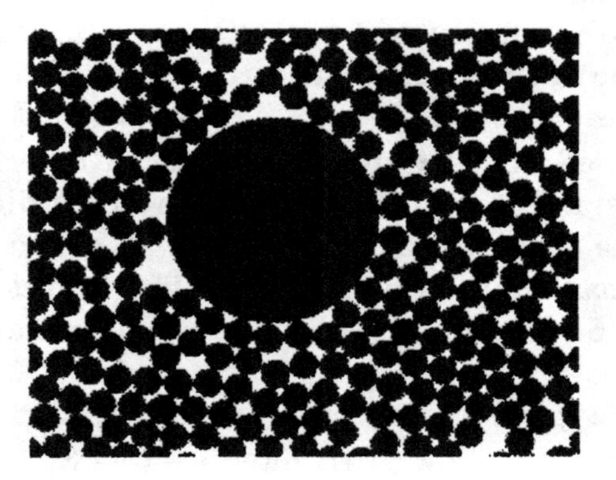

Figure 9. Typical grey-scale image from the Ektapro high speed video camera.

positions of the centres are then detected by Hough transformation of the resulting image; and (c) the position of all particles within the field of view are then tracked from frame to frame, from which speed and velocity distribution functions can be calculated. These operations are described in detail in [13].

4.1. VIBRO-FLUIDIZATION

The results of vibro-fluidization experiments on model two-dimensional powders consisting of 5-mm diameter steel spheres are described in this section. The experiments were performed using an electromagnetically driven shaker (Ling Dynamics Systems Model V650) driven by sine waves from a low-distortion signal generator and 1 kW power amplifier. The moving part of the shaker is a platform 156 mm in diameter which can attain a maximum peak to peak displacement of 25.4 mm and a maximum velocity and acceleration of 1.06 m/s and 70 g ($g = 9.81$ m/s^2), respectively. A cell made up of two glass plates 165 mm wide by 285 mm high was mounted on the moving platform. The width between the plates was controlled by spacers of varying thicknesses, to a resolution of 0.05 mm. By adjusting the plate spacing to exceed the particle size by 0.05 mm a close approximation to an idealized two-dimensional model powder was obtained. Vertical accelerations were monitored using piezoelectric accelerometers, one attached to each side of the support, and displacements of the vibrating cell were measured using a calibrated laser displacement meter. This allowed the horizontal acceleration to be checked at various points. The horizontal acceleration was found to be less than 2% of the vertical acceleration for the working range used in this paper, indicating an essentially one-dimensional acceleration field.

The number of spheres in the cell, N, took the values 27, 40, 60 and 90, and for each of these, experiments were carried out with amplitudes A_o of 0.5, 1.123, 1.84 and 2.12 mm. The base frequency was 50 Hz throughout. Camera runs were carried out at three different heights for each of these combinations of N and A_o, resulting in a total of nearly 80,000 frames of data [14].

Figure 10 shows the time-averaged three-dimensional packing fraction surface for the case $A_o = 2.12$ mm and $N = 90$. No significant side-wall effects are evident. Averaging the data across the width of the cell improves the signal-to-noise ratio; results for all four N values at $A_o = 2.12$ mm are shown in Figure 11(a). The $N = 90$ data are re-plotted on log-linear axes in Figure 11(b), together with a linear fit to the exponential tail of the distribution. The fact that the fit is reasonably good indicates that a Boltzmann distribution may provide a good approximation for the number

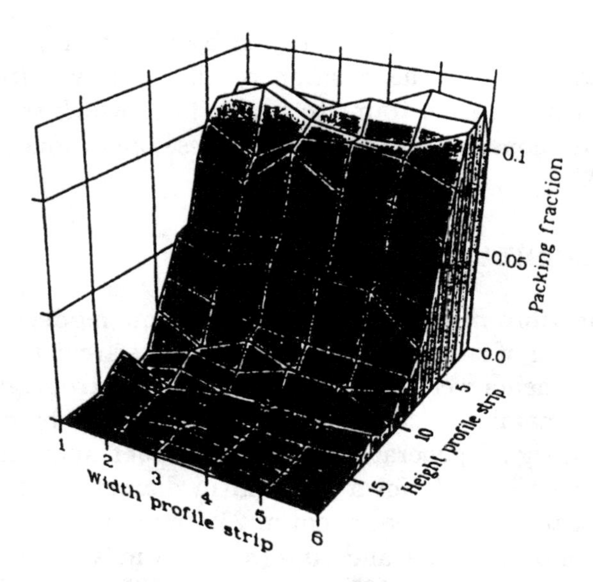

Figure 10. **Packing fraction distribution in fluidized powder ($N = 90$, $A_o = 2.12$ mm, $f = 50Hz$).**

density; the gradient of the best-fit line therefore provides one measure of the granular temperature, E_o, defined by

$$E_o = m\bar{c}^2/2 \tag{1}$$

where m is the particle mass, and c is its speed. The kinetic theory concept of granular temperature is widely used in theories of rapid granular flow. The reviews by Campbell [15] and Savage [16], on computer simulation and theoretical studies respectively, discuss the concept and its relevance to granular materials.

E_o can also be estimated directly from the width of the measured velocity distribution functions. Figure 12 shows the distribution for the horizontal component of velocity at a height in the cell of 65 mm, where $N = 90$ and $A_o = 2.12$ mm. Circles correspond to data points and the line is the best fit Gaussian curve.

The fluidization of granular materials has recently been investigated numerically by molecular dynamics and event-driven simulation techniques. Luding et al. [17, 18] obtained the conditions required to observe the condensed and fluidized regimes and showed scaling relations for the height of the center-of-mass for the system of beads in the fluidized regime in one- and two-dimensional powders. Previous experimental studies in two-dimensional systems have only considered surface fluidization whereby a condensed phase and fluidized phase coexist [19].

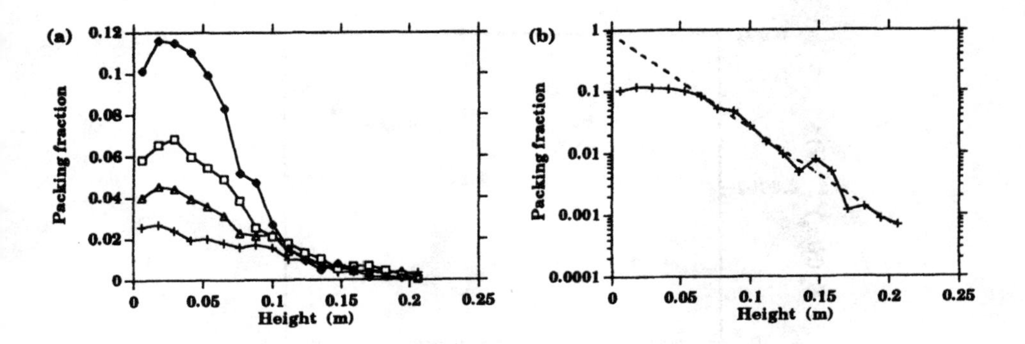

Figure 11. Effect of system size on the average packing fraction profiles. Crosses, triangles, squares and diamonds correspond to N = 27, 40, 60 and 90 respectively.

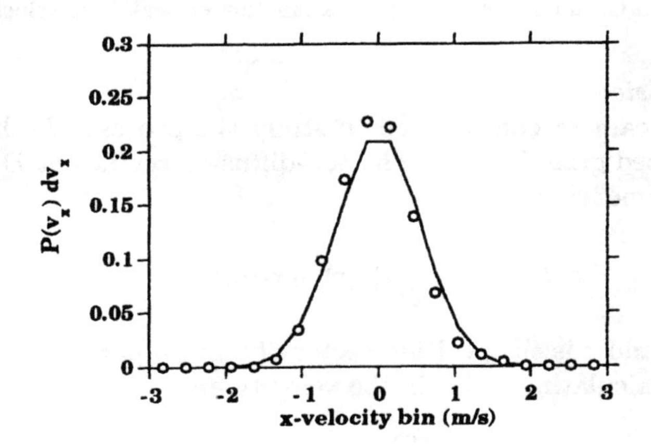

Figure 12. Horizontal velocity distribution: experimental (discrete points) with best fit Gaussian.

The simulations carried out by Luding et al. [17, 18] indicated that E_o scales with A_o and N as

$$E_o \propto (A_o\omega)^{\alpha}[N(1-\epsilon)]^{-\beta}, \tag{2}$$

where ω is the angular frequency of the base and ϵ is the restitution coefficient. The exponents α and β were found to take the values 2.0 and 1.0, respectively, for a one-dimensional powder, and the values 1.5 and 1.0 in two dimensions. Figure 13 shows the variation of E_o with $A_o\omega$ as calculated from the velocity distribution functions. The effects of system size have been scaled out using the average exponent $\beta = 0.60$. The data points fall on a line with a gradient $\alpha = 1.41 \pm 0.03$ which is close to the value α $= 1.5$ calculated from simulations.

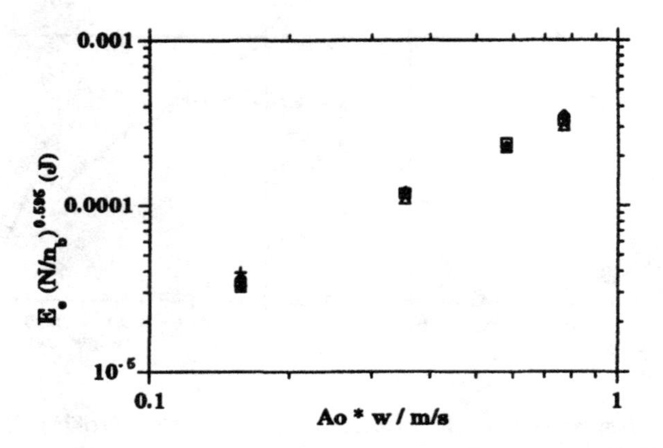

Figure 13. Granular temperature as a function of peak base velocity.

4.1.1. *Self Diffusion*

The high speed camera can be used to study the process of self-diffusion in a vibro-fluidized granular bed. The self-diffusion coefficient D takes the form (in three dimensions)

$$D = \lim_{t\to\infty} \frac{1}{6t} \langle |\mathbf{r}(t) - \mathbf{r}(0)|^2 \rangle, \tag{3}$$

where t is time, and $\mathbf{r}$ is the position vector of a given particle. An alternative method of calculating D is via the velocity auto-correlation function:

$$D = \frac{1}{3} \int_0^\infty \langle \mathbf{u}(t).\mathbf{u}(0) \rangle dt, \tag{4}$$

where $\mathbf{u}$ is the instantaneous velocity vector. Calculation of velocities amplifies measurement noise, and we therefore prefer to calculate D using Equation (3) rather than Equation (4). The particle tracking software allows many particle trajectories to be combined from a single high speed sequence to produce mean-square displacement plots with high signal-to-noise ratios.

Figures 14 and 15 show results from two separate experiments, one having a relatively small number of grains ($N = 90$) with low excitation ($A_o = 0.5$ mm), the other a larger number of grains ($N = 300$) and higher excitation ($A_o = 2.12$ mm). The mean particle speeds and mean packing fractions were respectively 0.27 m/s and 0.17 for Figure 14, and 0.37 m/s and 0.52 for Figure 15. Parts (a) and (b) show the long and short timescale behavior in each case. The horizontal and vertical contributions to the total mean-square displacement (marked as continuous lines) are shown separately as dashed and dotted lines, respectively. In both cases, there is

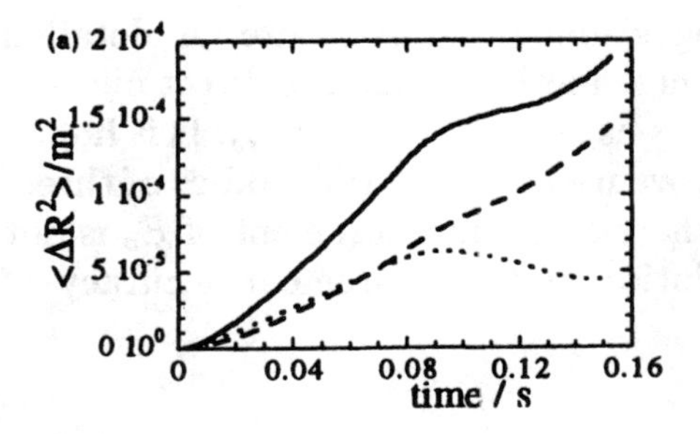

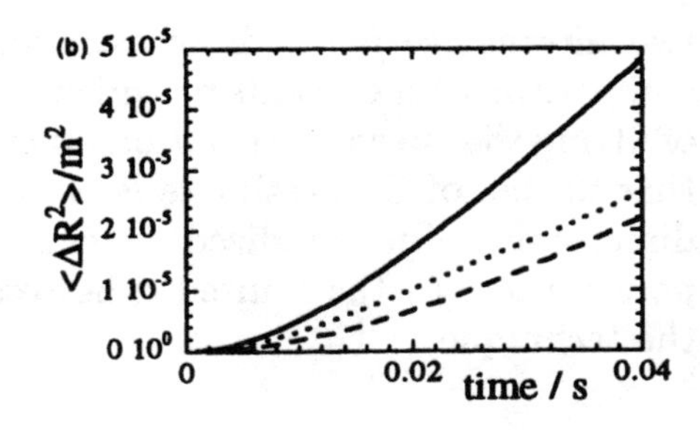

Figure 14. Mean-square displacement for small system ($N = 90$) under low excitation conditions ($A_o = 0.5$ mm). (a) Long time behavior; (b) short time behavior.

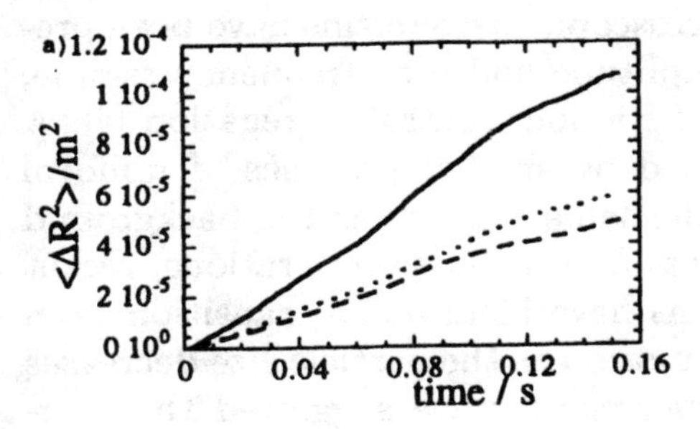

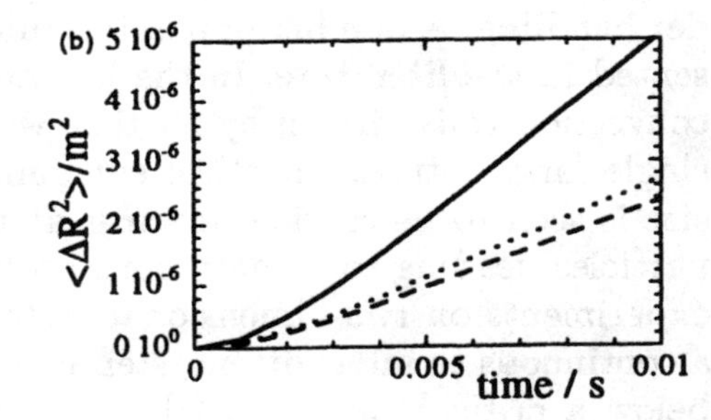

Figure 15. Mean-square displacement for large system ($N = 300$) under high excitation conditions ($A_o = 2.12$ mm). (a) Long time behavior; (b) short time behavior.

significant anisotropy in the diffusive behavior which can be attributed to larger velocity fluctuations in the vertical direction than in the horizontal direction. This is not surprising since energy is supplied to the system purely in the vertical direction.

In the case of Figure 14(a), the y-contribution is seen to saturate at large times due to the relatively low excitation, whereas the larger system (Figure 15(a)) shows an approximately linear response characteristic of diffusive behavior. At short times, (Figures 14(b) and 15(b)), the graphs have an approximately quadratic form. This is to be expected for times much shorter than the Enskog mean collision time, τ_E, (i.e. the mean time between two successive collisions suffered by any one sphere) where the particles are effectively moving freely. τ_E is approximately 33 ms for the experimental results in Figure 14, and 3 ms for those in Figure 15. The coefficient of the quadratic term can therefore provide a direct estimate of the mean particle speed, and hence the granular temperature through Equation (1). D is also related to the mean particle speed through the mean free path, which can be calculated for a given packing fraction. These approaches therefore provide two alternative methods of estimating granular temperature. A detailed comparison of the results provided by these methods, and the direct method of fitting the speed distribution functions, is currently underway. It is hoped that the use of D to estimate E_o will prove useful for PEPT studies of three-dimensional vibro-fluidized systems, where direct measurement of E_o is not possible due to the limited time resolution and measurement accuracy of the technique.

4.2. SIZE SEGREGATION

The segregation of powders under vibration according to particle size is a well-known phenomenon, and can cause significant problems during powder handling. A number of possible microscopic mechanisms have been presented in the literature. In the low amplitude and high frequency regime, convection rolls, driven by particle-wall friction, control segregation [4]. A single large intruder particle surrounded by smaller particles of uniform size is seen to be carried upwards at the same velocity as the background particles, leading to a continuous ascent. At very low accelerations, recent experiments on two-dimensional systems have identified a transition from a continuous to intermittent, step like, motion as the particle size decreases below a critical size ratio [5]. The experiments have suggested that segregation in this regime is no longer driven by convection, with the larger particles rising relative to the background particles. An arching effect model has been proposed [6] to explain this transition.

The ability to measure and track particles with the Ektapro camera has been exploited to investigate this transition between the continuous and intermittent regimes [20]. The experiments were carried out using a monolayer of approximately 5000 oxidized duralumin spheres (diameter = 2 mm) in the test cell. A set of intruders were made from 1 mm thick duralumin discs of various diameters, through which three 2-mm diameter chrome steel spheres were pressed to form an equilateral triangle. Γ is used to denote the peak acceleration normalised by g, the acceleration due to gravity. The vibration frequency throughout was 10 Hz.

Figure 16 shows three frames of an intruder with size ratio $\Phi = 7$, taken from the low acceleration regime ($\Gamma = 1.17$). Regions of disorder often appear around the intruder (a). Small gaps may open up below the intruder which particles can be pushed into by collective block motion. However, large gaps and avalanche events are not observed. In (b) a slip plane is captured below the intruder resulting in the upper block of particles moving upwards. Horizontal slip planes are also often observed (c). In Figure 17, the measured position of the intruder is plotted against time for a range of reduced accelerations. These plots are at a constant size ratio $\Phi = 7$.

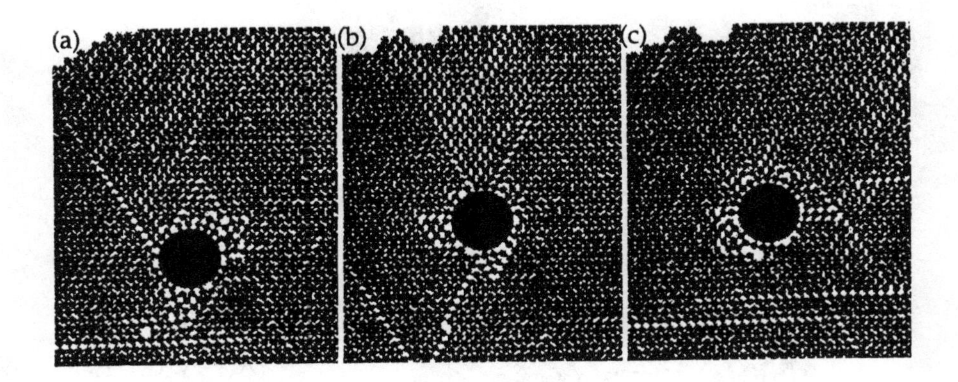

Figure 16. Images from size segregation experiments (see text for details).

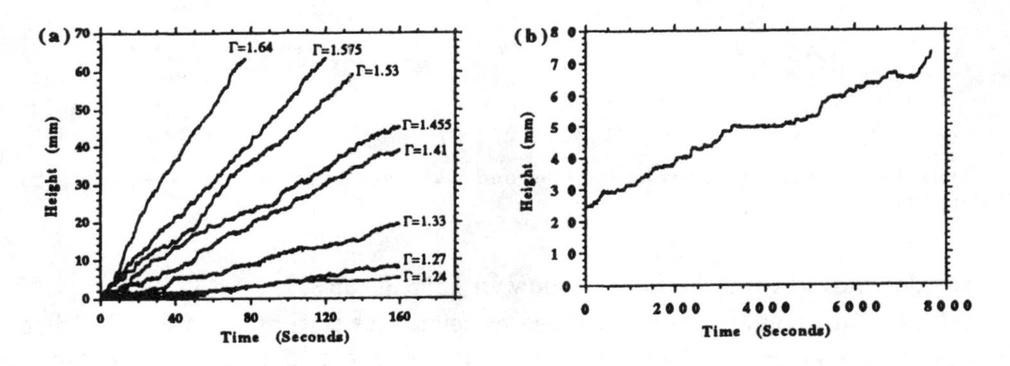

Figure 17. Intruder ascent diagrams for the (a) continuous and (b) intermittent regimes.

The results in (a) correspond to the continuous regime, whereas in (b) the acceleration is lower, ($\Gamma = 1.17$) and the intermittent motion is clearly visible. The results in (a) were obtained by filming at one frame per cycle; in (b) a frame was recorded every 48 cycles.

The use of Hough transforms and the particle tracking routines allowed trajectory maps of all the particles in the field to be calculated. Figure 18(a) shows results from one experiment with a size ratio $\Phi = 7$, and an acceleration of $\Gamma = 1.65$. This corresponds to a regime of quite strong convection; the intruder moved over the field of view in 266 frames as indicated by its trajectory. The other trajectories all show the position of the background particles **relative to the intruder** disc. A similar plot is shown in (b), but in this case the acceleration was reduced to $\Gamma = 1.32$ so that we are in the regime with intermittent rise characteristics. The trajectories still resemble those of Figure 18(a) showing that the intruder and background particles rise at the same rate in a collective motion; the

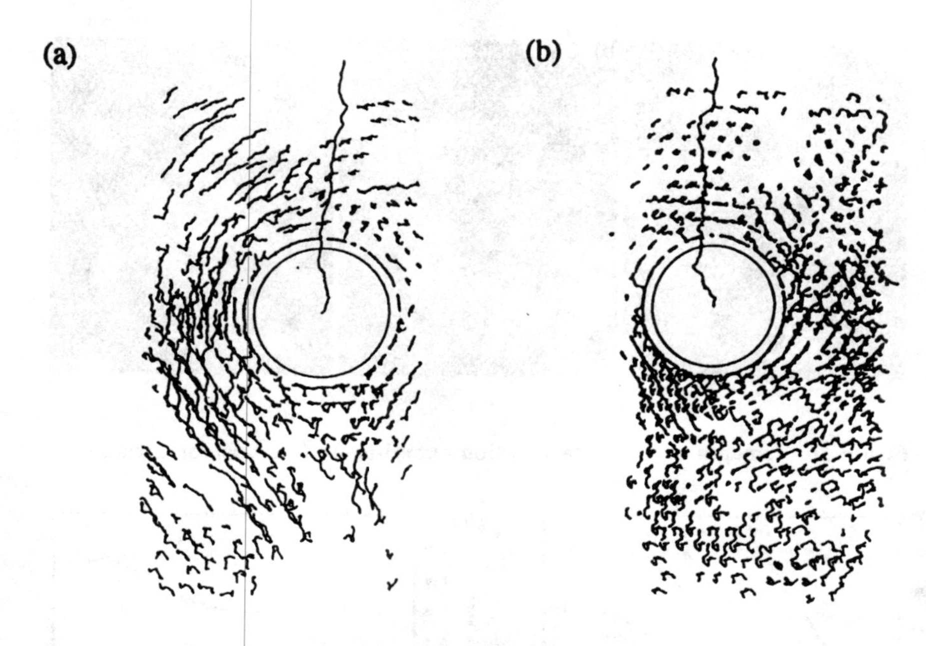

Figure 18. Trajectory maps for intruder and background particles: (a) $\Gamma = 1.65$; (b) $\Gamma = 1.32$.

intruder moved over the field of view in 1355 frames.

The main conclusion from these experiments is that the intruder disc rises at the same speed as the background particles over the entire range of accelerations studied. If we call such a phenomenon convective then the mechanism of segregation is driven by convection rolls over all accelerations. Previous studies which indicate that convection is absent at low accelerations (i.e. upwards motion of the intruder is not accompanied by upwards motion of the surrounding particles) and that geometrical effects strongly influence the segregation mechanism are not supported by our observations.

Acknowledgements

The research summarised in this paper benefited from experimental and programming input from R. Brockbank, W. Cooke, H. T. Goldrein and G.T.H. Jacques; technical support from D. Johnson, R. Flaxmann, and P. Bone; and useful discussions with J.-P. Hansen, C. C. Mounfield, S. F. Edwards, J. E. Field and E. M. Terentjev. Funding from the Department of Trade and Industry, Engineering and Physical Sciences Research Council, Unilever Plc, ICI Plc, Zeneca Plc, Schlumberger Cambridge Research, and Shell International Oil Products is also gratefully acknowledged. We thank Prof. S. Nagel for permission to reproduce Figures 3, 4 and 5, and the American Physical Society for permission to reproduce Figures 3 and 4. Figure 5 is reprinted (abstracted/excerpted) with permission from Reference 11. Copyright 1995 American Association for the Advancement of Science.

References

1. H. M. Jaeger and S. R. Nagel, Science **255**, 1523 (1992).
2. P. Evesque, Contemp. Phys. **33**, 245 (1992).
3. P. Evesque and J. Rajchenbach, Phys. Rev. Lett. **62**, 44 (1989).
4. J.B. Knight, H.M. Jaeger, and S.R. Nagel, Phys. Rev. Lett. **70**, 3728 (1993).
5. J. Duran, T. Mazozi, E. Clement, and J. Rajchenbach, Phys. Rev. E **50**, 5138 (1994).
6. J. Duran, J. Rajchenbach, and E. Clement, Phys. Rev. Lett. **70**, 2431 (1993).
7. S. Douady, S. Fauve, and C. Larouche, Europhys. Lett. **8**, 621 (1989).
8. H.K. Pak and R.P. Behringer, Phys. Rev. Lett. **71**, 1832 (1993).
9. L. Axel and L. Dougherty, Radiology **171** 841 (1989).
10. J.B. Knight, E.E. Ehrichs, V.Y. Kuperman, J.K. Flint, H.M. Jaeger, and S.R. Nagel, Phys. Rev. E **54**, 5726 (1996).
11. E.E. Ehrichs, H.M. Jaeger, G.S. Karczmar, J.B. Knight, V.Y. Kuperman, and S.R. Nagel, Science **267**, 1632 (1995).
12. D.J. Parker, A.E. Dijkstra, T.W. Martin, and J.P.K. Seville, Chem. Eng. Sci. **52**, 2011 (1997).
13. S. Warr, J.M. Huntley, and G.T.H. Jacques, Powder Technol. **81**, 41 (1995).
14. S. Warr, J.M Huntley, and G.T.H. Jacques, Phys. Rev. E. **52**, 5583 (1995).
15. C.S. Campbell, Ann. Rev. Fluid Mech. **22**, 57 (1990).
16. S.B. Savage, Adv. Appl. Mech. **24**, 289 (1984).
17. S. Luding, E. Clement, A. Blumen, J. Rajchenbach, and J. Duran, Phys. Rev. E **49**, 1634 (1994).
18. S. Luding, H.J. Herrmann, and A. Blumen, Phys. Rev. E **50**, 3100 (1994).
19. E. Clement and J. Rajchenbach, Europhys. Lett. **16**, 133 (1991).
20. W. Cooke, S. Warr, J.M. Huntley, and R.C. Ball, Phys. Rev. E **53**, 2812 (1996).

MODEL KINETIC EQUATIONS FOR RAPID GRANULAR FLOWS

J. JAVIER BREY
Física Teórica, Universidad de Sevilla
Apartado de Correos 1065. 41080 Sevilla. Spain

Abstract. Low density rapid granular flows are investigated by means of a combination of model kinetic equations and direct simulation Monte Carlo method. Both methods are based on the Boltzmann-Enskog equation for inelastic hard spheres, and provide a unique way to study far from equilibrium situations. Freely evolving systems and also steady driven states are considered.

1. Introduction

In this lecture the expression granular fluid will be used to refer to a system composed by identical smooth hard spheres or disks which collide inelastically. Moreover, we will restrict our considerations to relatively low densities, far from the close packing value. To describe and analyze the behavior of this kind of system, it is tempting to employ the same kind of techniques that have been successfully applied to ordinary (molecular) fluids over the years. In particular, standard kinetic theory can be straightforwardly adapted to incorporate the dissipation of energy in collisions. In this way, generalizations of the Boltzmann and Enskog equations for inelastic hard spheres have been derived, as discussed in detail in the lecture by Prof. Ernst in this volume. These equations provide, in principle, a firm starting point for the study of granular fluids. However, the lack of energy conservation complicates the equations in a nontrivial way, so that the derivation of exact solutions of the equations, describing specific physical situations, is even much harder than for elastic collisions. Inelasticity does not imply just a quantitative modification of the state of the system but a change in its own nature. For instance, boundary conditions leading to a homogeneous steady state for a molecular gas can drive a granular system

J. Karkheck (ed.),
Dynamics: Models and Kinetic Methods for Non-equilibrium Many Body Systems, 343–359.

into a highly inhomogeneous state (an example is given in Sec. 4). This is because dissipation in collisions and spatial inhomogeneities are coupled in a quite complicated way.

In the case of elastic fluids, the introduction of kinetic models has allowed obtaining relevant but otherwise inaccessible information about far from equilibrium states. In fact, the BGK model, originally proposed by Bhatnagar, Gross, and Krook [1], and its modifications, have provided the primary access to transport properties far from equilibrium [2]. The general idea of these models is to substitute the complicated collision operator appearing in the (exact) kinetic equations, by a much simpler form, usually a single relaxation term. Here we will show how the ideas behind the BGK approximation can be extended to granular systems, incorporating in an effective way the inelastic character of collisions. Let us mention from the very beginning that this is not a trivial task. The energy dissipation in collisions implies that there is no homogeneous steady state similar to the equilibrium state for molecular fluids. Also, the energy of a freely evolving system decreases monotonically in time, and this introduces another relevant time scale in the problem, associated with the cooling rate due to inelasticity. As a consequence of both facts, it is not clear whether the effect of collisions can be approximated by a tendency to approach a given velocity distribution and, if this were the case, which should be such a distribution and the characteristic time associated with the relaxation.

One of the most important applications of a description at the kinetic theory level is to investigate the possibility of a hydrodynamic-like characterization of the state of the system. This means going from a microscopic description to a macroscopic one with a drastic reduction in the number of variables used to identify the state of the system. It is known that on well defined time and space scales this reduction is possible for molecular fluids, but again this property can not be directly translated to granular media due to the modification of the time scales involved in the dynamical processes taking place in the system. The investigation of the relationship between the kinetic and hydrodynamic descriptions requires a detailed and controlled analysis of the spectrum of the linearization of the collision operator around some reference state. Model kinetic equations provide an appropriate way to carry out that analysis.

Far from equilibrium states of dilute molecular fluids described by the (elastic) Boltzmann equation have been investigated also by using the so-called direct simulation Monte Carlo method [3]. This is a numerical tool used to mimic the dynamic processes considered in the Boltzmann equation and has proven to be very useful to study a great variety of physical situations. The method can be easily adapted to the Boltzmann equation for inelastic hard spheres or disks since the only change needed refers to the collision rules, more concretely, to the expressions of the post-collisional velocities.

The aim of this lecture is to provide a revision of the work carried out along the above lines in the last few years. It will be shown that the combination of modeling and Monte Carlo simulation provides a unique tool to investigate granular flows in general situations.

2. The Model

The Boltzmann equation for the distribution function $f(\mathbf{r}, \mathbf{v}, t)$ of a low density inelastic gas of smooth hard spheres ($d = 3$) or disks ($d = 2$) of diameter σ, mass m, and coefficient of normal restitution α reads [4, 5]

$$\left(\frac{\partial}{\partial t} + \mathbf{v} \cdot \nabla\right) f(\mathbf{r}, \mathbf{v}, t) = J[f|f], \tag{1}$$

where $J[f|f]$ is the (inelastic) Boltzmann collision operator, whose explicit expression will not be given here. The important point for our present purposes is that it has the properties

$$\int d\mathbf{v} \begin{pmatrix} 1 \\ m\mathbf{v} \\ \frac{1}{2}mV^2 \end{pmatrix} J[f|f] = \begin{pmatrix} 0 \\ \mathbf{0} \\ -\frac{d}{2}nk_BT\zeta \end{pmatrix}, \tag{2}$$

with $\mathbf{V}(\mathbf{r}, t) = \mathbf{v} - \mathbf{u}(\mathbf{r}, t)$ the peculiar velocity, and the density $n(\mathbf{r}, t)$, flow velocity $\mathbf{u}(\mathbf{r}, t)$, and temperature $T(\mathbf{r}, t)$ defined in the standard way. The zeros on the right hand side of the above equation represent the conservation of mass and momentum in collisions. The term $-\frac{d}{2}nk_BT\zeta$ is the consequence of the energy loss due to dissipation, with ζ given by

$$\zeta[f] = (1-\alpha^2)\frac{m\pi^{\frac{d-1}{2}}\sigma^{d-1}}{4d\Gamma\left(\frac{d+3}{2}\right)nk_BT}\int d\mathbf{v}\int d\mathbf{v}_1\,|\mathbf{v}-\mathbf{v}_1|^3 f(\mathbf{r}, \mathbf{v}, t) f(\mathbf{r}, \mathbf{v}_1, t). \tag{3}$$

By using Eqs. (1) and (2), the following balance equations are obtained

$$\frac{\partial n}{\partial t} + \nabla \cdot (n\mathbf{u}) = 0, \tag{4}$$

$$\frac{\partial \mathbf{u}}{\partial t} + \mathbf{u} \cdot \nabla \mathbf{u} + (mn)^{-1} \nabla \cdot \mathsf{P} = 0, \tag{5}$$

$$\frac{\partial T}{\partial t} + \mathbf{u} \cdot \nabla T + \frac{2}{dnk_B}(\mathsf{P} : \nabla \mathbf{u} + \nabla \cdot \mathbf{q}) + T\zeta = 0. \tag{6}$$

The pressure tensor P and the heat flux $\mathbf{q}$ are defined as in the elastic case

$$\mathsf{P}(\mathbf{r},t)=\int d\mathbf{v} m\mathbf{V}\mathbf{V}f(\mathbf{v},\mathbf{r},t),\quad \mathbf{q}(\mathbf{r},t)=\int d\mathbf{v}\frac{m}{2}V^2\mathbf{V}f(\mathbf{r},\mathbf{v},t). \tag{7}$$

Equation (1) has been solved by means of a modified Chapman-Enskog procedure and closed transport equations have been derived to Navier-Stokes order with explicit expressions for the transport coefficients [6, 7]. Nevertheless, for systems with large gradients or outside the range of validity of the hydrodynamic description almost nothing is known about the solutions of the Boltzmann equation. For these situations it is convenient to consider kinetic model equations, as discussed in the Introduction. The idea is to replace the Boltzmann collision operator by a simpler form, but retaining those properties of the original operator considered as fundamental. This includes the ones which are responsible for the form of the balance equations, i.e. Eqs. (2).

The BGK model for the *elastic* Boltzmann equation is a one relaxation time model of the form [1]

$$\left(\frac{\partial}{\partial t}+\mathbf{v}\cdot\nabla\right)f=-\nu(f-f_M), \tag{8}$$

where f_M is the local equilibrium Maxwellian distribution and ν a parameter given by a functional of the local density and temperature. Therefore, the Boltzmann collision operator has been approximated by another one which, in addition to preserving the conservation of mass, momentum and energy, also preserves the equilibrium distribution function. For granular fluids, there is no equilibrium state. The basic solution of the (inelastic) Boltzmann equation playing a similar role, is the one describing the homogeneous cooling state (HCS). This is a distribution function $f_H(v,t)$ depending on the velocity and time only through the ratio v/v_0, where $v_0=(2k_BT(t)/m)^{1/2}$ is the thermal velocity. For this distribution function the Boltzmann equation reduces to

$$\frac{\zeta_H}{2}\frac{\partial}{\partial\mathbf{v}}\cdot(\mathbf{v}f_H)=J[f_H|f_H]. \tag{9}$$

Let us rewrite the Boltzmann equation in the form

$$\left(\frac{\partial}{\partial t}+\mathbf{v}\cdot\nabla\right)f-\frac{\zeta}{2}\frac{\partial}{\partial\mathbf{v}}\cdot(\mathbf{V}f)=J'[f|f], \tag{10}$$

with

$$J'[f|f]\equiv J[f|f]-\frac{\zeta}{2}\frac{\partial}{\partial\mathbf{v}}\cdot(\mathbf{V}f)\,. \tag{11}$$

The operator J' has the properties

$$J'[f_H|f_H] = 0, \quad \int d\mathbf{v} \begin{pmatrix} 1 \\ m\mathbf{v} \\ \frac{1}{2}mV^2 \end{pmatrix} J'[f|f] = 0, \tag{12}$$

i.e., it presents the same characteristic features as the usual Boltzmann collision operator for molecular gases. The representation in Eq. (10) is very suitable to formulate our model equation. We believe that the operator to be approximated by a single relaxation term is J' rather than J. Therefore, our model is defined by substituting $J'[f|f]$ by $-\nu(f - f_{\ell H})$, where $f_{\ell H}$ is the local version of f_H, with the local values of the temperature and density and with the velocity $\mathbf{v}$ replaced by the peculiar velocity $\mathbf{V}$,

$$\left(\frac{\partial}{\partial t} + \mathbf{v} \cdot \nabla\right) f = \mathcal{J}(\mathbf{v}|f), \tag{13}$$

$$\mathcal{J}(\mathbf{v}|f) = -\nu(f - f_{\ell H}) + \frac{\zeta}{2}\frac{\partial}{\partial \mathbf{v}} \cdot (\mathbf{V}f). \tag{14}$$

As in the elastic case, ν is a free parameter of the model, with a possible space and time dependence only through the low order moments of the distribution function f.

The model can be formulated in a less heuristic way by considering the linearization of the Boltzmann equation around f_H [8]. It is the spectrum of the linearization of J' that determines the relaxation modes for the distribution function in the homogeneous case. The term containing the derivative with respect to the velocity on the left hand side of Eq. (10) takes care of the time dependence of the temperature, i.e. of the time dependence of the reference state. In the kinetic model as formulated above, the spectrum of the linearization of J' is collapsed into the eigenvalue zero, corresponding to the conservation laws given in Eq. (12), and a single degenerate value $-\nu$, representing the relaxation in the velocity subspace orthogonal to 1, $\mathbf{v}$, and v^2.

In Eq. (13) appears the distribution function $f_{\ell H}$ that must be obtained from Eq. (9). Although the exact analytical solution of this equation is not known, it has been shown that it is very close to a Maxwellian even for rather small values of the coefficient of restitution α [4, 9, 10]. Therefore, for practical reasons we introduce one more approximation in the model, namely

$$f_{\ell H} \to f_\ell = n \left(\frac{m}{2\pi k_B T}\right)^{d/2} \exp\left(-\frac{mV^2}{2k_B T}\right). \tag{15}$$

The model still contains the free parameter ν that can be chosen to optimize the results in each specific situation. As a first application of the model let us consider the derivation of the hydrodynamic equations to first order in the gradient of the fluxes for a three-dimensional system. By using a modified Chapmann-Enskog expansion method, it is obtained

$$\mathsf{P} = p\mathsf{I} - \eta \left[\nabla \mathbf{u} + (\nabla \mathbf{u})^{+} - \frac{2}{d} \mathsf{I} \nabla \cdot u \right], \tag{16}$$

$$\mathbf{q} = -\kappa \nabla T - \mu \nabla n. \tag{17}$$

where I is the unit tensor, $p = nk_BT$ is the hydrostatic pressure, η is the shear viscosity, κ is the heat conductivity, and μ a new transport coefficient that has no analogue in molecular fluids. The above expressions have the same structure as those following from the (inelastic) Boltzmann equation. In order to compare the results for the transport coefficients, we must specify the choice for ν. In this context, we have fixed it by requiring that the viscosity from the model agrees as much as possible with that from the Boltzmann equation. A complete agreement can not be reached in a consistent way due to the difference in the expression of f_H. In Fig. 1 we compare the expressions for the heat conductivity from the model equation and from the Boltzmann equation. The latter has been obtained by using the first Sonine approximation [7]. A similar comparison for the coefficient μ is presented in Fig. 2. It is seen that the model provides a fairly good description of the α dependence of the transport coefficients.

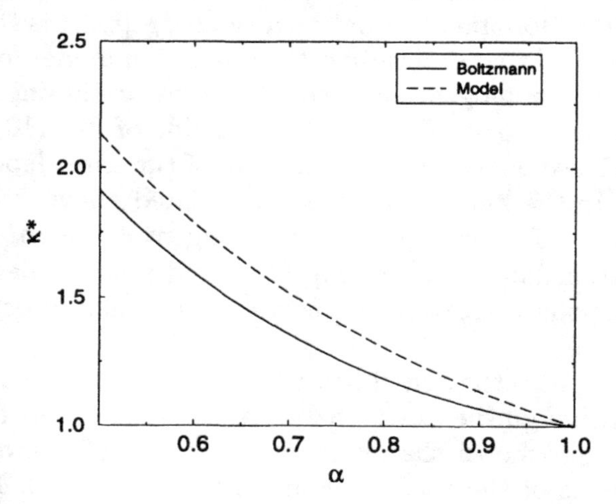

Figure 1. Thermal conductivity as a function of the coefficient of normal restitution α. The solid line is the result from the Boltzmann equation, and the dashed line is from the model equation. The heat conductivity is reduced in each case by its value in the elastic limit.

Up to now we have paid attention only to the one-particle distribution function, but it is possible to extend the model to provide also information

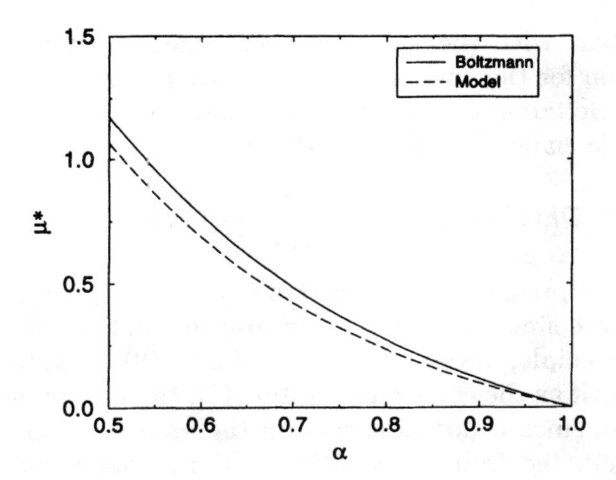

Figure 2. The transport coefficient μ as a function of the coefficient of restitution α. Symbols are the same as in Fig. 1. The transport coefficient is reduced by $\kappa_0 T/n$, where κ_0 is the elastic thermal conductivity.

about the spatial correlations. Let us consider the pair correlation function $g(\mathbf{r}_1, \mathbf{v}_1, \mathbf{r}_2, \mathbf{v}_2, t)$ defined by

$$g(\mathbf{r}_1, \mathbf{v}_1, \mathbf{r}_2, \mathbf{v}_2, t) = f_2(\mathbf{r}_1, \mathbf{v}_1, \mathbf{r}_2, \mathbf{v}_2, t) - f(\mathbf{r}_1, \mathbf{v}_1, t) f(\mathbf{r}_2, \mathbf{v}_2, t), \tag{18}$$

where f_2 is the two-particle distribution function. An exact equation for the function g can be obtained from the pseudo-Liouville equation for a gas of inelastic hard disks or spheres [5, 11]. In the low density limit it is

$$\begin{aligned}\left(\frac{\partial}{\partial t} + \mathbf{v}_1 \cdot \nabla_1 - \Omega_1 + \mathbf{v}_2 \cdot \nabla_2 - \Omega_2\right) g(\mathbf{r}_1, \mathbf{v}_1, , \mathbf{r}_2, \mathbf{v}_2, t) \\ = \delta(\mathbf{r}_{12}) \bar{T}_0(1,2) f(\mathbf{r}_1, \mathbf{v}_1, t) f(\mathbf{r}_2, \mathbf{v}_2, t),\end{aligned} \tag{19}$$

where Ω is the linearized (inelastic) Boltzmann collision operator and

$$\bar{T}_0(1,2) = \sigma^{d-1} \int d\widehat{\sigma}\, \theta(\mathbf{v}_{12} \cdot \widehat{\sigma}) |\mathbf{v}_{12} \cdot \widehat{\sigma}| [\alpha^{-2} b_\sigma^{-1}(1,2) - 1] \tag{20}$$

is a binary collision operator for inelastic hard spheres. Here $\widehat{\sigma}$ is a unit vector pointing from the center of particle 2 to that of particle 1 at contact, θ is the Heaviside step function, and $b_\sigma^{-1}(1,2)$ is an operator replacing all the velocities $\mathbf{v}_1$, $\mathbf{v}_2$ to its right by the precollisional velocities leading to them,

$$b_\sigma^{-1}(1,2)\mathbf{v}_1 = \mathbf{v}_1 - \frac{1+\alpha}{2\alpha}(\widehat{\sigma} \cdot \mathbf{v}_{12})\widehat{\sigma}, \tag{21}$$

$$b_\sigma^{-1}(1,2)\mathbf{v}_2 = \mathbf{v}_2 + \frac{1+\alpha}{2\alpha}(\widehat{\sigma} \cdot \mathbf{v}_{12})\widehat{\sigma}. \tag{22}$$

To be consistent with the approximation made to formulate the model kinetic equation for the one-particle distribution function, we approximate the linearized Boltzmann collision operators Ω_i in Eq. (19) by the linearization of the right hand side of Eq. (13), i.e.

$$\Omega h(\mathbf{v}) \to \int d\mathbf{v}' \left[\frac{\delta}{\delta f(\mathbf{r}, \mathbf{v}', t)} \mathcal{J}(\mathbf{v}|f) \right] h(\mathbf{v}'). \tag{23}$$

Of course, this expression can be written in a more explicit form, but it will not be done here since it does not add any relevant physical information. Although in principle, the right hand side of Eq. (19) could be approximated in the same spirit as the collision operator [12], there is no problem in using the exact form, since it does not contain the unknown function g and can be directly evaluated from the solution of the model equation for the one particle distribution function.

3. Homogeneous Cooling State

In many of the applications to far-from-equilibrium states of granular gases considered up to now, an even more simplified model has been considered. The term containing the velocity derivative in the expression of the model collision operator, Eq. (14), depends on the distribution function f both explicitly and through the bilinear functional dependence of ζ. These dependences greatly increase the complexity of the model, which is made much simpler if f is replaced everywhere in that term by the local equilibrium distribution f_ℓ, i.e. we approximate

$$\zeta[f] \to \zeta_\ell \equiv \zeta[f_\ell] = (1-\alpha^2) \frac{2\pi^{\frac{d-1}{2}} \sigma^{d-1}}{d\Gamma(d/2)} n \left(\frac{k_B T}{m} \right)^{1/2}, \tag{24}$$

$$\frac{\zeta}{2} \frac{\partial}{\partial \mathbf{v}} (\mathbf{V} f) \to \frac{\zeta_\ell}{2} \frac{\partial}{\partial \mathbf{v}} (\mathbf{V} f_\ell). \tag{25}$$

In summary, this simplified model kinetic equation reads

$$\left(\frac{\partial}{\partial t} + \mathbf{v} \cdot \nabla \right) f = -\nu (f - f_\ell) + \frac{\zeta_\ell}{2} \frac{\partial}{\partial \mathbf{v}} \cdot (\mathbf{V} f_\ell). \tag{26}$$

Let us consider a system in the HCS. We have carried out a linear stability analysis of this state based on the Navier-Stokes equations derived from the model kinetic equation. The effective collision frequency ν has been fixed now by requiring the model to reproduce the Boltzmann value of the shear viscosity in the elastic limit. This leads to

$$\nu = C n \sigma^{d-1} \left(\frac{\pi k_B T}{m} \right)^{1/2}, \tag{27}$$

with $C \simeq 16/5$ for $d = 3$ and $C \simeq 2$ for $d = 2$. The following general picture emerges from the analysis. There is a critical wavenumber

$$k_\perp^c = [\zeta^*(2 - \zeta^*)]^{1/2} l_0^{-1}, \tag{28}$$

separating two different regimes. Here $\zeta^* = \zeta_\ell/\nu$ and $l_0 = 2/C\sqrt{\pi}n\sigma^{d-1}$, which is proportional to the mean free path. All modes with wavenumber $k > k_\perp^c$ lead to a decay in time of any perturbation of the hydrodynamic variables, while for $k < k_\perp^c$ there is a $d - 1$ fold degenerated mode which decays slower than the thermal velocity associated with the temperature of the homogeneous reference state. This mode governs the time evolution of the transverse components of the velocity relative to $\mathbf{k}$, and it is usually referred to as the shear mode. The relationship of the behavior of this mode with the stability of the HCS has been discussed by Goldhirsch and Zanetti [13]. Let us point out that the dispersion relations obtained from this model for $d = 3$ are in very good agreement with the results from the Boltzmann equation [7].

In order to check the validity of the above linear analysis (based on a hydrodynamic description of the system) we have carried out direct Monte Carlo simulations of a initially homogeneous system of hard disks ($d = 2$) [14]. The system was a square of side L and periodic boundary conditions were used. Therefore, the smallest allowed wavenumber is $k_m = 2\pi/L$. Equation (28) implies that in systems with $L > L_c = 2\pi/k_\perp^c$ "unstable" shear modes are possible, while they cannot exist in smaller systems. Given a value of α, we have run simulations of systems of increasing size, starting from sizes for which the system remained homogeneous for the duration of the run. All runs lasted, at least, until the temperature decreased by a factor of the order of 10^{-12}.

By following the time evolution of the spatial average of density fluctuations and of the ratio of macroscopic kinetic energy to thermal kinetic energy, estimates of the critical size of the system L_c were made for different values of the coefficient of restitution α. For systems of sizes smaller than the critical one, the two monitored quantities stay within values corresponding to a homogeneous distribution. For larger systems, a clear separation between kinetic and thermal energy shows up at a given time, and later on there is a fast increase of density fluctuations, indicating the formation of cluster structures. As an example, a snapshot of the macroscopic velocity field for a given time in a system with $\alpha = 0.8$ is given in Fig. 3. It is clear that the size of the system is larger that the critical value, since fully developed velocity vortices are clearly identified. Nevertheless, a visualization of the density field shows no presence of significant density structures. They appeared in the system at later times. This chronological evolution seems to confirm the idea that the nonlinear effects associated with the transverse component of the flow field are responsible for the spatial inhomogeneities, and eventually for the formation of density clusters [13].

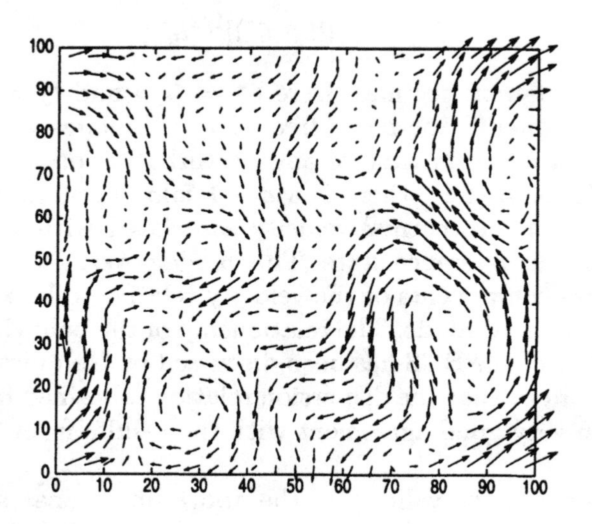

Figure 3. Snapshot of the momentum density in a system with $\alpha = 0.8$ and $L = 100\lambda$. The arrows represent the coarse grained momentum density in arbitrary units.

In Fig. 4 we have plotted the reduced critical size L_c/λ, where $\lambda = 1/2\sqrt{2}n\sigma$ is the mean free path, as a function of α. The continuous line is the prediction of the model, following from Eq. (28), the dotted line results from an expression obtained by Goldhirsch, Tan, and Zanetti [13] from the kinetic theory of Jenkins and Richmann [15]. It is seen that the model prediction fits very well the numerical results over the range of values of α considered. In fact, it seems to provide a more accurate description than the Jenkins and Richmann theory as the value of α decreases. Nevertheless, one must keep in mind the approximate character of the criterion used to determine the critical size of the system from the direct simulation Monte Carlo method. In any case, the results confirm the hydrodynamic nature of the cluster instability, at least in its initial set up.

We have also followed in the simulations the initial build-up of spatial correlations in the freely evolving system. In particular, we have focused on the two scalar correlation functions $C_{||}(r,t)$ and $C_{\perp}(r,t)$ defined through

$$\mathsf{C}(\mathbf{r},t) = \widehat{\mathbf{r}}\widehat{\mathbf{r}}C_{||}(r,t) + (\mathsf{I} - \widehat{\mathbf{r}}\widehat{\mathbf{r}})C_{\perp}(r,t), \tag{29}$$

where I is the unit tensor, $\widehat{\mathbf{r}}$ the unit vector parallel to $\mathbf{r}$, and

$$\mathsf{C}(\mathbf{r},t) = \frac{1}{L^2}\int d\mathbf{r}' \langle \mathbf{G}(\mathbf{r}+\mathbf{r}')\mathbf{G}(\mathbf{r})\rangle_t \tag{30}$$

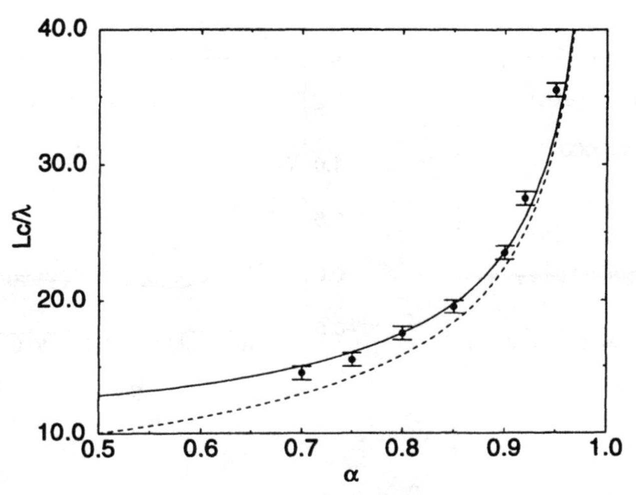

Figure 4. The critical size L_c for the shear mode as a function of the coefficient of normal restitution. The solid line is the theoretical prediction from the model, and the dashed line the low density limit of the results derived in Ref. [13]. The dots with the error bars have been obtained by direct simulation Monte Carlo method.

is the correlation tensor of the momentum fluctuations. Here $\mathbf{G}$ is the microscopic momentum density and the angular brackets denote ensemble average at time t. Expressions for $C_{||}$ $C_{\perp}$ have been derived from Eqs. (24) and (19) in Ref. [16]. They are based on the analysis of the spectrum of the linearized kinetic model equation. As an example, Fig. 5 compares the simulation results and the theoretical predictions for $\alpha = 0.85$ at two times. In the figures, the correlation functions have been scaled with $m^2 n_H^2 k_B T_H(t)$. Again, quite good agreement is found between the kinetic model and the Monte Carlo simulation results. The same problem has been studied in dense systems by using fluctuating hydrodynamics and Molecular Dynamics simulations [17]. Our results are consistent with theirs, in the sense that they tend to agree when the low density limit is considered.

4. Steady States of Driven Granular Systems

Most of the simplest far-from-equilibrium physical situations correspond to steady states. Nevertheless, as a consequence of the energy dissipation in collision, their nature is quite different in molecular and granular fluids. Consider for instance the uniform shear flow (USF). It is characterized by a linear profile of the velocity flow, $\mathbf{u}(\mathbf{r}) = \mathsf{a} \cdot \mathbf{r}$, where $a_{i,j} = a\delta_{i,x}\delta_{j,y}$, a being the constant shear rate. Otherwise, the density and the temperature are uniform. From the energy balance equation (6) it follows that the temperature changes in time due to the competition between viscous heating

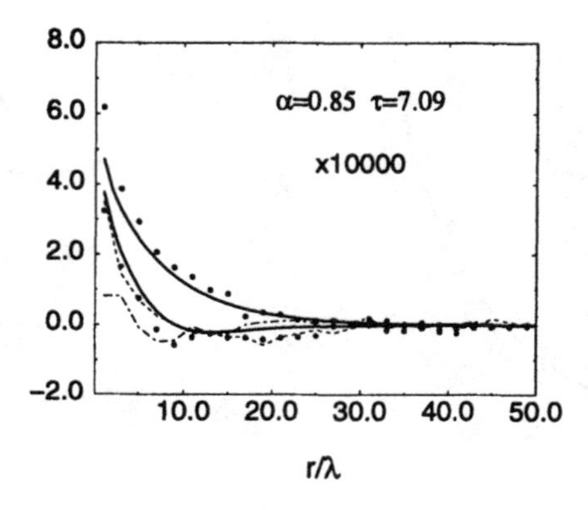

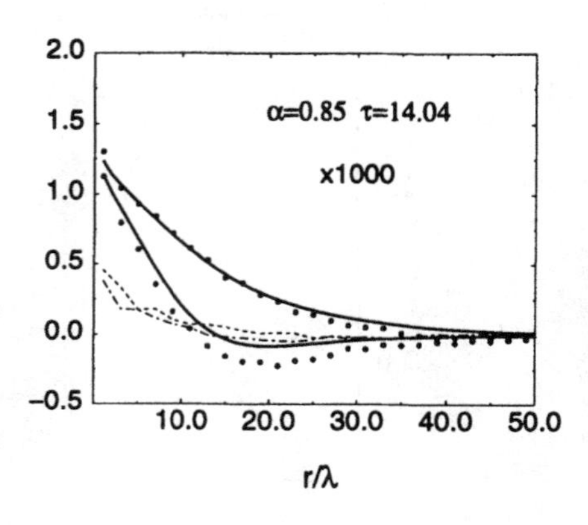

Figure 5. Spatial correlations of relative fluctuations around the HCS. They are scaled as indicated in the main text. Dots correspond to the parallel velocity and asterisks to the transverse velocity. The solid lines are the theoretical predictions from the model. The dashed lines and dot-dashed lines correspond to the density and energy respectively, not mentioned in the text.

and energy dissipation in collisions. Therefore, a steady state is possible when the effects cancel each other and the temperature remains constant. In that case, it is

$$aP_{xy} = -\frac{dnk_BT}{2}\zeta. \tag{31}$$

We have studied this steady simple shear flow, which does not exist for molecular fluids, by means of the kinetic model, as well as by performing direct Monte Carlo simulation of the Boltzmann equation [18]. Given that this state is already discussed elsewhere in this volume, we will limit ourselves to mention that also in this case the model provides an accurate description of what is observed in the numerical solution of the Boltzmann equation.

In this Section, we will consider another steady state: the one reached by a system confined between two parallel walls at the same temperature [19, 20]. For molecular fluids such a state is trivial since it is the uniform Maxwellian equilibrium one at the temperature of the walls, but for granular fluids this steady state is not homogeneous. When the granular balance equations given by Eqs. (4)–(6) are particularized for a steady state with no macroscopic velocity field and gradients in only one direction taken as the x axis, it is found

$$\frac{\partial}{\partial s}P_{xi} = 0 \tag{32}$$

for all i and

$$\frac{\partial}{\partial s}q_x = -T\frac{\zeta}{\nu}, \tag{33}$$

where we have introduced the new scale

$$s(x) = \int_0^x dx'\,\nu(x'). \tag{34}$$

If the Navier-Stokes approximation, Eqs. (17) and (18), is used, it is easily obtained that $P_{ij} = p\delta_{ij}$ and Eq. (32) implies that the pressure p is uniform. In a similar way, it is found that the temperature shows a parabolic profile in the variable s,

$$\frac{\partial^2}{\partial s^2}T(s) = \frac{\zeta^* d}{2a(\zeta^*)}. \tag{35}$$

with ζ^* defined below Eq. (28) and $a(\zeta^*)$ a certain function tending to $(d+2)k_B/m$ for $\zeta^* \to 0$. In the limit of elastic collisions, one recovers from Eq. (35) a temperature profile linear in the variable s that is characteristic of low density molecular gases. While the coefficient of the linear term in the temperature profile following from Eq. (35) is determined by the boundary conditions, the coefficient of the quadratic term is a given function of the coefficient of restitution. Therefore, the steady states of a granular fluid with no macroscopic fluxes are intrinsically inhomogeneous and gradients and inelasticity can not be considered as independent variables.

In Figs. 6 and 7 we present the results obtained from the direct Monte Carlo simulation of a system of hard disks for $\alpha = 0.99$. Note that the profiles are plotted as functions of the scaled distance s. Given the symmetry of the system, we have taken the origin to be the same distance from both walls. It is seen that, outside the boundary layers, the pressure is uniform and the temperature profile is accurately fitted by a parabola. Besides, the value of the slope of the numerical fit is in very good agreement with the prediction of Eq. (35) if the constant C in the effective collision frequency is chosen by requiring the model to give the same value for the Navier-Stokes thermal conductivity in the elastic limit $\alpha = 1$ as the Boltzmann equation [21]. Nevertheless, there is a clear anisotropy of the pressure tensor since P_{xx} is clearly larger than P_{yy}. This is contrary to the Navier-Stokes prediction. When the value of α decreases the anisotropy of the diagonal terms of the pressure tensor increases, and the hydrodynamic pressure becomes nonuniform in the bulk. Of course, the nonuniformity appears through P_{yy}, while P_{xx} remains homogeneous as required by the balance equations (see Eq. (32)). Finally, the slope of the temperature profile in the s variable becomes nonlinear showing the influence of higher-order terms.

Trying to describe the above behavior, we have looked for an exact solution of the model kinetic equation that would describe the steady state

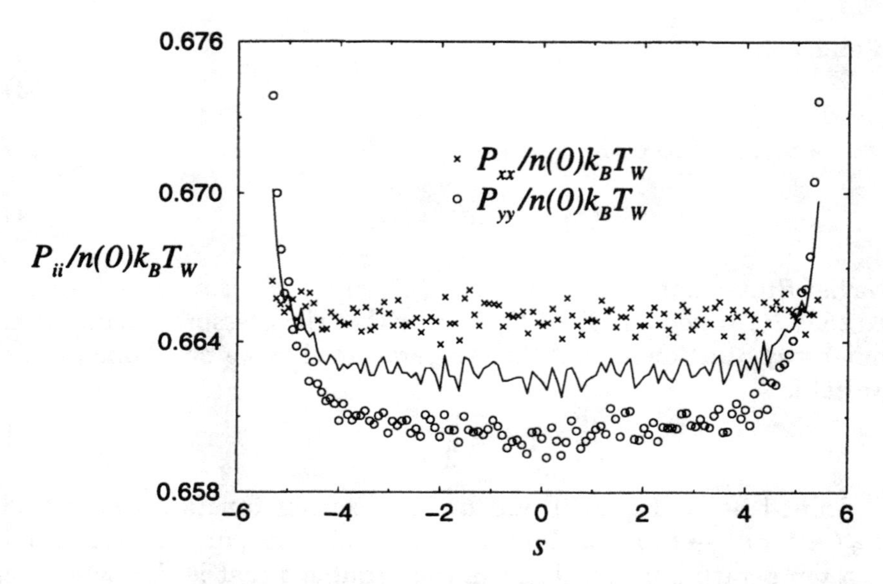

Figure 6. Profiles of the diagonal components of the pressure tensor obtained from Monte Carlo simulations in the steady state for $\alpha = 0.99$. The solid line is the hydrodynamic pressure p.

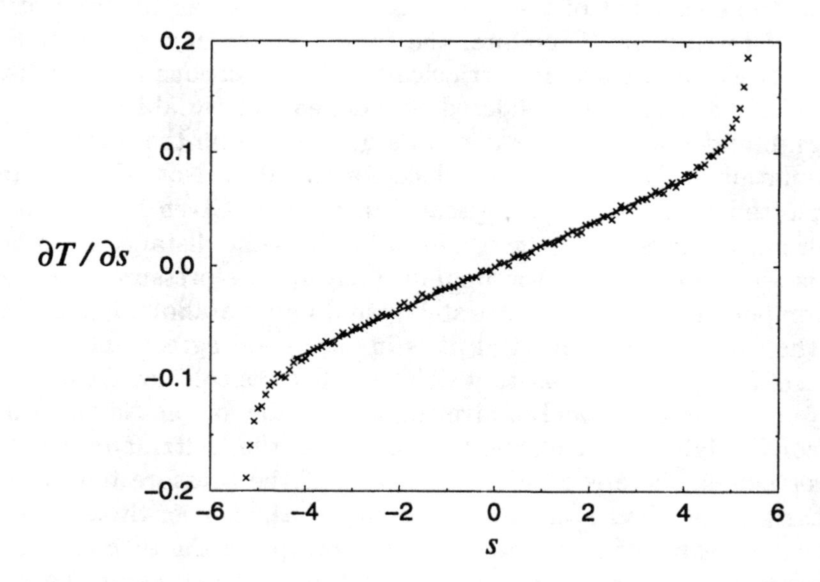

Figure 7. Slope of the temperature profile for the same state as in Fig. 6.

for arbitrary values of the coefficient of restitution. Since we were not able to construct it, we decided to follow a perturbative approach in the form of a series expansion around $\epsilon \equiv 1-\alpha^2 = 0$. The motivation for this expansion is that we are considering a system whose boundaries are kept at constant temperature, and we know that in such a situation the steady state for a molecular gas is the equilibrium one with no gradients. Therefore, all gradients must vanish in the limit $\epsilon \to 0$. Then, we have formally expanded

$$f = f_0 + \epsilon^{1/2} f_1 + \epsilon f_2 + \epsilon^{3/2} f_3 + \cdots, \tag{36}$$

and

$$\frac{\partial}{\partial s} = \epsilon^{1/2}\delta_1 + \epsilon\delta_2 + \epsilon^{3/2}\delta_3 + \cdots. \tag{37}$$

The expansion in powers of $\epsilon^{1/2}$ is motivated by the results obtained to Navier-Stokes order where, for instance, we saw that $\partial^2 T/\partial s^2$ is of order ϵ, i.e. $\partial T/\partial s \sim \epsilon^{1/2}$. However, self-consistency of the results obtained from the expansion is what indicates whether the assumed form is correct, at least up to the order considered in the calculations [21].

When the above expansion is carried out, the following expressions for the pressure tensor and the heat flux, valid up to order ϵ, are obtained

$$P_{ij} = \delta_{ij}\left[1 - (1 - \delta_{ix} d)\frac{2\zeta^*}{d+2}\right], \tag{38}$$

$$q_i = -\delta_{ix}\frac{(d+2)nk_B^2 T}{2m}\frac{\partial T}{\partial s}. \tag{39}$$

Thus, there are already normal-stress differences to this order. If the above expressions for the fluxes are used in the balance equations one gets

$$\frac{\partial p}{\partial s} = 0, \quad \frac{\partial^2 T}{\partial s^2} = \zeta^* \frac{md}{(d+2)k_B}. \tag{40}$$

For $\alpha > 0.99$ this temperature profile is practically equivalent to the one obtained in the Navier-Stokes approximation and, therefore, it fits very well the simulation results. The values of the ratio P_{xx}/P_{yy} as a function of ζ^* are plotted in Fig. 8. Also shown (solid line) is the theoretical prediction given by Eq. (38). It describes qualitatively well the asymmetry, but there is a clear discrepancy in the slope of the straight line. This is due to the criterion we have used to fix the collision frequency (to reproduce the elastic heat conductivity from the Boltzmann equation) and it is a consequence of the well known fact that single relaxation models of the Boltzmann equation can not describe in a quantitatively accurate way both the viscosity and the heat conductivity. The model result for the one-particle distribution function also agrees very well with the simulation data, at least for velocities in the thermal region. More details are given in Ref. [21].

Finally, let us mention that we have extended the perturbative calculations up to order ϵ^2 trying to explain the observed inhomogeneity for larger

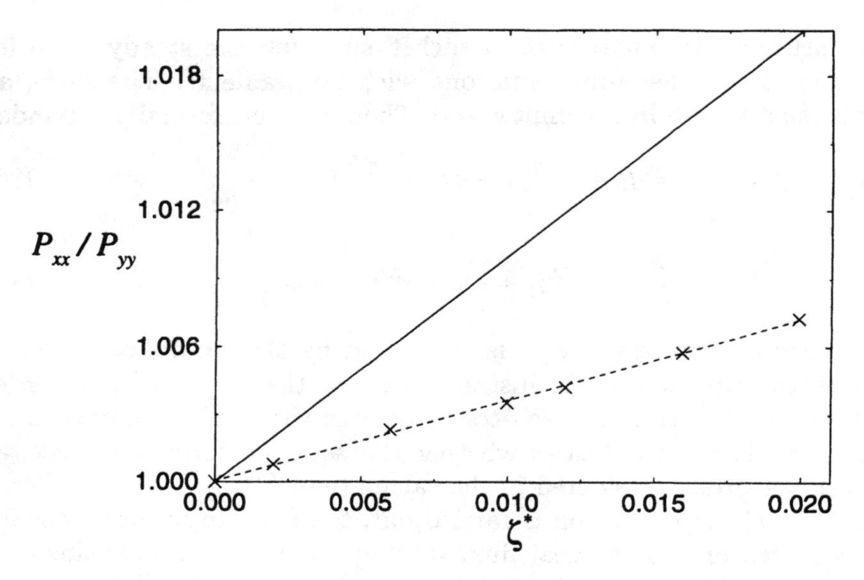

Figure 8. **Normal stresses ratio as a function of the dissipation parameter ζ^* for a system of hard disks. The solid line is from the model equation and the crosses from the Monte carlo simulation. The dotted line is a linear fit of the data.**

dissipation. Unfortunately, the results show quite clear evidence that the ϵ expansion, although probably asymptotic, is divergent. Further work is required to understand the reasons for this behavior.

5. Conclusion

I expect to have been able to translate the feeling that the combination of model kinetic equations and Monte Carlo simulation provides a unique way to study far-from-equilibrium situations also (and especially) for granular fluids. A general conclusion of the results presented here is that a hydrodynamic-like description is valid on the appropriate time and length scales.

The discussion here has been restricted to the low density limit for which the Boltzmann description is assumed to be appropriate. At higher densities an accurate description for elastic collisions is provided by the revised Enskog theory [22] whose generalization to the inelastic case is straighforward [23]. The kinetic modelling along the lines presented here has also been extended to this equation [5, 8]. Also a generalization of the direct simulation Monte Carlo simulation method to the revised Enskog theory has been proposed [24]. Both, model and Monte Carlo simulation, have been recently applied to the study of a simple granular shear flow and the results have been very promising [25].

Acknowledgements

The material presented in this work is based on published and unpublised work in collaboration with J.W. Dufty, A. Santos, M.J. Ruiz-Montero, F. Moreno, and D. Cubero to whom I am greatly indebted. This research was partially supported by Grant No. PB95-0534 from the Dirección General de Investigación Científica y Técnica (Spain).

References

1. C. Cercignani, *Theory and Application of the Boltzmann Equation* (Elsevier, New York, 1975).
2. A review is given by J. W. Dufty in *Lectures on Thermodynamics and Statistical Mechanics*, M. López de Haro and C. Varea, eds., (World Scientific, Singapore, 1990), pp. 166-181.
3. G. Bird, *Molecular Gas Dynamics and the Direct Simulation of Gas Flows* (Clarendon Press, Oxford, 1994)
4. A. Goldshtein and M. Shapiro, J. Fluid Mech. **361**, 41 (1998).
5. J. J. Brey, J. W. Dufty, and A. Santos, J. Stat. Phys. **87**, 1051 (1997).
6. N. Sela and Goldhirsch, J. Fluid Mech. **361**, 41 (1998).
7. J. J. Brey, J. W. Dufty, C. S. Kim, and A. Santos, Phys. Rev. E **58**, 4638 (1998).
8. J. J. Brey, J. W. Dufty, and A. Santos, J. Stat. Phys. **97**, 281 (1999).
9. T. P. C. van Noije and M. Ernst, Granular Matter, **1**, 57 (1998).
10. J. J. Brey, M.J. Ruiz-Montero, and D. Cubero, Phys. Rev. E, **54**, 3664 (1996).
11. T. P. C. van Noije and M. H. Ernst, Physica A, **251**, 266 (1998).
12. J. W. Dufty, M. Lee, and J. J. Brey, Phys. Rev. E **51**, 297 (1995).
13. I. Goldhirsch and G. Zanetti, Phys. Rev. Lett. **70**, 1619 (1993); I. Goldhirsch, M-L Tan, and G. Zanetti, J. Sc. Computing **8**, 1 (1993).
14. J. J. Brey, M. J. Ruiz-Montero, and F. Moreno, Phys. Fluids, **10**, 2976 (1998).
15. J. T. Jenkins and M. W. Richman, Arch. Ration. Mech. Anal. **87**, 355 (1985); Phys. Fluids **28**, 3485 (1986); J. Fluid Mech. **192**, 313 (1988).
16. J. J. Brey, F. Moreno, and M. J. Ruiz-Montero, Phys. Fluids, **10**, 2965 (1998).
17. T. P. C. van Noije, M. H. Ernst, R. Brito, and J. A. G. Orza, Phys. Rev. Lett. **79**, 411 (1997); T. P. C. van Noije, M. H. Ernst, and R. Brito, Phys. Rev. E **57**, R4891 (1998).
18. J. J. Brey, M. J. Ruiz-Montero, and F. Moreno, Phys. Rev. E **55**, 2846 (1997).
19. Y. Du, H. LI, and L. P. Kadanoff, Phys. Rev. Lett. **74**, 1268 (1995).
20. E. L. Grossman, T. Zhou, and E. Ben-Naim, Phys. Rev. E **55**, 4200 (1997).
21. J. J. Brey and D. Cubero, Phys. Rev. E, **57**, 2019 (1998).
22. H. van Beijeren and M. H. Ernst, Physica **68**, 437 (1973).
23. J. W. Dufty, J. J. Brey, and A. Santos, Physica A **240**, 212(1997).
24. J. M. Montanero and A. Santos, Phys. Rev. E **54**, 438 (1996); Phys. Fluids **9**, 2057 (1997).
25. J. M. Montanero, V. Garzó, A. Santos, and J. J. Brey, J. Fluid Mech. **389**, 391 (1999).

VELOCITY CORRELATIONS IN DRIVEN TWO-DIMENSIONAL GRANULAR MEDIA

C. BIZON, M.D. SHATTUCK, J.B. SWIFT AND
HARRY L. SWINNEY
Center for Nonlinear Dynamics and Department of Physics
University of Texas
Austin, TX 78712, USA

Abstract. Simulations of volumetrically forced granular media in two dimensions produce states with nearly homogeneous density. In these states, long-range velocity correlations with a characteristic vortex structure develop; given sufficient time, the correlations fill the entire simulated area. These velocity correlations reduce the rate and violence of collisions, so that pressure is smaller for driven inelastic particles than for undriven elastic particles in the same thermodynamic state. As the simulation box size increases, the effects of velocity correlations on the pressure are enhanced rather than reduced.

1. Introduction

In rapid flows of granular media, the mean time between collisions of grains is much longer than the duration of a collision [1]; for such flows, the machinery of kinetic theory is expected to apply. Continuum equations [2, 3] analogous to the Navier-Stokes equations can be produced, allowing quantitative analysis of flows. The simplest and most common formulations incorporate Boltzmann's assumption of molecular chaos: that particle velocities are uncorrelated.

While this assumption works well for low-density molecular gases, granular gases may not abide such a restriction because collisions between grains are inelastic. Inelastic collisions reduce relative velocities, so that post-collisional velocities are more parallel than pre-collisional velocities. Repeated inelastic collisions can lead to strong, long-range velocity correlations, which standard kinetic theory does not include. We will use molecular

J. Karkheck (ed.),
Dynamics: Models and Kinetic Methods for Non-equilibrium Many Body Systems, 361–371.

dynamics simulations to produce steady state granular gases and study the velocity correlations that develop.

The importance and intrinsic interest of velocity correlations in granular flows have been noted by a number of researchers. Two-dimensional simulations of an initially homogeneous distribution of inelastic disks without velocity correlations show that as time progresses, velocity correlations build in both strength and range [4]. These simulations are limited in time, however, because the homogeneous state is unstable to density fluctuations, and rapidly becomes inhomogeneous. Nevertheless, these simulations clearly displayed a characteristic vortex structure of the correlations. Based upon similar considerations, ring kinetic theory, which accounts for velocity correlations, has been applied to the cooling state [5]. One-dimensional simulations of stochastically forced point particles also show velocity correlations [6].

We apply stochastic forcing [6, 7] to two-dimensional event-driven simulations of inelastic disks. The forcing overcomes the tendency of the granular material to form density clusters, and approximately homogeneous steady states form. In an earlier study of these states [8], we found strong velocity correlations that extended throughout the entire simulation area. In the present work, we discuss the simulation method, show that the velocity correlations are essentially independent of the simulated area, and describe the vortex structure of the correlations.

2. Simulations of Driven Granular Gases

We treat collisions between molecules as instantaneous and binary. The collisions between grains conserve momentum but dissipate energy. Between collisions, particles travel along straight lines if unaccelerated, or along parabolas if accelerated. This model allows efficient simulation of collections of particles using event-driven molecular dynamics [9, 10].

When particles collide, the component of the relative particle velocity along the line joining particle centers, v_n, is reversed, and reduced by a factor e, the coefficient of restitution, which can take values between 1 for elastic particles and 0 for completely inelastic particles. We allow e to depend on v_n through

$$e(v_n) = \begin{cases} 1 - Bv_n^{\beta} & , v_n < v_o \\ \epsilon & , v_n > v_o \end{cases} , \tag{1}$$

where $B = (1-\epsilon)(v_o)^{-\beta}$, $\beta = 3/4$ and ϵ is a constant, chosen to be 0.7. These parameters give quantitative agreement to experiments on patterns in vertically oscillated granular media [11, 12]. The variation in e has the effect of removing inelastic collapse [13], which is a singularity in the inelastic hard-sphere model that produces an infinite number of collisions within a finite time [14, 15]. In general, colliding particles also exert frictional forces on one another; for this paper, we assume that the coefficient of friction is zero, so that we are studying only the effects of inelasticity.

Because of inelasticity, the energy of an unforced collection of grains inevitably decreases. To achieve steady states, then, we must force the granular material. Methods that force through boundaries, such as shaking, invariably produce strong inhomogeneities in the system; to achieve near-homogeneity, we force volumetrically, assuming the particles to be in contact with a white-noise heat bath [7]. Whenever two particles collide, the velocities of two other randomly selected particles are changed by amounts $|\delta \mathbf{v}|\hat{\mathbf{r}}_i$, where the magnitude of each kick, $|\delta \mathbf{v}|$, is always the same, but the direction vector, $\hat{\mathbf{r}}_i$ is randomly chosen for each kicked particle. In addition to the white noise heat bath, we perform a lesser number of runs with two other heat baths. To model the motions of pucks on an air table [16, 17], we can allow particles to accelerate randomly from collision to collision. Finally, we model the effects of a strong heat bath, which we denote the Boltzmann bath, by completely obliterating the velocities of randomly chosen particles, and giving new velocities based on a Boltzmann distribution. The details of all three forcing methods may be found in [8].

We perform simulations of N disks of diameter σ moving in a two-dimension square of side length L, which varies from 52.6σ to 420.8σ. The simulation box is periodic in both directions. The solid fraction, defined as $N\frac{\pi}{4}\frac{\sigma^2}{L^2}$, is 0.5 for all runs. Because of the variation of e with relative normal velocity, the velocity scale v_0 enters; we use v_0 to nondimensionalize velocities, and v_0^2 to nondimensionalize the granular temperature T. For T much larger than one, most particle collisions will occur with the high-velocity value of e, 0.7; for lower T, a range of e will occur.

3. Dependence of Correlations upon Simulation Area

We denote two particles 1 and 2, and $\hat{\mathbf{k}}$ the unit vector pointing from the center of 1 to the center of 2. The velocity of 1 then has components parallel to, $v_1^{||}$, and perpendicular to, $v_1^{\perp}$, $\hat{\mathbf{k}}$, as does particle 2. We define two correlation functions

$$\langle v_1^{||} v_2^{||} \rangle = \sum v_1^{||} v_2^{||} / N_r, \tag{2}$$

$$\langle v_1^{\perp} v_2^{\perp} \rangle = \sum v_1^{\perp} v_2^{\perp} / N_r, \tag{3}$$

where the sums are over the N_r particles such that the distance between the two particles is within δr of r. For uncorrelated particle velocities, $\langle v_1^{||} v_2^{||} \rangle$ and $\langle v_1^{\perp} v_2^{\perp} \rangle$ will both give zero.

In the smallest simulation area, where $L = 52.6\sigma$, correlations extend the full length of the computational cell. Cell filling structures may be divided into two cases: structures with a natural length that is larger than the box in which they exist and structures that will always grow to fill any finite box. To differentiate between the former and the latter, we performed four simulations with white noise forcing, quadrupling the area at each step, while holding the solid fraction, ν, fixed at 0.5. The granular temperature T is approximately 30, but varies between 28 in the smallest box and 32 in the largest. This variation in temperature is not important; for $T >> 1$, the coefficient of restitution is independent of collision velocity. In this limit, the role of the temperature is simply to set the velocity scale. The velocity correlation functions are shown in Fig. 1. Even in the largest simulation, composed of 112768 particles, the correlations fill the box. However, the correlation functions for the largest simulation are somewhat different from the smaller ones. This is probably due to poorer statistics; in terms of collisions per particle, this run lasted only one-half as long as the next largest.

Because velocity correlations are positive for small separations, inelastic particles collide less frequently and with less relative velocity than elastic particles at the same density, for which velocity correlations are much smaller. As a result, less momentum will be transferred through inelastic collisions than through elastic collisions, and the pressure, P, will be less.

Assuming that velocity correlations do not exist, the equation of state for dense granular gases is given by [3]

$$P = (4/\pi\sigma^2)\nu T(1 + (1 + e)G(\nu)). \tag{4}$$

The first term on the right hand side, $(4/\pi\sigma^2)\nu T$, accounts for momentum transfer due to particle streaming without collisions, while the second term, $(4/\pi\sigma^2)\nu T(1+e)G(\nu)$, accounts for the momentum transfer due to particle collisions [18]. In the absence of velocity correlations, $G(\nu)$ is defined as $\nu g(\nu, \sigma)$, where $g(\nu, \sigma)$ is the contact value of the radial distribution function for the particles. Calculation of P from simulation, via measurement of the virial [19], becomes a measurement of $G(\nu)$ which describes the collisional momentum transport. If velocity correlations exist, $G(\nu)$ will be reduced, since less momentum will be transported collisionally.

Figure 1 shows that the short range velocity correlations depend on the size of the box; therefore, $G(\nu)$ should also depend on L. Figure 2 displays $G(\nu)$ as a function of L for these four runs. Over about one decade, $G(\nu)$ scales with $\log L$. Clearly this scaling can not continue indefinitely, since unphysical negative values of $G(\nu)$ would result. Note also, that increasing the box size actually leads to values of $G(\nu)$ farther from the values for uncorrelated velocities.

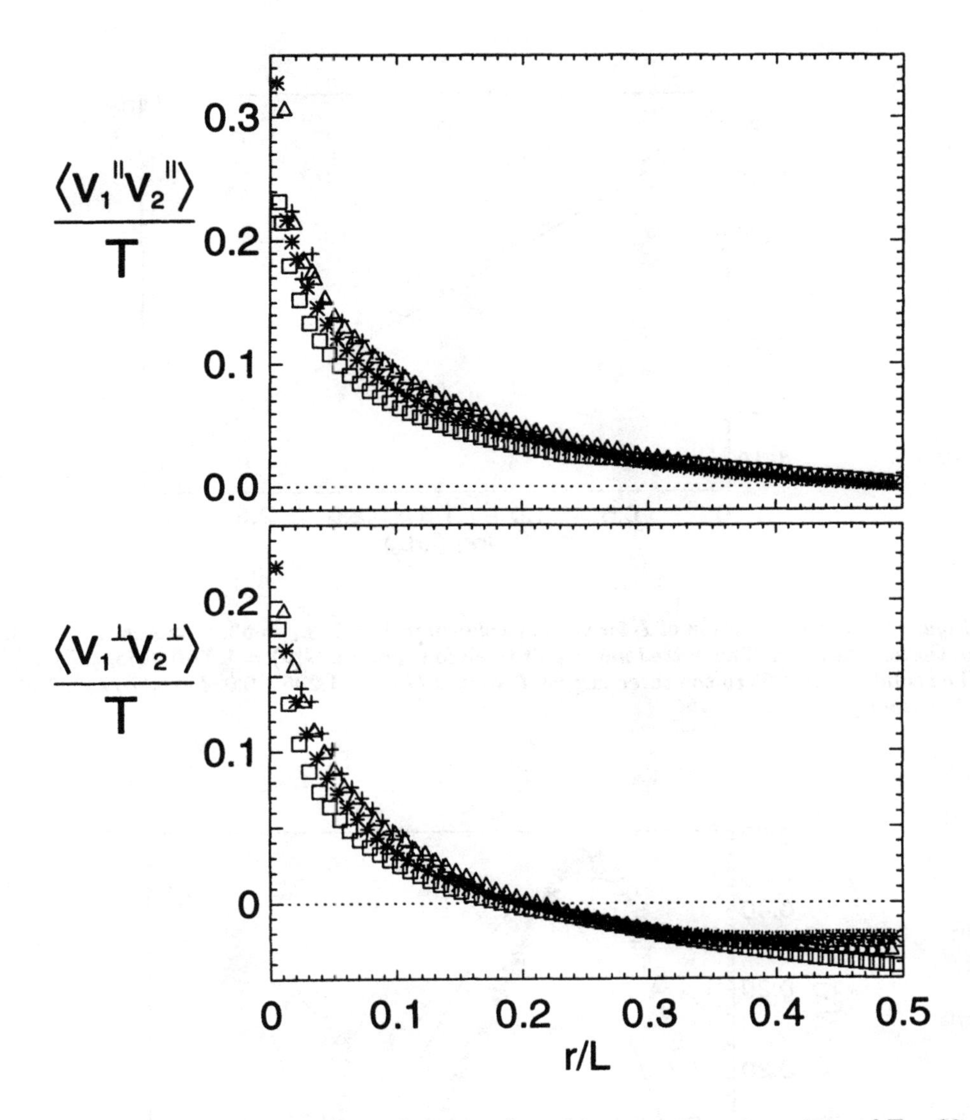

Figure 1. Velocity correlations as a function of particle separation at $\nu = 0.5$ and $T \approx 30$, for four different box sizes. $+$: $L = 52\sigma$, $\triangle$: $L = 105\sigma$, $*$: $L = 211\sigma$, $\square$: $L = 421\sigma$.

This unusual result, that the importance of velocity correlations increases with increasing computational area, can also be deduced from the distribution of collision velocities. Figure 3 exhibits these distributions for the runs displayed in Figures 1 and 2. As the computational area increases, so too does the deviation from the distribution predicted for particles chosen without correlation from a Boltzmann distribution, plotted as the solid curve in Figure 3.

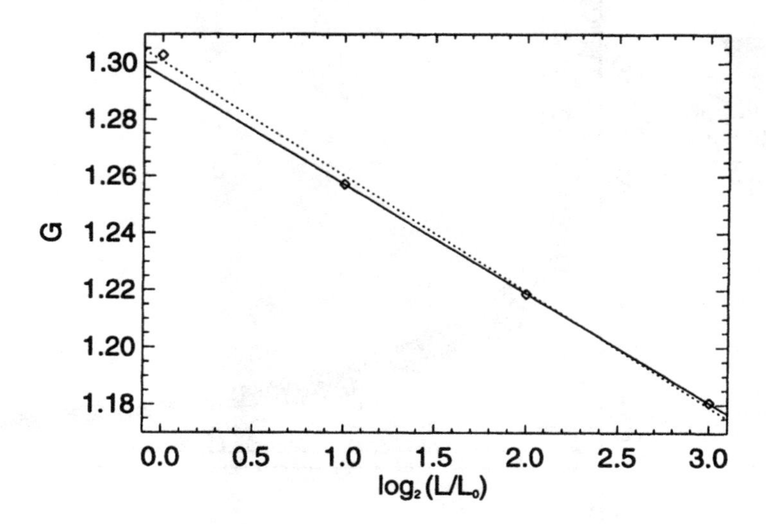

Figure 2. G as a function of L for the runs shown in Fig. 1. $L_o = 52\sigma$ denotes the length of the smallest box. The dotted line is a fit to all four points: $G(\nu) = 1.3 - 0.04\log_2(L/L_o)$, The solid line is a fit to the three largest L values: $G(\nu) = 1.295 - 0.038\log_2(L/L_o)$. Note that the log is base 2.

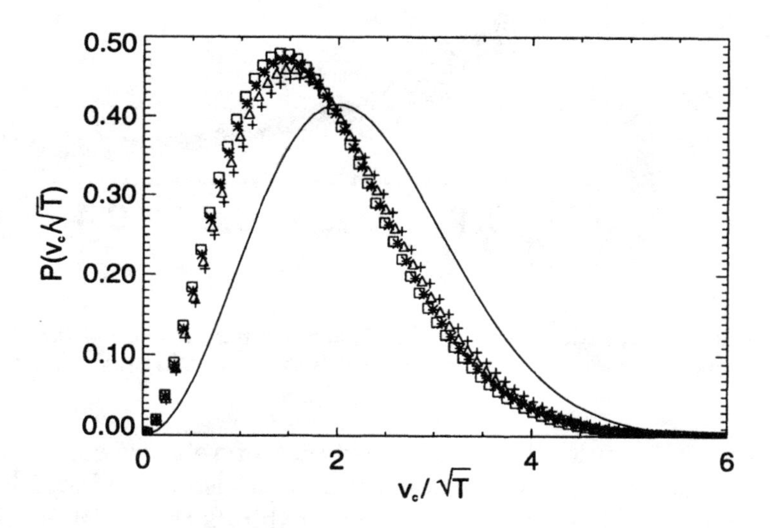

Figure 3. Probability distribution of collision velocities $v_c = |\mathbf{v}_1 - \mathbf{v}_2|$, for the data in Figs. 1 and 2. $+ : L = 52\sigma$, $\Delta : L = 105\sigma$, $* : L = 211\sigma$, $\Box : L = 421\sigma$. The solid curve is $P(v_c/\sqrt{T}) = (1/2\sqrt{\pi T^3})v_c^2 e^{-v_c^2/4T}$, which holds for elastic particles.

4. Vortex Structure

Inelasticity breeds velocity correlations; reduction of relative velocity in collisions leads to particles moving more alike after collisions than before. On average, then, particles will be surrounded by particles that are moving along with them. The structure of the velocity correlations can be elucidated by calculating this average flow around each particle.

For a single particle i, we can calculate the flow around it by translating it to the origin, and rotating so that its velocity lies along the positive x axis. If $\mathbf{v}(x, y)$ is the velocity field defined by the particles, then the flow around particle i is given by

$$\mathbf{u}_i = R_{\theta(i)}\mathbf{v}(x - x_i, y - y_i), \tag{5}$$

where $\theta(i)$ is the angle between the i-th particle velocity, $\mathbf{v}_i$, and the positive x axis, (x_i, y_i) is the position of the i-th particle, and R_θ is the operator that rotates vectors clockwise through angle θ. The average flow around particles, then, is first averaged over particles,

$$\mathbf{u} = \sum_{i=1}^{N} \mathbf{u}_i / N, \tag{6}$$

and then averaged over time. Finally, $\mathbf{u}$ is averaged over about 100 frames to reduce noise.

Figure 4 displays vector fields of the average flow around particles, $\mathbf{u}$, for the three types of forcing, as well as for unforced elastic particles, all at $\nu = 0.5$ and $T = 1.05$. In each case, the vector at the origin, which measures only the average particle speed, has been suppressed, and the longest remaining vector in each field has been scaled to unit length. In both the white noise and accelerated forcings, the average flow near the origin is along the positive x axis, *i.e.*, with the direction of the central particle's motion. The Boltzmann bath shows some indications of this effect close to the origin, but the correlations are destroyed by the strongly thermalizing forcing before they can propagate to larger length scale. For the elastic particles, there is no discernible flow, only noise.

Close to any particle, surrounding particles move along with it. Farther away, the correlations decay and cannot be seen on Fig 4, so the boxed regions for the white noise forcing and for elastic particles are expanded in Fig. 5. While expansion of the velocity field for elastic particles produces still more noise, the inelastic flow field reveals a highly ordered vortex structure. Along the direction of the central particle's motion, the velocities slowly drop to zero, while perpendicular to the original particle's motion, the velocities drop to zero and increase in the negative direction;

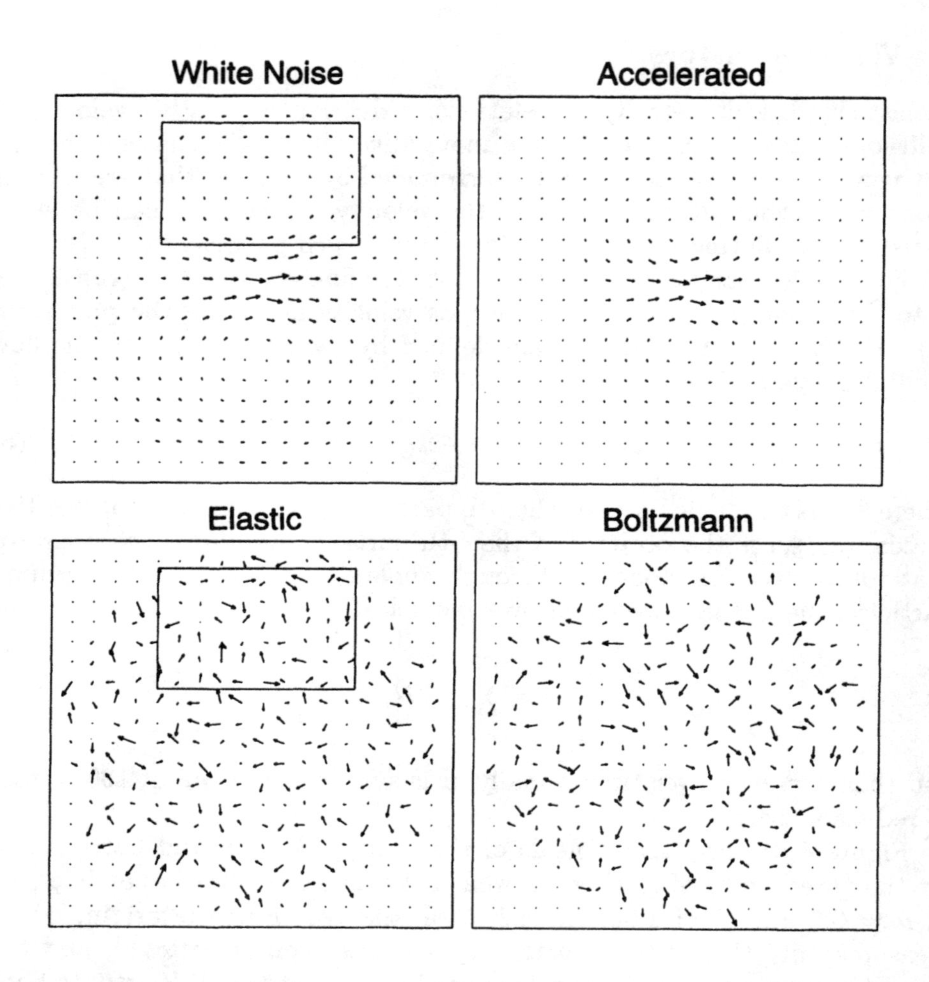

Figure 4. The average velocity fields around a particle centered in each cell and moving to the right, **u**, for elastic particles and for inelastic particles forced in three different ways (cf section 2). Each vector field is scaled separately so that its longest vector has length one. Compared to the (suppressed) central vector, these lengths are: White noise, 0.2; Accelerated, 0.27; Boltzmann, 0.008; Elastic 0.008. The boxed regions in the white noise and elastic flows are shown in Fig. 5.

this flow makes clear the structure of the velocity correlation functions in Fig. 1.

This vortical flow is reminiscent of similar structures produced in simulations of elastic particles [20, 21] by Alder and Wainwright. In their simulations, they discovered diffusive behavior different from that predicted by kinetic theory. The diffusion constant may be written in terms of the slope of the exponentially decaying autocorrelation function. However, Alder and Wainwright found deviations from exponential decay, and traced the devi-

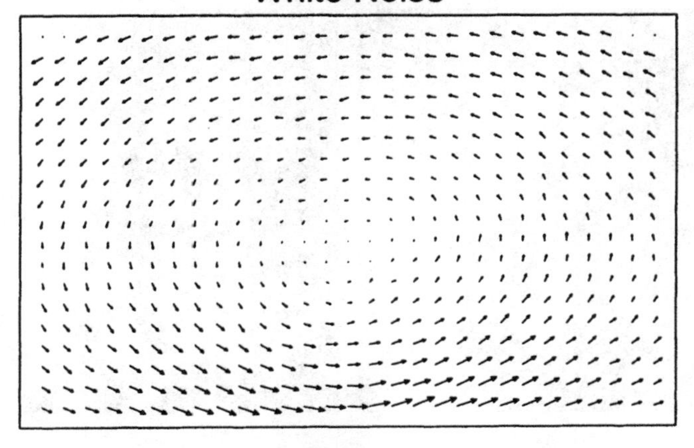

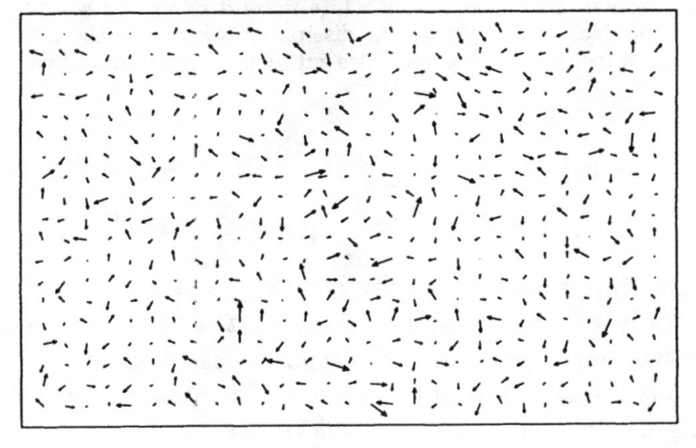

Figure 5. A close-up on the boxed regions in Fig. 4 with the velocities rescaled reveals that for inelastic particles, large vortices form, one on each side of the particle. The longest vector in the velocity field for inelastic particles represents a velocity nine times larger than that represented by the longest vector for elastic particles.

ations to a vortical flow. If particles a and b are initially uncorrelated, an elastic collision will correlate each particle's post collision velocity with the other particle's pre-collision velocity; both particles now have a correlation with the original velocity of particle a. As particle b collides with other particles, they gain information about particle a's initial velocity. Several collision times later, this information has been transmitted to many particles.

There are two main differences between the vortices in flows of elastic particles and those in flows of inelastic particles. Alder and Wainwright

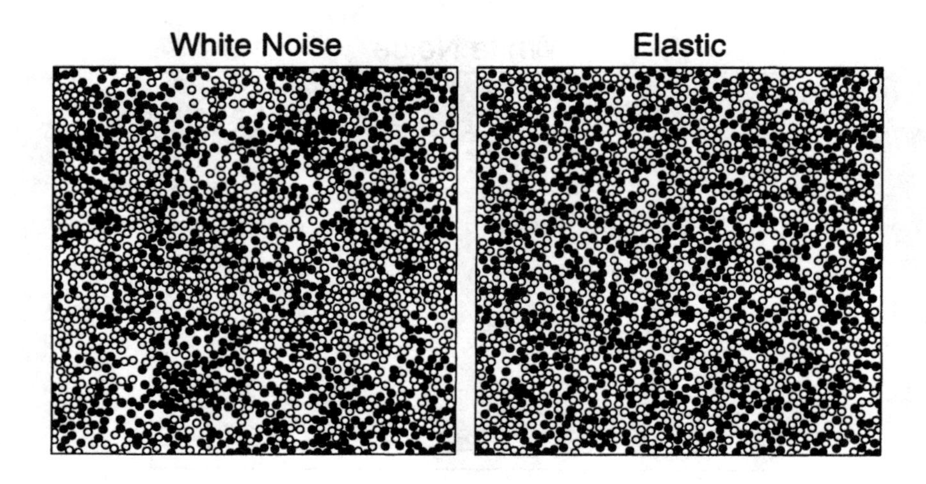

Figure 6. Snapshots of simulations with white noise driving and with elastic particles; large coherent structures are visible for the dissipative system on the left. Particles with positive horizontal velocity are black, particles with negative horizontal velocity are white. ($\nu = 0.5, T = 1.05$)

produced the flow field given by

$$\mathbf{u}(t')_i = R_{\theta(i,t)}\mathbf{v}(x - x_i(t), y - y_i(t), t'). \tag{7}$$

For $t' = t$, Eq. (5) is recovered; for elastic particles, no structure is apparent. It is only at later times, $t' > t$, that a vortex appears in $\mathbf{u}(t')$. For the inelastic particles, however, structure is clear at $t' = t$. The second difference is the strength of the vortex. The strongest velocity in Alder and Wainwright's vortex was about 2% of the original velocity, while for inelastic particles, the strongest velocity can be about 40% of the central velocity.

The inelastic vortex is so strong that hints of it are visible even in a single snapshot of particles. Figure 6 shows such a snapshot, with particles colored black if they have positive horizontal velocity and white if they have a negative horizontal velocity. For elastic particles, the black and white are well mixed, but in the inelastic case larger scale structure can be glimpsed. Black particles are concentrated along the top and bottom of the image, and white particles are concentrated along the central region.

5. Conclusion

The correlations we have found are consistent with those of simulations on the homogeneous cooling state [4]. In those simulations, the range of velocity correlations grew until the onset of large scale density variations.

The addition of forcing in our simulations suppresses the growth of density fluctuations, allowing the velocity correlations to continue to grow until they extend throughout the entire computational area.

The results we obtain are not particularly sensitive to the exact form of the forcing. In both the white noise and accelerated forcing schemes, vortical correlation structures form. Only when the bath explicitly destroys correlations, as in the Boltzmann bath, do the results differ.

Acknowledgements

We thank J. T. Jenkins, M. H. Ernst, and T. P. C. van Noije for useful discussions. This work was supported by the Engineering Research Program of the Office of Basic Energy Sciences of the U.S. Department of Energy.

References

1. C. S. Campbell, Ann. Rev. Fluid Mech. **2**, 57 (1990).
2. C. K. K. Lun, S. B. Savage, D. J. Jeffrey, and N. Chepurniy, J. Fluid Mech. **140**, 223 (1983).
3. J. T. Jenkins and M. W. Richman, Arch. Rat. Mech. Anal. **87**, 355 (1985).
4. J. A. G. Orza, R. Brito, T. P. C. van Noije, and M. H. Ernst, Int. J. Mod. Phys. C **8**, 953 (1998).
5. T. P. C. van Noije, M. H. Ernst, and R. Brito, Phys. Rev. E **57**, R4891 (1998).
6. M. R. Swift, M. Boamfă, S. J. Cornell, and A. Maritan, Phys. Rev. Lett. **80**, 4410 (1998).
7. D. R. M. Williams and F. C. MacKintosh, Phys. Rev. E **54**, R9 (1996).
8. C. Bizon, M. D. Shattuck, J. B. Swift, and H. L. Swinney, Phys. Rev. E, **60**, 4340 (1999).
9. B. D. Lubachevsky, J. Comp. Phys. **94**, 255 (1991).
10. M. Marín, D. Risso, and P. Cordero, J. Comput. Phys. **109**, 306 (1993).
11. C. Bizon, M. D. Shattuck, J. B. Swift, W. D. McCormick, and H. L. Swinney, Phys. Rev. Lett. **80**, 57 (1998).
12. J. R. de Bruyn, C. Bizon, M. D. Shattuck, D. Goldman, J. B. Swift, and H. L. Swinney, Phys. Rev. Lett. **81**, 1421 (1998).
13. D. Goldman, M. D. Shattuck, C. Bizon, W. D. McCormick, J. B. Swift, and H. L. Swinney, Phys. Rev. E **57**, 4831 (1998).
14. S. McNamara and W. R. Young, Phys. Fluids A **4**, 496 (1992).
15. S. McNamara and W. R. Young, Phys. Rev. E **50**, R28 (1994).
16. L. Oger, C. Annic, D. Bideau, R. Dai, and S. B. Savage, J. Stat. Phys. **82**, 1047 (1996).
17. I. Ippolito, C. Annic, J. Lemaître, L. Oger, and D. Bideau, Phys. Rev. E **52**, 2072 (1995).
18. S. Chapman and T. G. Cowling, *The Mathematical Theory of Non-uniform Gases* (Cambridge University Press, London, 1970).
19. D. C. Rapaport, *The Art of Molecular Dynamics Simulation* (Cambridge University Press, Cambridge, 1980).
20. B. J. Alder and T. E. Wainwright, J. Phys. Soc. Japan **26**, 267 (1968).
21. B. J. Alder and T. E. Wainwright, Phys. Rev. A **1**, 18 (1970).

THIRD
INTERLUDE

G. E. UHLENBECK ON PAUL EHRENFEST

E. G. D. COHEN
Center for the Studies in Physics and Biology
The Rockefeller University
1230 York Avenue, New York, NY 10021, USA

In 1875 Professor H. A. Lorentz was appointed at the University of Leiden as Professor of Theoretical Physics, I believe the first such position in Europe. When he chose to retire in 1911, he approached Ehrenfest as his successor, after having been unsuccessful to get Einstein to succeed him in Leiden. One could wonder why a world class physicist - by many considered the leading theoretical physicist of the day - would choose someone relatively unknown as Ehrenfest to succeed him. In his beautiful book on Ehrenfest [1], Martin Klein writes that "Lorentz had decided that Ehrenfest was the one he wanted, the one 'to dot the i's in physics'"[2]. This was, no doubt, largely based on the very critical and yet pedagogical article Ehrenfest had published, together with his wife, Tatiana Ehrenfest, in the Enzyclopädie der Mathematischen Wissenschaften, where they critically analyzed and summarized Boltzmann's ideas on kinetic theory and statistical mechanics [3].

Although I think this consideration of Lorentz is undoubtedly true, I wonder whether not another motive played a role as well. In the same book, Martin Klein writes [2] that Lorentz "always had 'the remarkable gift of maintaining a certain distance between his students and himself, without appearing to do so,'" and although Lorentz had had some brilliant Ph.D. students, like Miss van Leeuwen, L. S. Ornstein and A. D. Fokker, he never really created a school. I wonder, therefore, whether with his fine wisdom, he had not also spotted in Ehrenfest someone who would succeed there, where he might have felt that he failed himself. The future certainly bore him out: Ehrenfest did create a Dutch School of Physics, which was on the one hand sometimes called "the conscience of physics" and produced on the other hand scientists like J. Tinbergen, H. B. G. Casimir, J. M. Burgers, S. A. Goudsmit, G. E. Uhlenbeck and many others.

On the occasion of receiving the Oersted Medal for notable contributions to the teaching of physics in 1956, G. E. Uhlenbeck wrote a short paper,

J. Karkheck (ed.),
Dynamics: Models and Kinetic Methods for Non-equilibrium Many Body Systems, 375–378.

"Reminiscences of Professor Paul Ehrenfest", for the American Journal of Physics [4]. In the following I quote some passages of this paper, which illustrate the unique personality of Ehrenfest as a physics teacher and has not only historical but, in my opinion, also contemporary value. Here follow then some excerpts of Uhlenbeck's paper.

Let me begin with the art of giving a lecture. I can still hear Ehrenfest exclaim in his typical mixture of German and Dutch when one of us students gave a talk: "Please start writing on the upper left-hand corner of the blackboard; please do not erase before people have a chance to see what you wrote; please do not talk with your face towards the blackboard, etc." All perhaps rather trivial points, but one has only to go to a meeting of the American Physical Society to see how often people sin against these simple rules. But, of course, they touch only the surface of the technique of lecturing. With regard to the real problem of organizing the subject of your talk, we learned most by the example of Ehrenfest's lectures. It is difficult to say what made them so excellent. Partly, no doubt, it was their clarity. But that is not all; Lorentz for instance was also a wonderfully clear lecturer, but his lectures were often so "smooth" that it was difficult to recognize the real point of the argument, with the result that often at the end of the lecture one had forgotten what it was all about. This was never the case with Ehrenfest. He always told, and insisted that you told "der springende Punkt"[5] of the argument. "Was ist der Witz,"[6] he always used to ask, or he said: "Why do you say that now; is it the joke [7], or is it only because it happens to be true?" As a result, I still remember the main points of the many dificulties especially in statistical mechanics with which one struggled in those days, and often still struggles today. Ehrenfest's famous clarity must not be confused with rigor. In fact, only rarely would he present a precise formal proof. But he succeeded always to give an over-all view of the subject, to make clear what had been achieved and what remained dark. He used to say: "first the assertion and then the proof!" And he usually then only sketched the proof or made it plausible so that one understood it "with one's fingers". He was very clever in inventing simple models which showed the essential features of the argument (compare: the "wind-tree" model to explain the derivation of the H-theorem [8], the "dog-flea" model to explain the nature of irreversible processes [9]).

He used to say that one has not understood a derivation if one only could derive logically the result from the assumptions. Then one could "only dance on one leg". One should see all the connections, so that one's understanding becomes similar to a net.

In the second place let me say that the emphasis was always on the physical ideas and the logical structure of the theory. And I must say, although perhaps we did not learn how to compute, we certainly learned what the real problems were. It is difficult to say how this was accomplished. One reason was just the absence of technicalities. Only the fundamentals were carefully developed and drilled into one's mind; the rest of the time Ehrenfest would give wonderfully short bird's-eye views of various topics with a few characteristic results and with references, to whet the appetite of the student. In my opinion it is the best way to treat a subject, much better than the rigorous, complete, and systematic way which is in vogue in our universities. Even in undergraduate physics courses, the little I had to do with them, I always had most success with a combination of drill in the fundamentals and inspirational talks about subjects the students did not have to know.

In conclusion I would like to say a few words about what Ehrenfest did about perhaps the basic educational problem, namely the problem how to make a student do independent research. He worked essentially always only with one student, and that practically every afternoon during the week. He discussed with him either the problem on which he was working or recent papers in the literature which he wanted to understand in detail. It went fast, and one worked on the blackboards. When they were full, the main points were copied in little note books. I can personally testify that in the beginning, since one understood things so to say only with the tips of one's fingers, at the end of the afternoon one was dead tired. Especially because one had to follow in detail; the greatest sin was to say that one had understood the point if it was not the case. And it was always found out! The wonder was that after a while the tiredness disappeared and after a year one worked almost as equals. In fact, as a student you often had the sneaking suspicion that you really knew the things much better. At that point one stood on one's own legs and one had become a physicist!

I think that this method of Ehrenfest recognizes that one of the main requirements for the scientific investigator is confidence in himself, or if you want, courage. And Ehrenfest's method is the only one I know for the student to acquire this quality. It is essential that as a student one feels at least to be equal to one's teacher! Ehrenfest used to say: "Weshalb habe ich solche gute Studenten? Weil ich so dumm bin! [10]".

References

1. M. J. Klein, *Paul Ehrenfest*, Vol. 1, (North-Holland, Amsterdam, 1970). See also: M. J. Klein, "Physics in the Making in Leiden: Paul Ehrenfest as Teacher", *Physics in the Making*, A. Sarlemijn and M. J. Sparnaay, eds., (Elsevier, 1989), pp. 29-44.
2. Ref. 1, p. 189
3. (a) P. T. Ehrenfest, "Begriffliche Grundlagen der statistischen Auffassung in der Mechanik", Enz. d. Math. Wissensch. IV 32, pp. 3-90 Teubner (1912); (b) *Collected Works*, pp. 213-302; (c) English translation: M. J. Moravsik, *The Conceptual Foundations of the Statistical Approach in Mechanics*, (Cornell University Press, Ithaca,

NY 1959).
4. G. E. Uhlenbeck, "Reminiscences of Professor Paul Ehrenfest", Am. J. Phys. **24**, 431-433 (1956).
5. "the crucial step"
6. "What is the point"
7. "the joke" is a literal translation into English of the Dutch equivalent of the German "der Witz", used by Ehrenfest, which is best rendered in English here by "the point".
8. See ref. 3(a): pp. 19-20; 3(b): pp. 229-230; 3(c): pp. 11-13. For a modern account, see E. H. Hauge, "What can one learn from Lorentz models?" in *Transport Phenomena*, Sitges International School of Statistical Mechanics, G. Kirczenow and J. Marro, eds., (Springer, Berlin, 1974).
9. Also called the Ehrenfest urn model. See: P. and T. Ehrenfest, "Über zwei bekannte Einwände gegen das Boltzmannsche H-theorem", Phys. Zeit. **8**, 311-314 (1907). See also: M. Kac, "Random Walk and the Theory of Brownian Motion" in *Selected Papers on Noise and Stochastic Processes*, N. Wax, ed., (Dover, New York, 1954), pp. 311-317.
10. "Why do I have such good students? Because I am so dumb!".

QUANTUM KINETIC THEORY: THE DISORDERED ELECTRON PROBLEM

T.R.KIRKPATRICK
Institute for Physical Science and Technology and Department of Physics, University of Maryland College Park, MD 20742, USA

AND

D.BELITZ
Department of Physics and Materials Science Institute University of Oregon Eugene, OR 97403, USA

Abstract. These are notes for lectures delivered at the NATO ASI on Dynamics in Leiden, The Netherlands, in July 1998. The quantum kinetic theory for noninteracting electrons in a disordered solid is introduced and discussed. We first use many-body theory to derive the quantum Boltzmann equation that describes transport and time-correlation function in this system. Particular attention is paid to the calculation of the electrical conductivity σ, and the density response function χ_{nn}. We then consider corrections to the Boltzmann equation due to wave interference effects. The disorder expansion of the conductivity is addressed, and the so-called weak localization or long-time tail contribution to σ is discussed. We conclude with a brief discussion of the influence of electron-electron interactions on the properties of disordered electronic systems.

1. Introduction

Quantum kinetic theory has a long and interesting history. Shortly after the discovery of the Pauli exclusion principle, Sommerfeld applied it to electrons in metals and thereby resolved the most flagrant discrepancies between the observed thermal behavior of solids and the predictions of the classical Drude model of transport [1]. After the introduction of quantum mechanics,

J. Karkheck (ed.),
Dynamics: Models and Kinetic Methods for Non-equilibrium Many Body Systems, 379–398.

the new theory was incorporated into the standard Boltzmann description of transport. Like the classical Boltzmann equation, this quantum kinetic theory, also known as the Uhlenbeck-Uehling equation [2], is valid only for systems that are dilute in a sense to be explained below. Unlike the classical Boltzmann equation, it properly takes into account the statistics of the particles (fermionic or bosonic) as well as the quantum mechanical nature of the scattering process. However, it misses more subtle quantum mechanical effects. To get a feeling for what is missing, let us consider some length scales in the problem.

In a classical gas there are two length scales that occur in transport problems: The mean freee path ℓ, and the linear size of the particles or the impurities, which we denote by a. The Boltzmann transport theory is valid if the system is dilute in the sense that $a/\ell \ll 1$. In the quantum description of transport there is an additional length scale, namely the de Broglie wavelength λ of the scattered particle. Here we are interested in electron transport phenomena at temperatures that are low compared to the Fermi temperature, so the relevant length is $\lambda_{\rm F}$, the de Broglie wavelength at the Fermi surface. Due to approximations made in its derivation, the quantum Boltzmann equation or Uhlenbeck-Uehling equation is only valid if both $a/\ell \ll 1$ and $\lambda_{\rm F}/\ell \ll 1$. The latter expansion parameter indicates the relative importance of quantum mechanical wave interference effects.

In the first part of these lectures we introduce the standard Edwards model [3, 4] for noninteracting electrons in a disordered solid, and the two response functions that are most important for our purposes, viz. the electrical conductivity σ, and the density susceptibility $\chi_{nn}(\mathbf{k},\omega)$. We then use simple kinetic theory concepts in conjunction with the particle number conservation law to show that the behavior of the latter at small frequencies ω and small wavevectors $\mathbf{k}$ is dominated by the hydrodynamic density diffusion mode. Simple arguments lead to an estimate for the value of the diffusion coefficient D in terms of the collision mean-free time τ. The σ and D are related by an Einstein relation,

$$\sigma = e^2 \frac{\partial n}{\partial \mu} D, \tag{1}$$

with e the electron change, n the electron number density, and μ the chemical potential. For noninteracting electrons at zero temperature, $\partial n/\partial \mu = N_{\rm F}$, with $N_{\rm F}$ the single particle density of states at the Fermi surface. Following this, we discuss how to derive the quantum Boltzmann equation and how to compute σ from it.

As mentioned above, the quantum Boltzmann equation is valid only in the 'semi-classical' limit where $\lambda_{\rm F}/\ell \to 0$. In the remaining parts of these lectures we discuss some of the interesting phenomena associated with wave interference that appear at higher order in an expansion in powers of $\sim \lambda_F/\ell$. In particular, the nonanalytic nature of the disorder expansion of σ is discussed as well as weak-localization, or long-time tail, contributions to σ. We conclude with a few remarks on the combined effects of disorder and electron-electron interactions.

2. The Model, and Basic Concepts

2.1. THE MODEL

We will be concerned with the electronic transport properties of disordered solids. In order to keep things simple, we will mainly consider a model that ignores the electron-electron interaction. While this approximation has yielded many important insights, one should keep in mind that in general it is not justified. We will come back to this point at the end of the lectures. For now we consider a noninteracting electron gas model consisting of spin 1/2 fermions moving in a static random potential $u(\mathbf{x})$. Physically, $u(\mathbf{x})$ is produced by the impurities in the solid, and is therefore fixed for a given system, but we will assume that it is meaningful to consider an ensemble of systems and to compute averages of obervables over the random potential.[1] In the standard Edwards model [3, 4], $u(\mathbf{x})$ is taken to have zero mean and to be Gaussian distributed with the variance given by

$$\{u(\mathbf{x})u(\mathbf{y})\}_{\text{dis}} = \frac{\delta(\mathbf{x}-\mathbf{y})}{2\pi N_F \tau}. \tag{2}$$

Here $N_F = (m/2\pi)\,(k_F/\pi)^{d-2}$ is the free electron density of states per spin at the Fermi surface (in $d = 2, 3$ dimensions), τ is the elastic mean-free time between collisions, $\{\ldots\}_{\text{dis}}$ denotes the disorder average, k_F is the Fermi wavenumber, and m is the electron mass. Throughout the lectures we use units so that $\hbar = k_B = e = 1$, with $\hbar$ Planck's constant, k_B Boltzmann's constant, and e minus the electron charge.

To motivate this model we consider a quantum Lorentz gas, i.e. a system of moving electrons that interact with N stationary hard scatterers of radius a, but not with one another. The disorder average consists of averaging over the positions of the N scatterers in the system volume V,[2]

$$\{(\ldots)\}_{\text{dis}} = \frac{1}{V^N}\int d\mathbf{R}_1 \ldots d\mathbf{R}_N\ (\ldots), \tag{3}$$

[1] To what extent this assumption is justified is a very tricky question that we cannot get into. In principle one should calculate a whole distribution for every observable. Calculating the average is sufficient if this distribution is sharply peaked about its average.

[2] Here we neglect correlations between the scatterers, which are not important for our purposes.

and the electron-impurity potential can be written

$$v(\mathbf{x}) = \sum_{i=1}^{N} v(\mathbf{x} - \mathbf{R}_i), \tag{4}$$

with $\{\mathbf{R}_i\}$ the positions of the N scatterers or impurities. Assuming that $\lambda(\approx \lambda_F) \gg a$, with a the s-wave scattering length, we can use a pseudo-potential [1] for a single electron-impurity interaction. In $d = 3$,

$$v(\mathbf{x} - \mathbf{R}_i) = \frac{4\pi a}{m}\, \delta(\mathbf{x} - \mathbf{R}_i). \tag{5}$$

The relevant scattering potential is given by $u(\mathbf{x}) = v(\mathbf{x}) - \{v(\mathbf{x})\}_{\text{dis}}$. Since the mean, $\{v(\mathbf{x})\}_{\text{dis}}$, simply redefines the Fermi energy, we can ignore it and have

$$\{u(\mathbf{x})u(\mathbf{y})\}_{\text{dis}} = \left(\frac{4\pi}{m}\right)^2 a^2\, n_i\, \delta(\mathbf{x} - \mathbf{y}). \tag{2'}$$

Here $n_i = N/V$ is the impurity density. Using $\tau \sim (n_i\, a^2)^{-1}$, in $d = 3$, we see that Eq. (2') has the same structure as Eq. (2). The reason for the other factors in Eq. (2) will become clear below.

Finally, we write down the Hamiltonian for the Edwards model in second quantization [4],

$$\hat{H} = \int d\mathbf{x}\, \hat{\psi}^\dagger_\sigma(\mathbf{x}) \left(-\frac{\nabla^2}{2m}\right) \hat{\psi}_\sigma(\mathbf{x}) + \int d\mathbf{x}\, u(\mathbf{x})\, \hat{\psi}^\dagger_\sigma(\mathbf{x})\, \hat{\psi}_\sigma(\mathbf{x}). \tag{6}$$

Here $\hat{\psi}^\dagger_\sigma$ and $\hat{\psi}_\sigma$ are fermion creation and annihilation operators with σ denoting a spin label, and summation over repeated Greek labels is understood.

2.2. LINEAR DENSITY RESPONSE

A standard technique in nonequilibrium statistical mechanics is to apply an external field to equilibrium systems, and to calculate the response of the system to this field [5]. By expanding in powers of the external field, nonequilibrium quantities can be related to equilibrium time correlation functions. For example, let us apply an electric field $\mathbf{E}$ to an electron gas and expand to linear order in the field. Using Ohm's law, $\mathbf{J} = \sigma \cdot \mathbf{E}$, which relates the induced current $\mathbf{J}$ to the field, we obtain the so-called Kubo formula for the frequency dependent conductivity [6],

$$\sigma(\omega) = -\frac{n_e\, e^2}{m\, i\omega} + \frac{1}{\omega V} \int_0^\infty dt\, e^{i\omega t} \left\{ \left\langle 0 \middle| \left[\hat{J}_x(t), \hat{J}_x(t=0) \right] \middle| 0 \right\rangle \right\}_{\text{dis}}. \tag{7}$$

Here n_e is the electron density, $\hat{J}_x(t)$ is the total current operator at time t in the x-direction, $[\hat{a}, \hat{b}]$ denotes the commutator of operators $\hat{a}$ and $\hat{b}$, and $\langle 0 | \ldots | 0 \rangle$ denotes a ground state quantum mechanical expectation value.

Similarly, if we apply a potential $\varphi(\mathbf{x}, t)$ that couples to the density, then the change in the electronic density, δn, is given by

$$\delta n(\mathbf{k}, \omega) = \chi_{nn}(k, \omega)\, \varphi(\mathbf{k}, \omega), \tag{8}$$

with χ_{nn} the density response function and $k = |\mathbf{k}|$. The $\chi_{nn}(k, \omega)$ depends on k rather than on $\mathbf{k}$ because of rotational invariance once the disorder average has been performed. It is the Fourier transform of the density-density correlation function

$$\chi_{nn}(\mathbf{x}, t) = \Theta(t) \left\{ \left\langle 0 \middle| [\hat{n}(\mathbf{x} = 0, t = 0), \hat{n}(\mathbf{x}, t)] \middle| 0 \right\rangle \right\}_{\text{dis}}. \tag{9}$$

Here $\hat{n}$ is the density operator.

Equations (7) and (9) are valid for arbitrary systems. For noninteracting systems, some simplifications apply. In particular, the remaining many-body aspect of the problem (i.e., the correlations due to the Pauli principle) can be factored out, and the problem can be reduced to one of a *single* particle moving in a random potential. Technically, this is possible because the field operators ($\hat{\psi}^\dagger$ and $\hat{\psi}$) can be expanded in an exact eigenstate basis in terms of creation and annihilation operators $\hat{a}^\dagger$ and $\hat{a}$. The net result is that the quantum mechanical averages in Eqs. (7), (9) factorize. For example, the real part of the conductivity can be written

$$\begin{aligned} \sigma'(\omega) \equiv \mathrm{Re}\sigma(\omega) \quad = \quad & \frac{e^2}{2\pi V m^2} \mathrm{Re} \int d\mathbf{p}\, d\mathbf{p}_1\, p_x\, p_{1x} \Big[\Big\{ G^R(\mathbf{p}, \mathbf{p}_1, E_\mathrm{F} + \omega) \\ & \qquad \times G^A(\mathbf{p}_1, \mathbf{p}, E_\mathrm{F}) \Big\}_{\text{dis}} \\ & - \Big\{ G^A(\mathbf{p}, \mathbf{p}_1, E_\mathrm{F} + \omega)\, G^A(\mathbf{p}_1, \mathbf{p}, E_\mathrm{F}) \Big\}_{\text{dis}} \Big], \end{aligned} \tag{10}$$

with

$$G^{R,A}(\mathbf{p}_1, \mathbf{p}, E) = \left\langle \mathbf{p} \middle| \frac{1}{E - H \pm i0} \middle| \mathbf{p}_1 \right\rangle, \tag{11}$$

denoting retarded (R) and advanced (A) Green functions. $|\mathbf{p}\rangle$ denotes a single-particle plane wave state, and $\hat{H}$ is the single-particle Hamiltonian,

$$\hat{H} = -\frac{\nabla^2}{2m} + u(\mathbf{x}). \tag{12}$$

Similarly, the density response function can be written

$$\begin{aligned} \chi_{nn}(k, \omega) = N_\mathrm{F} \quad + \quad & \frac{i\omega}{2\pi} \int d\mathbf{p}\, d\mathbf{p}_1 \Big\{ G^R(\mathbf{p} + \mathbf{k}, \mathbf{p}_1 + \mathbf{k}, E_\mathrm{F} + \omega) \\ & \qquad \times G^A(\mathbf{p}_1, \mathbf{p}, E_\mathrm{F}) \Big\}_{\text{dis}}. \end{aligned} \tag{13}$$

2.3. SOME ADDITIONAL REMARKS

In the long-wavelength limit, the form of $\chi_{nn}(k,\omega)$ is determined by the diffusive nature of the electron dynamics in a disordered solid. This can be seens as follows. The conservation law for the particle number density is expressed by the continuity equation

$$\partial_t\, \delta n(\mathbf{x},t) + \nabla \cdot \mathbf{j}(\mathbf{x},t) = 0, \tag{14}$$

with $\mathbf{j}$ the number current density. In addition, Fick's law, valid in the long-wavelength limit, gives a relation between $\mathbf{j}$, δn, and φ,

$$\mathbf{j}(\mathbf{x},t) = -D\,\nabla\delta n + \mu_{\mathrm{e}}\nabla\varphi, \tag{15}$$

with D the diffusion coefficient and $\mu_{\mathrm{e}} = D\partial n/\partial\mu = DN_{\mathrm{F}}$ the electron mobility (not to be confused with the chemical potential μ). Using Eq. (15) in Eq. (14), performing a Fourier-Laplace transform, and comparing with Eq. (8) gives

$$\chi_{nn}(k,\omega) = \frac{N_{\mathrm{F}}Dk^2}{-i\omega + Dk^2}. \tag{16}$$

The above arguments tell us the functional form of χ_{nn} in the long-wavelength small-frequency limit, but they do not tell us the value of the diffusivity D. Rather, D enters as a purely phenomenological quantity. However, we can get a rough idea of the value of D, and of the closely related conductivity, by means of simple dimensional analysis. The units of D are L^2/T, with L a length scale and T a time scale. There are three length scales in the problem: The Fermi wavelength λ_{F}, the scattering length a, and the scattering mean-free path ℓ. Clearly, the relevant length for transport processes is ℓ. Similarly, the relevant time scale is the scattering mean-free time τ. For electrons on the Fermi surface, $\ell = v_{\mathrm{F}}\tau$, with $v_{\mathrm{F}} = k_{\mathrm{F}}/m$ the Fermi velocity. As an order of magnitude estimate of D we therefore expect

$$D \simeq v_{\mathrm{F}}^2\tau. \tag{17}$$

Using Eq. (1) and $v_{\mathrm{F}}^2 N_{\mathrm{F}} \simeq n_e/m$, the analogous estimate for the conductivity σ is

$$\sigma \simeq \frac{n_e e^2\tau}{m}. \tag{18}$$

This is the standard Drude result for σ.

Clearly, all of the above arguments and derivations are just phenomenological in nature, and therefore ultimately unsatisfactory, although they do contain the correct physics. In the following section we improve our treatment, and derive the Eqs. (16), (17), and (18) from the microscopic Hamiltonian.

3. Many-Body Perturbation Theory

3.1. THE DISORDER AVERAGED GREEN FUNCTION

For pedagogical reasons, let us first calculate the averaged Green function, although it is not directly needed to compute either Eq. (10) or Eq. (13). Since on average space is homogeneous, we have

$$\left\{G^{R,A}(\mathbf{p}, \mathbf{p}_1, E)\right\}_{\text{dis}} = \delta(\mathbf{p} - \mathbf{p}_1)\, G^{R,A}(\mathbf{p}, E). \tag{19}$$

A self-energy Σ can be defined in the usual way, leading to a Dyson equation for the Green function [5]. To be specific, we consider the retarded Green function,

$$\begin{aligned} G^R(\mathbf{p}, E) &= G_0^R(\mathbf{p}, E) + G_0^R(\mathbf{p}, E)\, \Sigma^R(\mathbf{p}, E)\, G^R(\mathbf{p}, E) \\ &= \frac{1}{E - \mathbf{p}^2/2m - \Sigma^R(\mathbf{p}, E) + i0}, \end{aligned} \tag{20}$$

with G_0 the free particle Green function (i.e., Eq. (20) with $\Sigma^R = 0$).

Standard methods lead to a diagrammatic theory for Σ^R [3]. In Fig. 1 (a), we show the lowest order, or Born, approximation for the self-energy. Here the straight line denotes a free particle Green function and the dashed line with a cross denotes a $\{u\, u\}_{\text{dis}}$ disorder correlation function. In this approximation,

$$\begin{aligned} \Sigma^R(\mathbf{p}, E) = \frac{1}{2\pi N_F \tau} \int_{\mathbf{q}} \frac{1}{[E - q^2/2m + i0]} \quad &\approx \quad \frac{-i\pi}{2\pi N_F \tau} \int_{\mathbf{q}} \delta\left(E - \frac{q^2}{2m}\right) \\ &= \quad -\frac{i}{2\tau}. \end{aligned} \tag{21}$$

Here $\int_{\mathbf{q}} \equiv \int d\mathbf{q}/(2\pi)^d$, and we have dropped the real part of Σ^R, which simply redefines the zero of the energy. In this approximation the average Green function is

$$G^R(\mathbf{p}, E) \approx \frac{1}{E - \mathbf{p}^2/2m + i/2\tau}. \tag{22}$$

We see that the factors in Eq. (2) were chosen so that the time Fourier transform of Eq. (22), $G^R(\mathbf{p}, t)$, decays exponentially with a relaxation time equal to τ. The physical meaning of this decay is the loss of phase correlations that is brought about by the electron-impurity scattering.

Figure 1. (a) Born approximation for the electron self-energy. (b) Perturbation theory for the two-particle correlation function Φ, Eq. (23). (c) The reducible vertex function Γ in terms of the irreducible vertex function U.

3.2. THEORY FOR THE CONDUCTIVITY AND THE DENSITY SUSCEPTIBILITY: GENERALIZED BOLTZMANN EQUATION

In order to calculate the conductivity and the density susceptibility, Eqs. (10) and (13), we introduce the two-particle correlation function,

$$\Phi_{\mathbf{p}\mathbf{p}_1}(\mathbf{k},\omega) = \left\{G^R\left(\mathbf{p}+\mathbf{k},\mathbf{p}_1+\mathbf{k},E_{\mathrm{F}}+\omega\right)\,G^A\left(\mathbf{p}_1,\mathbf{p},E_{\mathrm{F}}\right)\right\}_{\mathrm{dis}}. \qquad (23)$$

The advanced-advanced combination $G^A\,G^A$ also contributes to Eq. (10), but it is not important for the effects we are intererested in and we therefore do not dicuss it.

A diagrammatic expression for Φ is shown in Fig. 1 (b). Here the straight lines denote exact averaged Green functions, and the shaded box indicates disorder correlations between the top (retarded) and bottom (advanced) Green functions. A Bethe-Salpeter equation can be used to introduce a two-particle irreducible vertex, $U_{\mathbf{p}\mathbf{p}_1}$, which is the two-particle analog of the Dyson self-energy. The diagrammatic relationship between the reducible vertex Γ, and the irreducible one is shown in Fig. 1 (c). Analytically we have

$$\Phi_{\mathbf{p}\mathbf{p}_1}(\mathbf{k},\omega) = G^R(\mathbf{p}+\mathbf{k}, E_F+\omega)\, G^A(\mathbf{p}, E_F)\Big[\delta(\mathbf{p}-\mathbf{p}_1) + \int_{\mathbf{p}_2} U_{\mathbf{p}\mathbf{p}_2}(\mathbf{k},\omega)\Phi_{\mathbf{p}_2\mathbf{p}_1}(\mathbf{k},\omega)\Big]. \quad (24)$$

To put this into a more standard form we write [7]

$$G^R(\mathbf{p}+\mathbf{k}, E_F+\omega)\, G^A(\mathbf{p}, E_{rmF}) = -\frac{\Delta G_{\mathbf{p}}(\mathbf{k},\omega)}{\omega - \mathbf{k}\cdot\mathbf{p}/m - \Delta\Sigma_{\mathbf{p}}(\mathbf{k},\omega)}, \quad (25)$$

where

$$\Delta G_{\mathbf{p}}(\mathbf{k},\omega) = G^R(\mathbf{p}+\mathbf{k}, E_F+\omega) - G^A(\mathbf{p}, E_F), \quad (26)$$

$$\Delta\Sigma_{\mathbf{p}}(\mathbf{k},\omega) = \Sigma^R(\mathbf{p}+\mathbf{k}, E_F+\omega) - \Sigma^A(\mathbf{p}, E_F). \quad (27)$$

Defining $\widetilde{\Phi}$ through

$$\Phi_{\mathbf{p}\mathbf{p}_1}(\mathbf{k},\omega) = \Delta G_{\mathbf{p}}(\mathbf{k},\omega)\, \widetilde{\Phi}_{\mathbf{p}\mathbf{p}_1}(\mathbf{k},\omega), \quad (28)$$

and using the Ward identity [7]

$$\Delta\Sigma_{\mathbf{p}}(\mathbf{k},\omega) = \int_{\mathbf{p}_2} U_{\mathbf{p}\mathbf{p}_2}(\mathbf{k},\omega)\, \Delta G_{\mathbf{p}_2}(\mathbf{k},\omega), \quad (29)$$

the Eq. (24) can be written

$$\left\{\omega - \frac{\mathbf{k}\cdot\mathbf{p}}{m}\right\}\widetilde{\Phi}_{\mathbf{p}\mathbf{p}_1}(\mathbf{k},\omega) = -\delta(\mathbf{p}-\mathbf{p}_1) + \int_{\mathbf{p}_2} U_{\mathbf{p}\mathbf{p}_2}(\mathbf{k},\omega)\, \Delta G_{\mathbf{p}_2}(\mathbf{k},\omega) \times \left[\widetilde{\Phi}_{\mathbf{p}\mathbf{p}_1}(\mathbf{k},\omega) - \widetilde{\Phi}_{\mathbf{p}_2\mathbf{p}_1}(\mathbf{k},\omega)\right]. \quad (30)$$

Two comments are in order. First, for small wavevectors and frequencies,

$$\Delta G_{\mathbf{p}}(\mathbf{k}\to 0, \omega\to 0) \approx -i\pi\delta(E - p^2/2m)/\tau,$$

so that the ΔG in Eqs. (30) and (25) are delta-functions that confine the electrons to the energy shell. Second, Eq. (30) for Φ is exact, and has the form of a generalized Boltzmann equation. That is, the frequency term on the left hand side is the Fourier transform of a time derivative, and $\mathbf{k}\cdot\mathbf{p}/m$ is the spatial Fourier transform of $\mathbf{p}\cdot\nabla/m$. Therefore the left-hand side has the form of a standard free-streaming term. The right-hand side, on the other hand, is an initial condition term, plus a generalized collision operator. In the next subsection, we will see that the simplest approximation for U leads to the quantum Boltzmann equation.

In terms of $\widetilde{\Phi}$, the physical quantities of interest to us are

$$\sigma'(\omega) = \frac{e^2}{2\pi V m^2} \int d\mathbf{p}\, d\mathbf{p}_1\, p_x\, p_{1x}\, \Delta G_{\mathbf{p}}(\mathbf{0},\omega)\, \widetilde{\Phi}_{\mathbf{p}\mathbf{p}_1}(\mathbf{0},\omega) \tag{31}$$

and

$$\chi_{nn}(k,\omega) = N_{\mathrm{F}} + \frac{i\omega}{2\pi} \int d\mathbf{p}\, d\mathbf{p}_1\, \Delta G_{\mathbf{p}}(\mathbf{k},\omega)\, \widetilde{\Phi}_{\mathbf{p}\mathbf{p}_1}(\mathbf{k},\omega). \tag{32}$$

3.3. QUANTUM BOLTZMANN EQUATION

To lowest order in the disorder we have, cf. Fig. 2,

$$U_{\mathbf{p}\mathbf{p}_2}(\mathbf{k},\omega) = \frac{i}{2\pi N_{\mathrm{F}}\tau}. \tag{33}$$

Inserting Eq. (33) into Eq. (30) yields

$$\begin{aligned} \left\{-i\omega + i\frac{\mathbf{k}\cdot\mathbf{p}}{m}\right\} \widetilde{\Phi}_{\mathbf{p}\mathbf{p}_1}(\mathbf{k},\omega) &= i\delta(\mathbf{p}-\mathbf{p}_1) + \frac{1}{N_{\mathrm{F}}\tau} \int_{\mathbf{p}_2} \delta(E_{\mathrm{F}} - p_2^2/2m) \\ &\quad \times \left[\widetilde{\Phi}_{\mathbf{p}_2\mathbf{p}_1}(\mathbf{k},\omega) - \widetilde{\Phi}_{\mathbf{p}\mathbf{p}_1}(\mathbf{k},\omega)\right]. \end{aligned} \tag{34}$$

This is the quantum Boltzmann equation, which thus turns out to be the simplest nontrivial approximation within the formalism developed above. Higher order approximations for $U_{\mathbf{p}\mathbf{p}_2}$ lead to corrections of order $\lambda_{\mathrm{F}}/\ell$ and a/ℓ to the Boltzmann equation results.

Figure 2. Boltzmann approximation for the irreducible vertex function U.

The Eq. (34) can be easily solved exactly. One obtains

$$\begin{aligned} \widetilde{\Phi}_{\mathbf{p}\mathbf{p}_1}(\mathbf{k},\omega) &= i\delta(\mathbf{p}-\mathbf{p}_1)\, h_{\mathbf{p}}(\mathbf{k},\omega) \\ &\quad + \frac{i\,\delta(E_{\mathrm{F}} - p_1^2/2m)\, h_{\mathbf{p}}(\mathbf{k},\omega)}{N_{\mathrm{F}}\tau\left[1 - (1/N_{\mathrm{F}}\tau) \int_{\mathbf{p}_2} \delta(E_{\mathrm{F}} - p_2^2/2m)\, h_{\mathbf{p}_2}(\mathbf{k},\omega)\right]}, \end{aligned} \tag{35}$$

with

$$h_{\mathbf{p}}(\mathbf{k},\omega) = \left(-i\omega + i\,\frac{\mathbf{k}\cdot\mathbf{p}}{m} + \frac{1}{\tau}\right)^{-1}. \tag{36}$$

For general $\mathbf{k}$ and ω, Eq. (35) is quite complicated. However, using it in Eqs. (31) and (32) for small wavenumber and frequency immediately gives

$$\sigma_{\mathrm{B}} = n_e \frac{e^2\tau}{m} \tag{37}$$

and

$$\chi_{nn,\mathrm{B}}(\mathbf{k},\omega) \simeq \frac{N_{\mathrm{F}} D_{\mathrm{B}} k^2}{-i\omega + D_{\mathrm{B}} k^2}, \tag{38}$$

where the subscript B denotes the Boltzmann approximation. Notice that σ_{B} and D_{B} are related by the Einstein relation,

$$\sigma_{\mathrm{B}} = e^2 N_{\mathrm{F}} D_{\mathrm{B}}, \tag{39}$$

as they should be according to Eq. (1).

4. Corrections to Boltzmann Transport Theory: Wave Interference Effects

4.1. DISORDER OR DENSITY EXPANSION OF σ

Perhaps the most obvious theoretical problem concerning the wave interference effects is to characterize the corrections to σ_B that arise when $\lambda_{\mathrm{F}}/\ell$ is not zero. The first question is whether or not there is an analytic expansion in the small parameter $\epsilon = \lambda_{\mathrm{F}}/\ell$. Technically, this question can be answered by examining the coefficients a_i $(i = 1, 2, 3, \ldots)$ in the Taylor expansion

$$\frac{\sigma}{\sigma_{\mathrm{B}}} = 1 + a_1\,\epsilon + a_2\,\epsilon^2 + O(\epsilon^3). \tag{40}$$

Detailed calculations show that in general such an expansion does not exist, and that all the coefficients beyond a_{d-1} are infinite in d dimensions [8]. Although of different physical origin, a similar divergence is well known to exist in the kinetic theory of classical gases [9].

To see the origin of these infinities in the quantum case, consider the first few maximally crossed diagrams (MCD) in Fig. 3. The calculation gives

$$(3a) \propto \int_q G^R(\mathbf{q})\, G^A(\mathbf{p}+\mathbf{p}_1-\mathbf{q})\Big|_{\omega=1/\tau=0} \propto \frac{1}{|\mathbf{p}+\mathbf{p}_1|}, \tag{41}$$

$$(3b) \propto (3a)^2 \propto \frac{1}{|\mathbf{p}+\mathbf{p}_1|^2}. \tag{42}$$

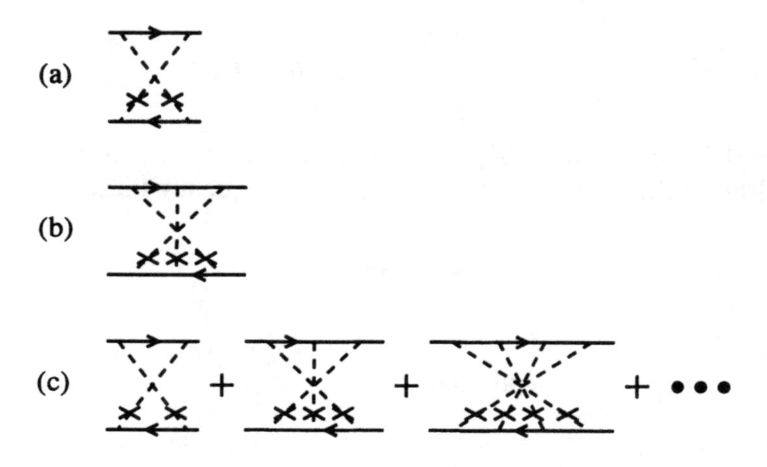

Figure 3. Maximally crossed diagrams.

That is, the MCD are singular for the momentum configuration that corresponds to backscattering, $\mathbf{p}_1 = -\mathbf{p}$. In any dimension, these singularities lead to infinite coefficients in Eq. (40), with the order in which the first infinity occurs depending on the dimension. The correct expansion of σ in powers of ϵ is therefore nonanalytic. In $d = 3$, it is of the form

$$\frac{\sigma}{\sigma_{\rm B}} = 1 + a_1\,\epsilon + a_2'\,\epsilon^2 \ln(1/\epsilon) + a_2\,\epsilon^2 + \ldots \quad . \tag{43}$$

In $d = 2$ it has been shown that power-law nonanalyticities appear in addition to logarithmic ones [10]. In the next subsection we further discuss the backscattering processes that cause these nonanalyticities.

For most systems, the values of the coefficients in Eq. (43) are of limited interest because the derivation has ignored electron-electron interactions, which are important in all solid state materials. However, there does exist a physical system for which the quantum Lorentz gas is a realistic model, namely electrons injected into He gas at low ($\leq$ 4°K) temperatures. In Ref. [11] detailed arguments have been given that on electronic time scales, the He atoms can be treated as stationary scatterers and that all other approximations used in our model are justified for describing transport in this system. The finite temperature electron mobility, μ, normalized to its Boltzmann value, can be expanded in the small parameter

$$\chi = \lambda_T/\pi\ell, \tag{44}$$

with $\lambda_T \equiv 2\pi/k_T = \sqrt{2\pi^2/mT}$ the thermal de Broglie wavelength. The result is

$$\frac{\mu}{\mu_{\rm B}} = 1 + \mu_1\,\chi + \mu_2'\,\chi^2 \ln\chi + \mu_2\,\chi^2 + \ldots, \tag{45}$$

with

$$\begin{aligned} \mu_1 &= -\pi^{3/2}/6 \quad , \\ \mu_2' &= (\pi^2 - 4)/32 \quad , \\ \mu_2 &= 0.236 \quad . \end{aligned} \tag{46}$$

In Fig. 4 this theoretical result, with no adjustable parameters, is compared to experimental data. We conclude that experiments are consistent with Eqs. (45) and (46). Further experiments are needed to unambigiously confirm the presence of the logarithm in these equations. For further discussions of this point, see Ref. [11].

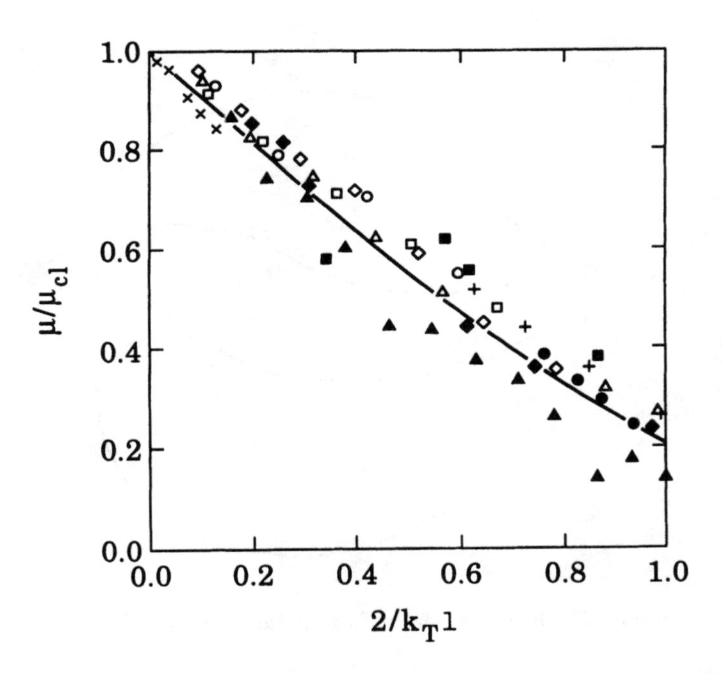

Figure 4. Mobility μ of electrons in dense gases, normalized to the Boltzmann value μ_{cl}, as a function of χ. The symbols represent experimental data as measured or quoted by Adams at al. [12]. The solid line is the theoretical result, Eqs. (45,46). From Ref. [13].

4.2. WEAK LOCALIZATION EFFECTS AND LONG-TIME TAILS

In classical transport theory it has proven to be very enlightening to examine the long-time properties of the time-correlation functions (TCF) that determine the transport coefficients. Generally, it has been found that these functions decay algebraically for long times and that therefore the transport coefficients themselves are nonanalytic functions of the frequency [9].

In quantum transport theory the decay of the TCF has proven to be even more interesting [13], and to be closely related to the phenomenon of Anderson localization [14]. Note that in either case, these algebraic tails are qualitatively different from the exponential decays predicted by both the Boltzmann equation and by simple theories for the Green function.

Here we mainly concentrate on a physical argument [15] that leads to these so-called weak localization effects. Some hints of their technical origin will also be given; details can be found elsewhere [14]. We will see that the weak localization effects are closely related to the same backscattering phenomena that are responsible for the nonanalytic disorder expansion discussed in the previous subsection.

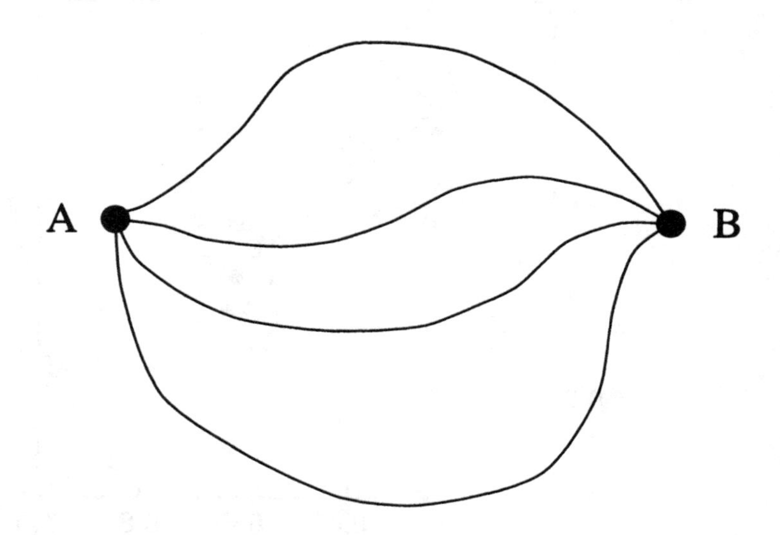

Figure 5. Feynman paths from point A to point B.

The physical argument starts by considering the different Feynman paths from, say, point A to point B, as illustrated in Fig. 5. Labeling the amplitude of path i by $\mathcal{A}_i$, the total probability to reach B from A is

$$w = \left|\sum_i \mathcal{A}_i\right|^2 = \sum_i |\mathcal{A}_i|^2 + \sum_{i \neq j} \mathcal{A}_i \mathcal{A}_j^*. \tag{47}$$

Usually, the different paths are uncorrelated and the second term in Eq. (47) averages to zero, leaving the first term, which is just the classical probability. The exception to this is if the points A and B coincide. In this case the path can be traversed in two opposite directions, namely, forward or backward. The amplitudes $\mathcal{A}_1$ and $\mathcal{A}_2$ for these two paths have a coherent phase relation, leading to constructive interference. If $\mathcal{A}_1 = \mathcal{A}_2 = \mathcal{A}$, then Eq. (47), for $i = 1, 2$, becomes

$$w = 2\,|\mathcal{A}|^2 + 2\mathcal{A}\,\mathcal{A}^* = 4\,|\mathcal{A}|^2\,, \tag{47'}$$

which is twice the classical probability. The important conclusion from this argument is that the quantum probabilities for time reversed, or intersecting, paths are enhanced compared to classical dynamics. Further, since paths that return to their starting point slow down the diffusion process, we expect these quantum corrections to lead to a decrease of σ compared to the semi-classical value.

To estimate the contribution of the closed loops to σ we argue as follows. If the particle is diffusing, then the probability of finding it at point $\mathbf{r}$ at time t, given that it was at point $\mathbf{r}_0$ at time $t = 0$ is

$$w(\mathbf{r}, t) = \frac{\exp[-(\mathbf{r} - \mathbf{r}_0)^2/4Dt]}{(4\pi Dt)^{d/2}}. \tag{48}$$

The $w(\mathbf{r}, t)$ is appreciably different from zero only within a diffusive volume V_{diff} determined by $(\mathbf{r} - \mathbf{r}_0)^2 < 4Dt$, or,

$$V_{\mathrm{diff}} \approx (Dt)^{d/2}. \tag{49}$$

Now, each Feynman path has a diameter proportional to $\lambda_{\mathrm{F}}^{d-1}$, and a

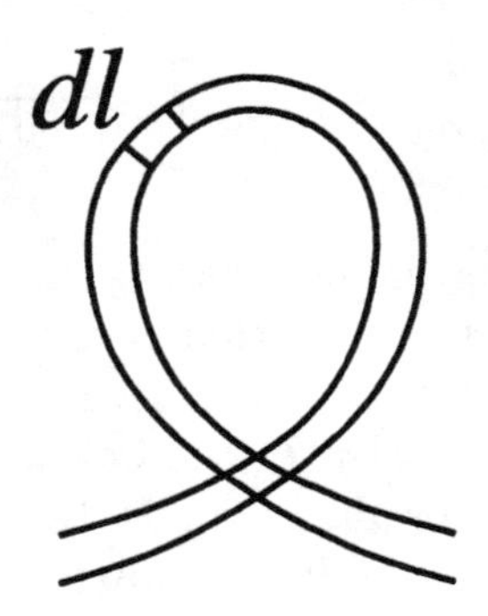

Figure 6. Geometry of a Feynman tube.

differential length $d\ell = v_{\mathrm{F}}\, dt$, cf. Fig. 6. The differential volume of the Feynman tube is $dV = v_{\mathrm{F}} \lambda_{\mathrm{F}}^{d-1} dt$, which should be compared to V_{diff}. If p is the probability that a closed path exists, then it is the ratio of these two volumes integrated over all times:

$$p = \int_{\tau'}^{T} \frac{dV}{V_{\mathrm{diff}}} = \lambda_{\mathrm{F}}^{d-1} v_{\mathrm{F}} \int_{\tau'}^{T} \frac{dt}{(Dt)^{d/2}}. \tag{50}$$

Here $\tau' = O(\tau)$ is a microscopic time where the diffusive description breaks down, and T is the observation time that we will equate with w^{-1}. The change in $\delta\sigma$ of the conductivity due to these wave interference effects is therefore

$$\frac{\delta\sigma}{\sigma_0} \propto -\begin{cases} 1/(\omega\tau)^{1/2}, & (d=1) \\ \ln(1/\omega\tau), & (d=2) \\ \text{const.} - (\omega\tau)^{1/2}, & (d=3). \end{cases} \tag{51}$$

Note that the low frequency correction to σ_0 in Eq. (51) diverges as $\omega \to 0$ for $d \leq 2$. Further, these contributions have a sign such as to decrease the conductivity. In a more complete theory that effectively resums all such divergent contributions, σ is identically equal to zero for $\omega = 0$ in $d \leq 2$ [14]. This phenomenon is known as Anderson localization. In $d > 2$ a metal-insulator transition, called the Anderson transition, occurs as a function of the disorder or the Fermi energy. In $d = 2$ there cannot be a metallic phase, as the electrons are always localized; $d = 2$ is therefore the lower critical dimension for the Anderson transition.

The same results can also be obtained from the generalized Boltzmann equation derived in Sec. 3. Resumming all of the backscattering or MCD diagrams shown in Fig. 3 leads to Eq. (51) with explicit coefficients. The crucial point is that the summations of the MCD can be related, in the absence of a magnetic field, to the summation of ladder diagrams that give χ_{nn}. Analytically, the MCD lead to a contribution of the form

$$\frac{\delta\sigma}{\sigma_0} \approx -c_d \int_{q<\ell^{-1}} \frac{1}{-i\omega + D_{\rm B}q^2}. \tag{52}$$

Here $\mathbf{q} = \mathbf{p} + \mathbf{p}_1$ is the backscattering momentum, a long-wavelength approximation has been made so that $q < \ell^{-1}$, and $c_d \sim \lambda_{\rm F}^{d-1}$ is a constant. Performing the integral gives Eq. (51). At finite temperatures the ω in Eq. (51) is effectively replaced by T (see below), so that σ depends nonanalytically, and in $d = 1, 2$ singularly, on T. In Fig. 7, the resistivity $\rho \sim 1/\sigma$ is plotted versus $\ln T$ for a two-dimensional amorphous film. The straight line confirms the existence of the logarithm in Eq. (51).

It is worth noting that the hydrodynamic pole in Eq. (52) occurs because the MCD are related to a spontaneously broken symmetry and a corresponding Goldstone mode [17]. This mode is soft or massless only at zero temperature. That is, the relationship of the MCD to $\chi_{\rm nn}$ that was noted above, is valid only for $T = 0$. At finite temperature, the ω in Eq. (52) is replaced by $\omega + 1/\tau_{\rm in}$, with $\tau_{\rm in}$ the temperature dependent inelastic mean-free time. The Goldstone modes thus acquire a mass, in contrast to $\chi_{\rm nn}$ which is always massless due to particle number conservation. In the literature, in analogy to particle physics terminology, these modes that are

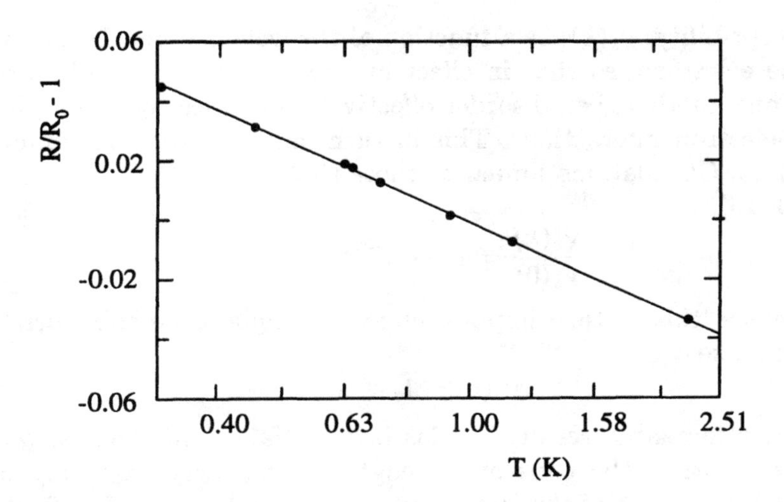

Figure 7. **Resistance, R, normalized to $R_0 = R(T = 1K)$, of a thin PdAu film plotted versus $\log T$ as measured by Dolan and Osheroff [16]. From Ref. [13].**

hydrodynamic only at $T = 0$ are often called either 'diffusons' or 'Cooperons', depending on how they arise.

4.3. THE EFFECTS OF ELECTRON-ELECTRON INTERACTIONS

In the absence of electron-electron interactions it is easy to show that thermodynamic quantities do not depend on the dynamical diffusive processes that lead to the low frequency nonanalyticities discussed in the previous subsection. However, when interactions are included this is no longer the case, and the static and dynamical properties of any quantum system are coupled together. From a technical point of view this is obvious within a many-body Matsubara formalism. From a scaling viewpoint, in fundamental units ($\hbar = 1 = k_{\mathrm{B}}$) T and ω have the same units, implying thermodynamics (T) and dynamics (ω) are coupled together.

To motivate the static results, we note that for diffusive systems, ω scales like k^2. For the wavenumber and frequency dependent conductivity this implies, for $2 < d < 4$,

$$\frac{\sigma(k=0,\omega)}{\sigma_0} = 1 + a\,\omega^{(d-2)/2} + \ldots, \tag{53}$$

$$\frac{\sigma(k,\omega=0)}{\sigma_0} = 1 + a'\,k^{d-2} + \ldots \quad . \tag{54}$$

Note that the coefficients a and a' in Eqs. (4.3) are positive since wave interference effects decrease $\sigma(0,0)$. Another interesting quantity is the static spin susceptibility, $\chi_s(k)$, as a function of the wavenumber. Disorder slows down the electrons, so that in effect any two electrons spend more time close to one another, i.e., disorder effectively increases the strength of the electron-electron interactions. This in turn implies a disorder induced increase in χ_s. Calculations similar to those that lead to Eqs. (53), (54) give, at $T = 0$ [18],

$$\frac{\chi_s(k)}{\chi_s(0)} = 1 - \tilde{a}\, k^{d-2} + 0(k^2). \tag{55}$$

In real space, this in turn implies long-range equal-time spin correlations that decay like

$$\chi_s(r) \sim 1/r^{2(d-1)}. \tag{56}$$

This is a remarkable result. In classical statistical mechanics, $\chi_s(r) \propto \exp(-r/\xi)$, with ξ the correlation length which, away from any critical point, is on the order of the lattice spacing. In contrast, at $T = 0$, the correlations in Eq. (55) are power-law like everywhere in the phase diagram. These long-range correlations are due to a coupling of the hydrodynamic spin transport to the charge diffusion processes which themselves are intrinsically of long range.

Physically, Eq. (55) implies that spin fluctuations effectively interact between themselves through long-range interactions. It is well known that long-range interactions have a profound effect on phase transitions and critical behavior. The implication of Eq. (55) for the paramagnetic to ferromagnetic quantum phase transition is discussed in the following lectures [19]. Note that long-range spatial correlations also exist in *nonequilibrium classical* systems. In that case, statics and dynamics are also coupled together, somewhat analogous to the *equilibrium quantum* statistical mechanical systems discussed here [13].

5. Final Remarks

We conclude with a few additional remarks. Although we have not emphasized this fact, the interesting phenomena in quantum kinetic theory discussed above have strong analogies in classical kinetic theory. Indeed, the discovery of the logarithmic singularity in the disorder expansion of σ, Eq. (43), was motivated by much earlier, related considerations in classical kinetic theory [9]. Similarly, long-time tail effects and low-frequency nonanalyticities were discussed already in 1967 for classical systems [21], whereas the weak localization effects discussed above were not discovered until 1979 [20].

Another aspect we have not stressed is that, even though the phenomena in classical and quantum systems are similar, the correlations in space and time are stronger in the quantum mechanical case. For example, in the classical Lorentz gas the low frequency nonanalytic contribution to the conductivity is proportional to $\omega^{d/2}$. Physically, this eventually leads to the conclusion that while classical particles are diffusive in $d = 2$, there is no diffusive phase in two-dimensional quantum systems [20, 14].

Finally, one other important feature in quantum systems is the coupling between statics and dynamics. In classical systems the thermodynamic correlation functions do not couple to the dynamical fluctuations that lead to the long-range correlations in time-correlation functions. In zero-temperature quantum systems the situation is different. As explained in the last section, equal-time correlations do couple to these dynamical fluctuations, and as a consequence of this, there are both long-ranged, power-law spatial correlations and long-ranged, power-law time correlations in quantum systems.

Acknowledgements

This work was supported by the NSF under grant numbers DMR-96-32978 and DMR-98-70597. This research was supported in part by the National Science Foundation under grant No. PHY94-07194.

References

1. See, e.g., N.W. Ashcroft and N.D. Mermin, *Solid State Physics* (Holt, Rinehart and Winston, New York, 1976).
2. E.A. Uehling and G. Uhlenbeck, Phys. Rev. **43**, 552 (1933).
3. S.F. Edwards, Philos. Mag. **3**, 1020 (1958).
4. A.A. Abrikosov, L.P. Gorkov, and I.E. Dzyaloshinski, *Methods of Quantum Field Theory in Statistical Physics* (Dover, New York, 1975).
5. See, e.g., A.L. Fetter and J.D. Walecka, *Quantum Theory of Many-Particle Systems* (McGraw-Hill, New York, 1971).
6. R. Kubo, J. Phys. Soc. Japan **12**, 570 (1957); see also G.D. Mahan, *Many-Particle Physics* (Plenum, New York, 1981), ch. 3.7.
7. D. Vollhardt and P. Wölfle, Phys. Rev. B **22**, 4666 (1980).
8. T.R. Kirkpatrick and J.R. Dorfman, J. Stat. Phys. **30**, 67 (1983); T.R. Kirkpatrick and J.R. Dorfman, Phys. Rev. A **28**, 1022 (1983); T.R. Kirkpatrick and D. Belitz, Phys. Rev. B **34**, 2168 (1986).
9. For a review see, e.g., J.R. Dorfman, T.R. Kirkpatrick, and J.V. Sengers, Ann. Rev. Phys. Chem. **45**, 213 (1994).
10. F. Evers, D. Belitz, and W. Park, Phys. Rev. Lett. **78**, 2768 (1997).
11. K.I. Wysokinski, W. Park, D. Belitz, and T.R. Kirkpatrick, Phys. Rev. E **52**, 612 (1995).
12. P. W. Adams, D. A. Browne, and M. A. Paalanen, Phys. Rev. B **45**, 8837 (1992).
13. For a review, see, e.g., T.R. Kirkpatrick and D. Belitz, J. Stat. Phys. **87**, 1307 (1997).
14. For a review, see, e.g., P.A. Lee and T.V. Ramakrishnan, Rev. Mod. Phys. **57**, 287 (1985).
15. G. Bergmann, Phys. Rep. **101**, 1 (1984).
16. G. J. Dolan and D. D. Osheroff, Phys. Rev. Lett. **43**, 721 (1979).
17. L. Schäfer and F. Wegner, Z. Phys. B **38**, 113 (1980).

18. D. Belitz, T.R. Kirkpatrick, and T. Vojta, Phys. Rev. B **55**, 9452 (1997).
19. D. Belitz and T.R. Kirkpatrick, this volume, p. 399.
20. E. Abrahams, P.W. Anderson, D.C. Licciardello, and T.V. Ramakrishnan, Phys. Rev. Lett. **42**, 673 (1979).
21. B.J. Alder and T.E. Wainwright, Phys. Rev. Lett. **18**, 988 (1967).

QUANTUM PHASE TRANSITIONS

D.BELITZ
Department of Physics and Materials Science Institute
University of Oregon
Eugene, OR 97403, USA

AND

T.R.KIRKPATRICK
Institute for Physical Science and Technology
and Department of Physics, University of Maryland
College Park, MD 20742, USA

Abstract. These are notes for lectures delivered at the NATO ASI on Dynamics in Leiden, The Netherlands, in July 1998. The main concepts relating to quantum phase transitions are explained, using the paramagnet-to-ferromagnet transition of itinerant electrons as the primary example. Some aspects of metal-insulator transitions are also briefly discussed. The exposition is strictly pedagogical in nature, with no ambitions with respect to completeness or going into technical details. The goal of the lectures is to provide a bridge between textbooks on classical critical phenomena and the current literature on quantum phase transitions. Some familiarity with the concepts of classical phase transitions is helpful, but not absolutely necessary.

1. Introduction to Continuous Phase Transitions

In these lectures we discuss so-called quantum phase transitions or zero-temperature phase transitions. Our motivation is the fact that these transitions provide a field where the long-range correlations in quantum systems that were introduced in the preceding lectures [1] have some spectacular consequences. Since quantum phase transitions have many features in common with ordinary classical or thermal phase transitions, we will first review the basic concepts of both, and then define and discuss the particular fea-

J. Karkheck (ed.),
Dynamics: Models and Kinetic Methods for Non-equilibrium Many Body Systems, 399–424.

tures of quantum transitions, before turning to some specific examples. We would like to warn the reader that it is not possible to do this subject justice within the time and space constraints of these lectures. Our aim can therefore only be to give some examples that highlight a few of the interesting features of the field. Thorough treatments of the phenomenology and theory of classical phase transitions can be found in Refs. [2], and aspects of quantum phase transitions are reviewed in Refs.[3, 4].

1.1. REVIEW OF BASIC CONCEPTS

1.1.1. *Continuous phase transitions*

Continuous phase transitions[1] are characterized by the considered system undergoing a transition from a symmetric or disordered state, which incorporates some symmetry of the Hamiltonian, to a broken-symmetry or ordered state, which does not have that symmetry, although the Hamiltonian still possesses it. A good example is given by a Heisenberg ferromagnet: The relevant symmetry of the Hamiltonian is the rotational symmetry in spin space, and the disordered and ordered states are represented by the paramagnetic and the ferromagnetic states, respectively. In the former, there is no preferred direction for the magnetic moments of the spins. The net magnetization is therefore zero, and the state has the same spin rotational symmetry as the Hamiltonian. That is, if we rotate the coordinate system in spin space, then the system will look the same. In the latter, on the other hand, there *is* a preferred direction for the spins, and this is the direction in which the overall magnetization will point. The state therefore no longer respects the spin rotational symmetry, and we say that the symmetry is *spontaneously broken* ('spontaneously', in order to distinguish this phenomenon from the explicit or external breaking that occurs if we apply an external magnetic field). In an ideal system the preferred direction is completely arbitrary, and in real systems it is determined by very small external fields, like e.g. the earth's magnetic field, or by a small breaking of the symmetry in the system's Hamiltonian as provided by, e.g., the ionic lattice structure. As a result, in an ideal system the magnetization in the ordered state is not only non-zero, but it is also non-unique. Such a thermodynamic quantitity that is zero in the disordered phase, and non-zero and non-unique in the ordered phase, is called an *order parameter*, a concept first introduced by Landau. The external field B that couples to the order parameter, in our example the magnetic field, is called the field *conjugate* to the order parameter.

[1]We will deal only with continuous transitions, and will simply refer to them as 'phase transitions' or 'critical points'.

As we approach the phase transition by changing some parameter, several remarkable phenomena are observed. For instance, the correlations of the order parameter become long-ranged. Let M be the order parameter, and let us consider the spatial correlations of M (in the case of a magnet, they can be measured by neutron scattering):

$$\langle M(\mathbf{x})\, M(\mathbf{y})\rangle = f(|\mathbf{x}-\mathbf{y}|/\xi). \tag{1}$$

Everywhere in parameter space except at the critical point, the function f decays in space on a length scale ξ called the *correlation length.*[2] The ξ diverges as the critical point is approached, usually via a power law that is characterized by the correlation length critical exponent ν,

$$\xi \propto t^{-\nu}, \tag{2}$$

where t denotes some dimensionless distance in parameter space from the critical point. For instance, if the transition occurs at a non-zero critical temperature T_c, we can use $t = |T - T_c|/T_c$. At criticality, i.e. at $t = 0$, the correlation length diverges, which indicates that the order parameter correlations decay only like a power law, which has no intrinsic scale.[3] It is customary to denote this power law by an exponent η:

$$\langle M(\mathbf{x})\, M(\mathbf{y})\rangle_{t=0} \propto |\mathbf{x}-\mathbf{y}|^{-d-2+\eta}, \tag{3}$$

where d denotes the spatial dimension of the system.

Apart from these long-range correlations in space, there are similar effects in the temporal behavior of the system. Let us denote by τ_c the equilibration time, i.e. the time scale for the system to return to equilibrium after it has been disturbed. This equilibration time diverges as criticality is approached, and it does so as a power of the correlation length, with the power law characterized by the exponent z,

$$\tau_c \propto \xi^z. \tag{4}$$

The inverse of τ_c defines a critical frequency scale ω_c that goes to zero as criticality is approached, a phenomenon called *critical slowing down*:

$$\omega_c(t \to 0) \propto 1/\tau_c \to 0. \tag{5}$$

The exponents ν, η, and z that we have defined so far are examples of *critical exponents*, which characterize the power law behavior of various

[2]For most phase transitions, the function f in Eq. (1) decays exponentially for large arguments: $\ln f(x \to \infty) = -x + 0(\ln x)$. However, in general the functional form can be more complicated.

[3]Mathematically, a power law is a homogeneous function, while an exponential, for instance, is not.

observables upon approach to the critical point. Three other important critical exponents are β, which describes the vanishing of the order parameter,

$$M(t \to 0) \propto t^{\beta}, \tag{6}$$

γ, which describes the divergence of the order parameter susceptibility, here in our example being the magnetic susceptibility $\chi = M/B$,

$$\chi(t \to 0) \propto t^{-\gamma}, \tag{7}$$

and the critical exponent δ, which describes the dependence of the order parameter on its conjugate field at criticality,

$$M(t = 0, B \to 0) \propto B^{1/\delta}. \tag{8}$$

Notice that a diverging susceptibility implies that there cannot be a region where the order parameter depends linearly on the conjugate field.

The set of critical exponents characterizes the critical behavior, and it turns out that all of the critical exponents are not independent. Rather, many of them are related to one another by *scaling laws* or *exponent relations.* Furthermore, it turns out that the complete set of critical exponents is the same for whole classes of phase transitions. For instance, the critical exponents for the ferromagnetic transitions in iron and nickel are the same, even though these metals have very different band structures and their Curie temperatures are very different. Perhaps even more surprising, the critical exponents observed at the critical point of water are the same as those at the liquid-gas critical points of the "quantum liquids" He3 and He4. Phase transitions that share the same critical behavior, as expressed by the critical exponents, are said to belong to the same *universality class*, and the existence of such universality classes is referred to as *universality* or *universal behavior.* It turns out that universality classes are determined by basic symmetries of the underlying Hamiltonian, and by the spatial dimensionality of the system. The fact that the critical behavior is independent of the microscopic details of the Hamiltonian is due to the diverging correlation length: Close to a critical point, the system performs an average over all length scales that are smaller than the (very large) correlation length. This observation also gives a clue as to how to theoretically deal with critical phenomena. In order to correctly describe the universal critical behavior it should be sufficient to work with an effective theory that keeps explicitly only the asymptotic long-wavelength or large-distance behavior of the original Hamiltonian. By contrast, observables that vary from system to system even within a given unversality class, such as the critical temperature, are called *non-universal* properties, and they are not accessible by means of such effective theories.

The theoretical understanding of these and other properties of phase transitions developed over a period of 100 years, starting with van der Waals, continuing with Landau, and ending with Wilson. Wilson's renormalization group, the essence of which is a set of scale transformations, provided us with a deep understanding of the origin of universality and the critical power laws. More recently, it has also turned out to be a very useful general tool for studying the statistical mechanics of many-body systems in general, not only close to some critical point. Here it is not our goal to describe the renormalization group, for this purpose we refer the reader to the many excellent expositions in the literature [2b, 2d, 6a]. Rather, we will take the critical phenomenology for granted, and on that basis discuss the role of quantum mechanics in the context of phase transitions. The only exception is Sec. 2.3, where we assume some familiarity with basic renormalization group arguments in order to derive some results. As an explicit example, we will focus on the ferromagnet, but in Sec. 3 we discuss the metal-insulator transition.

1.2. CLASSICAL VERSUS QUANTUM PHASE TRANSITIONS

A fundamental question is the following: To what extent is quantum mechanics necessary in order to understand the critical phenomena we have sketched in the previous subsection, and to what extent will classical physics suffice? One can get the correct answer by means of the following very simple considerations (which can be elaborated upon if that is considered desirable). Generally speaking, quantum mechanics is important whenever the temperature becomes lower than some characteristic energy of the system under consideration. For instance, in an atom that characteristic energy is the Rydberg energy. In our case, we have seen that there is a characteristic frequency, namely ω_c. Let us assume that the corresponding energy scale, $\hbar\omega_c$, is the smallest relevant energy scale. Since $\omega_c(t \to 0) \to 0$, this is a reasonable guess close to the transition. It then follows that quantum mechanics will be important whenever the temperature T obeys

$$k_B T < \hbar\omega_c. \tag{9}$$

Usually we think of phase transitions as taking place at some non-zero critical temperature T_c, e.g. at the Curie temperature in our magnet example. It then follows that sufficiently close to the transition, namely for

$$t < (T_c/T_0)^{1/\nu z} \tag{10}$$

we expect quantum mechanics not to be important for describing the system's behavior. Here T_0 is some microscopic temperature scale, e.g., the Fermi temperature in a metallic ferromagnet. It follows that for any phase transition that takes place at a non-zero critical temperature $T_c > 0$, the critical behavior asymptotically close to the transition can be described entirely by classical physics. This conclusion survives more rigorous arguments. These phase transitions are called *classical* or *thermal* transitions. What drives the correlation length to infinity are thermal fluctuation, which become very large close to criticality. In contrast, we might think of a transition that occurs at zero temperature, and that is triggered by varying some non-thermal parameter, e.g. the system's composition. For instance, if n is the concentration of some ingredient, and n_c is the critical concentration, we can choose $t = |n_c - n|/n_c$ as our dimensionless distance from criticality. Then it obviously follows that quantum mechanics *will* be important for describing the critical behavior. These transitions are called *quantum* or *zero-temperature* phase transitions, and the relevant fluctuatations are quantum fluctuations or zero-point motion. An example would be a magnetic system that is, at $T = 0$, continuously diluted with some non-magnetic material until it undergoes a transition to the paramagnetic state. (We will deal with the obviously relevant question of what happens at low, but non-zero, temperatures in the following subsection.) Notice that, according to this definition, some phase transitions in systems that are usually considered quintessentially quantum mechanical, like the superconducting transition in mercury at $T = 4.2\,\mathrm{K}$, or the λ-transition in helium at $T = 2.17\,\mathrm{K}$, are classical transitions. Indeed, in both cases the critical behavior (but *not* the physics that triggers the transition, or the properties of either phase) can be understood entirely by means of classical physics: Both transitions are in the universality class of a classical 3-d XY model.

1.2.1. *Classical phase transitions*

Let us now first consider some additional properties of classical phase transitions. Within classical statistical mechanics, consider a Hamiltonian

$$H(p, q) = K(p) + U(q), \tag{11}$$

where p and q are the generalized momenta and positions, and K and U are the kinetic and potential energy, respectively.[4] The partition function

$$Z = \int dp\, dq\; e^{-H/k_B T} = \int dp\; e^{-K/k_B T} \int dq\; e^{-U/k_B T} \tag{12}$$

then factorizes into a piece that depends only on K and one that depends only on U. As a result, one can study the system's static properties independently from its dynamical ones. In particular, the dynamical critical

[4]We exclude from our considerations systems of charged particles in magnetic fields, and other cases of velocity dependent potentials.

exponent z is independent from all of the other critical exponents, and the static critical behavior can be studied, following Landau, by means of an effective functional of a time-independent order parameter. One often expresses this by saying that 'statics and dynamics decouple'.

Close to the critical point, the free energy density, $f = (-1/V)\, k_{\rm B}T \,\ln Z$, obeys a generalized homogeneity law,

$$f(t,B,\ldots) = b^{-d}\, f(t\, b^{1/\nu}, B\, b^{x_B},\ldots). \tag{13}$$

Here V is the system volume, and t and B are the dimensionless distance from the critical point and the external field conjugate to the order parameter, respectively, as before. The b is an arbitrary positive real number called a scale parameter, and Eq. (13) holds for all $b > 0$. The ν is the correlation length critical exponent, and $x_B > 0$ is a critical exponent that is related to δ by $x_B = d\delta/(1+\delta)$. Since all thermodynamic quantities can be obtained from the free energy, Eq. (13) provides us with homogeneity laws for all of them. For instance, by differentiating f with respect to B we obtain a homogeneity law for the order parameter density, $M = \partial f/\partial B$,

$$M(t,B) = b^{x_B-d}\, M(t\, b^{1/\nu}, B\, b^{x_B}). \tag{14}$$

Since b is arbitrary, we can in particular put $b = B^{-1/x_B}$. At $t = 0$ we then recover the above relation between x_B and δ. By the same method, other exponent relations can be obtained, and it turns out that of all the static critical exponents, only two are independent, e.g. ν and x_B. Another useful substitution is to set $b = t^{-\nu} \propto \xi$. This makes it obvious that letting $b \to \infty$ is tantamount to approaching criticality. In this context, a remark about the suppressed variables in Eq. (13), which we denoted by "...", is in order. They all enter Eq. (13) analogously to t and B, but the exponents that characterize their 'scaled' entries on the right-hand side (i.e. the analogs of $1/\nu$ and x_B), turn out to be negative. As a result, these entries go to zero as one approaches criticality, and are called 'irrelevant variables' or 'irrelevant operators'. If the observable under consideration is a regular function of a particular variable for small values of the argument, then it follows that this variable becomes unimportant as we approach criticality and does not influence the critical behavior. Thus, it really becomes 'irrelevant' in the ordinary sense of the word. This is not the case, however, if the observable is a singular function of some argument for small values of the argument. In this case the irrelevant variable influences the critical behavior after all, and one speaks of a 'dangerous irrelevant variable'. Dangerous irrelevant variables can cause substantial complications in the technical analysis of scaling near critical points.

The homogeneity or scaling law, Eq. (13), was historically first postulated phenomenologically, and it turned out that all observed properties of phase transitions followed from it if appropriate values for ν and x_B were used, depending on the universality class under consideration. It was the triumph of Wilson's renormalization group [2b] that it allowed a derivation of the homogeneity law from first principles. This derivation is highly technical, and we cannot go into it. Instead, we refer the reader to the extensive literature on this subject [2b, 2d, 6a].

1.2.2. *Quantum phase transitions*

Let us now turn to the case of quantum phase transitions, i.e. transitions that occur at $T = 0$ and are triggered by some non-thermal control parameter. As we have seen above, in this case quantum mechanics is always important, and we need to employ quantum statistical mechanics in order to calculate the partition function and the free energy. Let the Hamiltonian operator of the system be $\hat{H}(\hat{a}^\dagger, \hat{a})$, with $\hat{a}^\dagger$ and $\hat{a}$ a collection of creation and annihilation operators. In the usual imaginary time formalism, the time evolution of $\hat{a}$ is given by $\hat{a}(\tau) = e^{\hat{H}\tau}\,\hat{a}\,e^{-\hat{H}\tau}$, with τ the imaginary time variable. A general theorem [5] then tells us that the partition function for any quantum many-body system can be written as a functional integral of the form

$$Z = \int D[\bar{\psi}, \psi]\; e^{S[\bar{\psi},\psi]}. \tag{15}$$

Here $\bar{\psi}$ and ψ are space and imaginary time dependent fields that are isomorphic to the sets of creation and annihilation operators in a second quantization formulation of the problem. Their nature depends on whether the quantum particles are fermions or bosons. For the latter, the fields are classical or bosonic (i.e., they commute), while for the former, they are fermionic (i.e., they anticommute). $D[\bar{\psi}, \psi]$ is an appropriate integration measure defined with respect to these fields [6]. The *action* $S[\bar{\psi}, \psi]$ is uniquely determined by the Hamiltonian operator, and is given by

$$\begin{aligned} S[\bar{\psi},\psi] &= \int d\mathbf{x} \int_0^{1/k_\mathrm{B}T} d\tau\; \bar{\psi}(\mathbf{x},\tau) \left[-\frac{\partial}{\partial\tau} + \mu\right] \psi(\mathbf{x},\tau) \\ &\quad - \int_0^{1/k_\mathrm{B}T} d\tau\; H\left(\bar{\psi}(\mathbf{x},\tau), \psi(\mathbf{x},\tau)\right). \end{aligned} \tag{16}$$

Here $H = \int d\mathbf{x}\; h\left(\bar{\psi}(\mathbf{x},\tau), \psi(\mathbf{x},\tau)\right)$, and the Hamiltonian density h as a function of $\bar{\psi}$ and ψ has the same functional form as does $\hat{H}$ as a function of $\hat{a}^\dagger$ and $\hat{a}$.

Since the Hamiltonian taken at some imaginary time does not commute with the Hamiltonian taken at another imaginary time, we see that for quantum systems the statics and the dynamics are intrinsically coupled and need to be treated together and simultaneously. All phenomena, quantum phase transitions in particular, that must be described by means of quantum statistical mechanics, therefore, automatically fall under the title of this NATO ASI. It further follows that for quantum phase transitions, in contrast to classical ones, there are three independent critical exponents, and the dynamical exponent z needs to be determined together with the static ones.

The homogeneity law for the free energy density, Eq. (13), can now easily be generalized to the quantum case. From Eqs. (2) and (4, 5) we see that, as a function of t, frequencies scale like $t^{\nu z}$. Furthermore, in the imaginary time formalism, temperature and Matsubara frequencies are directly proportional to one another, and it is therefore plausible that temperature and frequency will scale in the same way. We thus add T as an argument to our free energy density and acknowledge the explicit T in the definition of f to obtain

$$f(t, T, B, \ldots) = b^{-(d+z)} f(t\, b^{1/\nu}, T\, b^{z}, B\, b^{x_B}, \ldots). \tag{17}$$

Comparing Eqs.(13) and (17) we see that a quantum phase transition in d spatial dimensions resembles the corresponding classical transition in $d_{\text{eff}} = d + z$ spatial dimensions! This is also plausible from the point of view of the Landau functional (see below), where a spatial integral $\int d\mathbf{x}$ in the classical case gets replaced by a space-time integral $\int d\mathbf{x}\, d\tau$ in the quantum case. Early work on the subject suggested that this observation provides a fast and easy solution to the problem of quantum critical behavior. However, as we will see, the argument is too superficial to be reliable, and the extent to which it holds requires a careful and detailed discussion.

1.3. AN EXAMPLE: THE PARAMAGNET-TO-FERROMAGNET TRANSITION

Let us illustrate the concepts introduced above by means of a concrete example. For definiteness, we consider a metallic or itinerant ferromagnet.[5]

1.3.1. *The phase diagram*

Figure 1 shows a schematic phase diagram in the T-J plane, with J the strength of the exchance coupling that is responsible for ferromagnetism. The coexistence curve separates the paramagnetic phase at large T and small J from the ferromagnetic one at small T and large J. For a given J, there is a critical temperature, the Curie temperature T_c, where the phase transition occurs. This is the usual situation: a particular material has a given value of J, and the classical transition is triggered by lowering the temperature through T_c. Alternatively, however, we can imagine changing

[5]For the purposes of this subsection we might as well consider localized spins, but the theory discussed in Sec. 2 below applies to itinerant magnets only.

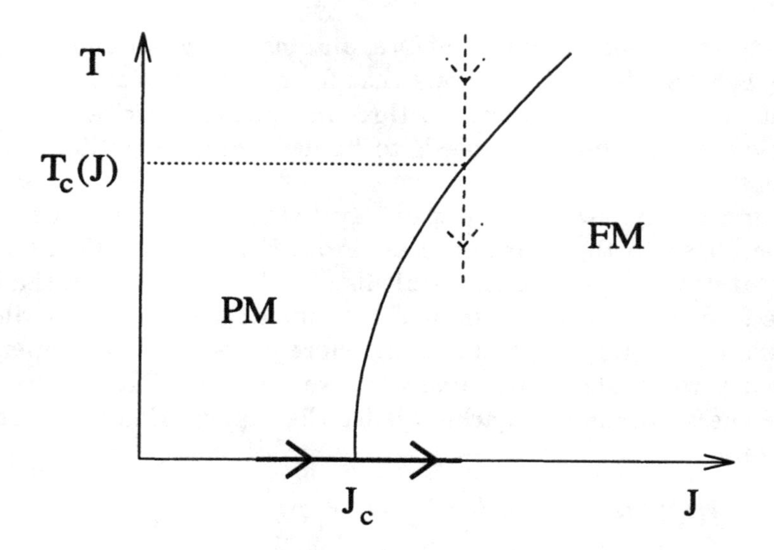

Figure 1. Schematic phase diagram showing a paramagnetic (PM) and a ferromagnetic (FM) phase. The path indicated by the dashed line path represents a classical phase transition, and the one indicated by the solid line a quantum phase transition.

J at zero temperature (e.g. by alloying the magnet with some non-magnetic material). Then we will encounter the paramagnet-to-ferromagnet transition at the critical value J_c. This is the quantum phase transition we are interested in. Since we have seen that the quantum transition is, loosely speaking, related to the classical one in a different spatial dimension, and since we know that changing the dimensionality usually means changing the universality class, we expect the critical behavior at this quantum critical point to be different from the one observed at any other point on the coexistence curve.

This brings us to the question of how continuity is guaranteed when one moves along the coexistence curve. The answer, which was found by Suzuki [7], also explains why the behavior at the $T = 0$ critical point is relevant for observations at small but non-zero temperatures. Consider Fig. 2, which shows an enlarged section of the phase diagram near the quantum critical point. The critical region, i.e. the region in parameter space where the critical power laws can be observed, is bounded by the two dashed lines. It then turns out that the critical region is divided into two subregions, denoted by 'QM' and 'classical' in Fig. 2, in which the observed critical behavior is predominantly quantum mechanical and classical, respectively. The division between these two regions (shown as a dotted line in Fig. 2) is not sharp (and neither is the boundary of the critical region), but rather a smooth *crossover* from predominantly quantum mechanical to predominantly classical critical behavior is observed as J is increased at low but

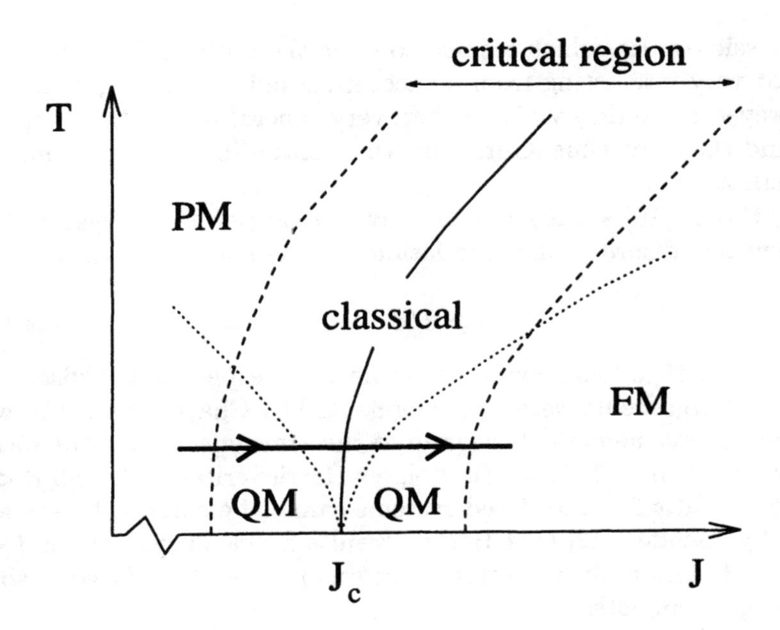

Figure 2. Schematic phase diagram as in Fig. 1 showing the vicinity of the quantum critical point ($J = J_c, T = 0$). Indicated are the critical region, and the regions dominated by the classical and quantum mechanical critical behavior, respectively. A measurement along the path shown will observe the crossover from quantum critical behavior away from the transition to classical critical behavior asymptotically close to it.

non-zero T. The figure illustrates that the asymptotic critical behavior, very close to the transition, is classical for all non-zero T, but that for small T there nevertheless is a sizeable region where quantum critical behavior is observable. It also makes it clear that the abovementioned continuity is realized by means of a distribution limit.

1.3.2. *Classic results for itinerant ferromagnets*

The subject of quantum magnetic phase transitions was pioneered by Hertz [8], who built on earlier work by Suzuki [7] and others [9]. Hertz's main result can be recovered by combining the above observation that the quantum phase transition should correspond to the classical one in $d_{\text{eff}} = d+z$ dimensions, with the fact that for classical magnets the upper critical dimension d_c^+, i.e. the dimensionality above which the critical behavior is mean-field like, is $d_c^+ = 4$. It then follows that for the quantum transition, $d_c^+ = 4 - z$. Hertz further found that $z = 3$ for clean itinerant ferromagnets, and $z = 4$ for disordered ones (we will reproduce his results in Sec. 2 below). It then seemed to follow that $d_c^+ = 1$ for clean magnets, and $d_c^+ = 0$ for disordered ones, so that the quantum critical behavior would be mean-field like in all physical dimensions ($d = 2, 3$).

These classic results, which seemed to solve the problem (albeit the answer was not very interesting from a theoretical point of view), cannot be correct, however, since they violate other, very general results that were obtained around the same time. Curiously, this contradiction went unnoticed for more than 20 years.

In 1974, Harris [10] argued that at any critical point in a system that contains quenched disorder, the correlation length exponent must obey the inequality

$$\nu \geq 2/d \tag{18}$$

in order for the critical behavior to be stable with respect to the disorder.[6] Harris' physical arguments were later augmented by Chayes et al. [11], who proved a rigorous mathematical theorem to the same effect. Now the mean-field value of ν is $\nu_{\rm MF} = 1/2$, which violates Harris' criterion for all $d < 4$. Hence Hertz's results for disordered systems cannot be correct. It was also pointed out by Sachdev [12] that Hertz's results for clean systems in $d < 1$ (an academic, but nonetheless interesting case) were at odds with some general scaling arguments.

It finally turned out that the classic results for both the clean and the disordered case are indeed incorrect, for rather subtle and interesting reasons. It also was shown that the theory can be salvaged with relatively little effort, provided that one acknowledges the long-range correlations in itinerant electron systems that were explained in the preceding lectures [1]. This theory was developed in a series of papers [13, 14], the main arguments and results of which we reproduce in the next section.

2. Quantum Critical Behavior of Itinerant Ferromagnets

We now sketch the derivation of an effective field theory for itinerant ferromagnets, starting from a microscopic Hamiltonian. As we will see, the derivation makes massive use of our knowledge of the properties of *non-magnetic* electron systems. It is this knowledge that makes the task of determining the correct quantum critical behavior much easier than it would have been in the 1970s.

2.1. LANDAU-GINZBURG-WILSON THEORY

Let us consider a microscopic model for an itinerant ferromagnet. What we have in mind is to develop an effective theory from such a microscopic model that will be capable of correctly describing the quantum critical behavior without being more detailed than necessary in other respects. To

[6]This is not how Harris formulated his criterion, but it can be brought into this form by using suitable exponent relations.

this end we follow Landau's strategy [15] of formulating a theory in terms of an appropriate order parameter, and integrating out all other degrees of freedom. For a ferromagnet, the relevant observable is the spin density

$$\mathbf{n}_s(\mathbf{x},\tau) = \bar{\psi}(\mathbf{x},\tau)\,\vec{\sigma}\,\psi(\mathbf{x},\tau), \tag{19}$$

where the vector $\vec{\sigma}$ denotes the Pauli matrices. The relevant electron-electron interaction is the spin-triplet interaction between the spin density fluctuations,

$$S_{\text{int}} = J\int d\mathbf{x}\int_0^{1/k_B T} d\tau\ \mathbf{n}_s(\mathbf{x},\tau)\,\cdot\,\mathbf{n}_s(\mathbf{x},\tau). \tag{20}$$

The complete action is

$$S = S_0 + S_{\text{int}}. \tag{21}$$

Here S_0 comprises all parts of the action other than S_{int} defined in Eq. (20). It contains a term describing free electrons (or lattice electrons[7]) as well as electron-electron interactions in all channels other than the spin-triplet one. In particular, S_0 contains the direct or Coulomb interaction between the electronic charge densities. We will refer to the fictitious system that is described by the action S_0 as the *reference ensemble*, and it will play an important role in what follows.

Our goal is to calculate the partition function

$$Z = \int D[\bar{\psi},\psi]\ e^{S_0[\bar{\psi},\psi]}\ e^{J\int dx\ \mathbf{n}_s(x)\cdot\mathbf{n}_s(x)}. \tag{22}$$

Here and in what follows we adopt a four-vector notation $x \equiv (\mathbf{x},\tau)$ for space-time integrals. We now concentrate on the term S_{int}, and decouple it by means of a Gaussian or Hubbard-Stratonovich transformation [16]. The latter consists of introducing a classical (i.e. commuting) auxiliary field $M(x)$ that couples linearly to the spin density, and rewriting $e^{S_{\text{int}}}$ as a Gaussian integral,

$$Z = \int D[\bar{\psi},\psi]\ e^{S_0[\bar{\psi},\psi]}\int D[M]\ e^{-J\int dx\ M^2(x)+2J\int dx\ M(x)n_s(x)}. \tag{23}$$

From now on, for simplicity we ignore the vector nature of the spin density. Taking it into account just complicates the notation without leading to qualitatively important effects. Next, we interchange the M and ψ-integrations, and formally carry out the latter,

$$Z = \int D[M]\ e^{-J\int dx\ M^2(x)}\ Z_0[M], \tag{24}$$

[7]For our purposes, which are aimed at long-wavelength phenomena, it is irrelevant whether one considers a lattice model or a continuum one.

where we recognize

$$Z_0[M] = \int D[\bar{\psi},\psi]\; e^{S_0[\bar{\psi},\psi]+2J\int dx\; M(x)\, n_s(x)} \tag{25}$$

as the partition function of the reference ensemble in an external 'magnetic field' that is proportional to $M(x)$. From the linear coupling between $M(x)$ and the spin density it is clear that the expectation values of these two quantities are proportional to one another. We thus identify $M(x)$ as the order parameter field whose expectation value is the magnetization. It is customary to define a Landau-Ginzburg-Wilson or LGW functional $\Phi[M]$,

$$\Phi[M] = J\int dx\; M^2(x) - \ln Z_0[M], \tag{26}$$

where

$$Z_0[M] = \langle e^{2J\int dx\; M(x)\, n_s(x)}\rangle_0. \tag{27}$$

Here $\langle\ldots\rangle_0$ denotes an average with respect to the reference ensemble action S_0, and the partition function can be expressed in terms of the LGW functional as

$$Z = \int D[M]\; e^{-\Phi[M]}. \tag{28}$$

The above derivation of an order parameter theory for quantum ferromagnets is due to Hertz [8]. The salient point is that the LGW functional is given in terms of the free energy of the reference ensemble in a (space and time dependent) 'external field' that is given by the order parameter field. A more technical way to say this is that the LGW is given by the generating functional for connected spin density correlation functions in the reference ensemble, viz. $\ln Z_0$. Notice that all of the above have been exact manipulations. The strategy underlying this exact rewriting of the partition function is to express the free energy of the full system, which undergoes a phase transition, in terms of that of the reference ensemble, which does not.[8] The price one pays is that one needs to know the free energy of the latter in the presence of an arbitrary space and time dependent magnetic field. As we will see, however, enough is known about correlated electrons in magnetic fields to successfully implement this strategy.

2.2. THE LANDAU EXPANSION

Follwing Landau, the next step is to expand the LGW functional in powers of M. Remembering that the reference ensemble has no spontaneous

[8]Remember that the reference ensemble is missing the spin-triplet part of the electron-electron interaction that is responsible for triggering ferromagnetic phase transitions.

magnetization, and hence $\langle n_s \rangle_0 = 0$, we have

$$Z_0 = 1 + 2J^2 \int dx\, dy\; M(x)\, M(y)\, \langle n_s(x)\, n_s(y) \rangle_0 + \ldots, \qquad (29)$$

where the terms not shown explicitly are of $O(M^4)$ or higher. Taking a Fourier transform, we thus obtain for the LGW functional

$$\begin{aligned} \Phi[M] \quad = \quad & \frac{1}{V} \sum_{\mathbf{q}} T \sum_{\omega} M(\mathbf{q}, \omega)\; [1 - J\chi_s(q, \omega)]\; M(-\mathbf{q}, \omega) \\ & + \sum_{\{\mathbf{q}, \omega\}} u_4(\{\mathbf{q}, \omega\})\, M^4(\mathbf{q}, \omega) + \ldots, \end{aligned} \qquad (30)$$

where $q = |\mathbf{q}|$, V is the system volume, T is the temperature,

$$\chi_s(q, \omega) = \langle n_s(\mathbf{q}, \omega)\, n_s(-\mathbf{q}, -\omega) \rangle_0^c \qquad (31)$$

is the connected two-point spin-density correlation function of the reference ensemble, i.e., its spin susceptibility, and

$$u_4 = \langle n_s\, n_s\, n_s\, n_s \rangle_0^c \qquad (32)$$

is the corresponding connected four-point function, etc. We have used an obvious schematic notation for the quartic terms, since we will not study them in detail here.

The salient point of this formal development is that the LGW functional is given in terms of the connected spin density correlation functions of the reference ensemble. The reference ensemble, however, is a Fermi liquid, or its generalization to the case of quenched disorder, and so its correlation functions are known! Indeed, the preceding lectures [1] has explained in some detail what they are, and we can now draw on that knowledge. For the sake of definiteness we will discuss only disordered systems,[9] where the effects we want to demonstrate are most pronounced. Qualitatively similar, albeit weaker, phenomena are present in clean systems as well, as has been discusssed in the original literature [13, 14]. The appropriate limit to study the correlation functions is that of small wavenumbers and frequencies,

[9]Strictly speaking, we should have used a replicated theory to deal with the quenched disorder. In the interest of keeping our pedagogical discussion simple we have suppressed this technical point. For an introductory discussion of the replica trick, see Ref. [17]

$q, \omega \to 0$, with $\omega \ll q$ [10] in suitable units.[11] As we have seen in the preceding lectures [1], see also [18], the spin susceptibility of a disordered Fermi liquid in this limit reads

$$\chi_s(q,\omega) = \chi_s^0(q)\left[1 - |\omega|/Dq^2 + \ldots\right], \tag{33}$$

and the long-wavelength expansion of the static spin susceptibility χ_s^0 reads

$$\chi_s^0(q) = \text{const.} - q^{d-2} - q^2 + o(q^2), \tag{34}$$

where $o(q^2)$ denotes terms that are smaller than q^2, and we have omitted positive prefactors of all terms in the expansion, as they will not be important for what follows. The LGW functional now takes the form

$$\begin{aligned}\Phi[M] &= \frac{1}{V}\sum_{\mathbf{q}} T \sum_{\omega} M(\mathbf{q},\omega)\left[t + q^{d-2} + q^2 + |\omega|/q^2\right] M(-\mathbf{q},-\omega) \\ &\qquad + O(M^4).\end{aligned} \tag{35}$$

The four-point correlation function u_4 also contains nonanalyticities, which show up the in the quartic term in $\Phi[M]$, and the same is true for all higher order terms. While a careful study of these terms is necessary for a complete treatment of the problem, we suppress them here for brevity and simplicity, and refer the interested reader to the original literature where they have been discussed in detail [13].

Before we analyze the LGW functional, Eq.(35), in the next subsection, two remarks are in order. First, if we had used a reference ensemble of non-interacting electrons, rather than our more realistic one that includes electron-electron interactions, then we would have missed the non-analytic terms proportional to q^{d-2} in Eqs.(34) and (35). As a consequence, we would have recovered Hertz's results which, as we have seen earlier, cannot be correct. While it seems at this point as if a realistic choice of the reference ensemble were crucial, this is disturbing from some fundamental theoretical points of view. Indeed, a careful investigation reveals that the precise choice of the reference ensemble is *not* important. We will come back to this point in Sec.2.4 below. Second, the physical origins of the q^{d-2} term are long-range

[10] Otherwise one does not reach criticality, which can be seen as follows. Since the magnetization is conserved, ordering on a length scale L requires some spin density to be transported over that length, which takes a time $t \propto L^2/D$, with D the spin diffusion coefficient. Now look at the system at a momentum scale q or a length scale $L \propto 1/q < \xi$, with ξ the coherence length. Because of the time it takes the system to order on that scale, the condition for criticality is $L^2 < \text{Min}(Dt, \xi^2)$. In particular, one must have $L^2 < Dt$, or $\omega \propto 1/t < Dq^2$.

[11] For instance, one can measure q in units of the Fermi wavenumber k_F, and ω in units of Dk_F^2, with D the spin diffusion coefficient of the reference ensemble.

spatial correlations in the reference ensemble of interacting electrons that underlies our effective LGW action, as has been explained in the preceding lectures [1]. These long-range spatial correlations are in turn a consequence of the coupling of statics and dynamics in quantum systems. We now see what has happened: In addition to the order parameter fluctuations that develop a long range near criticality, our electron systems also contains long-range correlated degrees of freedom (viz. the 'diffusons' of the preceding lectures) that have nothing to do with phase transition physics, and that are present even far away from the transition. These degrees of freedom have been integrated out in our derivation of the LGW functional, which by definition is a functional of the order parameter field only. The inevitable consequence of this integrating out of slow modes is nonanalyticities in the resulting LGW functional. Such nonanlyticities violate the very spirit of the LGW concept, and to perform explicit calculations for the non-local field theory given by Eq.(35) would be very difficult indeed. An obvious way to avoid these problems would be to *not* integrate out the diffusons, but rather derive a generalized LGW theory in terms of *all* the soft modes in the systems. However, as we will see in the next subsection, for the purpose of determining the critical behavior the non-local theory can be handled, and the present route is the fastest one to answer the questions we have asked.

2.3. RENORMALIZATION GROUP ANALYSIS

To finish our treatment of itinerant ferromagnets, we now analyze the LGW functional, Eq.(35), by means of power counting or a tree-level renormalization group (RG) analysis. Space and time constraints do not allow us to explain this technique here. Readers not familiar with it can find excellent and very accessible treatments in Refs. [2].

Let us restrict ourselves to spatial dimensions $d < 4$, where the q^{d-2} term dominates over the analytic q^2 term, and let us look for a Gaussian RG fixed point where neither the q^{d-2} term nor the $|\omega|/q^2$ term in Eq.(35) are renormalized. At such a fixed point, the critical order parameter correlation function will behave like

$$\langle M\,M\rangle_{t=0} \propto 1/q^{d-2} \equiv q^{2-\eta}, \tag{36}$$

where the last relation reflects the definition of the critical exponent η, see Eq. (3). We thus have

$$\eta = 4 - d. \tag{37}$$

Furthermore, at such a Gaussian fixed point the frequency clearly scales with the wavenumber like $\omega \sim q^d \equiv q^z$,[12] which yields a dynamical critical exponent

$$z = d. \tag{38}$$

Finally, the exponent γ for the Gaussian fixed point can also just be read off the Gaussian action: At zero frequency and wavenumber, we have

$$\langle M\, M \rangle_{q=\omega=0} \propto 1/t \equiv t^{-\gamma}, \tag{39}$$

so that we have

$$\gamma = 1. \tag{40}$$

In order to determine the correlation length exponent ν, we define the scale dimension $[q]$ of a wavenumber q to be $[q] = 1$. The requirement that the q^{d-2} term in the LGW functional be dimensionless then leads to a scale dimension of the order parameter field of $[M] = -(d-2)/2$. Since the t term must also be dimensionless, this yields

$$t \sim q^{d-2} \sim \xi^{-(d-2)} \equiv \xi^{-1/\nu}, \tag{41}$$

from which we read off ν as

$$\nu = 1/(d-2). \tag{42}$$

All of these results obviously hold only for $2 < d < 4$. For $d \leq 2$ the electrons in the reference ensemble become localized (see the preceding lectures) and our theoretical framework breaks down, while for $d > 4$ the q^2 term dominates over the q^{d-2} term, and we recover Hertz's mean-field exponents, i.e. $\eta = 0$, $z = 4$, $\gamma = 1$, and $\nu = 1/2$.

A quick check shows that our result fulfills the Harris criterion, $\nu \geq 2/d$ (see Sec.1.3.2 above) for all values of d. While this is encouraging, it is of course only a necessary criterion for our results representing the correct quantum critical behavior, not a sufficient one. To establish the latter, one needs to consider the higher order terms in the LGW functional, and establish that our Gaussian fixed point is stable. At tree level, this has been done in the original literature [13], and the result was that the fixed point is indeed stable. The analysis of the quartic term in Φ in particular also yields the equation of state, and thus the critical behavior of the order parameter itself, i.e. the critical exponents β and δ. The result is

$$\beta = 2/(d-2) \quad , \quad \delta = d/2, \tag{43}$$

[12] We use $\propto$ for 'proportional to', and $\sim$ for 'scales like' or 'has the same scale dimension as'.

for $2 < d < 6$, while for all $d > 6$ these exponents have their mean-field values $\beta = 1/2$ and $\delta = 3$, respectively. The analysis of the higher order terms thus establishes $d = 6$ as another upper critical dimensionality, in addition to $d = 4$. In $d = 4$ and in $d = 6$ logarithmic corrections to scaling occur, as is usually the case at an upper critical dimensionality.

Finally, we mention again that an analogous analysis of clean systems yields qualitatively very similar results. Essentially, the exponent $d - 2$ in the non-analytic terms gets replaced by $d - 1$, which leads to a single upper critical dimension $d_c^+ = 3$. Contrary to the disordered case, the critical behavior in the most interesting dimension $d = 3$ is therefore mean-field like with logarithmic corrections to scaling [13].

2.4. FINAL REMARKS

We finish this section with a few additional remarks.[13] As we have sketched in the preceding subsection, and as has been more carefully established in the original literature, we have managed to determine the critical behavior at the quantum ferromagnetic transition exactly, yet this behavior is not mean-field like. This is surprising, as usually non-mean field like critical behavior cannot be obtained exactly, save for a very few models. To find out what has enabled us to do so, let us look again at our LGW functional, Eq.(35). One way to state the effect of the non-critical soft modes that have led to the q^{d-2} term is to say that they have established an effective long-range interaction between the order parameter fluctuations. Indeed, a Fourier transform of the non-analytic term yields an interaction that falls off like $1/r^{2(d-1)}$. It is well known that such long-range interactions stabilize Gaussian critical behavior that is not mean-field like [19]. What is remarkable here is that this long-range interaction is not put in by hand, but rather is generated by the system itself via the non-critical slow modes.

These considerations raise the question of whether the phenomenon discussed here is germane to quantum phase transitions. In principle, it is not. Whenever there are slow modes in addition to the order parameter fluctuations that couple to the latter, one will obtain nonanalyticities in the LGW functional, and resulting unusual critical behavior, irrespective of whether one deals with a classical or a quantum phase transition. However, quantum transitions are much more susceptible to this mechanism, for the simple reason that there are many modes that are soft only at $T = 0$ and acquire a mass at non-zero temperature. Our diffusons are a good example: as was discussed in the preceding lectures, they are indeed massive at $T > 0$.

[13] Like the preceding subsection, some of these remarks require familiarity with renormalization group techniques.

Finally, we come back to a point raised at the end of Sec.2.2. There we mentioned that whether or not one obtains the correct critical behavior seems to depend crucially on the choice of the reference ensemble. While one might argue that a non-interacting reference ensemble is simply not a realistic model, this leads to the following paradox. Suppose we consider a model whose action consists only of a free electron part, and the spin-triplet interaction S_{int} of Eq.(20). Then a decoupling of S_{int}, as performed above, seems to lead to mean-field critical behavior. However, if we considered some fraction of S_{int} a part of S_0, and decoupled the rest, then we would have an interacting reference ensemble and would obtain the above non-mean field like critical behavior. Clearly, both procedures are equally valid and should lead to the same result. The resolution of this paradox is as follows. In the case of an interacting reference ensemble, already the bare LGW functional contains the crucial nonanalyticities. Therefore, an RG analysis at tree level is sufficient to obtain the correct critical behavior. In the case of a non-interacting reference ensemble, on the other hand, the bare action does not contain the crucial non-analytic terms, but they are generated if the RG analysis is carried to higher order in the loop expansion. Indeed, an inspection of Hertz's model shows that the q^{d-2} term is indeed generated by the RG, starting at one-loop order. Since the generated term is relevant with respect to the mean-field fixed point, it invalidates the zero-loop analysis. These observations serve as a reminder of a fact that is well-known in principle, but occasionally forgotten. Any RG analysis to a given order in a loop expansion gives the correct answer only if no relevant new terms in the action are generated at higher order. Unfortunately, *proving* that no such terms are generated at any order in the loop expansion amounts to proving that the theory is renormalizable, a task that is very difficult and has been done only for a few select models.

3. The Anderson-Mott Transition

At the end of these lectures, we would like to briefly touch upon some very different quantum phase transitions, namely the types of metal-insulator transition known as Anderson, and Anderson-Mott transitions, respectively. Our motivation for doing so is chiefly to dispel any possible misconception that the concepts developed so far apply only to quantum phase transitions that are magnetic in nature, or that are represented as the $T = 0$ end point of a line of classical phase transitions. Indeed, the Anderson-Mott transition is neither. We would also like to illustrate the point that the scaling ideas that have been so successful in connection with phase transitions, classical and quantum mechanical, may be applied to transport coefficients as well as to thermodynamic quantities [20]. Apart from making these two points, the time and space we devote to this subject are grossly inadequate. Extensive reviews can be found in Refs. [3, 21, 22, 23].

3.1. THE SIMPLEST METAL-INSULATOR TRANSITION: THE CLASSICAL LORENTZ MODEL

The simplest metal-insulator transition occurs in a classical model, viz. the classical Lorentz gas. It is well known that with increasing scatterer density n, the diffusivity D of the moving particle decreases (see the preceding lectures), until it finally reaches zero at a critical scatterer density n_c. This behavior can be seen in the numerical data shown in Fig. 3. With $t =$

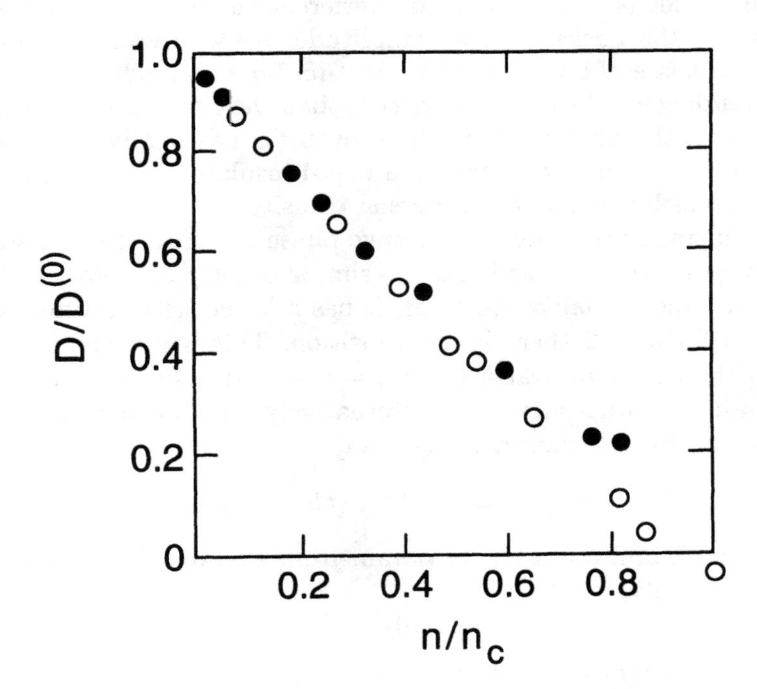

Figure 3. Numerical simulation data for the diffusion coefficient D versus the scatterer density n of a 2-d classical Lorentz model. Full symbols represent data by Bruin [28], and open ones data by Alder and Alley [29], and by Alley [30]. D is normalized by its Boltzmann value $D^{(0)}$, and n by its critical value n_c. From Ref. [21].

$|n - n_c|/n_c$ being the dimensionless distance from the critical point, one has

$$D(t \to 0) \propto t^s, \tag{44}$$

with s a critical exponent. The reason for the localization of the diffusing particle is that with increasing scatterer density it is increasingly likely to get trapped in cages of scatterers from which it cannot escape.

3.2. ELECTRONS: THE ANDERSON TRANSITION, AND THE MOTT TRANSITION

Let us now add quantum mechanics to these considerations, and consider the quantum Lorentz model that was discussed in detail in the preceding lectures. Quantum mechanics has two competing effects on the transport properties: On the one hand, it should increase the diffusivity, since it allows the particle to tunnel out of cages it would be trapped in classically. On the other hand, it leads to the quantum interference or weak-localization effects that enhance the backscattering amplitude. As we saw in the preceding lectures, the latter effect wins. The quantum Lorentz model therefore has a stronger tendency to localize the particle than the classical one, to the point that in $d = 2$ the particle is localized even for arbitrarily small scatterer density. In $d > 2$, however, there is a metal-insulator transition at a finite value of n, which is called an Anderson transition.

The Anderson transition is a strange phase transition from a statistical mechanics point of view, as it has no simple order parameter, and no upper critical dimensionality. However, it has a lower critical dimensionality, $d_c^- = 2$, as for $d \leq 2$ there is no transition. This has been exploited for studies of the Anderson transition in $\epsilon = d - 2$ expansions. More generally, the dynamical conductivity σ (or, alternatively, the diffusion coefficient D) obeys a generalized homogeneity law [24]

$$\sigma(t, \omega) = b^{-(d-2)}\, \sigma(t\, b^{1/\nu}, \omega\, b^z). \tag{45}$$

Putting $\omega = 0$, and $b = t^{-\nu}$, we obtain from Eq.(45) the behavior of the static conductivity,

$$\sigma(t, 0) \propto t^s, \tag{46}$$

with a conductivity exponent

$$s = \nu(d-2) \geq 2(d-2)/d. \tag{47}$$

The equality in Eq.(47) is known as Wegner's scaling law, and the inequality results from the Harris criterion, Eq.(18). Due to the poor convergence properties of the $2 + \epsilon$ expansions, no reliable theoretical values for s or ν are available. Numerical calculations in $d = 3$ yield values for ν in the range 1.3 - 1.5 [23].

The quantum Lorentz model does not contain any electron-electron interaction, and the Anderson transition is therefore entirely driven by disorder. The opposite case, namely a metal-insulator transition that is entirely driven by interactions, with no quenched disorder present, was proposed by Mott to explain why certain materials with one electron per unit cell, e.g. NiO, are insulators. Mott's original idea hinged on the long-range nature of the Coulomb interaction, and in this case the metal-insulator transition comes about by means of a breakdown of screening and is of first order [25]. A similar, albeit continuous, transition is believed to occur in a model with a short-ranged electron-electron interaction known as the Hubbard model. This Mott-Hubbard transition is still not well understood in $d = 3$, although much progress has been made recently on high-dimensional models [26].

3.3. THE ANDERSON-MOTT TRANSITION

An Anderson-Mott transition results if one combines the driving forces for the Anderson and Mott transitions and considers the case of interacting electrons in the presence of disorder. This is a problem of great experimental interest. For instance, the metal-insulator transition that is observed in doped semiconductors as a function of the dopant concentration is believed to be of that type. Fig. 4 shows the infrared absorption spectrum of phosphorus-doped silicon. For very low dopant concentrations one sees a hydrogen-like spectrum that is produced by isolated phosphorus atoms. With increasing dopant concentration these 'atoms' start to overlap, which leads to a broadening of the spectral features. At the highest concentration shown in the figure, the spectrum is smooth, but it still represents an insulator (no absorption at zero energy). With further increasing donor concentration, the system undergoes a quantum phase transition to a metal, as can be seen in Fig. 5, where both the static conductivity and the dielectric susceptibility are plotted versus the phosphorus concentration. This system, and some other doped semiconductors, are, experimentally, the best-studied examples of metal-insulator transitions. Notice that this is a quantum phase transition that has no classical counterpart, as there can be no true insulator at any non-zero temperature.

Much theoretical effort has been devoted to the Anderson-Mott transition. These approaches fall into two distinct classes. The first one contains ϵ-expansions about the lower critical dimension $d_c^- = 2$ [21]. These theories lead again to Wegner scaling, Eq.(45), as in the case of the Anderson transition, and Wegner's scaling law, Eq.(47), still holds. This leads to $s = \nu \geq 2/3$ in $d = 3$, which contradicts the experimental result that the value of s is close to $1/2$ [32].[14]

The second class is formed by an order parameter theory [3] that uses the fact that the density of states at the Fermi level may be considered

[14]Not all experiments agree with this result. Ref. [27] has reported $s > 1$ for Si:P.

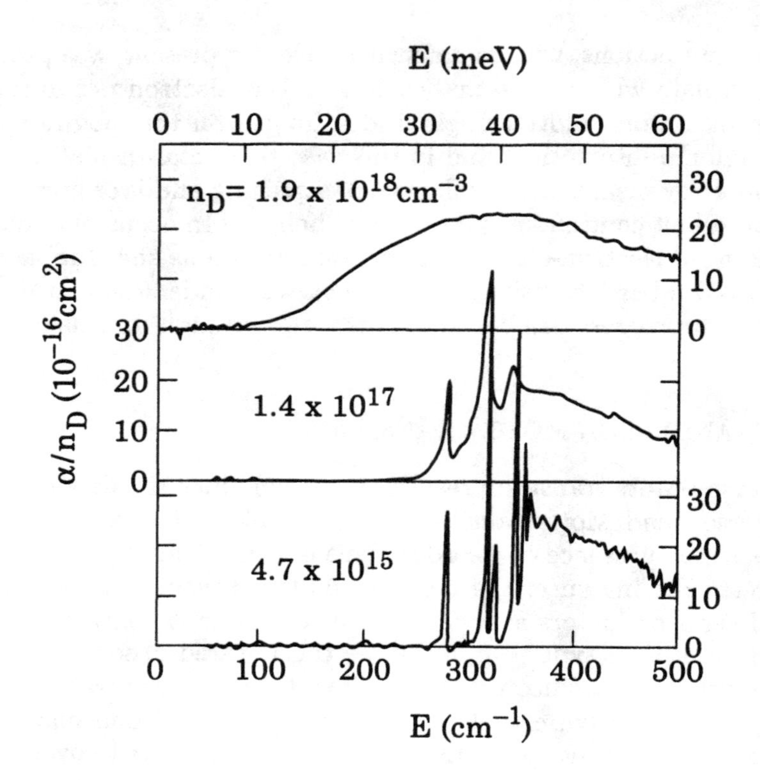

Figure 4. Infrared absorption coefficient α for three different donor concentrations n_D in Si:P as measured by Thomas at al. [31]. The critical concentration in this system is $n_c \approx 3.7 \times 10^{18}\ \mathrm{cm}^{-3}$. From Ref.[21].

an order parameter for the Anderson-Mott transition (in contrast to the case of the Anderson transition, where the density of states is uncritical.) This theory has established $d_c^+ = 6$ as the upper critical dimension for the Anderson-Mott transition (again in contrast to the Anderson transition, where $d_c^+ = \infty$), and the critical behavior for $d > 6$ is exactly known and mean-field like. Certain technical analogies between this theory and theories for magnets in random magnetic fields have led to the suggestion that in $d < 6$, and in particular in $d = 3$, the Anderson-Mott transition has aspects that are reminiscent of a glass transition, with exponential rather than power-law critical behavior for many observables. A scaling theory has been developed for this unorthodox critical behavior [3], which is at least consistent with existing experimental results, but so far no microscopic theory exists.

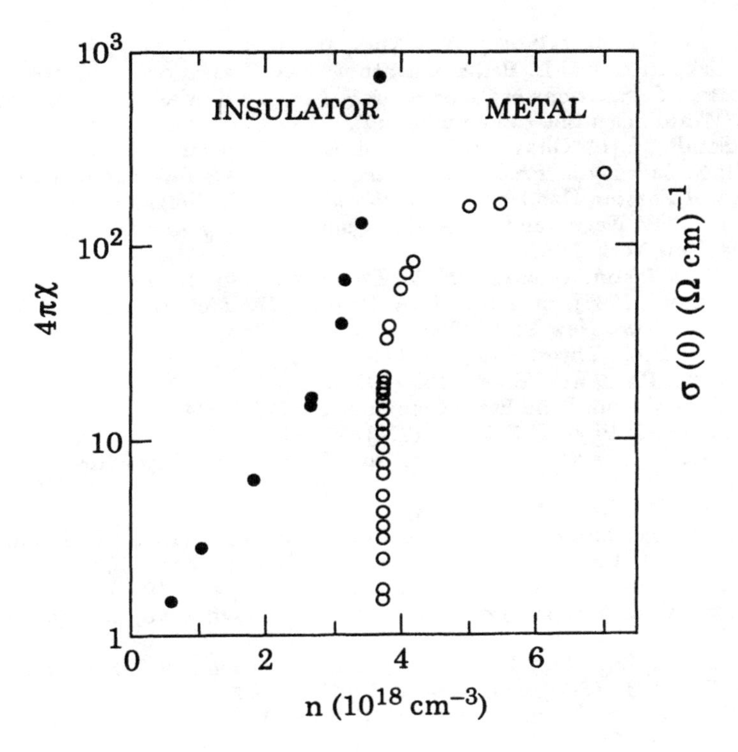

Figure 5. Divergence of the dielectric susceptibility χ (full symbols), and vanishing of the static conductivity σ (open symbols), both extrapolated to zero temperature, at the metal-insulator transition in Si:P as measured by Rosenbaum et al. [32]. From Ref.[21].

Acknowledgements

We would like to thank our collaborators on some of the work on ferromagnetic systems reviewed above, Thomas Vojta, Rajesh Narayanan, and Andy Millis. This work was supported by the NSF under grant numbers DMR–96–32978 and DMR–98–70597. This research was supported in part by the National Science Foundation under grant No. PHY94-07194.

References

1. T.R. Kirkpatrick and D. Belitz, this volume, p. 379.
2. (a) For a description of classical phase transition phenomenology, see, H.E. Stanley, *Introduction to Phase Transitions and Critical Phenomena*, (Oxford University Press, Oxford, 1971). For the modern theoretical treatment of the subject, see, e.g., (b) K. G. Wilson and J. Kogut, Phys. Rep. **12**, 75 (1974); (c) S. K. Ma, *Modern Theory of Critical Phenomena*, (Benjamin, Reading, MA, 1976); (d) M. E. Fisher in *Advanced Course on Critical Phenomena*, F. W. Hahne, ed., (Springer, Berlin, 1983), p.1; (e) N. Goldenfeld, *Lectures on Phase Transitions and the Renormaliza-*

tion Group, (Addison-Wesley, New York, 1992).

3. T.R. Kirkpatrick and D. Belitz, Quantum Phase Transitions in Electronic Systems, in *Electron Correlations in the Solid State*, Norman H. March, ed., (Imperial College Press/World Scientific, to be published).
4. S. L. Sondhi, S. M. Girvin, J. P. Carini, and D. Shahar, Rev. Mod. Phys. **69**, 315 (1997); S. Sachdev in *Proceedings of the 19th IUPAP International Conference on Statistical Physics*, Hao Bailin, ed., (World Scientific, Singapore, 1996), p. 289.
5. See, e.g., J. W. Negele and H. Orland, *Quantum Many–Particle Systems*, (Addison-Wesley, New York, 1988).
6. (a) J. Zinn-Justin, *Quantum Field Theory and Critical Phenomena*, (Clarendon Press, Oxford, 1989), ch. 27; (b) F. A. Berezin, *The Method of Second Quantization*, (Academic Press, New York, 1966).
7. M. Suzuki, Prog. Theor. Phys. **56**, 1454 (1976).
8. J. A. Hertz, Phys. Rev. B **14**, 1165 (1976).
9. M. T. Beal-Monod, Solid State Commun. **14**, 677 (1974).
10. A. B. Harris, J. Phys. C **7**, 1671 (1974).
11. J. Chayes, L. Chayes, D. S. Fisher, and T. Spencer, Phys. Rev. Lett. **57**, 2999 (1986).
12. S. Sachdev, Z. Phys. **94**, 469 (1994).
13. T.R.Kirkpatrick and D.Belitz, Phys. Rev. B **53**, 14364 (1996); D. Belitz and T. R. Kirkpatrick, J. Phys. Cond. Matt. **8**, 9707 (1996); Thomas Vojta, D. Belitz, R. Narayanan, and T. R. Kirkpatrick, Z. Phys. B **103**, 451 (1997).
14. D. Belitz, T. R. Kirkpatrick, A. J. Millis, and Thomas Vojta, Phys. Rev. B, **58**, 14155 (1998).
15. L. D. Landau, Zh. Eksp. Teor. Fiz. **7**, 19 (1937); *Collected Papers of L.D. Landau*, D. Ter Haar, ed., (Pergamon, Oxford, 1965), p. 193.
16. R. L. Stratonovich, Dokl. Akad. Nauk. S.S.S.R. **115**, 1907 (1957) [Sov. Phys. Doklady **2**, 416 (1957)]; J. Hubbard, Phys. Rev. Lett. **3**, 77 (1959).
17. G. Grinstein, in *Fundamental Problems in Statistical Mechanics VI*, E.G.D. Cohen, ed., (North Holland, Amsterdam, 1985), p. 147.
18. D. Belitz, T.R. Kirkpatrick, and T. Vojta, Phys. Rev. B **55**, 9452 (1997).
19. M. E. Fisher, S. K. Ma, and B. G. Nickel, Phys. Rev. Lett. **29**, 917 (1972).
20. P.C. Hohenberg and B.I. Halperin, Rev. Mod. Phys. **49**, 435 (1977).
21. For a review, see, e.g., D. Belitz and T. R. Kirkpatrick, Rev. Mod. Phys. **66**, 261 (1994).
22. For a review, see, P. A. Lee and T. V. Ramakrishnan, Rev. Mod. Phys. **57**, 287 (1985).
23. For a review, see, e.g., B. Kramer and A. MacKinnon, Rep. Progr. Phys. **56**, 1469 (1993).
24. F. Wegner, Z. Phys. B **25**, 327 (1976).
25. N.F. Mott, *Metal-Insulator Transitions*, (Taylor & Francis, London, 1990).
26. See, e.g., A. Georges, G. Kotliar, W. Krauth, and M. J. Rozenberg, Rev. Mod. Phys. **68**, 13 (1996).
27. H. Stupp, M. Hornung, M. Lakner, O. Madel, and H. v. Löhneysen, Phys. Rev. Lett. **71**, 2634 (1993). See also T.F. Rosenbaum, G.A. Thomas, and M.A. Paalanen, Phys. Rev. Lett. **72**, 2121 (1994); H. Stupp, M. Hornung, M. Lakner, O. Madel, and H.v. Löhneysen, Phys. Rev. Lett. **72**, 2122 (1994).
28. C. Bruin, Phys. Rev. Lett. **29**, 1670 (1972); Physica (Utrecht) **72**, 261 (1974); Ph.D. Thesis (Delft University 1978).
29. B. J. Alder and W. E. Alley, J. Stat. Phys. **19**, 341 (1978).
30. W. E. Alley, Ph.D. Thesis (UC Davis, 1979).
31. G. A. Thomas, M. Capizzi, F. DeRosa, R. N. Bhatt, and T. M. Rice, Phys. Rev. B **23**, 5472 (1981).
32. T. F. Rosenbaum, R. F. Milligan, M. A. Paalanen, G. A. Thomas, R. N. Bhatt, and W. Lin, Phys. Rev. B **27**, 7509 (1983).

SCATTERING, TRANSPORT & STOCHASTICITY IN QUANTUM SYSTEMS

PIERRE GASPARD
Université Libre de Bruxelles
Center for Nonlinear Phenomena and Complex Systems
& Service de Chimie Physique,
Campus Plaine, Code Postal 231,
B-1050 Brussels, Belgium

1. Introduction

Recent work has shown the importance of chaotic behavior and of the sensitivity to initial conditions to understand how irreversible processes such as diffusion, viscosity, heat conductivity or reactions may arise in classical Hamiltonian systems [1, 2]. Indeed, the exponential separation between classical trajectories which are infinitesimally close can be shown to have for consequence exponential decays or relaxations in the Hamiltonian scattering by a hill, in some chaotic scattering systems, or in simple fully chaotic maps such as the multibaker map [2].

Moreover, in the escape-rate formalism, the transport coefficients have been shown to be related to the difference between the sum of positive Lyapunov exponents – which characterize the sensitivity to initial conditions – and the Kolmogorov-Sinai (KS) entropy – which characterizes the dynamical randomness of the trajectories that are trapped forever in the scattering region and that form a so-called fractal repeller [3, 4, 5, 6].

These results and others [7, 8, 9, 10, 11] suggest that the microscopic chaos in the motion of atoms or molecules composing matter plays an important role in the transport and reaction properties of classical dynamical systems.

However, the question arises whether these considerations about classical chaos are still relevant at the quantum-mechanical level of description. The reason is that quantum mechanics is a linear theory although classical mechanics is nonlinear. This problem has been the subject of much work for about two decades and many results have accumulated which provide

J. Karkheck (ed.),
Dynamics: Models and Kinetic Methods for Non-equilibrium Many Body Systems, 425–456.

a rich and detailed picture of the properties of classically chaotic quantum systems.

In order to connect quantum to classical mechanics and to study how chaotic behavior may emerge at the classical level, a method of choice is the semiclassical theory in which the quantum-mechanical properties are expanded asymptotically in the formal limit where the Planck constant vanishes $\hbar \to 0$. A central result of modern semiclassical theory is the famous Gutzwiller trace formula [12] which allows the periodic-orbit quantization of classically chaotic systems and, in particular, of scattering systems such as the disk scatterers [13], the helium atom [14, 15] and the hydrogen negative ion [16], as well as metastable triatomic molecules [17]. Although practically limited to few-body systems, the semiclassical quantization method already allows us to study irreversible decay processes such as electronic conductance, unimolecular reactions or atomic auto-ionizations.

Spatially extended scattering systems have also been studied in order to characterize the transport across these systems from the viewpoint of scattering theory. In particular, Landauer proposed a scattering theory of electronic conductance which is powerful for the mesoscopic semiconductor circuits. Besides conductance which is obtained from the transmission coefficient in Landauer's theory, time-dependent properties of relaxation type are also of great interest because they are related to the diffusion coefficient in the quasiclassical limit.

In Sections 2 and 3, we shall give an overview of the methods of semiclassical quantization and of some of its applications to scattering systems.

In Section 4, we shall be concerned with relaxation and dynamical randomness in many-body quantum systems. It turns out that quantum systems acquire some of the dynamical properties of classical chaotic systems in the large-system limit or thermodynamic limit. In this limit, several types of properties may be studied such as the statistical properties of the energy eigenvalues and eigenfunctions. Much work has been devoted to these spectral properties showing that many quantum systems with a sufficient degree of genericity behave like random matrices on the small energy scale where individual eigenvalues are resolved [18, 19]. Such considerations were initiated by Wigner and others in the fifties in the context of nuclear physics [18]. Today, these random-matrix properties have been experimentally observed also in atomic, molecular, electromagnetic, and acoustic systems. Besides, several numerical calculations have shown that many-body quantum systems of solid-state physics behave similarly. We shall here review such results for many-spin quantum systems [20]. These random-matrix properties are also observed in classical chaotic systems and they allow to justify some of the basic laws of thermostatics. In particular, Srednicki has shown that some of the statistical properties of eigenfunctions of classically chaotic systems imply thermalization and the quantum equilibrium thermal distributions [21]. These properties can also justify dynamical properties such as the decay of time-correlation functions evaluated with typical eigenfunctions. In this context, a nonMarkovian stochastic Schrödinger equation has recently been derived on the basis of similar and related considerations [22].

Finally, we shall be concerned with the characterization of dynamical randomness in large quantum systems and with the definition of quantum analogues of the dynamical entropies per unit time.

2. Semiclassical Quantization

2.1. QUANTUM TIME EVOLUTION OF PURE STATES

In quantum mechanics, a system is described by a wavefunction which evolves in time according to the linear Schrödinger equation

$$i\hbar\, \partial_t\, \psi \;=\; H\, \psi\; . \tag{1}$$

If the Hamiltonian H is time independent, the wavefunction at the current time t is given by applying the time evolution operator to the initial wavefunction

$$\psi_t \;=\; \exp(-iHt/\hbar)\; \psi_0\; . \tag{2}$$

In the position representation, this equation has the following integral form

$$\psi_t(\mathbf{q}) \;=\; \int d\mathbf{q}_0\; K(\mathbf{q},\mathbf{q}_0,t)\; \psi_0(\mathbf{q}_0)\; , \tag{3}$$

where $K(\mathbf{q},\mathbf{q}_0,t)$ is the propagator that is the probability amplitude for the particle to move between the positions $\mathbf{q}_0$ and $\mathbf{q}$ during the time t.

Feynman has shown that the quantum propagator – which is the time evolution operator in the position representation – can be obtained by a path integral involving the Lagrangian function $L(\mathbf{q},\dot{\mathbf{q}},t)$ of the system as

$$K(\mathbf{q},\mathbf{q}_0,t) = \langle \mathbf{q}|\exp(-iHt/\hbar)|\mathbf{q}_0\rangle = \int \mathcal{D}\mathbf{q}(t)\; \exp\frac{i}{\hbar}\int_0^t L(\mathbf{q},\dot{\mathbf{q}},\tau)d\tau \tag{4}$$

for the propagation from $\mathbf{q}_0$ to $\mathbf{q}$ during the time t.

If the propagation proceeds in the semiclassical regime, the action may be supposed to be much larger than the Planck constant,

$$W = \int_0^t L(\mathbf{q},\dot{\mathbf{q}},\tau)d\tau \gg \hbar\; . \tag{5}$$

In this semiclassical limit, the path integral can be evaluated by the method of stationary phases, which selects the preferred paths for the propagation as the classical trajectories which are the rays obeying the Hamiltonian equations

$$\dot{\mathbf{q}} = + \frac{\partial H_{\rm cl}}{\partial \mathbf{p}} , \qquad \text{and} \qquad \dot{\mathbf{p}} = - \frac{\partial H_{\rm cl}}{\partial \mathbf{q}} . \tag{6}$$

The propagator can thus be asymptotically approximated by a sum over all the classical trajectories ℓ which connects $\mathbf{q}_0$ to $\mathbf{q}$ during the time t

$$K(\mathbf{q}, \mathbf{q}_0, t) \simeq_{\hbar \to 0} \sum_{\ell} \mathcal{A}_{\ell}(\mathbf{q}, \mathbf{q}_0, t) \ \exp \frac{i}{\hbar} W_{\ell}(\mathbf{q}, \mathbf{q}_0, t) . \tag{7}$$

Since the propagator $K(\mathbf{q}, \mathbf{q}_0, t)$ is still a quantum amplitude it is given by the linear superposition of the quantum amplitudes of the different classical trajectories. W_ℓ is the action of the classical trajectory and $\mathcal{A}_\ell$ is an amplitude associated with the classical trajectory, the expression of which can be found elsewhere [23]. This amplitude behaves differently depending on the stability properties of the classical trajectory. Globally, over a long time interval, we may say that the more unstable the trajectory is the faster its amplitude decays.

2.2. QUANTUM TIME EVOLUTION OF MIXED STATES

In order to consider a statistical ensemble of pure quantum states, we introduce the density matrix $\rho = \sum_i |\psi_i\rangle p_i \langle\psi_i|$, where p_i are the probabilities of occurrence of the states ψ_i in the statistical ensemble. The average of an observable D is then given in terms of this density matrix by $\langle D \rangle = \mathrm{tr}\rho D$. The time evolution of the density matrix is governed by the Landau-von Neumann or quantum Liouvillian operator

$$\partial_t \, \rho = \frac{1}{i\hbar}[H, \rho] = \mathcal{L}\rho . \tag{8}$$

An important related superoperator is the energy superoperator

$$\mathcal{H}(\cdot) = \frac{1}{2}(H \cdot + \cdot H) . \tag{9}$$

The energy superoperator always commutes with the Liouvillian superoperator $[\mathcal{L}, \mathcal{H}] = 0$, so that they have common eigenstates such that

$$\begin{cases} \mathcal{L}\rho = s\rho , \\ \mathcal{H}\rho = E\rho , \end{cases} \tag{10}$$

where s is a complex frequency and E an energy. For a bounded quantum system, these common eigenstates are given by $\rho_{mn} = |E_m\rangle\langle E_n|$ in terms of the energy eigenstates, $H|E_n\rangle = E_n|E_n\rangle$.

In order to consider a representation which is close to the classical phase-space representation, we can introduce the Wigner transform of an operator X by

$$X_{\rm W}(\mathbf{q},\mathbf{p}) = \int d^f r \, \exp(i\mathbf{p}\cdot\mathbf{r}/\hbar) \left\langle \mathbf{q}-\frac{\mathbf{r}}{2}\Big|X\Big|\mathbf{q}+\frac{\mathbf{r}}{2}\right\rangle , \tag{11}$$

where f equals the number of degrees of freedom. In the Wigner representation, the Liouvillian and energy superoperators are given by [24]

$$\begin{cases} (\mathcal{L}\rho)_{\rm W} = \frac{2}{\hbar}\, H_{\rm W}\, \sin\frac{\hbar\hat{\Lambda}}{2}\, \rho_{\rm W} = \left(\mathcal{L}_{\rm cl} + \hbar^2\mathcal{L}^{(2)} + \hbar^4\mathcal{L}^{(4)} + \cdots\right)\rho_{\rm W} , \\ (\mathcal{H}\rho)_{\rm W} = H_{\rm W}\, \cos\frac{\hbar\hat{\Lambda}}{2}\, \rho_{\rm W} = \left(H_{\rm cl} + \hbar^2\mathcal{H}^{(2)} + \hbar^4\mathcal{H}^{(4)} + \cdots\right)\rho_{\rm W} , \end{cases} \tag{12}$$

where $\hat{\Lambda} = \overleftarrow{\partial_{\mathbf{q}}}\,\overrightarrow{\partial_{\mathbf{p}}} - \overleftarrow{\partial_{\mathbf{p}}}\,\overrightarrow{\partial_{\mathbf{q}}}$, $H_{\rm W} = H_{\rm cl}$ is the classical Hamiltonian, while

$$\mathcal{L}_{\rm cl}(\cdot) = \{H_{\rm cl}, \,\cdot\,\}_{\rm Poisson} \tag{13}$$

is the classical Liouvillian operator given by the Poisson bracket with the Hamiltonian.

These results lead to the most interesting observation that the energy superoperator becomes an operator of multiplication with the classical Hamiltonian at the leading order of an expansion in powers of the Planck constant. In the quasiclassical limit, if we consider the eigenstates of the energy superoperator they must therefore satisfy

$$\left[H_{\rm cl} + \mathcal{O}(\hbar^2)\right]\, \rho_{\rm W} = E\, \rho_{\rm W} . \tag{14}$$

If the corrections in $\hbar^2$ can be neglected, we have an equation of the form $xf(x) = 0$ for an unknown density $f(x)$ depending on the energy variable $x = E - H_{\rm cl}$. Such an equation has a solution in the theory of generalized functions or Schwartz distributions, which is given by the Dirac distribution $f(x) = \delta(x)$ up to a constant factor. We can therefore conclude that, in the quasiclassical limit, the eigenstate is defined on the energy shell $H_{\rm cl} = E$ by

$$\rho_{\rm W} = g\, \delta(E - H_{\rm cl}) + \mathcal{O}(\hbar^2) , \tag{15}$$

where $g(\mathbf{q},\mathbf{p})$ is some density which may be proportional to other delta distributions involving further constants of motion.

If we consider an eigenstate with the eigenvalue $s = 0$ for the Liouvillian operator and if the system is ergodic, the function g is constant and we recover the microcanonical ensemble in the quasiclassical limit $\hbar \to 0$. This result is related to a conjecture by Berry and Voros [25] that averages over the Hamiltonian eigenstates, $\langle E_n|A|E_n\rangle$, tend to the microcanonical average in the limit $\hbar \to 0$. This conjecture has been proved for certain classically chaotic systems which are ergodic, mixing and hyperbolic of Anosov type [26].

The time-dependent properties of a system can be characterized by the time-correlation functions of the various observables of the system, such as the autocorrelation function of D,

$$C(t) = \langle D(0)D(t)\rangle , \quad \text{with} \quad D(t) = \exp(iHt/\hbar)\, D\, \exp(-iHt/\hbar) , \tag{16}$$

where the average is taken over a time-invariant state which is either a Hamiltonian eigenstate or a mixed state given by a density matrix which is a function of the Hamiltonian operator: $\rho = \mathcal{P}(H)$. It is also interesting to introduce the spectral functions that are the Fourier transforms of the correlation functions

$$S(\omega) = \int_{-\infty}^{+\infty} dt\, \exp(-i\omega t)\, \langle D(0)D(t)\rangle . \tag{17}$$

An example of such a spectral function is the cross-section for photoabsorption that is given by the Fourier transform of the time-autocorrelation function of the dipole electric moment of the system [27]. The average is taken over the quantum state of the system prior to the absorption. If the system is at a low temperature with respect to the energy of the absorbed photon, the initial system can be considered in its ground state. This case is treated in the following Subsection 2.3. However, if the temperature is higher, the initial state is a thermal state described by a density matrix and a different treatment is required which is described in Subsection 2.4.

2.3. TRACE FORMULA FOR PURE-STATE AVERAGING

In this subsection, our aim is to obtain a semiclassical expression for the spectral function (17) in the case where the average is carried out over a pure state of density matrix $\rho = |\varphi_0\rangle\langle\varphi_0|$ which projects on the ground Hamiltonian eigenstate: $H|\varphi_0\rangle = E_0|\varphi_0\rangle$.

In this case, the spectral function (17) becomes

$$S(\omega) = \text{tr}\ A\ \delta(E_0 + \hbar\omega - H) \qquad \text{with} \qquad A = 2\pi\hbar\ D|\varphi_0\rangle\langle\varphi_0|D . \tag{18}$$

Therefore, at low temperatures, the spectral function is expressed as a trace of an operator directly involving a lone resolvent of the Hamiltonian. Such an expression can be calculated semiclassically and expanded in terms of periodic orbits with the Gutzwiller trace formula.

We owe to Gutzwiller [12] the first derivation of a semiclassical trace formula including the oscillating contributions of the classical periodic orbits for quantities like (18). Under certain circumstances, the periodic-orbit contributions are able to approximate semiclassically the effect of quantization. Previously, the work by Weyl and Wigner showed how to approximate quasiclassically expressions given by traces [24]. Such quasiclassical expansions are based on the use of the Wigner transform (11). However, such quasiclassical expansions do not reproduce the effect of quantization. Gutzwiller derived a periodic-orbit trace formula for the level density which is defined by Eq. (18) with A replaced by the identity operator I. The trace has the effect that the selected classical trajectories are closed, i.e., are periodic orbits. However, Gutzwiller's treatment is general and extends also to the spectral functions for which we can obtain the following semiclassical approximation [23, 28]:

$$\begin{aligned} S(\omega) &= \int \frac{d^f q d^f p}{(2\pi\hbar)^f} A_W(\mathbf{q},\mathbf{p})\, \delta\left[E - H_{\mathrm{cl}}(\mathbf{q},\mathbf{p})\right] + \mathcal{O}(\hbar^{-f+1}) \\ &+ \frac{1}{\pi\hbar} \sum_p \sum_{r=1}^{\infty} \left(\oint_p A_W dt \right) \frac{\cos\left[\frac{r}{\hbar} S_p(E) - r\frac{\pi}{2}\mu_p\right]}{|\det(\mathbf{m}_p^r - \mathbf{I})|^{1/2}} + \mathcal{O}(\hbar^0)\,, \end{aligned} \tag{19}$$

with $E = E_0 + \hbar\omega$. In this formula, A_W denotes the Wigner transform (11) of the operator A defined in Eq. (18);

$$S_p(E) = \oint_p \mathbf{p} \cdot d\mathbf{q} \tag{20}$$

is the reduced action of the periodic orbit p; μ_p is its Maslov index which characterizes the winding of the trajectories around the periodic orbit and $\mathbf{m}_p$ is the matrix of the linearized Poincaré map near the periodic orbit. The sum is carried out over all the prime periodic orbits and their repetition $r = 1, 2, 3, \ldots$ The period of the periodic orbit is given in terms of the reduced action (20) according to

$$T_p(E) = \frac{\partial S_p(E)}{\partial E}\,. \tag{21}$$

With the formula (19), the spectral function is decomposed into a smooth quasiclassical background given by the first term and oscillating contributions from the periodic orbits which are superposed on top of the smooth background. The peaks of the spectral function may allow us to identify the energy eigenstates of the system.

It turns out that the contribution of the unstable periodic orbits to the spectral function can be rewritten as

$$S(\omega)\Big|_{\mathrm{po}} = \frac{1}{\pi}\, \mathrm{Im} \frac{\partial}{\partial\lambda} \ln \tilde{Z}(E,\lambda)\Big|_{\lambda=0} + \mathcal{O}(\hbar^0)\,, \tag{22}$$

with $E = E_0 + \hbar\omega$, in terms of the so-called Selberg Zeta function

$$\tilde{Z}(E,\lambda) = \prod_p \prod_{m_1,\ldots,m_{f-1}=0}^{\infty} \left\{ 1 - \frac{\exp\left[\frac{i}{\hbar}\tilde{S}_p(E,\lambda) - i\frac{\pi}{2}\mu_p\right]}{\prod_{k=1}^{f-1} |\tilde{\Lambda}_p^{(k)}(E,\lambda)|^{\frac{1}{2}} \tilde{\Lambda}_p^{(k)}(E,\lambda)^{m_k}} \right\}, \quad (23)$$

which is a product over the periodic orbits of the perturbed Hamiltonian system, $\tilde{H}_{\rm cl}(\mathbf{q},\mathbf{p},\lambda) = H_{\rm cl}(\mathbf{q},\mathbf{p}) + \lambda A_{\rm W}(\mathbf{q},\mathbf{p})$ [23, 29]. In the Zeta function (23), the stability factors $\tilde{\Lambda}_p^{(k)}(E,\lambda)$ are the eigenvalues of the linearized Poincaré map $\tilde{\mathbf{m}}_p(E,\lambda)$ and they satisfy $|\tilde{\Lambda}_p^{(k)}| > 1$. If the classical system is structurally stable, we have that $S_p(E) = \lim_{\lambda\to 0} \tilde{S}_p(E,\lambda)$. The product over the periodic orbits can be expanded into a sum over topological combinations of all the periodic orbits, called the cycle expansion [30, 31]. By regrouping terms of high period, this series can be reordered into terms of lower and lower magnitudes as the period increases. Truncation can be carried out for numerical computations [31].

The presence of several periodic orbits leads to interferences between their different quantum amplitudes in the cycle-expanded Zeta function. These interferences cause irregularities in the structures of the spectral function. The greater is the number of periodic orbits with different reduced actions, the greater are the irregularities in the spectral function [32].

2.4. TRACE FORMULA FOR MIXED-STATE AVERAGING

Let us now consider that the spectral function involves an average over a statistical mixture described by a density matrix which is a function of the Hamiltonian operator $\rho = \mathcal{P}(H)$. This is the case for a canonical ensemble for which $\mathcal{P}(E) = \exp(-\beta E)/Z$ with the inverse temperature β.

The spectral function (17) can be expressed as

$$S(\omega) = \int dt\, dE\, \mathrm{e}^{-i\omega t}\, \mathcal{P}(E)\, \mathrm{tr}\, \delta(E-H) D(0) D(t) \ , \quad (24)$$

in terms of a time-autocorrelation function averaged over a microcanonical ensemble. By using the Wigner transform and the Gutzwiller trace formula, the spectral function can be semiclassically approximated as [33, 34]

$$\begin{aligned} S(\omega) &= \int dt \int \frac{d^f x d^f p}{(2\pi\hbar)^f}\, \mathcal{P}(H_{cl})\, D_{\rm W}\, \mathrm{e}^{(\mathcal{L}_{cl}-i\omega)t} D_{\rm W} + \mathcal{O}(\hbar^{-f+1}) \\ &+ \frac{2}{\hbar} \sum_{p,r,n} \mathcal{P}_{p,n} |D_{p,n}|^2 \left|\frac{dS_{p,n}}{d\omega}\right| \frac{\cos\left(\frac{r}{\hbar}S_{p,n} - r\frac{\pi}{2}\mu_p\right)}{|\det(\mathbf{m}_{p,n}^r - \mathbf{I})|^{1/2}} + O(\hbar^0), \quad (25) \end{aligned}$$

where all the quantities in the periodic-orbit contributions are evaluated at the energies $E = E_{p,n}(\omega)$ at which there is resonance between the driving frequency ω and the intrinsic frequency of the periodic orbit p, $\omega = 2\pi n/T_p(E)$. The first term of (25) is the classical expression involving the classical Liouvillian operator. $D_W = D_W(\mathbf{q}, \mathbf{p})$ is the Wigner transform of the operator D while H_{cl} is the classical Hamiltonian. The second term contains the oscillating contributions of each unstable periodic orbit p at repetition r. The magnitude of each periodic-orbit contribution is proportional to the square of the coefficient of the Fourier expansion of the observable D_W evaluated at the periodic orbit p,

$$D_{p,n} = \frac{1}{T_p} \int_0^{T_p} D_W \left[\mathbf{q}(t), \mathbf{p}(t)\right] \exp\left(-i\frac{2\pi n}{T_p} t\right) dt \ . \tag{26}$$

The $\mathbf{m}_{p,n}$ is the same matrix of the linearized Poincaré map of the periodic orbit p at the energy $E = E_{p,n}(\omega)$ [33].

We observe that, here again, the spectral function has a smooth background given by the quasiclassical expression while the oscillating contributions from the periodic orbits are superposed on top of this smooth background because the system is finite. The same semiclassical calculation can be carried out for the dynamic susceptibility of a non-interacting system of Fermions in a trapping potential [33]. In this example, the oscillating contributions are due to the Fermionic shells which exist in atoms, in nuclei or in metallic clusters. These periodic-orbit corrections prevent the existence of normal dissipation or transport, but since they only appear $\hbar^{f-1}$ away from the leading quasiclassical term these oscillating contributions tend to disappear when the number of degrees of freedom f increases indefinitely so that we may expect that normal dissipation or transport is restored when $f \to \infty$.

The comparison with Eq. (19) shows that a very different expression for the spectral function is obtained here in the case of a state given by the statistical mixture. Here, we find a non-trivial time dependence under the classical dynamics involving a statistical ensemble of trajectories according to the density $\mathcal{P}(H_{cl})$. For a classically mixing system, the autocorrelation function is expected to decay for $t \to \pm\infty$. This decay is controlled by classical resonances called the Pollicott-Ruelle resonances which are the generalized eigenvalues of the classical Liouvillian operator [35]. These resonances are given by taking the trace of the Frobenius-Perron operator $\exp(\mathcal{L}_{cl}t)$ over the energy shell $H_{cl} = E$. We can view this trace of a classical operator as the quasiclassical limit of an appropriate trace of the quantum Liouvillian evolution operator restricted to the eigenstates of the energy superoperator. Using a result by Cvitanovic and Eckhardt for the trace of the classical Frobenius-Perron operator in a classically hyperbolic system with unstable periodic orbits [36], we find that the trace of the quantum

Liouvillian evolution operator is approximated quasiclassically by

$$\mathrm{Tr}_E \, \exp(\mathcal{L}t) \;=\; \sum_p \sum_{r=1}^{\infty} T_p \, \frac{\delta(t - rT_p)}{|\det(\mathbf{m}_p^r - \mathbf{I})|} \;+\; \mathcal{O}(\hbar^2) \;, \tag{27}$$

where T_p is the period of the prime periodic orbit p and $\mathbf{m}_p$ is the same matrix of the linearized Poincaré map as in Eq. (19).

The Laplace transform of this evolution operator gives the 'trace' of the resolvent of the Liouvillian operator [2]

$$\int_0^{\infty} dt \, \exp(-st) \, \mathrm{Tr}_E \, \exp(\mathcal{L}t) \;=\; \mathrm{Tr}_E \, \frac{1}{s - \mathcal{L}} \;=\; \frac{\partial}{\partial s} \, \ln \, Z_{\mathrm{cl}}(s; E) + \, \mathcal{O}(\hbar^2) \;, \tag{28}$$

in terms of the classical Zeta function

$$Z_{\mathrm{cl}}(s; E) = \prod_p \prod_{m_1, \ldots, m_{f-1}=0}^{\infty} \left\{ 1 - \frac{\exp\left[-sT_p(E)\right]}{\prod_{k=1}^{f-1} |\Lambda_p^{(k)}(E)| \Lambda_p^{(k)}(E)^{m_k}} \right\}^{(m_1+1)\cdots(m_{f-1}+1)} . \tag{29}$$

The zeros of this classical Zeta function, $Z_{\mathrm{cl}}(s; E) = 0$ give minus the classical decay rates for the dynamics on each energy shell $H_{\mathrm{cl}} = E$. We notice that both the periods T_p and the stability eigenvalues Λ_p depend on the energy E, as expected. At this level, we see the importance of introducing the energy superoperator and its eigenstates, which allows us to obtain a classical dynamics restricted to one energy shell as it should be. This classical behavior is in contrast with the quantum-mechanical behavior in which the quantization of energy selects the eigenenergies. In a classical scattering system, the leading Pollicott-Ruelle resonance gives the so-called classical escape rate $\gamma_{\mathrm{cl}}(E) > 0$ of trajectories out of the interacting region

$$s \;=\; s_0(E) \;=\; -\,\gamma_{\mathrm{cl}}(E) \;. \tag{30}$$

We remark that the classical Zeta function (29) has similarities with the semiclassical quantum Zeta function (23), but they have also important differences because they are concerned with the time evolution of different types of quantities. The semiclassical quantum Zeta function is concerned with quantum amplitudes while the classical Zeta function is concerned with classical probabilities. Since the probabilities are essentially given by the squares of the quantum amplitudes, we may explain the differences as follows. Let us consider the quantum amplitude associated with an unstable periodic orbit of a two-degrees-of-freedom system

$$\frac{\exp\left[\frac{i}{\hbar} S_p(E) - i\frac{\pi}{2}\mu_p\right]}{|\Lambda_p(E)|^{\frac{1}{2}}} \, . \tag{31}$$

This amplitude depends on the quantum phase of the orbit and on the square root of its stability eigenvalue. At the level of the density matrix, the relevant quantity is the amplitude (31) multiplied by the complex conjugate of another quantum amplitude, which should give a classical probability. Since we consider in Eq. (29) a time-dependent process with a decay rate $-s$, the other amplitude should be taken at the energy $E' = E - i\hbar s$ with respect to the energy E of (31). The difference between E' and E is an imaginary energy corresponding to the imaginary frequency $-s$. Accordingly, in the classical expression, we expect to find the factor

$$
\begin{aligned}
& \frac{\exp\left[\frac{i}{\hbar}S_p(E) - i\frac{\pi}{2}\mu_p\right]}{|\Lambda_p(E)|^{\frac{1}{2}}} \, \frac{\exp\left[-\frac{i}{\hbar}S_p(E') + i\frac{\pi}{2}\mu_p\right]}{|\Lambda_p(E')|^{\frac{1}{2}}} \\
\simeq \quad & \frac{\exp \frac{i}{\hbar}\left[S_p(E) - S_p(E) + i\hbar s T_p(E) + \mathcal{O}(\hbar^2)\right]}{|\Lambda_p(E)|} \\
\simeq \quad & \frac{\exp\left[-sT_p(E)\right]}{|\Lambda_p(E)|} \qquad (32)
\end{aligned}
$$

because of the classical formula (21). This argument explains that:

1. The stability eigenvalues themselves appear in the classical Zeta function (29) instead of their square roots which appear in the semiclassical Zeta function (23).

2. The period multiplied by the decay rate $-s$ appears in (29) instead of the quantum phase as in (23).

A last difference comes from the exponents $(m_k + 1)$ of the periodic-orbit factors in (29), which has its origin also in the fact that the stability eigenvalues are classically involved instead of their square roots.

2.5. CHARACTERIZATION OF CLASSICAL CHAOS

Nonlinear classical systems governed by Hamilton's equations generate a dynamics of trajectories. In the phase space of the system, these trajectories may be stable or unstable. The possible dynamical instability may generate dynamical randomness. Under such circumstances, the time average of an observable may be equivalent to an average over a statistical ensemble of points which are distributed in the phase space according to an invariant probability measure.

Such invariant measures can be constructed from the knowledge of the instability of the trajectories visiting successively different cells $\omega_1\omega_2\cdots\omega_n$ in the phase space. In a two-degrees-of-freedom system, the dynamics stretches the phase-space cells by a so-called stretching factor $\Lambda_{\omega_1\cdots\omega_n}$ which corresponds to a time interval $T_{\omega_1\cdots\omega_n}$. An invariant measure can be defined by assuming that the probability weight of the trajectories visiting the cells $\omega_1\omega_2\cdots\omega_n$ is smaller if this phase-space region is more unstable. Different invariant measures can be constructed depending on an exponent β given to each stretching factor [37]. In the limit $n \to \infty$, the probability weight of the successive cells $\omega_1\omega_2\cdots\omega_n$ is thus defined by

$$\mu_\beta(\omega_1\omega_2\cdots\omega_n) \simeq \frac{|\Lambda_{\omega_1\omega_2\cdots\omega_n}|^{-\beta}}{\sum_{\omega_1\omega_2\cdots\omega_n}|\Lambda_{\omega_1\omega_2\cdots\omega_n}|^{-\beta}} , \tag{33}$$

where the denominator guarantees that the invariant measure is normalized to unity. The classical dynamics naturally induces the invariant measure with $\beta = 1$, in which case the probability weight is inversely proportional to the stretching factor. This result is consistent with the fact that each periodic orbit has a probability weight which is inversely proportional to its stability eigenvalue as seen in Eq. (32).

In analogy with equilibrium statistical mechanics, Ruelle has introduced an associated pressure function as [38]

$$P(\beta) \equiv \lim_{\substack{t\to\infty \\ \delta\to 0}} \frac{1}{t} \ln \sum_{\substack{\omega_1\cdots\omega_n \\ t < T_{\omega_1\cdots\omega_n} < t+\Delta t}} |\Lambda_{\omega_1\omega_2\cdots\omega_n}|^{-\beta} , \tag{34}$$

where δ is the diameter of the cells ω. For a two-degrees-of-freedom system, the invariant measure μ_β is equivalently characterized by the KS entropy per unit time and by the mean Lyapunov exponent, or by the pressure function.

For an open chaotic system with two degrees of freedom (2F), the dynamics may select a set of trajectories which are forever trapped in the interacting region. This set is called the repeller and it is characterized by an escape rate as well as by partial fractal dimensions. All the different characteristic quantities of the repeller can be obtained from the pressure function as shown in Table 1 [4, 6, 39, 40]. In Table 1, the four first quantities are evaluated for the natural invariant measure with $\beta = 1$ which is directly induced by the classical dynamics. The next quantity characterizes the topological chaos which is probed with the invariant measure with $\beta = 0$ for which all the trajectories have an identical probability weight.

The pressure function can be evaluated from the unstable periodic orbits of the system as the leading zero $s = P(\beta; E)$ of the following inverse Ruelle zeta function [2],

$$\zeta_0^{-1}(s;\beta;E) = \prod_p \left\{ 1 - \frac{\exp[-sT_p(E)]}{\prod_{k=1}^{f-1}|\Lambda_p^{(k)}(E)|^\beta} \right\} . \tag{35}$$

TABLE 1. Characteristic quantities of chaos in an open 2F system

Escape rate	$\gamma_{\text{cl}} = -P(1)$
Lyapunov exponent	$\lambda = -P'(1)$
KS entropy	$h_{\text{KS}} = \lambda - \gamma_{\text{cl}} = P(1) - P'(1)$
Partial information dimension	$d_{\text{I}} = \frac{h_{\text{KS}}}{\lambda} = 1 - \frac{P(1)}{P'(1)}$
Topological entropy	$h_{\text{top}} = P(0)$
Partial Hausdorff dimension	$P(d_{\text{H}}) = 0$

The pressure function depends on the energy shell where the invariant measure (33) is defined.

3. Scattering Theory of Decay Processes

3.1. SCATTERING RESONANCES

Decay processes occur in finite systems which are excited by collisions with particles or waves coming from the exterior of the system. This is the case for a vast set of experimental situations which can therefore be studied from the viewpoint of scattering theory.

If the Hamiltonian of the system is time independent, the time evolution of the system can always be described by the time evolution operator (2) or the propagator (3), which admit a decomposition on the eigenfunctions of the Hamiltonian. The quantum-mechanical time evolution is thus completely determined by the energy spectrum of the system.

For the purpose of obtaining the asymptotic time evolution for either $t \to \pm\infty$, a remarkable method is given by the analytic continuation of the resolvent of the Hamiltonian toward either the lower-half complex energy surface or the upper-half surface, respectively. In this way, the unitary time evolution is asymptotically approximated by either the evolution operator of the forward semigroup or the one of the backward semigroup [23]. This method is most natural and convenient to derive the asymptotic irreversible time evolution for the quantum wavefunction. By analytic continuation, we can obtain the contributions from the complex singularities of the Hamiltonian resolvent, such as the poles and the branch cuts which are issued from the energy thresholds of each scattering channel. The poles of the resolvent are located in the lower second Riemann sheet of the complex energy surface at the complex energies

$$z_r = E_r - i\,\frac{\Gamma_r}{2}\,, \tag{36}$$

and are called the scattering resonances of the system. These poles of the Hamiltonian resolvent are also poles of the S-matrix describing the scattering of plane waves in the system. If a resonance is sufficiently isolated and is excited individually, its contribution to the wavepacket decays in time with damped oscillations and the corresponding probability density decays exponentially with a lifetime given by the imaginary part of the complex energy (36) as

$$\tau_r = \hbar/\Gamma_r \, . \tag{37}$$

Long-time tails with power-law decays of the wavefunction also exist at energies near the thresholds where new scattering channels open but they play a minor role [23].

This framework of the scattering and resolvent theories provides a firm basis to study the decay properties of open few-body quantum systems. In this framework, the resonances give direct information on the decay because each of them determines a lifetime. In particular, reaction rates can be determined from the knowledge of the scattering resonances. Moreover, we can associate some Hamiltonian eigenstates with the resonances, in a similar way as eigenstates are associated with the bound states.

In many systems, the scattering resonances form a dense spectrum which has to be analyzed statistically. For this purpose, the semiclassical theory is an important method where the quantum-mechanical quantities such as the Hamiltonian resolvent or its trace are expanded around the classical trajectories in the limit $\hbar \to 0$. Whether the states are pure or mixed, different expressions are obtained in the semiclassical limit. For both pure and mixed states, the time evolution can be decomposed onto the quantum scattering resonances. However, many systems evolve in a semiclassical regime involving numerous quantum scattering resonances. The effect of the accumulation of many quantum scattering resonances can be studied with the classical Liouvillian theory. In this regard, the classical escape rate can be considered as an average over the multiple quantum decays due to the many individual quantum scattering resonances.

3.2. DISTRIBUTIONS OF SCATTERING RESONANCES AND BOUND ON THE QUANTUM LIFETIMES

As we have noted above, the quantum scattering resonances are given by the poles of the resolvent of the Hamiltonian operator. In the semiclassical approximation, the periodic-orbit contributions of this resolvent are given as [23]

$$\mathrm{tr}\frac{1}{z-H}\bigg|_{\mathrm{po}} = \frac{\partial}{\partial z}\ln Z(z) \; + \; \mathcal{O}(\hbar^0) \, , \tag{38}$$

in terms of the semiclassical quantum Zeta function (23) with $\lambda = 0$ for a classically hyperbolic system with unstable periodic orbits such as a classically chaotic system. As a consequence of (38), the poles of the resolvent are semiclassically obtained from the zeros of the Zeta function: $Z(z) = 0$.

For a system with two degrees of freedom, the periodic-orbit contributions to the resolvent is a series which can thus be evaluated as

$$\mathrm{tr}\, \frac{1}{z-H}\bigg|_{\mathrm{po}} \sim \sum_p \sum_{r=1}^{\infty} T_p(z) \frac{\exp\left[\frac{i}{\hbar} r S_p(z) - i\frac{\pi}{2} r \mu_p\right]}{|\Lambda_p(z)|^{\frac{r}{2}}} . \tag{39}$$

Taking a complex energy

$$z = E - i\, \frac{\hbar}{2\tau} , \tag{40}$$

the reduced action can be expanded in powers of $\hbar$ as

$$S_p(z) = S_p(E) - i\, \frac{\hbar}{2\tau}\, T_p(E) + \mathcal{O}(\hbar^2) \tag{41}$$

because of Eq. (21). Whereupon the series becomes

$$\mathrm{tr} \frac{1}{z-H}\bigg|_{\mathrm{po}} \sim \sum_p \sum_{r=1}^{\infty} T_p(E)\, \exp\left[\frac{i}{\hbar} r S_p(E) - i\frac{\pi}{2} r \mu_p\right] \frac{\exp\left[r T_p(E)/2\tau\right]}{|\Lambda_p(E)|^{\frac{r}{2}}} . \tag{42}$$

This series converges absolutely under the condition that the sum of the absolute values of the terms with their period T_p in the time interval $t < T < t + \Delta t$ vanishes. An upper bound on this sum is given in terms of the pressure function by

$$\sum_{t<T<t+\Delta t} T\, \frac{\exp\left(T/2\tau\right)}{|\Lambda_T|^{\frac{1}{2}}} \sim_{t\to\infty} t\, \exp(t/2\tau)\, \exp\left[tP(\beta = 1/2; E)\right] \to_{t\to\infty} 0 , \tag{43}$$

with $T = rT_p$ and $\Lambda_T = \Lambda_p^r$. We find the pressure function evaluated at $\beta = 1/2$ because the quantum amplitudes are essentially the square roots of the classical probabilities except for a quantum phase, which is not relevant for absolute convergence. We infer from (43) that the sum vanishes and the series converges absolutely under the condition that the imaginary part of the complex energy (40) satisfies $(1/2\tau) + P(1/2; E) < 0$ at the real energy E. Therefore, the poles of the resolvent may only appear at the complex energies (36) such that

$$\frac{\Gamma_r}{\hbar} = \frac{1}{\tau_r} \geq -2\, P\left(\frac{1}{2}; E_r\right) \tag{44}$$

in the asymptotic limit $\hbar \to 0$ [13, 23, 32].

In this way, we observe that the pressure at $\beta = 1/2$ determines a boundary on the complex energy surface above which no resonance is expected. As long as the pressure at $\beta = 1/2$ is positive, this inequality does not impose a bound on the resonances. However, the inequality becomes effective when the pressure is negative. In this case, there is a gap below the real energy axis which is empty of scattering resonance. Since the pressure function vanishes at the partial Hausdorff dimension of the repeller in a two-degrees-of-freedom system, this gap appears when the corresponding classical dynamics has an invariant measure supported by a fractal set with a partial dimension less than one half. If $d_{\rm H} < 1/2$, we may speak about a filamentary fractal as opposed to a bulky fractal with $d_{\rm H} > 1/2$.

The above result is remarkable in many respects if we compare with the mean lifetime expected from the classical Liouvillian dynamics. Besides the bound (44), the semiclassical quantum lifetimes can also be estimated with the escape rate given by the quasiclassical Liouvillian theory. Since the classical escape rate is given by minus the pressure function at $\beta = 1$, we find that, quasiclassically, the lifetime is estimated as

$$\gamma_{\rm cl} = \frac{1}{\tau_{\rm cl}} = -P(1;E) \tag{45}$$

instead of (44). The properties of the pressure function imply that

$$P\left(\frac{1}{2};E\right) \geq \frac{1}{2} P(1;E) . \tag{46}$$

The equality occurs if the pressure function is linear, i.e., if the scattering system has a periodic unstable motion on its repeller. If the motion on the repeller is chaotic, the pressure function is convex and a strict inequality occurs in Eq. (46). In this chaotic case, the mean classical lifetime will thus be shorter than the largest possible quantum lifetimes which may approach the semiclassical bound (44). Consequently, we find this important difference between the classically periodic and chaotic systems [13]:

$$\text{periodic}: \quad \tau_{\rm cl}(E) = \tau_{\rm q}(E) , \tag{47}$$

$$\text{chaotic}: \quad \tau_{\rm cl}(E) < \tau_{\rm q}(E) , \tag{48}$$

where the quantum lifetime at energy E is defined as the longest possible lifetime of the quantum scattering resonances: $\tau_{\rm q}(E) = \max\{\tau_r\}_{E_r \simeq E}$. On the other hand, the classical lifetime is defined on the decay of a quantum statistical mixture involving many individual quantum resonances, as explained in Subsection 2.4. In this sense and as suggested by comparing Eq. (45) with Eq. (44), the inverse of the classical lifetime gives some kind of average value for the imaginary parts of the energies of the quantum scattering resonances. The classical lifetime acquires its importance in the case of bulky fractal repellers, for which the bound (44) is no longer directly useful. In the case of a bulky repeller, the classical lifetime may give in some systems an estimation of the imaginary part of the energy of a typical scattering resonance.

3.3. APPLICATION TO DISK SCATTERERS AND UNIMOLECULAR REACTIONS

The resonance spectrum has been studied in detail for the scattering of a wave on several disks with different boundary conditions [30, 41, 42]. These scattering systems have also been the object of scattering experiments with acoustic waves [42] and electromagnetic microwaves [43].

The scattering resonances can also be used to characterize unimolecular reactions of molecules excited by absorption of a photon. This excitation may bring the molecule up to an upper electronic Born-Oppenheimer potential energy surface where the motion may be dissociative with a filamentary classical repeller. This is the case in HgI_2 and CO_2 [17], for which agreement has been found with the bound on the quantum lifetimes described here above [32].

In many molecules, the classical repeller is bulky and it traps the trajectories in a quasibounded phase-space region before escaping above a barrier which forms a bottleneck. In this bulky case, the classical escape rate becomes very useful to evaluate the unimolecular reaction rate by assuming a quasi-equilibrium in the quasibounded region. Indeed, if the repeller is bulky, the fractal dimension of the invariant measure is close to the phase-space dimension and it is reasonable to assume a quasi-equilibrium distribution in the quasibounded region. Under these assumptions of RRKM theory, the reaction rate can be calculated which gives an expression for the classical escape rate [44]. In the simplest model, the scattering resonances are supposed to be distributed around the classical escape rate according to the χ^2-distribution. This statistical theory has been applied to many experimental and numerical data on unimolecular reactions as well as to nuclear reactions.

3.4. APPLICATION TO SPATIALLY EXTENDED SCATTERERS

Landauer's model of electronic conductance is based on the idea that the electron undergoes a scattering process inside the electronic circuit connected to external reservoirs by wires which behave as waveguides. The circuit may be localized as in the case of a quantum constriction but it may also be spatially extended if it is a regular or a disordered chain composed of many successive units.

Recently, we have analyzed the spectrum of resonances for the scattering of a particle in a one-dimensional potential formed by many identical units [45]. We have shown that the resonances form band structures which are reminiscent of the Bloch energy bands of the infinite periodic potential. Actually, if the potential contains N units satisfying certain conditions of regularity, $(N-1)$ resonances accumulate just below each energy band.

If the one-dimensional chain is disordered, the Anderson localization is known to happen for the infinite system. For a finite disordered chain, the spectrum of resonances and of bound states is irregular [46]. For energies where the localization length is smaller than the total length of the scatterer, the transmission coefficient vanishes and the scattering resonances have long lifetimes. At higher energies where the localization length is larger than the system size, the transmission coefficient fluctuates just below the value unity and the lifetimes of the scattering resonances are consistently much shorter [46].

In spatially extended systems with a chaotic classical dynamics, diffusion may occur as in the Lorentz gas or in the multibaker map. In such systems, the classical escape rate decreases as $\gamma_{\rm cl}(E) \simeq D(E)(\pi/L)^2$ with the system size L, where $D(E)$ is the energy-dependent diffusion coefficient [13]. A similar behavior is expected for the distribution of the imaginary parts of the quantum scattering resonances in such systems. In this context, the Pollicott-Ruelle resonances and the escape rate have been shown to play an important role in the semiclassical quantum theory of diffusion [47].

4. Many-Body Quantum Systems

4.1. TRANSPORT IN MANY-BODY QUANTUM SYSTEMS

Transport properties such as diffusion, viscosity, heat or electric conductivities can be described quantitatively within the Green-Kubo theory. For a quantum system in a thermal state, a transport coefficient is given by the integral of the time-autocorrelation function of the microscopic current associated with the transport process according to the Kubo formula [48], which can be transformed into an Einstein-type formula by introducing the associated Helfand moment [49]. In this way, a simple proof can be given that the transport coefficients are non-negative [50].

If the many-body system is bounded, its energy spectrum is discrete and, as a consequence, its time-correlation functions are almost-periodic functions of time. Such an almost-periodic function of time has the property to be recurrent, which is problematic because the recurrences spoil the Kubo formula. The way out of this difficulty is to take a thermodynamic limit where the number of particles in the system increases with the volume of the system, keeping the density constant. In this limit, the time-correlation function is expected to converge to a function which decays. Indeed, for a finite system, the almost-periodic oscillations and the recurrences only appear for times longer than the Heisenberg time defined by the Planck constant multiplied by the mean level density $d_{\rm av}(E)$

$$t_{\rm Heisenberg} \equiv \hbar\, d_{\rm av}(E) \sim (\hbar/k_{\rm B}T)\,(n\,\lambda^3_{\rm de\ Broglie})^{-N} , \qquad (49)$$

where $\lambda_{\rm de\ Broglie} \equiv 2\pi\hbar/\sqrt{mk_{\rm B}T}$ is the thermal de Broglie wavelength of the particles of mass m and n is the number of particles per unit volume. The Heisenberg time grows exponentially with the number N of particles in the system, under the condition that the temperature is large enough with respect to the particle density in order that $n\,\lambda^3_{\rm de\ Broglie} < 1$. Since the Heisenberg time becomes astronomical even for a few thousand particles under standard conditions, the Kubo formula can be justified thanks to a rapidly attained thermodynamic limit.

4.2. SPECTRAL RANDOMNESS IN MANY-BODY QUANTUM SYSTEMS

In spite of the fact that for many-body systems the energy eigenvalues become too dense to be resolved, their spectral properties still govern the time evolution, which has motivated the statistical analysis of the energy spectrum.

A property of interest is the degeneracy of the eigenvalues, which determines the dimension of the nullspace of the quantum Liouvillian operator and, therefore, the number of constants of motion, i.e., of operators which commute with the Hamiltonian. The larger is the nullspace dimension, the greater is the number of constants of motion which may restrict the normal transport properties.

In this context, we have compared the spectral properties of different systems with many spins $S = 1/2$ forming a square two-dimensional lattice, namely, the Ising, the Heisenberg and the dipolar spin systems. This comparison has revealed strikingly different spectral properties [20].

In the Ising system, the degeneracy is the highest as expected for this exactly solvable system.

For the Heisenberg system, the degeneracy is intermediate with high degeneracies for the states associated with the ferromagnetic spin waves. Geometric and some dynamic symmetries could be used to block diagonalize the Hamiltonian. Within the remaining blocks of the Hamiltonian, the eigenvalues follow a Poisson-type spacing statistics, suggesting the existence of further symmetries besides the one considered.

However, for the dipolar spin system, the degeneracy was the smallest due only to the known geometric symmetries. After block diagonalization, the eigenvalues of each remaining block follow a Wigner-type spacing statistics, precluding the existence of further symmetries. The results show that the dipolar system has the less regular dynamics. The transport of heat has also been studied numerically [52].

The statistics of the curvature of the energy levels, i.e. of the second derivative of each eigenvalue with respect to an external field, has also been studied for the dipolar spin system [51]. The probability density of the level curvatures presents the universal power law $\mathcal{P}(K) \sim K^{-3}$ for $K \to \infty$, confirming the existence of a Wigner spacing statistics in the dipolar spin system. The curvature distribution has also been considered in the context of electronic states in disordered media in relation to the problem of electronic conductance [53].

4.3. NONMARKOVIAN STOCHASTIC SCHRÖDINGER EQUATION FOR A SLOW SUBSYSTEM COUPLED TO A FAST BATH

Among the possible many-body quantum systems, we often find a small and slow subsystem which interacts with a fast thermal bath. Examples of such systems are a spin or a cluster of spins coupled to the vibrations of their host molecule, liquid or solid, which is common in NMR. Effective two-level subsystems also occur in electron- or charge-transfer reactions in condensed phases. The total Hamiltonian of such systems has the form

$$H = H_{\rm s} + H_{\rm b} + \lambda V \qquad \text{with} \qquad V = \sum_{\alpha} S_{\alpha} B_{\alpha} , \tag{50}$$

where $H_{\rm s}$ is the Hamiltonian of the isolated subsystem, $H_{\rm b}$ is the one of the bath, while the interaction between the subsystem and the bath is described by the subsystem operators S_α and the bath operators B_α, respectively.

The kinetics of such open systems may be described by a quantum master equation which governs the time evolution of the subsystem density matrix [54]. If we make a formal analogy with a classical process such as Brownian motion, the master equation would be the Fokker-Planck equation but we should also expect the existence of an equivalent stochastic equation such as the Langevin stochastic differential equation. Its quantum analogue would be a stochastic Schrödinger equation, as recently introduced and studied [55, 56]. However, the recently introduced stochastic Schrödinger equations are restricted to be equivalent to Markovian quantum master equations of Lindblad type [57]. The Markovian assumption requires that the time correlation functions of the bath,

$$C_{\alpha\beta}(t) = \langle B_\alpha(t) B_\beta(0) \rangle , \tag{51}$$

be proportional to the Dirac delta distribution $\delta(t)$, which is a good approximation for light-matter interaction because of the fast velocity of light.

However, most of the thermal baths have a nonvanishing relaxation time for their time-correlation functions. For such systems, we have recently derived a nonMarkovian stochastic Schrödinger equation in the weak coupling limit [22]. The derivation is based on three main ideas:

1. A suitable definition is given to the concept of 'wavefunction' for a subsystem. The stochastic Schrödinger equation is supposed to govern the time evolution of such a 'subsystem wavefunction'. For this purpose, the total wavefunction is decomposed in the basis of the eigenfunctions $\{\chi_n(x_{\rm b})\}$ of the bath Hamiltonian:

$$\Psi(x_{\rm s}, x_{\rm b}; t) \;=\; \sum_n \phi_n(x_{\rm s}; t)\; \chi_n(x_{\rm b}) \;. \tag{52}$$

The coefficients $\phi_n(x_{\rm s}; t)$ of this linear decomposition still depend on the subsystem coordinates $x_{\rm s}$ and they play the role of the 'subsystem wavefunctions'. Since the bath is very large, there is an enormous number of such coefficients $\phi_n(x_{\rm s}; t)$, many of them behaving statistically in a similar way. Accordingly, we suppose here that they form a statistical ensemble. One typical coefficient may be expected to evolve stochastically because it is driven by all the other coefficients.

2. In order to obtain the equation of motion for the 'typical' coefficient $\phi_l(x_{\rm s}; t)$, we use the Feshbach projection-operator method. This projection method is used in the Hilbert space, which fulfils our need to remain with a quantum-state description. An operator P of projection onto the 'typical' coefficient $\phi_l(x_{\rm s}; t)$ as well as its complement $Q = I - P$ are introduced by

$$(P\,\Psi)(x_{\rm s}, x_{\rm b}) \;\equiv\; \phi_l(x_{\rm s})\; \chi_l(x_{\rm b}) \;, \tag{53}$$

$$(Q\,\Psi)(x_{\rm s}, x_{\rm b}) \;\equiv\; \sum_{n\,(\neq l)} \phi_n(x_{\rm s})\; \chi_n(x_{\rm b}) \;, \tag{54}$$

which obey the standard relations for projector operators: $P^2 = P$, $Q^2 = Q$, and $QP = PQ = 0$. Using the projection method by Feshbach, the equation for the typical coefficient $P\Psi(t)$ is obtained as

$$\begin{aligned} i\,\partial_t\, P\Psi(t) \;&=\; PHP\; P\Psi(t) \;+\; PHQ\; e^{-iQHQt}\; Q\Psi(0) \\ &-\; i \int_0^t d\tau\; PHQ\; e^{iQHQ(\tau-t)}\; QHP\; P\Psi(\tau) \;. \end{aligned} \tag{55}$$

In the right-hand side, the first term is essentially the subsystem Hamiltonian, the second is the stochastic forcing of the subsystem by the thermal bath and the third is the dissipative loss from the subsystem toward the bath.

3. In order to close the equation for the typical coefficient $P\Psi(t)$ or $\phi_l(x_{\rm s}; t)$, a triple hypothesis is introduced. Firstly, the average over a typical

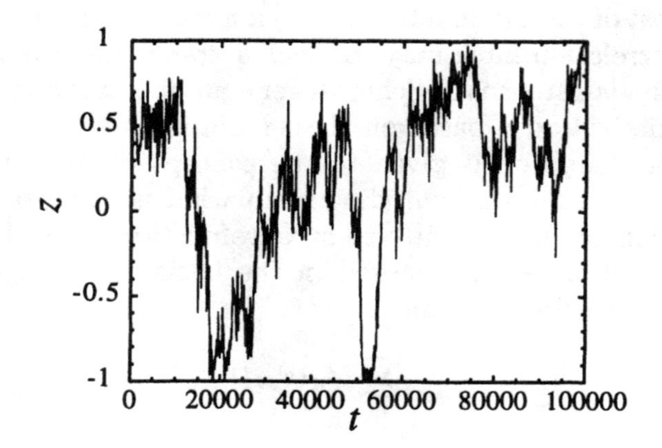

Figure 1. **Bloch component $z = \langle \sigma_z \rangle$ versus the time t for an individual trajectory of the nonMarkovian stochastic Schrödinger equation for the spin-boson model.**

bath eigenfunction $\chi_l(x_b)$ is assumed to be equivalent to an average over a thermal statistical mixture for the bath. Secondly, the different subsystem wavefunctions $\phi_m(x_s; t)$ are assumed to differ by a random phase. Thirdly, the noises are assumed to be Gaussian. The first part of the hypothesis can be justified if the bath is classically chaotic and has eigenfunctions $\chi_n(x_b)$ which satisfy the Berry-Voros conjecture [25]. In this regard, Srednicki has shown that the quantum thermal equilibrium distributions for Bosons and Fermions are consequences of the Berry-Voros conjecture [21]. The second part of the hypothesis allows us to fix the form of the initial wavefunction.

By using this triple hypothesis as well as a perturbative expansion up to second order in the coupling parameter λ, we have been able [22] to derive the following nonMarkovian stochastic Schrödinger equation for a wavefunction $\psi(x_s; t)$ which is proportional to a typical coefficient $\phi_l(x_s; t)$ of the decomposition (52):

$$\begin{aligned} i\,\partial_t\,\psi(t) &= H_s\,\psi(t) + \lambda \sum_\alpha \eta_\alpha(t)\,S_\alpha\,\psi(t) \\ &- i\,\lambda^2 \int_0^t d\tau \sum_{\alpha\beta} C_{\alpha\beta}(\tau)\,S_\alpha\,e^{-iH_s\tau}\,S_\beta\,\psi(t-\tau) + \mathcal{O}(\lambda^3)\,, \end{aligned} \tag{56}$$

where $C_{\alpha\beta}(t)$ are the bath correlation functions (51) and $\eta_\alpha(t)$ are Gaussian noise functions that satisfy

$$\overline{\eta_\alpha(t)} = 0\,, \qquad \overline{\eta_\alpha(t)\eta_\beta(0)} = 0\,, \qquad \text{and} \qquad \overline{\eta^*_\alpha(t)\eta_\beta(0)} = C_{\alpha\beta}(t)\,. \tag{57}$$

We have shown that this stochastic equation has for master equation the nonMarkovian quantum master equation of second-order perturbation theory [22]. In the long-time limit, this nonMarkovian master equation tends

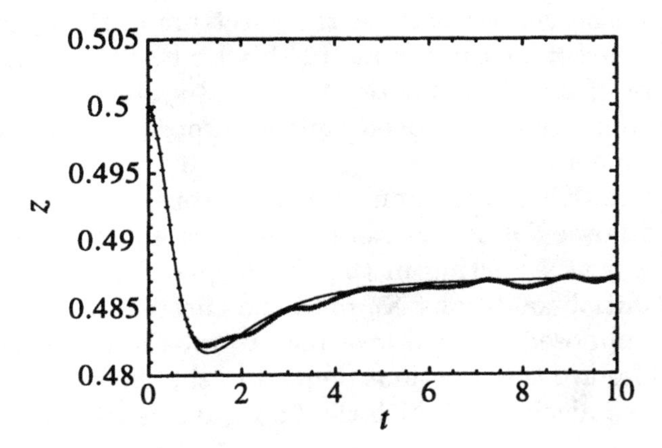

Figure 2. For the spin-boson model under the same conditions as in Fig. 1, comparison between a simulation using the nonMarkovian stochastic Schrödinger equation (56) and the prediction of the corresponding nonMarkovian master equation at the early stage of the time evolution of the Bloch variable z in order to show the slippage of initial conditions over the bath relaxation time $t_b \sim 1$. The average (pluses) is here taken over a statistical ensemble of 2×10^5 individual trajectories.

to the famous Redfield master equation [54] supplemented by a slippage of initial conditions [58]. This slippage of initial conditions can solve the well-known problem of the violation of positivity by the Redfield master equation [58]. The nonMarkovian stochastic Schrödinger equation also avoids this problem of positivity and consistently reproduces the slippage of initial conditions [22].

Our stochastic Schrödinger equation has been applied to the spin-boson model of a spin $S = 1/2$ coupled to a thermal bath of bosonic harmonic oscillators. Fig. 1 depicts the spin variable $z = \langle \sigma_z \rangle$ for a long trajectory of an individual system taken in the statistical ensemble. Fig. 2 shows an average over the statistical ensemble and the agreement between the simultation with our stochastic Schrödinger equation and the calculation based on the quantum master equation just after the initial condition, demonstrating the slippage of initial conditions on the time scale of relaxation of the bath correlation function (51).

4.4. QUANTUM ENTROPIES PER UNIT TIME AND THE CHARACTERIZATION OF DYNAMICAL RANDOMNESS

Stochastic behavior is expected to occur in large many-body quantum systems. In contrast, few-body quantum systems which are bounded have a discrete energy spectrum and, therefore, an almost-periodic time evolution. In the thermodynamic limit, the almost-periodic time evolution could be expected to become stochastic because the spectrum of the time evolution operator typically becomes continuous in this limit. Such a stochasticity is also the feature of classical chaotic dynamical systems where dynamical entropies per unit time have been defined in order to characterize the dynamical randomness.

In the late fifties, Kolmogorov and Sinai rigorously defined an entropy per unit time as follows [59]. For classical dynamical systems, suppose that the dynamics is a flow $\mathbf{\Phi}^t$ acting in the phase space Γ of the system and which maps the initial conditions $\mathbf{X}_0$ onto the current point $\mathbf{X}_t = \mathbf{\Phi}^t\mathbf{X}_0$. The dynamics is supposed to leave invariant a probability measure μ. The classical observables are the functions defined in the phase space and they form a commutative algebra in which the flow induces the time evolution

$$\phi^t\,[A(\mathbf{X})] \;=\; A(\mathbf{\Phi}^t\mathbf{X})\;. \tag{58}$$

Among the observables, we find the indicator functions of cells ω defined in the phase space: $A(\mathbf{X}) = I_\omega(\mathbf{X})$. We define a denumerable partition $\mathcal{P} = \{\omega\}$ of the phase space Γ into non-overlapping cells covering the whole phase space: $\omega \cap \omega' = \emptyset$ and $\cup\omega = \Gamma$. The classical system is observed stroboscopically at regular time intervals $t_n = n\Delta t$ with $n = 0, 1, 2, ...$ The probability that the trajectories visit successively the cells $\omega_0\omega_1\omega_2\cdots\omega_{n-1}$ at the times $t_0, t_1, t_2, ..., t_{n-1}$ is $\mu(\omega_0\omega_1\omega_2\cdots\omega_{n-1})$. The entropy per unit time of the partition $\mathcal{P}$ is defined by

$$h(\mathcal{P}) \;\equiv\; \lim_{n\to\infty}\; -\frac{1}{n\Delta t}\sum_{\omega_0\omega_1\cdots\omega_{n-1}} \mu(\omega_0\omega_1\cdots\omega_{n-1})\;\ln\mu(\omega_0\omega_1\cdots\omega_{n-1})\;. \tag{59}$$

The KS entropy per unit time is the supremum of the entropy over all the partitions [59]

$$h_{\mathrm{KS}} \;=\; \sup_{\mathcal{P}}\, h(\mathcal{P})\;. \tag{60}$$

The entropy per unit time is the rate of decrease of the multiple-time probabilities because of the following theorem.

Shannon-McMillan-Breiman theorem: If $(\mathbf{\Phi}^t, \Gamma, \mu)$ is an ergodic classical dynamical system and $\mathcal{P}$ is a partition of Γ, then

$$\mu(\omega_0\omega_1\cdots\omega_{n-1}) \;\sim\; \exp\left[-n\;\Delta t\; h(\mathcal{P})\right] \tag{61}$$

for almost all initial conditions $\mathbf{X} \in \Gamma$ such that $\mathbf{\Phi}^{k\Delta t}\mathbf{X} \in \omega_k$ [59].

We observe that the entropy per unit time is a property of n-time probabilities, i.e., of n-time correlation functions, in contrast to the transport, Burnett and super-Burnett coefficients which are properties of the 2-, 3- and 4-time correlation functions. This observation gives us a hint about how the concept of KS entropy can be generalized to quantum systems.

In 1987, Connes, Narnhofer and Thirring (CNT) made a first proposal for an extension of the concept of dynamical entropy to noncommutative algebras [60]. The CNT quantum entropy reduces to the KS entropy when the algebra of operators becomes commutative. The notion of partition was replaced by a set of noncommutative subalgebras and the entropy was defined for this set of subalgebras. The CNT quantum entropy per unit time was defined by considering the subalgebras generated by shifting successively in time an initial subalgebra. The CNT entropy per unit time vanishes as expected for few-body quantum systems with a discrete energy spectrum. It may be positive for large quantum systems with infinitely many particles in a thermal state. For noninteracting Bosons or Fermions, a closed expression was derived for the CNT entropy per unit time [61]. As expected again by correspondence with the classical entropy, this entropy per unit time is proportional to the surface crossed by the ideal gas of Bosons or Fermions. The corresponding classical entropy per unit time of an ideal gas is recovered in the limit $\hbar \to 0$, as shown elsewhere [62].

More recently, another definition of a quantum entropy per unit time was proposed by Alicki and Fannes [63]. It is based on a decomposition $\mathcal{X}$ of the identity operator as

$$\sum_{\omega} X_\omega^\dagger \, X_\omega \;=\; I \;, \tag{62}$$

where X_ω are operators in the quantum algebra. These operators are shifted in time by steps Δt according to

$$\phi^t(X_\omega) = X_\omega(t) = \exp(iHt/\hbar) \; X_\omega \;\; \exp(-iHt/\hbar) \; . \tag{63}$$

A multiple-time matrix $\mathbf{D}_n$ of time-correlation functions is defined by the matrix elements

$$D_n\,[\Omega,\Omega'] \;\equiv\; \mathrm{tr}\Big[\underbrace{X_{\omega_{n-1}}(t_{n-1})\cdots X_{\omega_1}(t_1)X_{\omega_0}(t_0)}_{\Omega} \times\rho_{\mathrm{eq}}\,\underbrace{X^\dagger_{\omega'_0}(t_0)X^\dagger_{\omega'_1}(t_1)\cdots X^\dagger_{\omega'_{n-1}}(t_{n-1})}_{\Omega'}\Big] \;, \tag{64}$$

where $\Omega = \omega_0\omega_1\cdots\omega_{n-1}$ and $\Omega' = \omega'_0\omega'_1\cdots\omega'_{n-1}$ are two arbitrary sequences of labels of the operators X_ω and where ρ_{eq} is the equilibrium density matrix. If there are M operators X_ω in the decomposition (62), the matrix (64) has the size $M^n \times M^n$. This matrix is Hermitian $D_n[\Omega,\Omega'] = D_n[\Omega',\Omega]^*$ and non-negative.

An example of decomposition is given by taking the operators X_ω as the projection operators P_ω, which satisfy (62) because $P_\omega P_{\omega'} = P_\omega \delta_{\omega\omega'}$, $P_\omega =$

$P_\omega^\dagger$, and $\sum_\omega P_\omega = I$. With this assumption, the sequences Ω and Ω' define two quantum histories of Griffiths, Gell-Mann, Hartle and Omnès and the matrix (64) characterizes the decoherence between the quantum histories Ω and Ω' [64]. In particular, each diagonal element of the decoherence matrix is non-negative, $D_n[\Omega, \Omega] \geq 0$, and defines a probability for the quantum history Ω. If the histories Ω and Ω' have no common coherence the matrix element $D_n[\Omega, \Omega']$ vanishes.

If the algebra is commutative as in classical mechanics, the projection operators P_ω may be taken as the indicator functions $I_\omega(\mathbf{X})$ of the cells of a partition $\mathcal{P} = \{\omega\}$ of the classical phase space. In this case, the different factors composing (64) commute and $D_n[\Omega, \Omega'] = 0$ if $\Omega \neq \Omega'$ because the cells of the partition do not overlap. Therefore, the decoherence matrix is diagonal for a commutative algebra as in classical dynamical systems,

$$\text{commutative algebra:} \qquad D_n[\Omega, \Omega'] \;=\; D_n[\Omega, \Omega]\; \delta_{\Omega\Omega'} \; . \tag{65}$$

The Alicki-Fannes (AF) entropy per unit time of a quantum dynamical system is defined by [63]

$$h_{\text{AF}} \;\equiv\; \sup_{\mathcal{X}} \lim_{n\to\infty} \; - \; \frac{1}{n\Delta t} \; \text{Tr} \; \mathsf{D}_n \ln \mathsf{D}_n \; . \tag{66}$$

For a commutative algebra, the AF entropy reduces to the KS entropy by the property (65).

A Renyi-type quantum entropy per unit time can similarly be defined as

$$h(q) \;\equiv\; \sup_{\mathcal{X}} \lim_{n\to\infty} \; - \; \frac{1}{n\Delta t} \; \frac{\ln \text{Tr} \; \mathsf{D}_n^q}{q-1} \; , \tag{67}$$

with a Renyi parameter q. For $q = 1$, $h(q) = h_{\text{AF}}$. The quantum entropy (67) reduces to the classical Renyi-type entropy per unit time in the classical limit $\hbar \to 0$ by (65).

The above entropies per unit time are expected to be nonvanishing only for infinite quantum systems defined by a thermodynamic limit. We may conjecture that, for a quantum system with a positive entropy per unit time, a matrix element of (64) would decay exponentially in a similar way as predicted by the classical Shannon-McMillan-Breiman theorem (61). Already the diagonal matrix elements of (64) have been shown to decay exponentially for the system of quantum spins coupled by a dipolar interaction [65].

The AF entropy has already been evaluated numerically for finite quantum systems confirming the vanishing of the entropy per unit time in these systems. Many questions remain open concerning its value in infinite quantum systems where it is conjectured to be positive.

5. Conclusions and Perspectives

We have given a general overview of recent results on the decay, transport and stochastic properties of quantum systems.

In Section 2, we have described the quasiclassical (Liouvillian) and the periodic-orbit semiclassical approaches to quantum systems. We have shown that, generally, a spectral function can be decomposed into a quasiclassical smooth contribution plus semiclassical periodic-orbit contributions which oscillate with the frequency or the energy and which are supposed to approximate the effect of quantization of energy. The smooth quasiclassical contribution is obtained in the Wigner representation by a Weyl series in powers of the Planck constant. The leading term is the classical expression. The semiclassical periodic-orbit contributions are of Gutzwiller-type and their precise form depends on the stability of the periodic orbits. Unstable periodic orbits only contribute by terms of order $\hbar^{-1}$ while the leading classical term is of order $\hbar^{-f}$ where f is the number of degrees of freedom. This result is of special importance in our context because it shows that the periodic-orbit contributions will in general become negligible in many-body quantum systems with $f \to \infty$. Therefore, we may conclude that the Weyl series would give the essential contribution in interacting many-body quantum systems at nonvanishing temperatures such as fluids. The periodic-orbit contributions are nevertheless important in microscopic and mesoscopic quantum systems like atoms, nuclei, molecules, atomic clusters, or in solid-state systems where the quantum coherence of the few-body dynamics remains important.

The contributions of the unstable periodic orbits can be expressed as the logarithmic derivative of a semiclassical quantum Zeta function which is a product over all the periodic orbits. This result appears to be general in the Gutzwiller semiclassical theory which is based on the periodic-orbit trace formula. In this regard, it should be emphasized that the periodic orbits are selected in the semiclassical approximation because the quantum expression is defined by the trace of an operator involving an evolution operator or a resolvent operator. The trace is given by a Feynman path integral where the paths are closed on themselves so that, in the semiclassical limit, the paths reduce to closed classical trajectories, i.e., to periodic orbits (and also to stationary points).

Furthermore, we have pointed out that different expressions are obtained for the spectral function whether it is defined by an average over a pure or a mixed state. In the case of a pure state, the leading quasiclassical term is a microcanonical average of a static quantity involving the pure state and the operator of the spectral function. In the case of a mixed state, the leading quasiclassical term is the Fourier transform of the classical autocorrelation function. The decay of such classical autocorrelation functions is often controlled by classical resonances called the Pollicott-Ruelle resonances which can be calculated thanks to a classical trace formula derived by Cvitanović and Eckhardt [36]. This classical trace formula is essentially the trace of the resolvent of the classical Liouvillian operator and, for a chaotic system, it is given by the logarithmic derivative of a classical Zeta function which is also a product over the unstable periodic orbits of the system. The differences and similarities between the semiclassical quantum Zeta function and the classical Zeta function have been explained in Subsection 2.4. The large-deviation formalism to characterize classical chaos was also summarized in Subsection 2.5.

In Section 3, the semiclassical methods have been applied to quantum scattering systems in order to characterize the decay processes which usually take place in scattering systems. The decay can be described thanks to the quantum scattering resonances which are the poles of the scattering matrix or also the poles of the resolvent of the Hamiltonian operator. This approach is of prime importance, in particular, in chemical kinetics in order to study unimolecular reactions.

The semiclassical approach becomes of interest in systems where the scattering resonances accumulate and where the resonance spectrum needs to be characterized by its statistical properties. The periodic-orbit semiclassical method turns out to be powerful in this context to describe the structure of the resonance spectrum at high energies where the scattering resonances can be approximated by the zeros of the semiclassical quantum Zeta function.

Moreover, a bound on the quantum lifetimes can be obtained which is useful when the classical repeller, made of trajectories trapped in the interacting region, is filamentary. On the other hand, we have shown that a decay process involving many scattering resonances may behave quasiclassically and obey the Liouvillian dynamics. In this case, the Pollicott-Ruelle resonances become relevant and, in particular, the leading resonance which is the classical escape rate. In this regard, the classical escape rate appears as a kind of average over the many quantum rates defined by the individual scattering resonances. For classically chaotic systems, we have shown that the classical lifetime corresponding to the escape rate is shorter than the longest quantum lifetimes. This result is compatible with the interpretation of the classical escape rate as an average over the quantum rates associated with the scattering resonances. The classical behavior predicted by the escape rate is thus expected to occur on an intermediate time scale, after which the quantum scattering resonances with the longest lifetimes would dominate the decay process. The theory is illustrated by applications to the disk scatterers, to ultrafast unimolecular reactions, and also to spatially extended scatterers.

In Section 4, we have been concerned by systems with many degrees of freedom taken in the thermodynamic limit. It is in this limit that the time evolution of a bounded quantum system may have a continuous spectrum and, thus, become mixing. The mixing property is indeed required in order to guarantee the relaxation of the time-correlation functions and is, therefore, a necessary condition for the existence of positive and finite transport coefficients according to the Green-Kubo formula, as explained in Subsection 4.1.

In Subsection 4.2, we have presented several spectral properties of systems of increasing size in the case of spin systems. In particular, the number of constants of motion has been studied as well as the Wigner spacing statistics [20]. Different spectral statistics are observed for systems with or without dynamical constants of motion, which are expected to determine the transport properties of the infinite system. Here, the observation of a Wigner statistics is evidence for the absence of extra constants of motion. Similar results have been obtained for electronic systems. Such properties of spectral statistics are thus expected to be general and of importance for many-body quantum systems.

Subsection 4.3 contains a summary of a recent derivation of a non-Markovian stochastic Schrödinger equation, which describes the dynamics of a slow quantum subsystem interacting with a faster thermal bath [22]. This stochastic Schrödinger equation is associated with the Redfield quantum master equation with a slippage of initial conditions [58] and is thus relevant for the relaxation of spins or other two-level systems in condensed phases as in NMR. This stochastic Schrödinger equation may turn out to be useful for the simulation of quantum subsystems with a large state space of dimension $N \gg 1$, because the stochastic equation only requires to integrate the $2N$ real components of the quantum state, while the associated master equation requires the simultaneous integration of the N^2 variables of the density matrix.

Finally, in Subsection 4.4, we have presented methods to characterize dynamical randomness, i.e., stochasticity, in large quantum systems. We have argued that such a characterization can be performed with quantum entropies per unit time which are the Connes-Narnhofer-Thirring entropy and the Alicki-Fannes entropy. This latter entropy can be interpreted as the rate of decay of the multiple-time decoherence matrices of the quantum histories introduced by Griffiths, Gell-Mann, Hartle, and Omnès. We should note that we are here concerned with the decay of the decoherence matrices as the number of times increases. In a certain sense, these multiple-time decoherence matrices form the quantum generalization of the multiple-time probabilities introduced by Onsager and Machlup for classical Gaussian irreversible processes [66]. However, methods to evaluate conveniently the quantum dynamical entropy are still missing. Such methods are desirable, especially, in order to calculate the quantum corrections to the KS entropy recently obtained by kinetic theory for the hard-sphere gas [11].

Acknowledgements

The author thanks Professor G. Nicolis for support and encouragement in this research. He is supported by the National Fund for Scientific Research (F. N. R. S. Belgium). This work is financially supported by the Training and Mobility Program of the European Commission, by the Belgian government under the Pôles d'Attraction Interuniversitaires Program, and by the F. N. R. S. (Belgium).

References

1. J. R. Dorfman, *An Introduction to Chaos in Nonequilibrium Statistical Mechanics*, (Cambridge University Press, Cambridge, 1999).
2. P. Gaspard, *Chaos, Scattering, and Statistical Mechanics*, (Cambridge University Press, Cambridge, 1998).
3. P. Gaspard and G. Nicolis, Phys. Rev. Lett. **65**, 1693 (1990).
4. P. Gaspard and F. Baras, Phys. Rev. E **51**, 5332 (1995).
5. J. R. Dorfman and P. Gaspard, Phys. Rev. E **51**, 28 (1995).
6. P. Gaspard and J. R. Dorfman, Phys. Rev. E **52**, 3525 (1995).
7. H. A. Posch and W. G. Hoover, Phys. Rev. A **38**, 473 (1988); *ibid.* **39**, 2175 (1989).
8. D. J. Evans, E. G. D. Cohen, and M. P. Morriss, Phys. Rev. A **42**, 5990 (1990).
9. N. I. Chernov, G. L. Eyink, J. L. Lebowitz, and Ya. G. Sinai, Phys. Rev. Lett. **70**, 2209 (1993).
10. G. Gallavotti and E. G. D. Cohen, Phys. Rev. Lett. **74**, 2694 (1995); J. Stat. Phys. **80**, 931 (1995).
11. H. van Beijeren, J. R. Dorfman, H. A. Posch, and Ch. Dellago, Phys. Rev. E **56**, 5272 (1997).
12. M. C. Gutzwiller, *Chaos in Classical and Quantum Mechanics* (Springer, New York, 1990).
13. P. Gaspard and S. A. Rice, J. Chem. Phys. **91**, 2225, 2242, 2255 (1989).
14. D. Wintgen, K. Richter, and G. Tanner, Chaos **2**, 19 (1992).
15. B. Grémaud and P. Gaspard, J. Phys. B: At. Mol. Phys. **31**, 1671 (1998).
16. P. Gaspard and S. A. Rice, Phys. Rev. A **93**, 54 (1993).
17. I. Burghardt and P. Gaspard, J. Chem. Phys. **100**, 6395 (1994); J. Phys. Chem. **99**, 2732 (1995); Chem. Phys. **225**, 259 (1997).
18. C. E. Porter, ed., *Statistical Theories of Spectra: Fluctuations* (Academic Press, New York, 1965).
19. M. J. Giannoni, A. Voros, and J. Zinn-Justin, eds., *Chaos and Quantum Physics* (North-Holland, Amsterdam, 1991).
20. P. van Ede van der Pals and P. Gaspard, Phys. Rev. E **49**, 79 (1994).
21. M. Srednicki, Phys. Rev. E **50**, 888 (1994); *ibid.* **54**, 954 (1996).
22. P. Gaspard and M. Nagaoka, J. Chem. Phys. **111**, 5676 (1999).
23. P. Gaspard, D. Alonso, and I. Burghardt, Adv. Chem. Phys. **90**, 105 (1995).
24. P. Carruthers and F. Zachariasen, Rev. Mod. Phys. **55**, 245 (1983).
25. M. V. Berry, J. Phys. A: Math. Gen. **10**, 2083 (1977); A. Voros, Ann. Inst. Henri Poincaré A **24**, 31 (1976); *ibid.* **26**, 343 (1977).

26. V. Colin de Verdière, Commun. Math. Phys. **102**, 497 (1985); S. Zelditch, Duke Math. J. **55**, 919 (1987).
27. E. J. Heller, J. Chem. Phys. **68**, 2066, 3891 (1978).
28. B. Eckhardt, S. Fishman, K. Müller, and D. Wintgen, Phys. Rev. A **45**, 3531 (1992).
29. A. Voros, J. Phys. A: Math. Gen. **21**, 685 (1988).
30. P. Cvitanović and B. Eckhardt, Phys. Rev. Lett. **63**, 823 (1989).
31. R. Artuso, E. Aurell, and P. Cvitanović, Nonlinearity **3**, 325, 361 (1990).
32. P. Gaspard and I. Burghardt, Adv. Chem. Phys. **101**, 491 (1997).
33. P. Gaspard and S. R. Jain, Pramana-J. of Phys. **48**, 503 (1997).
34. B. Mehlig and K. Richter, Phys. Rev. Lett. **80**, 1936 (1998).
35. M. Pollicott, Invent. Math. **81**, 413 (1985); D. Ruelle, Phys. Rev. Lett. **56**, 405 (1986); J. Stat. Phys. **44**, 281 (1986); J. Diff. Geom. **25**, 99, 117 (1987).
36. P. Cvitanović and B. Eckhardt, J. Phys. A: Math. Gen. **24**, L237 (1991).
37. Ya. G. Sinai, Russ. Math. Surv. **27**, 21 (1972); R. Bowen and D. Ruelle, Invent. Math. **29**, 181 (1975).
38. D. Ruelle, *Thermodynamic Formalism* (Addison-Wesley, Reading, 1978).
39. D. Bessis, G. Paladin, G. Turchetti, and S. Vaienti, J. Stat. Phys. **51**, 109 (1988).
40. Z. Kovács and T. Tél, Phys. Rev. Lett. **64**, 1617 (1990).
41. A. Wirzba, Phys. Rep. **309**, 1 (1999).
42. Y. Decanini, A. Folacci, E. Fournier, and P. Gabrielli, J. Phys. A: Math. Gen. **31**, 7865, 7891 (1998).
43. W. Lu, M. Rose, K. Pance, and S. Sridhar, Phys. Rev. Lett. **82**, 5233 (1999).
44. K. A. Holbrook, M. J. Pilling, and S. H. Robertson, *Unimolecular Reactions*, 2nd ed., (Wiley, Chichester, 1996).
45. F. Barra and P. Gaspard, J. Phys. A: Math. Gen. **32**, 3357 (1999).
46. P. Gaspard, in *Quantum Chaos*, G. Casati, I. Guarneri, and U. Smilansky, eds., (North-Holland, Amsterdam, 1993), p. 307.
47. B. D. Simons, Physica A **263**, 148 (1999).
48. R. Kubo, J. Phys. Soc. Jpn. **12**, 570 (1957).
49. E. Helfand, Phys. Rev **119**, 1 (1960); Phys. Fluids **4**, 681 (1961).
50. P. Gaspard, in *Nonlinear Dynamics and Computational Physics*, V.B. Sheorey, ed., (Narosa Publishing House, New Delhi, 1999), p. 54.
51. P. Gaspard and P. van Ede van der Pals, Chaos, Solitons, and Fractals **5**, 1183 (1995).
52. P. van Ede van der Pals, *Etude de systèmes quantiques dans les limites semi-classique et thermodynamique*, Thèse de doctorat, Université Libre de Bruxelles (1997).
53. V. E. Kravtsov, I. V. Yurkevich, and C. M. Canali, in *Supersymmetry and Trace Formulae: Chaos and Disorder*, I.V. Lerner, J.P. Keating, and D.E. Khmelnitskii, eds., (Kluwer/Plenum, New York, 1999), p. 269.
54. A. G. Redfield, IBM J. Research Devel. **1**, 19 (1957); Adv. Magn. Reson. **1**, 1 (1965).
55. N. van Kampen, *Stochastic Processes in Physics and Chemistry*, 2nd ed., (North-Holland, Amsterdam, 1992).
56. M. B. Plenio and P. L. Knight, Rev. Mod. Phys. **70**, 101 (1998).
57. G. Lindblad, Commun. Math. Phys. **48**, 119 (1976).
58. P. Gaspard and M. Nagaoka, J. Chem. Phys. **111**, 5668 (1999).
59. I. P. Cornfeld, S. V. Fomin, and Ya. G. Sinai *Ergodic Theory* (Springer, Berlin, 1982).
60. A. Connes, H. Narnhofer, and W. Thirring, Commun. Math. Phys. **112**, 691 (1987).
61. H. Narnhofer and W. Thirring, Lett. Math. Phys. **14**, 89 (1987).
62. P. Gaspard, Prog. Theor. Phys. Suppl. **116**, 369 (1994).
63. R. Alicki and M. Fannes, Lett. Math. Phys. **32**, 75 (1994).
64. R. B. Griffiths, J. Stat. Phys. **36**, 219 (1984); M. Gell-Mann and J. B. Hartle, in *Foundations of Quantum Mechanics in the Light of New Technology*, S. Kobayashi, H. Ezawa, Y. Murayama, and S. Nomura, eds., (Physical Society of Japan, Tokyo,

1990); R. Omnès, Rev. Mod. Phys. **64**, 339 (1992).
65. P. van Ede van der Pals and P. Gaspard, in *Quantum Classical Correspondence*, D. H. Feng and B. L. Hu, eds., (International Press, Cambridge MA, 1997), p. 483.
66. L. Onsager and S. Machlup, Phys. Rev. **91**, 1505 (1953).

SOME PROBLEMS OF THE KINETIC THEORY OF MESOSCOPIC SYSTEMS

V.L. GUREVICH
Solid State Physics Division, A.F. Ioffe Institute
194021 Saint Petersburg, Russia

Abstract. We discuss the influence of phonons on ballistic transport in a two-terminal nanostructure consisting of a spatially uniform semiconductor quantum wire connected to two large classical reservoirs, each in thermal equilibrium with itself. The diagrammatic technique is applied to calculate the phonon-assisted variation of ballistic resistance of such a structure.

This approach is compared with another approach based on iteration of the quasiclassical Boltzmann equation. Several physical phenomena are discussed in this context, such as d.c. current produced in a nanostructure by a travelling acoustic wave (acoustoelectric current), Coulomb drag under ballistic conditions in quantum wires and generation of heat by ballistic current-carrying nanostructures.

1. Introduction

The advances in semiconductor nano-fabrication and material science in recent years have made available materials of great purity and crystalline perfection. The electrical conduction and some other transport phenomena in such nanoscale structures have been a focus of numerous investigations, both theoretical and experimental, with a number of important discoveries and even patent applications.

The essence of electrical conduction in these structures is that the quantum nature of the electron leaves its distinct trace in a macroscopic measurement. Namely, electrons move through nanostructures phase coherently so that the phase difference of an electron's wave function between any two spatial points remains constant. Also, nanostructures act effectively as waveguides. This is possible since the largest dimension of the structure is smaller than the electron mean free path as well as the coherence break-

J. Karkheck (ed.),
Dynamics: Models and Kinetic Methods for Non-equilibrium Many Body Systems, 457–471.

ing length in the problem (typically a few μm). These nanoscale systems are characterized by low electron densities, which may be easily varied by means of gate voltage. It also possible to make these nanostructures in such a way that the impurity scattering can be ignored. The transport of electrons in such a regime is called *ballistic*.

In these lectures we give a discussion of the effects of electron-phonon scattering on ballistic electron transport and of the ways of treating transport phenomena in semiconductor nanostructures.

2. Phonon-Assisted Ballistic Resistance. Formulation of the Problem

We will make use of the physical picture developed by Landauer-Büttiker-Imry [1, 2], i.e. we consider a semiconductor quantum nanostructure (quantum wire) connecting the two reservoirs, each in equilibrium with itself. Both are described by the equilibrium Fermi functions, however at different values of chemical potentials, $\mu^{(+)}$ and $\mu^{(-)}$, their difference being equal to eV,

$$\mu^{(+)} - \mu^{(-)} = eV, \tag{1}$$

where V is the voltage applied. This means that in a ballistic situation we have in a quantum wire two counterflows of electrons moving in the opposite directions [3].

In the linear response regime we have the Ohm's law

$$J = G^{(0)}V. \tag{2}$$

It is well known (see, for instance, the review [3]) that in the absence of electron scattering, the conductance $G^{(0)}$ is a step-like function of the Fermi level or the gate voltage. Each step corresponds to the inclusion of a new mode of transverse quantization to the conduction process. The height of each step is equal to the quantum of conductance,

$$G_0 = 2e^2/h, \tag{3}$$

multiplied by a prefactor whose physical interpretation is that of a transmission probability. It is unity in the case of uniform 1D conductors which we consider here.[1] We consider the simplest case of a uniform mesoscopic conductor of constant transverse cross section (a quantum wire). In such a case no redistribution of the electrostatic potential with the onset of the phonon scattering should be expected.

[1] To obtain the classical ballistic conductance the sum over a large number of modes of transverse quantization should be replaced by an integral and one gets a result analogous to the 3D Sharvin's [4] point contact conductance.

The electron wave functions within the wire have the form

$$\mathcal{V}^{-1/2}\exp(ipx/\hbar)\phi_n(\mathbf{r}_\perp). \tag{4}$$

Here $\mathcal{V}$ is the normalization volume, $\mathbf{r}_\perp = y, z$ are the coordinates perpendicular to the x-axis (along the wire), $\phi_n(\mathbf{r}_\perp)$ are the wave functions of transverse quantization. They are determined by the equation

$$\left[-\frac{\hbar^2}{2m}\left(\frac{\partial^2}{\partial y^2}+\frac{\partial^2}{\partial z^2}\right)+U(\mathbf{r}_\perp)\right]\phi_n(\mathbf{r}_\perp) = \epsilon_n(0)\phi_n(\mathbf{r}_\perp), \tag{5}$$

where $U(\mathbf{r}_\perp)$ is the confining potential for the conduction electrons while $n = 1, 2, 3, \ldots$ enumerate the eigenvalues of transverse quantisation, $\epsilon_n(0)$.

The electron spectrum consists of a set of 1D subbands. Each is characterized by the dispersion law

$$\epsilon_n(p) = \epsilon_n(0) + p^2/2m, \tag{6}$$

where m is the effective mass of an electron, while p is the x-component of the electron quasimomentum.

Due to the electron-phonon scattering we have for a full conductance $G = G^{(0)} - \Delta G$. Our immediate purpose will be to calculate the phonon-assisted part, ΔG. During the intermediate calculations we will often omit the factor $\hbar$ but will restore it in the final formulas. To apply a perturbation theory we assume that the change of the current ΔJ due to electron-phonon interaction is relatively small.

3. Diagrammatic-Technique Approach

In this section and the next two sections we will follow the paper [5]. We will discuss how a diagrammatic technique can be applied for calculation of ΔG. As the initial state of the electron system (unperturbed by the electron-phonon interaction) is a nonequilibrium one we resort to the Keldysh diagrammatic technique (see Ref. [6]) in order to calculate the phonon-controlled part of the current.

The Keldysh Green function is (apart from the spin variable — see below) a 2×2 matrix. There are two representations for the Green function, quadratic and triangular. In the quadratic representation the electron Green function matrix [6] is

$$\hat{\mathcal{G}} = \begin{pmatrix} \mathcal{G}^c & \mathcal{G}^+ \\ \mathcal{G}^- & \tilde{\mathcal{G}}^c \end{pmatrix}. \tag{7}$$

The matrix elements $\mathcal{G}_{\alpha\beta}$ are defined in the following way

$$\mathcal{G}^c = -i\left\langle \mathrm{T}\psi(r)\psi^+(r')\right\rangle_0, \quad \tilde{\mathcal{G}}^c = -i\left\langle \tilde{\mathrm{T}}\psi(r)\psi^+(r')\right\rangle_0, \tag{8}$$

where $r = (\mathbf{r}, t)$, T and $\tilde{\mathrm{T}}$ are the time ordering and the reverse time ordering operators, ψ^+ and ψ are the creation and annihilation electron field operators in the Heisenberg representation, and $|\ \rangle_0$ is the state of the system at $t \to -\infty$ before the electron-phonon interaction is turned on (we assume that it is turned on adiabatically). In the absence of the interaction we have ψ_0^+ and ψ_0 instead of ψ^+ and ψ, so that

$$\mathcal{G}_0^c(r, r') = -i \left\langle \mathrm{T}\psi_0(r)\psi_0^+(r') \right\rangle_0 \tag{9}$$

and

$$\tilde{\mathcal{G}}_0^c(r, r') = -i \left\langle \tilde{\mathrm{T}}\psi_0(r)\psi_0^+(r') \right\rangle_0 . \tag{10}$$

Now,

$$\mathcal{G}_0^-(r, r') = -i \left\langle \psi_0(r)\psi_0^+(r') \right\rangle_0 , \tag{11}$$

while

$$\mathcal{G}_0^+(r, r') = -i \left\langle \psi_0^+(r')\psi_0(r) \right\rangle_0 . \tag{12}$$

A typical term in the perturbative expansion is of the form

$$(\mathcal{G}_0)_{lj}\, \gamma_{ij}^k\, (\mathcal{G}_0)_{jm}\, (\mathcal{D}_0)_{kn} , \tag{13}$$

$\mathcal{D}_0$ being the zero-order phonon Green function, where the subscripts refer to the electron lines while the superscripts refer to the phonon lines and the summation is assumed over the repeated indices. For each vertex we use a matrix of a third rank of the form

$$\gamma_{ij}^k = \delta_{ij}(\sigma_3)_{jk}, \tag{14}$$

where σ_3 is Pauli spin matrix in the z direction and δ_{ij} is Kronecker's delta function. We assume that the Green function matrix $\mathcal{G}$ is diagonal with respect to the spin variables and neglect the spin-orbit interaction. Therefore, in what follows we omit the spin indices altogether.

To the zeroth order in the electron-phonon interaction, the Green functions are

$$\mathcal{G}_0^-(\mathbf{r}, \mathbf{r}'; \epsilon) = -2\pi i \cdot 2 \sum_n \int \frac{dp}{2\pi\hbar} (1 - f) e^{ip(x-x')} \phi_n^*(\mathbf{r}'_\perp) \phi_n(\mathbf{r}_\perp) \delta[\epsilon - \epsilon_n(p)], \tag{15}$$

$$\mathcal{G}_0^+(\mathbf{r}, \mathbf{r}'; \epsilon) = 2\pi i \cdot 2 \sum_n \int \frac{dp}{2\pi\hbar} f e^{ip(x-x')} \phi_n^*(\mathbf{r}'_\perp) \phi_n(\mathbf{r}_\perp) \delta[\epsilon - \epsilon_n(p)]. \tag{16}$$

The electron distribution function, $f = f_n(p)$, is, in general, a nonequilibrium function. The subscript 0 here denotes zeroth order in electron-phonon coupling in the nanostructure. Now,

$$\mathcal{G}_0^c(\mathbf{r},\mathbf{r}';\epsilon) = 4\pi \sum_n \int \frac{dp}{2\pi\hbar} \left[\frac{f}{\epsilon - \epsilon_n(p) - i0} + \right. \tag{17}$$

$$\left. \frac{1-f}{\epsilon - \epsilon_n(p) + i0} \right] e^{ip(x-x')} \phi_n^*(\mathbf{r}'_\perp)\phi_n(\mathbf{r}_\perp),$$

$$\tilde{\mathcal{G}}_0^c(\mathbf{r},\mathbf{r}';\epsilon) = 4\pi \sum_n \int \frac{dp}{2\pi\hbar} \left[\frac{f}{\epsilon - \epsilon_n(p) + i0} + \right. \tag{18}$$

$$\left. \frac{1-f}{\epsilon - \epsilon_n(p) - i0} \right] e^{ip(x-x')} \phi_n^*(\mathbf{r}'_\perp)\phi_n(\mathbf{r}_\perp).$$

The current density, j_i can be expressed through $\mathcal{G}^+$ as

$$j_i = \frac{e\hbar}{m} \lim_{\mathbf{r}\to\mathbf{r}'} \left(\frac{\partial}{\partial x_i} - \frac{\partial}{\partial x'_i} \right) \mathcal{G}^+(\mathbf{r},\mathbf{r}'), \tag{19}$$

where the limit is to be taken only after the derivatives have been evaluated. One can show that by neglecting the electron-phonon scattering entirely one can get Eq.(2), and, if necessary, also its nonlinear generalization.

In the same manner one can write for the matrix of phonon Green functions

$$\hat{\mathcal{D}}(\mathbf{r},\mathbf{r}',\omega) = \int \frac{d^3q}{(2\pi)^3} e^{i\mathbf{q}(\mathbf{r}-\mathbf{r}')} \hat{\mathcal{D}}(\mathbf{q},\omega), \tag{20}$$

where

$$\left. \begin{matrix} \mathcal{D}^c(\mathbf{q},\omega) \\ \tilde{\mathcal{D}}^c(\mathbf{q},\omega) \end{matrix} \right\} = W_\mathbf{q} \left(\frac{N_\mathbf{q}+1}{\omega \mp \omega_\mathbf{q} + i0} + \right. \tag{21}$$

$$\left. \frac{N_\mathbf{q}}{\omega \pm \omega_\mathbf{q} + i0} - \frac{N_\mathbf{q}+1}{\omega \pm \omega_\mathbf{q} - i0} - \frac{N_\mathbf{q}}{\omega \mp \omega_\mathbf{q} - i0} \right)$$

while

$$\mathcal{D}^\pm(\mathbf{q},\omega) = -2\pi i W_\mathbf{q}[(N_\mathbf{q}+1/2\pm 1/2)\delta(\omega\mp\omega_\mathbf{q}) + (N_{-\mathbf{q}}+1/2\mp 1/2)\delta(\omega\pm\omega_\mathbf{q})]. \tag{22}$$

Here $N_\mathbf{q}$ is the phonon distribution function. We assume it to be the equilibrium Bose function.

4. Electron-Phonon Interaction

We assume that the factor describing the electron-phonon interaction, $W_{\mathbf{q}}$, is incorporated into the phonon Green functions. For the deformation potential interaction with longitudinal acoustic phonons in an isotropic elastic medium

$$W_{\mathbf{q}} = \frac{\pi \Lambda^2 q^2}{\rho \omega_{\mathbf{q}}}, \tag{23}$$

where Λ is the deformation potential constant for the longitudinal phonons, and ρ is the mass density. For interaction with acoustic phonons of any branch one should make a substitution

$$\Lambda^2 \to \Lambda_a^2. \tag{24}$$

Here Λ_a denotes the deformation potential constant for the acoustic wave belonging to branch a. For the piezoelectric unscreened electron-phonon interaction we have

$$W_{\mathbf{q}} = \frac{\pi}{\rho \omega_{\mathbf{q}}} \left[\frac{4\pi e \beta_{q,lq} \nu_l(\mathbf{q}, a)}{\varepsilon_{qq}} \right]^2. \tag{25}$$

Here $\beta_{i,ln}$ is the tensor of piezoelectric moduli (which is symmetric in the last two indices), ε_{il} is the tensor of dielectric susceptibility, $\nu_l(\mathbf{q}, a)$ is the polarization vector (i.e. a unit vector along the elastic displacement $\mathbf{u}$) of the phonon with wave vector $\mathbf{q}$, belonging to the branch a. Index q indicates the projection of a tensor on $\mathbf{q}$ direction. Tensor $\beta_{i,ln}$ enters the equation for the electrostatic induction $\mathbf{D}$ (see, for instance, Ref. [7]):

$$D_i = D_{0i} + \varepsilon_{il} E_l - 4\pi \beta_{i,ln} u_{ln}, \tag{26}$$

where $\mathbf{D}_0$ is a constant vector (that does not enter the equation for the electron-phonon interaction), $\mathbf{E}$ is the a.c. piezoelectric field while u_{ln} is the strain produced by the acoustic wave. For the polar optical scattering [8],

$$W_{\mathbf{q}} = 2\pi (e/q)^2 \omega_o (\varepsilon_0 - \varepsilon_\infty)/\varepsilon_0 \varepsilon_\infty. \tag{27}$$

Here ε_0 and ε_∞ are the lattice dielectric susceptibilities for $\omega \to 0$ and $\omega \to \infty$ respectively. We ignore the dispersion of the optical phonons, namely, $\omega_{\mathbf{q}} = \omega_o$.

5. Perturbation Theory

The easiest way to calculate the diagram is to use the p-representation. This means that we use the fact that the Green function is a sharp function of the difference $x_d = x - x'$ and a smooth function of $x_s = (x + x')/2$. The dependence on x_s should take place because $\Delta\mathcal{G}_+$ should satisfy certain boundary conditions at $x_s = \pm L_x/2$. Therefore the problem is not entirely homogeneous along the x-direction.

To first order in the electron-phonon interaction, the perturbation theory diagram is

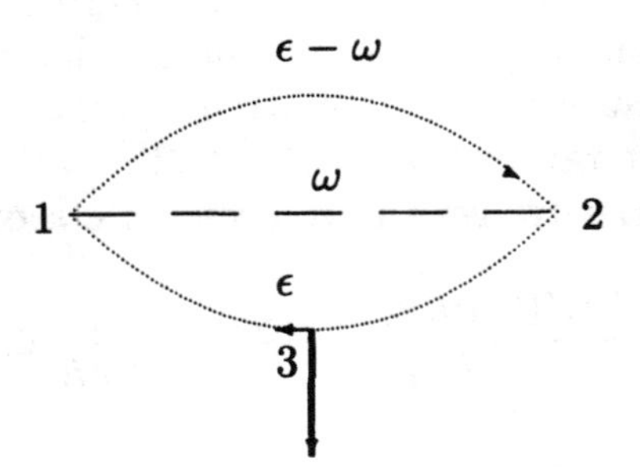

The vertical arrow indicates the current vertex, i.e. it represents the operation

$$e\sum_n \int \frac{d\epsilon}{2\pi i} \int \frac{dp}{2\pi\hbar} v. \tag{28}$$

The dashed line represents the phonon propagator. Each electron-phonon vertex implies summation over the Keldysh matrix indices.

The calculation associated with the diagram yields $\Delta\mathcal{G}^+$. The diagram represents a contribution of several terms that are obtained by summation over the matrix indices of the vertices in the figure. Our purpose is to calculate the total current. Since the electron velocity is diagonal in regard to the matrix index n, we calculate the diagonal part, $\Delta\mathcal{G}^+_{nn}$ which is actually the double projection of $\Delta\mathcal{G}^+$ as a function of $\mathbf{r}_\perp$ and $\mathbf{r}'_\perp$ on the states n. So we transform away the confined degrees of freedom which are perpendicular to the orientation of the wire,

$$\mathcal{G}^+_{nn'}(r, r') = \int\int d\mathbf{r}_\perp d\mathbf{r}'_\perp \phi^*_n(\mathbf{r}_\perp)\mathcal{G}^+(\mathbf{r}, t, \mathbf{r}', t')\phi_{n'}(\mathbf{r}_\perp), \tag{29}$$

and take a Fourier transform with respect to the rapidly varying x_d coordinate.

Let us ascribe quasimomentum $p - \kappa/2$ to the electron line going to the current vertex and quasimomentum $p + \kappa/2$ to the outgoing electron line. The $1/\kappa$ determines the scale of the spatial variation of the electron's distribution function with respect to the variable x_s. Hence, in our case it should be considered as a large parameter as compared to the de Broglie wavelength of an electron. Each item of the diagram contains the following product of pole denominators

$$\int \frac{d\epsilon}{2\pi i} \frac{1}{\epsilon - \epsilon_{p-\hbar\kappa/2} \pm i0} \frac{1}{\epsilon - \epsilon_{p+\hbar\kappa/2} \mp i0} = \pm \frac{1}{v\kappa}. \tag{30}$$

We see that $\Delta\mathcal{G}^+$ is proportional to $1/\kappa v$ where $v = \partial\epsilon_n/\partial p$. This large factor disappears in the expression for the spatial derivative, $\partial\Delta\mathcal{G}^+/\partial x_s$. This means that what we are calculating by the perturbation theory is actually the derivative rather than $\Delta\mathcal{G}^+$ itself.

Performing a Fourier transform over variable κ one can write

$$\frac{\partial\Delta(\mathcal{G}^+(p, x_s, \epsilon))_{nn}}{\partial x_s} = -\frac{i}{2\pi\hbar}\mathcal{R}, \tag{31}$$

where

$$\mathcal{R} = \frac{1}{2\pi\hbar v}\delta[\epsilon - \epsilon_n(p)] \sum_{n'} \int_{-\infty}^{\infty} dp' \int \frac{d^2\mathbf{q}_\perp}{(2\pi)^d} C_{nn'} W_{\mathbf{q}} (B^{(+)} + B^{(-)}). \tag{32}$$

Here

$$C_{nn'} = \left|\langle n'| \exp\left(i\mathbf{q}_\perp \cdot \mathbf{r}_\perp\right)|n\rangle\right|^2,$$

$$B^{(\pm)} = [f'(1-f)\left(N_{\mathbf{q}} + 1/2 \pm 1/2\right) - f(1-f')(N_{\mathbf{q}} + 1/2 \mp 1/2)]\delta(\epsilon' - \epsilon \mp \hbar\omega_{\mathbf{q}}), \tag{33}$$

where the distribution functions are given by $f = f(p)$ and $f' = f(p')$ and the summation over the phonon branches is implied. The integrations in Eq. (32) are over the three components of the phonon wave vector. The two transverse wave vector components are indicated by $\mathbf{q}_\perp$; the third integration is equivalent to the integration over the electron quasimomentum, p', because of the conservation of quasimomentum.

The boundary condition on $\Delta\mathcal{G}^+$ is such that this function must vanish for each electron population at the point of their injection into a mesoscopic wire,

$$\Delta\mathcal{G}^+(p, x_s, \epsilon) = 0 \text{ at } x_s = \begin{cases} -L_x/2 & p > 0 \\ +L_x/2 & p < 0. \end{cases} \tag{34}$$

This must happen since at such a location it is expected that no appreciable phonon mediated interactions between the two populations of electrons, each out of equilibrium with the other and yet in equilibrium with itself, has taken place yet. It vanishes for all collisions involving strictly one population by an explicit assumption that electrons in each contact are in thermal equilibrium with themselves and hence obey Fermi-Dirac statistics.

To this order in the perturbation theory, the Green's functions are diagonal with respect to the subband indices.

The solution of Eq. (31) satisfying these boundary conditions is

$$\Delta\mathcal{G}^{+}_{nn}(p, x_s, \epsilon) = -\left(x_s \pm \frac{L}{2}\right)\frac{i}{2\pi\hbar}\mathcal{R}. \tag{35}$$

Taking this into consideration and making use of Eqs. (19) and (35) one gets the following relation between $\Delta\mathcal{G}^{+}_{nn}(p, x_s, \epsilon)$ and the variation of current ΔJ

$$\Delta J = 2e\int\frac{d\epsilon}{2\pi i}\sum_{n}\int_{-\infty}^{\infty}\frac{dp}{2\pi\hbar}v\Delta\mathcal{G}^{+}_{nn}(p, x_s, \epsilon). \tag{36}$$

Combining Eqs. (19) and (35) we get as a result the following general expression for the variation ΔG due to the electron-phonon scattering,

$$\Delta G = -\frac{G_0 L_x}{\pi\hbar k_{\mathrm{B}}T}\sum_{nn'}\int\frac{d^2q_{\perp}}{(2\pi)^2}\int_0^{\infty}dp\ \times$$

$$\int_{-\infty}^{0}dp' C_{nn'}N_{\mathbf{q}}W_{\mathbf{q}}f^{(F)}(\epsilon)[1-f^{(F)}(\epsilon')]\delta(\epsilon'-\epsilon-\hbar\omega_{\mathbf{q}}). \tag{37}$$

Here we made use of the identities satisfied by the phonon frequencies and the matrix elements of the electron-phonon interaction that are consequences of time reversal symmetry:

$$\omega_{\mathbf{q}} = \omega_{-\mathbf{q}} \text{ and } |\langle n|\exp(i\mathbf{q}_{\perp}\cdot\mathbf{r}_{\perp})|n'\rangle|^2 = |\langle n|\exp(-i\mathbf{q}_{\perp}\cdot\mathbf{r}_{\perp})|n'\rangle|^2, \tag{38}$$

as well as of the identities for the Fermi and Bose functions:

$$1-f^{(F)}(\epsilon-\mu) = \exp\left(\frac{\epsilon-\mu}{k_{\mathrm{B}}T}\right)f^{(F)}(\epsilon-\mu);\, N(\hbar\omega)+1 = \exp\left(\frac{\hbar\omega}{k_{\mathrm{B}}T}\right)N(\hbar\omega). \tag{39}$$

6. Quasi-Classical Perturbative Approach

Before we start analysis of Eq.(37) it is worthwhile to discuss an alternative method to obtain the same result which, incidentally, permits one to get a better physical insight into the problem. The classical Boltzmann-Drude conductivity, σ, is proportional to the average time, τ, that an electron travels between two successive collisions. Up to a multiplicative constant of the order unity,

$$\sigma \approx ne^2\tau/m, \tag{40}$$

where n is the number of conduction electrons per unit volume. Consequently, even in the weak limit of electron-phonon coupling W the perturbation theory cannot be employed directly for the calculation of the conductivity. The first term in the expansion of τ and therefore of σ in powers of W scales with W as W^{-1} and the calculation of σ typically requires solving the Boltzmann integral equation.

There exist cases, however, where the expansion of σ in positive powers of W is possible. One such case is the transverse magnetoconductivity

$$\sigma \approx \frac{ne^2\tau/m}{1+(\omega_c\tau)^2}, \tag{41}$$

where $\omega_c = eB/mc$ is the cyclotron frequency. When $\omega_c\tau \gg 1$, $1/\tau$ in Eq.(41) can be expanded in powers of W by making use of the perturbation theory applied, for example, to the Kubo formula (see, for instance [9]). An analogous situation is found in the calculation of the a.c. conductivity in a high-frequency electric field.

These examples show that the applicability of the perturbation theory depends on values assumed by the conductance, G, in the limit of $W \to 0$. If G approaches infinity as $W \to 0$ the perturbation theory cannot be directly applied. This means that one should solve an integro-differential Boltzmann equation to obtain the conductance. If, on the other hand, G remains finite as $W \to 0$, there exists an interval of values of W where the perturbation theory is valid.

It is well known that (see for instance [2, 3]) in the absence of scattering G is a step-like *finite* function of the Fermi level. Therefore it is permissible to use the perturbation theory for calculation of the variation in the ballistic conductance, ΔG, due to the interaction with phonons provided that the electron-phonon coupling constant is sufficiently small. It means, in other words, that only a small fraction of the total number of electrons within the quantum wire suffers any collisions with the phonons. Eq.(37) is nothing more than the leading term of the perturbation theory expansion in powers of the small parameter proportional to W. Before making an analysis of Eq.(37) it is instructive to describe a different approach to obtain this result.

7. Calculation of the Current

The influence of the phonons manifests itself, then, in two ways. (*i*) a background resistance appears in addition to the ballistic resistance. The background depends on temperature, chemical potential, and, in general (nonlinear case) on the applied voltage, V. (*ii*) the form of the conductance steps and their heights may be altered, due to redistribution of the electrostatic potential, $\phi(x)$, within the conductor which is a consequence of phonon scattering. Due to the complexity of the problem, we consider a uniform conductor only, i.e. a transverse cross section independent of the coordinate x along the conductor. Then the redistribution of $\phi(x)$ due to the scattering does not contribute to the current and the phonons only add a background conductance.

According to Ref. [3], the distribution function of the electrons, $f_n(p)$ in the absence of the electron-phonon interaction is given by

$$f^{(0)}(p) = f^{(F)}(\epsilon \mp eV/2), \tag{42}$$

where one should take the upper (lower) sign in Eq.(42) for $p > 0 (p < 0)$.

A weak electron-phonon interaction creates a small variation of the electron distribution function $f = f^{(0)} + \Delta f$ with Δf satisfying the equation

$$v\frac{\partial \Delta f}{\partial x} = I\,[f]\,. \tag{43}$$

For $p > 0 (p < 0)$ the solution of Eq. (43) is

$$\Delta f(x) = \left(x \pm \frac{L}{2}\right)\frac{1}{v} I\,[f]\,. \tag{44}$$

Here we have assumed that the boundary condition $\Delta f = 0$ is satisfied at $x = \pm L/2$ and have neglected the small term Δf as compared to $f^{(0)}$ on the right-hand side of Eq. (44).

Variation of the current, ΔJ, due to the nonequilibrium electrons is given by (cf. Ref. [10], [11])

$$\Delta J = 2e\left(x + \frac{L}{2}\right)\sum_n \int_0^\infty \frac{dp}{2\pi\hbar} I\left[f^{(0)}\right] + 2e\left(x - \frac{L}{2}\right)\sum_n \int_{-\infty}^0 \frac{dp}{2\pi\hbar} I\left[f^{(0)}\right]. \tag{45}$$

The summation over all the phonon branches is implied, as above.

The explicit form of the collision term reads:

$$I[f] = \sum_n \int \frac{dp'}{2\pi\hbar} \int \frac{d^2 q_\perp}{(2\pi)^2} |\langle n'| e^{i\mathbf{q}_\perp \mathbf{r}_\perp} |n\rangle|^2 W_{\mathbf{q}} (B^{(+)} + B^{(-)}), \tag{46}$$

where the functions $B^{(\pm)}$ were introduced above in Eq. (33).

We will be interested in the acoustic phonon scattering. The integrations in Eq.(37) are over the three components of the phonon wave vector. $\mathbf{q}_\perp$ indicates the two transverse wave vector components; the third integration is equivalent to the integration over the electron quasimomentum, p', because of the conservation of the quasimomentum. Within the method of successive approximations we insert the zeroth approximation, Eq.(42), for the distribution function into Eq.(46). We assume that the phonons are in equilibrium. Then N is the Bose function. Detailed balance results in a vanishing of the collision term for the equilibrium distribution functions depending on the same temperature and chemical potential. This means that the distribution functions (42) give finite contribution in the collision term if and only if p and p' are of opposite sign, so that their chemical potentials are different. In other words, only those phonons contribute that can *backscatter* the electrons. It is readily seen that this condition should impose a certain limitation from below on the phonon wave vectors. For such phonons to be excited the temperature should be sufficiently high. At lower temperatures $|\Delta G|$ should be *exponentially small.*

With these considerations and expanding the Fermi distribution functions in powers of the small parameter eV/k_BT we arrive, after some algebra, at the following result for the variation of conductance collision term

$$\Delta G = -\frac{4e^2L}{k_BT}\sum_{nn'}\int\frac{d^2q_\perp}{(2\pi)^2}C_{nn'}\int_0^\infty\frac{dp}{2\pi\hbar}\int_{-\infty}^0\frac{dp'}{2\pi\hbar}\times$$
$$N_{\mathbf{q}}W_{\mathbf{q}}f^{(F)}(\epsilon)[1-f^{(F)}(\epsilon')]\delta(\epsilon-\epsilon'-\hbar\omega_{\mathbf{q}}). \qquad (47)$$

In the course of the derivation we have made use of Eqs. (38). This result coincides with Eq. (37) obtained by diagrammatic techniques. In essence, the coincidence is due to the fact that the longitudinal motion of electrons in the quantum wire is classical and therefore the problem can be treated quasi-classically.

To begin the analysis of Eq. (47), we consider a case where one channel takes part in the conduction. On the one hand, the initial and final electron states should be within the energy interval of the width k_BT near the Fermi level; on the other hand, these two states should have p of opposite sign. This means that the electron should be *backscattered* and the quasimomentum variation in the course of such a transition should be $2p_F$. This is the quasimomentum of the phonon that is involved in such a transition. Its energy, $\hbar\omega_{\mathbf{q}}$, for acoustic phonons should be equal to $2wp_F$ where w is the sound velocity. If k_BT is smaller than this energy these phonons are not excited and the phonon-assisted part of the resistance is exponentially small. This example supports the statement that a simple comparison of the length L with the electron mean free path does not, in general, give an estimate of the relative magnitude of the phonon-assisted resistance.

If several channels are involved these considerations may be still valid to some extent. However, one should take into consideration that there are different values of p_F for different channels and that the interchannel transitions are possible. For a very great number of channels the classical result of the type obtained by Yanson and Kulik in Ref. [11] where the backscattering does not play such a pronounced role should be obtained. One should, however, keep in mind that the classical ballistic transport can be fully restored provided not only μ but also $k_B T$ is bigger than the distances between the bands' bottoms.

One can easily do numerical estimates for ΔG for quasi-1D GaAs quantum wires. Using the following typical numbers $L_z = 100\mathring{A}$, $L_y = 1000\mathring{A}$ and L_x=1μm (we remind that the x-axis is along the propagation direction while the effective width of the channel is L_y) we can check that the condition ensuring applicability of the perturbation theory,

$$|\Delta G|/G_0 \ll 1, \tag{48}$$

is fulfilled for a wide range of transverse wire dimensions and temperatures. More detailed estimates as well as detailed results of numerical calculations can be found in Ref. [17].

8. Final Remarks

The methods developed here can also be applicable for other problems. I would like to name here several such problems. One example is the acoustoelectric effect in nanostructures under the ballistic conductance regime. A travelling acoustic wave propagating in the semiconductor creates a net drag of the electrons and hence a d.c. acoustoelectric current. Such current can be calculated by the methods described above (see Refs. [12, 13]).[2]

A lot of interest has been attracted to the problem of Coulomb drag in various systems of low dimensionality. The methods described in the present lecture can be applied to calculate the Coulomb drag current created under ballistic transport regime in a one-dimensional nanowire by a ballistic current in a near-by nanowire [14]. These results may be of a relevance to the issue of the cross-wire talk which is of pivotal importance to the proper operation of the scaled-down devices and VLSI circuits.

A problem also attracting much interest is a shot noise in semiconductor nanowires under the condition of non-Ohmic phonon-assisted quasiballistic transport. An expression for the shot noise under conditions of predominant electron-phonon scattering can also be derived using the methods described above [15].

Yet another such problem is the spatial distribution of the heat generated by current through a semiconductor microstructure [16]. It turns out that the heat is spread over the length of the electron mean free path in the reservoirs while the amount of heat generated per second in both reservoirs that are joined by the nanostructure is the same.

[2]I take this opportunity to correct some misprints in Ref. [12]. In the caption to Fig. 2 one should replace the inequality $\hbar\omega \gg k_B T$ by $\hbar\omega \ll k_B T$ *et vice versa*. In the numerical estimates of acoustoelectric current J and acoustoelectric voltage V per unit of sound intensity one should use the units A$\cdot$ cm^2/W and mV$\cdot$ cm^2/W, respectively.

In the present paper we discuss only the Ohmic conduction. The simplest example of non-Ohmic conduction is a ballistic resistance for $eV \gg k_BT$ [5, 17, 18]. Again the heat generation takes place well outside the nanostructure, within the contacts. Its calculation can be also based on application of the Boltzmann equation.

More complicated are various aspects of the non-Ohmic phonon-assisted ballistic resistance [17, 19]. In the classical regime, nonlinear phenomena in the current-voltage characteristics of point contacts between normal metals were observed and discussed in a pioneering work by Yanson [20]. Here we mean a quantum situation where the conduction electrons experience some scattering within the nanostructure. As indicated in [21], the situation here is quite unlike the usual collisionless transport where the electron-phonon (as well as the electron-electron and electron-impurity) interactions are restricted to the contacts and hence all the heat is released in the contacts only. In the mentioned case some energy is transferred to phonons and may be released as heat by the phonon system outside both the wire and the contacts.

I have discussed two methods to solve kinetic problems for quantum mesoscopic systems. One of them is a direct perturbative approach. It can be very effective for solution of a number of particular problems related to the transport phenomena in mesoscopic systems. The other method is based on application of diagrammatic techniques. This method is more general and, as a rule, more involved.

References

1. R. Landauer, IBM J. Res. Develop. **1**, 233 (1957); **32**, 306 (1989).
2. Y. Imry, in *Directions in Condensed Matter Physics*, G. Grinstein and G. Mazenko, eds., (World Scientific, Singapore, 1986) p. 101; M. Büttiker, Phys. Rev. Lett. **57**, 1761 (1986).
3. C.W.J. Beenakker and H. van Houten, Solid State Phys. **44**, 1 (1991).
4. Yu.V. Sharvin, Zh. Eksp. Teor. Fiz. **48**, 984 (1965) [Sov. Phys. — JETP **21**, 655 (1965)].
5. V.L. Gurevich and V.B. Pevzner (unpublished).
6. L.V. Keldysh, Zh. Eksp. Teor. Fiz. **47**, 1515 (1964) [Sov. Phys. — JETP **20**, 1018 (1965)].
7. V.L. Gurevich, *Transport in Phonon Systems* (North-Holland, Amsterdam, 1986).
8. H. Frölich, Proc. Roy. Soc. **A160**, 230 (1937).
9. V.L. Gurevich and Yu.A. Firsov, Zh. Eksp. Teor. Fiz. **40**, 198 (1961) [Sov. Phys. – JETP **13**, 137 (1961)].
10. I.O. Kulik, R.I. Shekhter, and A.N. Omelyanchuk, Solid State Comm. **23**, 301 (1977).
11. I.K. Yanson and I.O. Kulik, Journal de Physique **39**, 1564 C6 (1978).
12. V.L. Gurevich, V.B. Pevzner, and G.J. Iafrate, Phys. Rev. Lett. **77**, 3881 (1996).
13. V.L. Gurevich, V.I. Kozub, and V.B. Pevzner, Phys. Rev. B **58**, 13088 (1998).
14. V.L. Gurevich, V.B. Pevzner, and E.W. Fenton, J. Phys.: Condens. Matter **10**, 2551 (1998).

15. V.L. Gurevich and A.R. Rudin, Phys. Rev. B **53**, 10078 (1996).
16. V.L. Gurevich, Phys. Rev. B **55**, 4522 (1997).
17. V.L. Gurevich, V.B. Pevzner, and K. Hess, J. Phys.: Condens. Matter **6**, 8363 (1994); Phys. Rev. B **51**, 5219 (1995).
18. V.L. Gurevich, V.B. Pevzner, and G.J. Iafrate, Phys. Rev. Lett. **75**, 1352 (1995).
19. V.B. Pevzner, V.L. Gurevich, and E.W. Fenton, Phys. Rev. B **51**, 9465 (1995).
20. I. K. Yanson, Zh. Eksp. Teoret. Fiz. **66**, 1035 (1974) [Sov. Phys. — JETP **39**, 506 (1974)].
21. V.L. Gurevich, V.B. Pevzner, and G.J. Iafrate, J. Phys.: Condens. Matter **7**, L445 (1995).

KINETIC THEORY FOR ELECTRON DYNAMICS IN SEMICONDUCTORS AND PLASMAS

JAMES W. DUFTY
Department of Physics, University of Florida
Gainesville, FL 32611, USA

1. Introduction

The developing technology of femtosecond pulsed lasers has allowed experimental access to conditions that provide theoretical challenges for the exploration of charge carrier dynamics. New features of these experiments include states driven far from equilibrium and probes on time scales short compared to carrier scattering or other characteristic times. Traditional many-body approximations for electron dynamics in solids must be generalized to accommodate these features. Two general approaches have been pursued: the Keldysh / Kadanoff-Baym method for nonequilibrium Green's functions, and the quantum BBGKY hierarchy for reduced density operators. The latter method has a closer relationship to the extensive body of literature on classical kinetic theory for fluids and plasmas developed over the past forty years, and will be emphasized here. A primary objective of this review is to introduce the theoretical challenges of these new experimental tools in a language more familiar to the statistical mechanics and kinetic theory community, to facilitate access to a literature dominated by the tools and language of solid state physics. Accordingly, an attempt is made to connect the dynamical description for laser-pulsed electron dynamics in semi-conductors to those for simple classical fluids and plasmas. Following a formal review of the quantum BBGKY hierarchy, an approximation is proposed that is appropriate on all time scales and which preserves symmetries necessary for the exact local conservation laws. The approximation results from a physically transparent closure of the hierarchy and incorporates the dominant mechanisms for quantum exchange symmetry for the electrons, as well as the charge correlations present for both classical and quantum systems. Furthermore, this approximation is implemented in a representation-independent form so the results can be applied uniformly

J. Karkheck (ed.),
Dynamics: Models and Kinetic Methods for Non-equilibrium Many Body Systems, 473–490.

to both electrons in solids (Bloch representation for band occupation and polarization densities) and quantum plasmas (momentum or Wigner representation). In the final Discussion, some comments on the Green's function approach are offered for comparison.

2. The Challenge

In the past decade the application of short-pulsed lasers to investigate properties of semiconductors has provided a wealth of information not previously available by other means. The most significant feature is the ability to probe charge carrier dynamics on a wide range of time scales. Currently, pulses as short as 10 femtoseconds ($1fs = 10^{-15}$ seconds) can be produced for a number of different experiments: e.g., pump/probe, continuum probe, luminescence, reflectivity, four-wave mixing, photon echoes. From these experiments information is obtained about such physical properties as scattering rates, band structure, optical matrix elements, dephasing times, and phonon lifetimes. Some recent texts describing the physics of these phenomena are given in references [1-3].

Consider as an important illustration a pump-probe experiment on a semiconductor initially at equilibrium with all electrons in a single valence band. The initial pump excites the valence electrons to various conduction bands during an initial interval of $10 - 50$ fs, for example. At this point the electrons are very far from equilibrium and proceed to thermalize through various scattering mechanisms during a time of the order of picoseconds. Due to the initial pump, the resulting electron temperature is considerably higher than the lattice temperature, so there is a cooling of the electrons over the next hundred picoseconds to reach the lattice temperature. Finally, on a scale of $1 - 100$ nanoseconds there is a recombination of the electrons into the equilibrium band structure. A delayed laser probe can be used to study this dynamics at any point after the initial pump.

This separation of the dynamics into four stages corresponding to different effective mechanisms is an oversimplification. In fact the stages overlap considerably, pairwise. For example the first stage is dominated by the external driving field with a coherent dynamics similar to that for a multilevel atom in an external field. As indicated below, the approximate description of this coherent dynamics is quite similar to the optical Bloch equations of atomic physics. However, during the latter part of this coherent stage, electron-electron interactions and inelastic scattering events begin to occur as well. Thus, the separation of the theory into Bloch equations on one time scale followed by a Boltzmann equation for electron scattering on another time scale is a coarse approximation that misses the physics of their correlated mechanisms. For similar reasons the electron-electron scattering cannot be separated from the late stage electron-phonon dynamics. Instead, it is desired to have a theoretical formulation that is capable of uniformly treating all relevant mechanisms across the entire time interval.

The coherent dynamics is described by a mean-field kinetic equation, the time dependent Hartree-Fock equation, that is accurate at asymptotically short times. The corresponding density matrix equations in a Bloch representation are called the semiconductor Bloch equations and have the same structure as the optical Bloch equations. Electron-electron interactions appear only via a mean field which is responsible for dephasing the "Rabi oscillations" and for quasi-bound states (excitons). Thus, even at this simplest level the dynamics is rich and complex. Corrections to this description require dynamical correlations that eventually lead to scattering events on time scales long compared to the scattering times. At this longer time, the dynamical correlations provide collisional relaxation via a Boltzmann-like collision operator. Due to the long range nature of the Coulomb interaction the electron-electron scattering rates at long times must also account for many-body effects that provide screening. Consequently, the proper description must include a transition domain during which incomplete (off energy shell) scattering and the build-up of screening must be treated properly. For example, in this domain the kinetic energy of scattered states is no longer relevant and a detailed accounting of the transfer between kinetic and potential energy must be made.

The theoretical challenge is to formulate a kinetic theory that is accurate on short as well as long time scales, that preserves the exact conservation laws (particularly that for energy), that applies for arbitrary quantum degeneracy of the electrons and includes the mechanism for dynamical screening (polarization). In the following, this problem is approached from the exact BBGKY hierarchy for the electron reduced density operator [4, 5]. The hierarchy is transformed to make explicit the effects of exchange symmetry [5, 6]. An approximate closure is proposed that accounts for all residual two particle correlations and which is consistent with the exact local conservation laws [6]. The resulting kinetic theory is appropriate for all time scales and retains the relevant physical mechanisms of Hartree-Fock renormalized single particle energies, strong collisions, and dynamical polarization. Some limiting forms of this kinetic equation are described to connect with other results from semiconductor theory and the theory of classical plasmas.

3. Phenomenological Bloch Equations

Before discussing the theoretical formulation from the BBGKY hierarchy it is useful to recall the first order phenomenological equations for electron dynamics in semiconductors driven by an external laser field. To simplify the discussion, all of the following assumes N electrons interacting with a rigid ionic lattice, with overall charge neutrality, in the presence of an external classical transverse electric field, $\mathbf{E}(t)$, via a dipole interaction. Thus, electron-phonon interactions and quantum features of the laser field are not considered. In this case it is appropriate to use a Bloch representation, in terms of single electron states in the extended crystal with energy eigenvalues $\varepsilon_\alpha(\mathbf{k})$ labeled by a band index α and wavevector $\mathbf{k}$. The relevant physical properties in most experiments are determined from the diagonal and off diagonal matrix elements of the one-particle reduced density operator, defined in the next section, respectively,

$$n_\alpha(\mathbf{k};t) =< \alpha\mathbf{k}\, | f(t) \mid \alpha\mathbf{k} > \qquad p_{\alpha\alpha'}(\mathbf{k};t) \equiv< \alpha\mathbf{k}\, | f(t) \mid \alpha'\mathbf{k} >, \tag{1}$$

where $n_\alpha(\mathbf{k};t)$ are the band occupation densities and $p_{\alpha\alpha'}(\mathbf{k};t)$ are the polarization densities (for $\alpha \neq \alpha'$). The Bloch equations are an approximate set of coupled equations for these two sets of densities. As an example, consider an idealized case of only a single conduction band and a single valence band. Then the phenomenological equations have the form

$$\partial_t n_c(\mathbf{k};t) - 2Im\left(\mathbf{E}(t)\cdot \mu_{\mathbf{cv}} p^*(\mathbf{k};t)\right) = 0, \tag{2}$$

$$\left[\partial_t - i\left(\varepsilon_c(\mathbf{k}) + \varepsilon_h(\mathbf{k})\right) - \gamma\right] p(\mathbf{k};t) + i\mathbf{E}(t)\cdot \mu_{\mathbf{cv}}\left[n_v - n_c(\mathbf{k};t)\right] = 0. \tag{3}$$

Here, $\mu_{\mathbf{cv}}$ is the dipole matrix element between the conduction and valence states. The equation for the valence occupation numbers, $n_v(\mathbf{k};t)$, has not been written since in the absence of collisions it is the same as for $n_c(\mathbf{k};t)$. These equations describe the rich "coherent" phenomena mentioned above of Rabi oscillations, dephasing (via the parameter γ) and bound states (excitons). They are effectively the same as the optical Bloch equations for a two level atom in an external field. The solution to these equations for physically relevant initial conditions and external fields has been the subject of many fruitful investigations over the past decade. The new theoretical challenges are to include corrections to these equations for the additional mechanisms expected to be important after the very short intial stage:

- correlations (exchange, Coulomb)
- collisions (non-Markovian, energy conserving)
- screening / polarization (dynamical)

4. BBGKY Hierarchy and Exchange Effects

Most physical properties of interest can be obtained from the reduced one-particle density operator $f^{(1)}(1;t)$. It is related to the exact BBGKY (Born, Bogoliubov, Green, Kirkwood, Yvon) hierarchy [4] for the reduced

s-particle density operators defined in terms of the N-particle density operator $\rho(t)$ according to

$$f^{(m)}(1,..,m;t) = N^m Tr_{m+1..N}\,\rho(t). \tag{4}$$

The notation on the right side indicates a trace over the degrees of freedom associated with particles $m+1$ through N. More precisely, in matrix representation it denotes summation over the diagonal quantum numbers for these particles. The reduced density operators $f^{(m)}$ inherit an $m-$ particle symmetrization operator from the corresponding N particle symmetrization operator inherent in ρ. The Hamiltonian has the form

$$H(t) = H_0(t) + U \tag{5}$$

$$H_0(t) = \sum_{i=1}^{N} H(i;t), \qquad U = \frac{1}{2}\sum_{i\neq j}^{N} V(i,j), \tag{6}$$

where $H(1;t) = h_0(1) - \mu(1)\cdot\mathbf{E}(t)$ is the single particle Hamiltonian in the external semi-classical laser dipole field, and $V(i,j)$ is the Coulomb interaction for particles i and j. The Liouville-von Neumann equation for the $N-$ particle density operator is

$$\partial_t\rho(t) + i[H(t),\rho(t)] = 0. \tag{7}$$

Units such that $\hbar = 1$ are used. The BBGKY hierarchy equations for the reduced density operators for $m = 1,2$ follow directly from the definition (4)

$$\partial_t f(1;t) + i\,[H(1;t), f(1;t)] + Tr_2\, i\left[V(1,2), f^{(2)}(1,2;t)\right] = 0 \tag{8}$$

$$\partial_t f^{(2)}(1,2;t) + i\left[H(1,2;t), f^{(2)}(1,2;t)\right]$$
$$+\sum_{i=1}^{2} Tr_3\, i\left[V(i,3), f^{(3)}(1,2,3;t)\right] = 0, \tag{9}$$

where $H(1,2) = H(1;t) + H(2;t) + V(1,2)$ is the two-particle Hamiltonian. Here and in the following the simplified notation $f(1;t) = f^{(1)}(1;t)$ is used.

An approximate kinetic equation is obtained from a *closure approximation* expressing $f^{(2)}(1,2;t)$ as a functional of $f(1;t)$. Any such approximation in (8) yields a closed kinetic equation for $f(1;t)$. The choice of the approximation involves an analysis of correlations in the two- and three-pariticle distribution functions. These correlations are due to two different physical effects: quantum statistics and forces between the particles.

The former are always present at low and moderate temperatures for the electrons and it is useful to extract them explicitly in the hierarchy. This is done by recognizing that the reduced density operators have implicit symmetrization operators and writing this explicitly by

$$f^{(2)}(1,2;t) \equiv [f(1;t)f(2;t) + \overline{g}(1,2;t)]\,\Lambda_{12}, \tag{10}$$

$$\begin{aligned} f^{(3)}(1,2,3;t) \equiv [&f(1;t)f(2;t)f(3;t) + f(1;t)\overline{g}(2,3;t) + f(2;t)\overline{g}(1,3;t) \\ &+f(3;t)\overline{g}(1,2;t) + \overline{g}(1,2,3;t)]\,\Lambda_{123}. \end{aligned} \tag{11}$$

The correlation operators $\overline{g}(1,2;t)$ and $\overline{g}(1,2,3;t)$ describe correlations due primarily to the Coulomb forces, whereas the two and three particle (anti) symmetrization operators $\Lambda_{12} = 1 - P_{12}$ and $\Lambda_{123} = 1 - P_{12} - P_{13} - P_{23} + P_{13}P_{12} + P_{23}P_{12}$ generate correlations due to the quantum statistics (where P_{12} is the permutation operator for quantum numbers 1 and 2). Separation of these two physically different mechanisms is useful for introducing the approximations below where three-particle quantum correlations are important but three-particle force correlations are weak. Substitution of these definitions into (8) and (9) leads to the corresponding equations for the density and correlation operators

$$i\frac{\partial}{\partial t}f(1;t) - [H^{HF}(1;t), f(1;t)] = \mathrm{Tr}_2\,[V_s(1,2), \overline{g}(1,2;t)]\,, \tag{12}$$

$$\begin{aligned} (i\partial_t + \mathcal{L}_{12}(t))\,\overline{g}(1;t) &= \theta_{12}(t)f(1;t)f(2;t) \\ -Tr_3\,[(V_{13} + V_{23})\,&, \overline{g}(1,2,3;t)\,(1 - P_{13} - P_{23})]\,, \end{aligned} \tag{13}$$

where $V_s(1,2) = V(1,2)\Lambda_{12}$ is the pair potential for anti-symmetrized two particle states. The generator for the effective pair dynamics in (13) is

$$\mathcal{L}_{12}(t)\overline{g}(1,2;t) \equiv (\mathcal{L}_1(t) + \mathcal{L}_2(t) - \theta_{12}(t))\,\overline{g}(1,2;t), \tag{14}$$

$$\mathcal{L}_1(t)\overline{g}(1,2;t) \equiv -\left[H^{HF}(1;t), \overline{g}(1,2;t)\right] - \mathcal{V}_1\overline{g}(1,2;t), \tag{15}$$

$$\theta_{12}(t)\overline{g}(1,2;t) = \left[\hat{V}(1,2;t)\overline{g}(1,2;t) - \overline{g}(1,2;t)\hat{V}^{\dagger}(1,2;t)\right]. \tag{16}$$

These two equations are still exact, but a number of important physical effects have been made explicit by this transformation. First, the single-particle Hamiltonian has been renormalized to the Hartree-Fock Hamiltonian $H^{HF}(1;t)$

$$H^{HF}(1;t) = H(1;t) + \mathrm{Tr}_2 f(2;t)V_s(1,2). \tag{17}$$

Next, dynamical Coulomb polarization is described by the quantum Vlasov operators

$$\mathcal{V}_1\overline{g}(1,2;t) = Tr_3\,[\,V_s(1,3), f(1;t)]\,\overline{g}(3,2;t)\Lambda_{23}. \tag{18}$$

This operator is the generator for dynamical screening and polarization ("rings" or RPA summation in diagrammatic language), including exchange. Finally, the exchange effects have led to a modification of the pair potential $\hat{V}(1,2;t)$ due to Pauli blocking

$$\begin{aligned}\hat{V}(1,2;t) &\equiv (1 - f(1;t) - f(2;t))V(1,2)\\ &= (f^{-}(1;t)f^{-}(2;t) - f(1;t)f(2;t))\, V(1,2),\end{aligned} \tag{19}$$

where $f^{-}(1;t) = 1 - f(1;t)$ is the "hole" occupation operator. This means that the pair potential will have matrix elements between quantum states with proper accounting for their exact nonequilibrium occupancy.

In summary, the first two hierarchy equations have been transformed to make explicit the effects of pair correlations (Hartree-Fock and Vlasov terms) and the effects of both pair and triplet correlations due to exchange effects (Pauli blocking). All residual correlations reside in $\overline{g}(1,2,3;t)$ which is associated with strong Coulomb correlations. The classical limit of these equations has been the starting point for most kinetic theories of simple plasmas. In this respect, the problem of describing degenerate electrons in semiconductors is seen to be very close to the older and well-studied problems of classical plasma physics.

5. Closure Approximation and Kinetic Theory

Equations (12) and (13) are exact but do not constitute a kinetic theory since they describe $f(1;t)$ and $\overline{g}(1,2;t)$ in terms of the unknown $\overline{g}(1,2,3;t)$. A closed description requires a formal or approximate representation of $\overline{g}(1,2,3;t)$ in terms of $f(1;t)$ and $\overline{g}(1,2;t)$. Given such a representation Eqs. (12) and (13) can be solved self-consistently for $f(1;t)$ and $\overline{g}(1,2;t)$, or Eq. (13) can be solved as a functional of $f(1;t)$ and inserted in the right side of Eqs. (12) to obtain a closed equation for $f(1;t)$. The latter is what is usually referred to as a kinetic equation.

5.1. EXAMPLE: TIME-DEPENDENT HARTREE-FOCK EQUATION

The simplest example of a kinetic equation results from neglecting $\overline{g}(1,2;t)$ on the right side of (12)

$$i\frac{\partial}{\partial t}f(1;t) - [H^{HF}(1;t), f(1;t)] = 0. \tag{20}$$

This result is deceptively simple. In the classical limit it becomes the nonlinear Vlasov equation, providing a mean-field description of plasma dynamics in an external field. It has proved particularly useful for states far from equilibrium in the analysis of plasma instabilities, for example. In this respect it is expected to play a similar role for electrons in laser-pulsed semiconductors since the initial states are also far from equilibrium and the non-linear dynamics is expected to be an important feature. In fact this is the case, and Eq. (20) provides a good description of the dominant coherent effects at short times. The representation independent operator form given here emphasizes the close relationship of the two fields and also exposes the essential mean field physics of this approximation. In practice, a Bloch state representation of (20) is used and comprises a coupled set of equations for diagonal and off-diagonal matrix elements in terms of the relevant quantum numbers (band index and momentum). Further details of this representation are given below. If the number of bands and momentum states required is not too large, (20) can be solved numerically to give a detailed description of the electron dynamics in the presence of the laser pulse.

The form of the time-dependent Hartree-Fock kinetic equation in Bloch representation is very similar to that for the density matrix of an atom in a time dependent electric field, with the band indices playing the role of the atomic states. The latter are the optical Bloch equations of laser physics, and for this reason they are referred to as the semiconductor Bloch equations in the present context. There are no Coulomb correlations in this approximation, but there are two-particle exchange correlations (the Fock terms). Thus, there is no a priori limitation on the degree of degeneracy but there are limitations to weak Coulomb forces.

5.2. PAIR-CORRELATION APPROXIMATION

To account for the important Coulomb correlations that dominate after the short pulse period it is necessary to have an adequate description of $\overline{g}(1,2;t)$. This is provided by an appropriate closure of Eq. (13). Thus the problem reduces to a choice for $\overline{g}(1,2,3;t)$ which represents the residual three-particle correlations beyond those due to three-particle exchange effects. The transformation leading to this equation has extracted explicitly the mechanisms for exchange and polarization so any choice for $\overline{g}(1,2,3;t)$ can only effect the quantitative description of these mechanisms. The primary suggestion here is an approximation resulting from the complete neglect of $\overline{g}(1,2,3;t)$, referred to below as the pair-correlation approximation [7, 8, 9]. This terminology is somewhat misleading since all three-particle exchange correlations are included in this approximation (see Eq. (11)).

There are other important constraints that are satisfied by this approximation. The definition of reduced distribution functions (4) implies

$$Nf(1;t) = Tr_2 f^{(2)}(12;t), \qquad Nf^{(2)}(12;t) = Tr_3 f^{(3)}(123;t). \tag{21}$$

Therefore, the first hierarchy equation must result from a partial trace of the second hierarchy equation [10]. A second constraint is imposed by the exact local conservation laws for mass, energy, and momentum density [5, 6]. The conservation laws for energy are of particular importance for the electron dynamics as the system evolves from short times with weak correlation to long times with strong correlations. During this evolution there is a conversion between kinetic and potential energy that must be described accurately in the kinetic equation for the correct system properties at later times (e.g., kinetic temperature). It is possible to show that the approximation $\overline{g}(1,2,3;t) \rightarrow 0$ preserves these important properties of the conservation laws and the exact form of the first hierarchy equation. The proposed approximation is therefore

$$i\frac{\partial}{\partial t}f(1;t) - [H^{HF}(1;t), f(1;t)] = \mathrm{Tr}_2\left[V_s(1,2), \overline{g}(1,2;t)\right], \tag{22}$$

$$(i\partial_t + \mathcal{L}_{12}(t))\,\overline{g}(1,2;t) = \theta_{12}(t)f(1;t)f(2;t). \tag{23}$$

These are first order differential equations in time and must be supplemented with the initial conditions $f(1;0)$ and $\overline{g}(1,2;0)$ or, equivalently, $f(1;0)$ and $f^{(2)}(1,2;0)$. The formulation as an initial value problem is particularly important for the laser-pump problems of interest here. Equations (22) and (23) will be referred to below as the pair-correlation approximation and are the main result of this discussion. The terminology is somewhat misleading since all three-particle exchange correlations are included in this approximation (see Eq. (11)). If instead, *all* three-particle correlations are neglected the Fock energies, ring exchange effects, and Pauli blocking would be lost. Thus, the extraction of all exchange correlations prior to neglecting $\overline{g}(1,2,3;t)$ is an important feature of this analysis, and assures that the pair approximation is valid for all degrees of electron degeneracy. Of course, the neglect of $\overline{g}(1,2,3;t)$ assumes weak three-particle Coulomb correlations but these conditions are probably relevant for the semiconductor problem. This approximation would not be appropriate for strongly coupled plasmas. The description of dynamical phenomena in degenerate, strongly coupled plasmas is still an open problem.

It is remarkable that (22) and (23) retain all the most important strutural features of the exact hierarchy [5, 6, 10]:

- Exact symmetries of the Galilei group
- Time reversal invariance
- Exact local conservation laws
- All exchange effects (statistics)
- Polarization effects (rings)
- Strong collision effects (ladders)

– First hierarchy equation exact at short times

5.3. KINETIC EQUATION

The solution to (22) and (23) can be approached as a self-consistent pair of equations, with a numerical iterative method: Set $\overline{g}(1,2;t) = \overline{g}(1,2;0)$ in (22) and solve for $f(1;t)$; use this solution in (23) and solve for $\overline{g}(1,2;t)$; with this improved estimate for $\overline{g}(1,2;t)$ repeat the process to find an improved $f(1;t)$; iterate until convergence. This may in fact be the most efficient method to determine $f(1;t)$, and yields as a byproduct the information in $\overline{g}(1,2;t)$ as well. However, the more usual approach is to use (22) and (23) to determine a single, closed equation for $f(1;t)$. The latter is referred to as "the kinetic equation". To obtain it, first solve (23) as a functional of $f(1;t)$:

$$\overline{g}(1,2;t) = \mathcal{U}_{12}(t,0)\overline{g}(1,2;0) + \int_0^t \mathcal{U}_{12}(t,\tau)\theta_{12}(t)f(1;t)f(2;t)d\tau, \tag{24}$$

where $\mathcal{U}(1,2;t,\tau)$ is the *superoperator* defined by

$$(i\partial_t + \mathcal{L}_{12}(t))\mathcal{U}_{12}(t,\tau) = 0, \qquad \mathcal{U}_{12}(t,t) = \mathcal{I}_{12}, \tag{25}$$

and $\mathcal{I}_{12}$ is the two particle identity operator. Substitution of this formal result into (22) gives the kinetic equation for $f(1;t)$,

$$i\frac{\partial}{\partial t}f(1;t) - [H^{HF}(1;t), f(1;t)] = I_0(1;t) + I(1;t \mid f). \tag{26}$$

The first term on the right side is due to the initial correlations of $\overline{g}(1,2;0)$

$$I_0(1;t) = \mathrm{Tr}_2\left[V_s(1,2), \mathcal{U}_{12}(t,0)\overline{g}(1,2;0)\right]. \tag{27}$$

The second term on the right side is the non-Markovian collision operator

$$I(1;t \mid f) = \int_0^t \mathrm{Tr}_2\left[V_s(1,2), \mathcal{U}_{12}(t,\tau)\theta_{12}(\tau)f(1;\tau)f(2;\tau)\right]d\tau. \tag{28}$$

The contribution $I_0(1;t)$ from initial correlations is expected to vanish at long times for physically realistic initial states. However, it can be an important modification of the time dependent Hartree-Fock dynamics at short times. The collision operator describes the build up of correlations due to Coulomb interactions. It is vanishingly small at short times and eventually transforms to a Boltzmann-like completed-collision description for times long compared to the electron-electron scattering times. At intermediate times it describes the dynamical conversion between kinetic and potential energy, and the transformation from bare to screened Coulomb interactions. Both the fundamental approximation leading to (22) and (23), and the formal solution (24) are applicable on all time scales, so the kinetic equation (26) provides an accurate description from the coherent initial laser-pump stage, through the build-up of correlations, to the kinetic stage of elastic collisions. Hence the primary challenges outlined above have been met.

6. Plasma and Semiconductor Bloch Representations

An important advantage of the above analysis based on the operator form of the BBGKY hierarchy is that no reference has been made to a particular representation. The physical consequences of the transformation from (8) and (9) to (12) and (13), and the physical conditions behind the pair approximation are the same for electrons in a plasma and for electrons in semiconductors. In fact these equations are applicable for neutral particles as well. Some illustrations are provided by the one component plasma (OCP) and the two band semiconductor.

6.1. ONE COMPONENT PLASMA

Consider an idealized system of electrons in a uniform neutralizing background. There are no bound states possible so this is a prototype fully ionized plasma. The simplest case is for spatially homogeneous states. An appropriate representation is given by the single particle momentum states, $\mid \mathbf{p}\rangle$, and $f(1;t)$ is diagonal in this case. Equations (26) to (28) give

$$i\frac{\partial}{\partial t}n(p;t) = I_0(p;t) + I(p;t \mid f), \tag{29}$$

where $n(p;t) = \langle \mathbf{p} \mid f(1;t) \mid \mathbf{p}\rangle$. Here

$$I_0(1;t) = \sum_{p_2} \left[V_s(\mathbf{p},\mathbf{p}_2)\langle \mathbf{p},\mathbf{p}_2 \mid \mathcal{U}_{12}(t,0)\overline{g}(1,2;0) \mid \mathbf{p},\mathbf{p}_2\rangle - hc\right], \tag{30}$$

$$I(1;t \mid f) = \int_0^t d\tau \times$$

$$\sum_{p_2} \left[V_s(\mathbf{p},\mathbf{p}_2)\langle \mathbf{p},\mathbf{p}_2 \mid \mathcal{U}_{12}(t,\tau)\theta_{12}(\tau) f(1;\tau) f(2;\tau) \mid \mathbf{p},\mathbf{p}_2\rangle - hc\right]. \tag{31}$$

The notation hc denotes the Hermitian conjugate and $V_s(\mathbf{p},\mathbf{p}_2) = \langle \mathbf{p},\mathbf{p}_2 \mid V_s(1,2) \mid \mathbf{p},\mathbf{p}_2\rangle$. These equations constitute a generalization of the classical Lennard-Balescu kinetic equation to include strong collisions, degeneracy effects, and short times. In contrast to the Lennard-Balescu equation, the total energy is conserved by this kinetic equation. The $\mathcal{U}_{12}(t,\tau)$ is governed by an effective two-particle equation, except for the quantum Vlasov operators $\mathcal{V}_1$ and $\mathcal{V}_2$. The latter contain the many-body effects, but these are within the familiar random phase approximation (generalized to nonequilibrium states) so the analysis of the collision operator is brought under control. For a detailed analysis of the corresponding neutral case and the relationship to the quantum Boltzmann equation at long times see [5].

6.2. TWO-BAND SEMICONDUCTOR

Consider an idealized semiconductor with a conduction band and a valence band. The Bloch states are denoted by $\mid \alpha\mathbf{k} >$ where α denotes the band and $\mathbf{k}$ is a wavevector. The Coulomb matrix elements in this representation become

$$< \alpha_1\mathbf{k}_1;\alpha_2\mathbf{k}_2|V \mid \alpha_2'\mathbf{k}_2';\alpha_1'\mathbf{k}_1'>=$$

$$\delta_{\mathbf{k}_1'+\mathbf{k}_2',\mathbf{k}_1+\mathbf{k}_2}\sum_{\mathbf{q}}\widetilde{V}(q)\delta_{\mathbf{k}_1',\mathbf{k}_1+\mathbf{q}}K_{\alpha_1\alpha_1'}(\mathbf{k}_1,-\mathbf{q})K_{\alpha_2\alpha_2'}(\mathbf{k}_2,\mathbf{q}), \qquad (32)$$

where $\widetilde{V}(q)$ is the Fourier transform of the Coulomb potential and the overlap integral is given by

$$K_{\alpha\alpha'}(\mathbf{k},\mathbf{q}) \equiv \int d\mathbf{r}\, e^{i\mathbf{q}\cdot\mathbf{r}}\psi^*_{\alpha\mathbf{k}}(\mathbf{r})\psi_{\alpha'\mathbf{k}-\mathbf{q}}(\mathbf{r}). \qquad (33)$$

In many cases of practical interest the excitations are all near the band edge and it is possible to consider the small q limit $K_{\alpha,\nu}(\mathbf{k},\mathbf{q}) \to K_{\alpha\nu}(\mathbf{k},0) = \delta_{\alpha,\nu}$. Taking various matrix elements of equation (26) and using the definitions (1) leads to the coupled set of equations

$$\partial_t \begin{pmatrix} n_c(\mathbf{k}) \\ n_v(\mathbf{k}) \end{pmatrix} + 2Im\,(\mathbf{E}\cdot\mu_{cv}p_{vc}(\mathbf{k})) + 2Im\sum_{\mathbf{q}}\widetilde{V}(q)p_{cv}(\mathbf{k}+\mathbf{q})p_{vc}(\mathbf{k})$$

$$= \begin{pmatrix} < c\mathbf{k}\,|I_0(1;t)\mid c\mathbf{k} > + < c\mathbf{k}\,|I(1;t\mid f)\mid c\mathbf{k} > \\ < v\mathbf{k}\,|I_0(1;t)\mid v\mathbf{k} > + < v\mathbf{k}\,|I(1;t\mid f)\mid v\mathbf{k} > \end{pmatrix}, \qquad (34)$$

$$\partial_t p_{cv}(\mathbf{k}) + i\left(\varepsilon_c(\mathbf{k}) - \varepsilon_v(\mathbf{k}) - \mathbf{E}\cdot[\mu_{cc}-\mu_{vv}]\right)p_{cv}(k)$$

$$-i\sum_{\mathbf{q}}\widetilde{V}(q)\left[n_c(\mathbf{k}+\mathbf{q}) - n_v(\mathbf{k}+\mathbf{q})\right]p_{cv}(\mathbf{k})$$

$$-i\left\{\sum_{\mathbf{q}}\widetilde{V}(q)p_{cv}(\mathbf{k}+\mathbf{q}) + \mathbf{E}\cdot\mu_{cv}\right\}(n_v(\mathbf{k}) - n_c(\mathbf{k}))$$

$$=< c\mathbf{k}\,|I_0(1;t)\mid v\mathbf{k} > + < c\mathbf{k}\,|I(1;t\mid f)\mid v\mathbf{k} >. \qquad (35)$$

The left sides of these equations give the "coherent" dynamics from the time dependent Hartree-Fock terms and correspond to the phenomenological Bloch equations (2) and (3). The right sides include the effects of initial correlations, important on the same time scale as the coherent dynamics, plus the non-Markovian collisional effects that grow from zero to dominance on longer time scales. It is straightforward to work out the matrix elements of the three collision matricies using (28), in terms of the effective pair dynamics of $\mathcal{U}_{12}(t,\tau)$.

Of course, these equations for the band occupation and polarization densities look very different from the kinetic equation for the OCP, but the underlying physics of both is seen to be the same and considerably simpler when expressed in the common operator form (26).

7. Polarization Approximation

The main result of this presentation is the pair approximation leading to equations (22) and (23). However, in some cases additional practical approximations may be possible. One of these is the neglect of strong collisions through the replacement $\mathcal{L}_{12}(t) \to \mathcal{L}_1(t) + \mathcal{L}_2(t)$ in (23), where $\mathcal{L}_1(t)$ is the generator of single-particle mean-field dynamics given by (15). This will be called the polarization approximation. The formal solution for the correlation function in (24) then simplifies to

$$\overline{g}(1,2;t) = \mathcal{U}_1(t,0)\mathcal{U}_2(t,0)\overline{g}(1,2;0)$$
$$+ \int_0^t \mathcal{U}_1(t,\tau)\mathcal{U}_2(t,\tau)\theta_{12}(\tau)f(1;\tau)f(2;\tau)d\tau, \tag{36}$$

which has been reduced to the single-particle problem

$$\left(i\partial_t + \mathcal{L}_\infty(t)\right)\mathcal{U}_1(t,\tau) = 0 \qquad \mathcal{U}_1(t,t) = \mathcal{I}_1. \tag{37}$$

The generator is the sum of that for Hartree-Fock dynamics and polarization effects in the quantum Vlasov operator. Formally, the Hartree-Fock dynamics can be extracted in the second term of (36) to display the polarization effects

$$\mathcal{U}_1(t,\tau)\mathcal{U}_2(t,\tau)\theta_{12}(\tau) = \mathcal{U}_1^{HF}(t,\tau)\mathcal{U}_2^{HF}(t,\tau)\widetilde{\theta}_{12}(t,\tau). \tag{38}$$

The new interaction operator $\widetilde{\theta}_{12}(t,\tau)$ is given by (16) except that the interaction potential $\hat{V}(1,2;t)$ is replaced with a dynamically screened potential due to the polarization effects [11, 8, 9].

The polarization approximation retains most of the physical mechanisms of the pair approximation. In particular, it is valid on all time scales and therefore properly describes both coherent and collisional dynamics in the short- and long-time limits, respectively. A further simplification, obtained by using only the Hartree-Fock dynamics in (16) or by using a statically screened potential in $\widetilde{\theta}_{12}(t, \tau)$, is a generalization of the familiar Born approximation to include short time scales. However, this improper treatment of polarization effects leads to internal inconsistencies. A detailed application of the polarization approximation to semiconductor dynamics has not yet been carried out.

8. Discussion

The primary feature of this review has been to use the operator form of the BBGKY hierarchy for reduced density operators to address the significant new problems encountered in laser-pulsed semiconductors. The electron distributions are far from equilibrium so that familiar linear response methods do not apply and the full nonlinear kinetic equation is required. The electrons are at room temperature and therefore strongly degenerate, requiring a full quantum mechanical description. The external driving field persists over a short initial time and must be included explicitly to understand the "preparation" of the electron distribution in conduction bands, so that short-time dynamics must be accurately described. The long range Coulomb potential becomes screened at long times due to the build-up of correlations but is "bare" at short times, requiring a complete treatment of the dynamics of screening. It has been shown here that a simple closure approximation for the first two hierarchy equations incorporates all of these features. The approximation is to neglect three-particle Coulomb correlations while retaining the corresponding three-particle exchange correlations. The result is a pair of coupled operator equations for the one-particle density operator and the pair-correlation operator. In this way the many-body problem is reduced to the controlled level of an effective two-particle problem.

The analysis in terms of operators provides a unified treatment of several different problems: a gas of neutral particles, plasmas, and electrons in solids. The approximation is applicable for all degrees of degeneracy and is easily modified for Bose statistics. While attention here has been focused on degenerate electrons, the classical limit of the pair approximation provides a generalization of familiar kinetic equations (Boltzmann, Lennard-Balescu) to short times as well. Such non-Markovian effects are essential for underlying symmetries such as time reversal invariance and local conservation of energy. The closure is phenomenological but physically transparent. However, while improvements to describe strong coupling have been discussed recently for classical systems [10], their extension to the quantum case has not been considered yet in this context.

Finally, some comments on the more traditional approach using Green's functions are in order. Since the systems considered are far from equilibrium an appropriate nonequilibrium Green's function formalism must be used (e.g., Keldysh Green's functions). The application of this approach in the present context is discussed in references [12 - 14]. The primary advantage of the Green's function methods is a diagrammatical or functional representation that allows structural renormalization to include the desired physical mechanisms. Ultimately it is still uncontrolled since the renormalization entails a phenomenological selection of diagrams to be included. Still, the context of the approximations often preserves essential symmetries and invariances. To obtain a kinetic equation the matrix element of the one body distribution is related to the Green's function according to $< \alpha|f(1;t) \mid \alpha'>=<\psi^*_\alpha(t)\psi_{\alpha'}(t) >_{ne}$ where ψ_α is the creation operator for a single-particle state with quantum number α. However, the formal equations govern the more general *two-time* Green's functions $g^{\rangle}_{\alpha\alpha'}(t,t') = <\psi_\alpha(t)\psi^*_{\alpha'}(t') >_{ne}$, $g^{\langle}_{\alpha\alpha'}(t,t') = <\psi^*_{\alpha'}(t')\psi_{\alpha'}(t) >_{ne}$, and the corresponding retarded and advanced functions on a complex Keldysh-Schwinger contour. An approximation to these equations still leaves a coupled set of equations for the two time quantities, and typically an additional phenomenological approximation is introduced to obtain an equation for the desired density matrix. This latter consideration is referred to as the "reconstruction" problem. The original ansatz of Kadanoff and Baym is not correct for the nonequilibirum states of interest here [15]. An improved version of this ansatz has been proposed and in the polarization approximation leads to the same results as obtained here. However, it entails an assumed relationship of $g^{\langle}_{\alpha\alpha'}(t,t')$ to the density matrix via a spectral density plus an approximation for the spectral density. In this respect the Green's function approach must be considered phenomenological just as the hierarchy approach proposed here.

An alternative Green's function approach avoiding the reconstruction problem would be to solve the two time Kadanoff-Baym equations for $g^{\rangle}_{\alpha\alpha'}(t,t')$ and $g^{\langle}_{\alpha\alpha'}(t,t')$ for a chosen approximation, and then determine $< \alpha|f(1;t) \mid \alpha'>$ from the solution at $t = t'$. Recently, significant progress on this difficult numerical problem has been made for simple approximations to the Kadanoff-Baym equation [9]. It is a promising approach limited only by the practical (but considerable) numerical problems.

Acknowledgements

The author is indebeted to M. Bonitz, University of Rostock, C-S Kim, Chonnam University, and C. Stanton, University of Florida for introducing him to this field and for their collaboration on the topics discussed here.

This research was supported by grants from the National Science Foundation (INT 9414072 and NSF 9722133).

Appendices

A. Exchange Effects

The evaluation of the three-particle exchange effects leading to (12) and (13) are illustrated in this Appendix. Important identities for the permutation operators are

$$P_{ij}P_{ij} = 1, \qquad Tr_j P_{ij} = 1, \qquad P_{ij}X(i,j,k)P_{ij} = X(j,i,k). \tag{39}$$

Consider the three-particle contribution to the equation for $\overline{g}(1,2;t)$:

$$\begin{aligned} Tr_3 V(1,3)\overline{g}(1,3;t)f(2;t)P_{23} &= f(2;t)Tr_3 P_{23}P_{23}V(1,3)\overline{g}(1,3;t)P_{23} \\ &= f(2;t)Tr_3 P_{23}V(1,2)\overline{g}(1,2;t) = f(2;t)V(1,2)\overline{g}(1,2;t). \end{aligned} \tag{40}$$

This is one of the Pauli-blocking terms leading to $\widehat{V}(1,2;t)$. As a second example consider another three-particle contribution to the equation for $\overline{g}(1,2;t)$:

$$\begin{aligned} &Tr_3 V(1,3)\left[\overline{g}(1,2;t)f(3;t) - \overline{g}(2,3;t)f(1;t)P_{13}\right] \\ &= Tr_3 V(1,3)\left[\overline{g}(1,2;t)f(3;t) - P_{13}P_{13}\overline{g}(2,3;t)f(1;t)P_{13}\right] \\ &= Tr_3 f(3;t)V(1,3)\left(1 - P_{13}\right)\overline{g}(1,2;t). \end{aligned} \tag{41}$$

This gives the Hartee-Fock shift in the single-particle energy.

B. Mean-Field Dynamics

The solution to the mean-field equation, $\mathcal{U}_1(t,\tau)$, in (36) and (37) is expressed in terms of single-particle superoperators. This can be transformed to a representation in terms of normal operators in the two-particle Hilbert space. The solution operator obeys the equation

$$(i\partial_t + \mathcal{L}_1)\mathcal{U}_1(t,\tau) = 0 \qquad \mathcal{U}_1(t,t) = \mathcal{I}_1. \tag{42}$$

The operators with label 1 are defined over a single particle Hilbert space $\mathcal{H}_1(1)$. In all cases of interest these are linear operators, e.g., the superoperator $\mathcal{L}_1$ defines a linear map of the operators of $\mathcal{H}_1$ into themselves. It has the representation

$$\mathcal{L}_1 A(1) = Tr_2 K(1,2)A(2), \tag{43}$$

where $K(1,2)$ is an operator in the two particle Hilbert space $\mathcal{H}_2(1,2) \equiv \mathcal{H}_1(1) \otimes \mathcal{H}_1(2)$. The action of $\mathcal{U}_1(t,\tau)$ on a single particle operator $y(1)$ is

$$y(1;t) = \mathcal{U}_1(t,\tau)y(1). \tag{44}$$

Next, note the relation for arbitrary operators $A(1)$,

$$A(1) = Tr_2\Delta(1,2)A(2), \tag{45}$$

where $\Delta(1,2)$ is the operator in $\mathcal{H}_2(1,2)$ given by

$$\Delta(1,2) = \sum_{\alpha\beta} P_{\alpha\beta}(1)P_{\beta\alpha}(2), \qquad P_{\alpha\beta} \equiv\mid \alpha\rangle\langle\beta\mid. \tag{46}$$

Thus, $\Delta(1,2)$ is analogous to a delta function, except with respect to the trace instead of an integral. In matrix representation it is

$$\langle\alpha\mu \mid \Delta \mid \beta\nu\rangle = \delta_{\alpha,\nu}\delta_{\beta,\mu}. \tag{47}$$

Use of the relationship (47) in the formal solution (46) gives

$$y(1;t) = Tr_2 U(1,2;t,\tau)y(2), \tag{48}$$

where $U(1,2;t)$ is an operator in $\mathcal{H}_2(1,2)$ formally given by

$$U(1,2;t,\tau) = \mathcal{U}_1(t,\tau)\Delta(1,2). \tag{49}$$

The subscript on $\mathcal{L}_1$ indicates it is defined over the single particle operators of $\mathcal{H}_1(1)$. The ordinary operator $U(1,2;t)$ is determined from the equation

$$(i\partial_t + \mathcal{L}_1)\,U(1,2;t,\tau) = 0, \qquad U(1,2;t,t) = \Delta(1,2), \tag{50}$$

or more specifically

$$i\frac{\partial}{\partial t}U(1,2;t,\tau) - [H^{HF}(1;t), U(1,2;t,\tau)]$$
$$-Tr_3\,[V_s(1,3), f(1;t)]\,U(3,2;t,\tau)\Lambda_{23} = 0. \tag{51}$$

References

1. H. Haug and S. Koch, *Quantum Theory of the Optical and Electronic Properties of Semiconductors* (World Scientific, New Jersey, 1990).
2. W. Knox, *Hot Carriers in Semiconductor Nanostructures: Physics and Applications* (Academic Press, Boston, 1992).
3. R. Phillips, ed., *Coherent Optical Interactions in Semiconductors* (Plenum, New York, 1995).

4. N. Bogoliubov, *Lectures in Quantum Statistics* (Gordon and Breach, New York, 1967).
5. D. Boercker and J. Dufty, Ann. Phys. **119**, 43 (1979).
6. J. Dufty and D. Boercker, J. Stat. Phys. **57**, 827 (1989).
7. J. Dufty, C. Kim, M. Bonitz, and R. Binder, Int. J. Quant. Chem. **65**, 929 (1997).
8. M. Bonitz, J. Dufty, and C. Kim, Phys. Stat. Sol. (b) **206**, 181 (1998).
9. M. Bonitz, *Quantum Kinetic Theory* (TEUBNER-TEXTE zur Physik, Bd 33, Stuttgart, 1998).
10. J. Dufty, Contrib. Plasma Phys. **37**, 129 (1997).
11. H. Hohenester and W. Potz, Phys. Rev. B **56**, 13177 (1997).
12. R. Binder and S. Koch, Prog. Quant. Elect. **19**, 307 (1995).
13. H. Haug and A. Jauho, *Quantum Kinetics in Transport and Optics of Semiconductors* (Springer-Verlag, New York, 1996).
14. D. Kremp, W. Kraeft, and M. Schlanges, *Quantum Statistics of Strongly Coupled Plasmas* (Springer-Verlag, New York, 1998).
15. P. Lipavsky, V. Spicka, and B. Velicky, Phys. Rev. B **34**, 6933 (1986).

QUANTUM KINETIC THEORY OF TRAPPED ATOMIC GASES

H.T.C. STOOF
Institute for Theoretical Physics, University of Utrecht
Princetonplein 5, 3584 CC Utrecht, The Netherlands

Abstract. We present a general framework in which we can accurately describe the non-equilibrium dynamics of trapped atomic gases. This is achieved by deriving a single Fokker-Planck equation for the gas. In this way we are able to discuss not only the dynamics of an interacting gas above and below the critical temperature at which the gas becomes superfluid, but also during the phase transition itself. The last topic cannot be studied on the basis of the usual mean-field theory and was the main motivation for our work. To show, however, that the Fokker-Planck equation is not only of interest for recent experiments on the dynamics of Bose-Einstein condensation, we also indicate how it can, for instance, be applied to the study of the collective modes of a condensed Bose gas.

1. Introduction

The most important reason for the present interest in Bose condensed atomic gases is the possibility to study in detail the dynamics of a superfluid system in this case. In particular, it is possible to compare *ab initio* many-body theories for the non-equilibrium dynamics directly to experiment, which is for instance not possible for liquid helium. Two important issues that are of interest to us here are the dynamics of condensate formation and the eigenfrequencies of the collective modes of the Bose condensed gas cloud. In both of these problems it is important to realize that the gas consists of two components, which, due to the harmonic confinement of the gas, are roughly speaking also spatially separated. More precisely, the density profile of the gas can be viewed as a relatively narrow condensate peak on top of a broad thermal background.

J. Karkheck (ed.),
Dynamics: Models and Kinetic Methods for Non-equilibrium Many Body Systems, 491–502.

Theoretically, we anticipate that the noncondensate or thermal cloud behaves similarly as a gas above the critical temperature for Bose-Einstein condensation. Its dynamics is therefore accurately described by an appropriate kinetic equation. However, the condensate is a macroscopic quantum object and the dynamics of the condensate must therefore be determined by an equation for its wavefunction. On the basis of these arguments we see that to describe the coupled dynamics of the thermal and condensate clouds, we need a quantum kinetic theory that is capable of simultaneously treating both the incoherent as well as the coherent processes taking place in the gas. It is the main purpose of this lecture to explain as simple as possible how such a quantum kinetic theory can be derived from first principles [1]. Moreover, as an application and illustration of the theory, we also present the first results on the formation of the condensate and the collective modes that we have recently obtained.

It should be noted that in this lecture we use only physical arguments to explain and motivate the final structure of our quantum kinetic theory. For a detailed derivation by means of field-theoretical methods we refer to the existing literature [2, 3]. In addition, we mainly restrict ourselves here to a discussion of the so-called weak-coupling limit in which interactions have only a relatively small, but nevertheless crucial, effect on the behavior of the gas. This restriction is again made for reasons of clarity only, because in the weak-coupling limit the theory is most transparent and therefore most easily understood. Moreover, once this limit is well understood, in principle it is straightforward to generalize and treat also the strong-coupling limit. However, before we can start to consider the effect of interactions, we first need to reformulate the theory of the ideal Bose gas in a somewhat unusual way, that nevertheless turns out to be very convenient for our purposes.

2. Ideal Bose Gas

Let us therefore consider an ideal Bose gas in an external trapping potential $V^{\text{trap}}(\mathbf{x})$ with one-particle energy levels ϵ_α and corresponding wavefunctions $\chi_\alpha(\mathbf{x})$ that can be found from the Schrödinger equation

$$\left\{-\frac{\hbar^2\nabla^2}{2m} + V^{\text{trap}}(\mathbf{x})\right\}\chi_\alpha(\mathbf{x}) = \epsilon_\alpha\chi_\alpha(\mathbf{x}) \ . \tag{1}$$

Using the methods of second quantization, in the Heisenberg picture the non-equilibrium dynamics of the gas is fully determined by the initial density matrix $\hat{\rho}(t_0)$ and the hamiltonian

$$\hat{H} = \sum_\alpha \epsilon_\alpha \hat{N}_\alpha(t) = \sum_\alpha \epsilon_\alpha \hat{\psi}^\dagger_\alpha(t)\hat{\psi}_\alpha(t) \ , \tag{2}$$

where $\hat{\psi}^{\dagger}_{\alpha}(t)$ and $\hat{\psi}_{\alpha}(t)$ create and annihilate at time t a particle in the state $\chi_{\alpha}(\mathbf{x})$, respectively.

If we would follow the usual treatment of an ideal Bose gas, we would at this point introduce the basis $|\{N_{\alpha}\};t\rangle$ in which the occupation numbers of all the one-particle state $\chi_{\alpha}(\mathbf{x})$ are specified, and proceed to calculate the probability distribution

$$P(\{N_{\alpha}\};t) = \mathrm{Tr}\left[\hat{\rho}(t_0)\,|\{N_{\alpha}\};t\rangle\langle\{N_{\alpha}\};t|\right] \; . \tag{3}$$

For an ideal gas the hamiltonian commutes with the number operators $\hat{N}_{\alpha}(t)$ and the above probability distribution is in fact independent of time. In particular, this implies that the average occupation numbers $\langle\hat{N}_{\alpha}(t)\rangle \equiv \mathrm{Tr}\left[\hat{\rho}(t_0)\hat{N}_{\alpha}(t)\right]$ are constant, which physically makes sense because without any interactions there is no way in which the particles can scatter from one state to another.

As mentioned in the introduction, we are not only interested in the occupation numbers of the gas, but also in the condensate wavefunction. Therefore we do not want to use as a basis the eigenstates of the number operators $\hat{N}_{\alpha}(t)$, but instead the eigenstates of the annihilation operators $\hat{\psi}_{\alpha}(t)$. More precisely, we introduce the so-called coherent states $|\phi;t\rangle$, which are (properly normalized) eigenstates of the field operator

$$\hat{\psi}(\mathbf{x},t) = \sum_{\alpha} \chi_{\alpha}(\mathbf{x})\hat{\psi}_{\alpha}(t) \tag{4}$$

with complex eigenvalue $\phi(\mathbf{x})$ [4], and consider the corresponding probability distribution

$$P[\phi^*,\phi;t] = \mathrm{Tr}\left[\hat{\rho}(t_0)\,|\phi;t\rangle\langle\phi;t|\right] \; . \tag{5}$$

We now need to determine the equation of motion, i.e., the appropriate Fokker-Planck equation, for this probability distribution. This can be achieved most easily as follows. We know that by definition the annihilation operators $\hat{\psi}_{\alpha}(t)$ obey the Heisenberg equation

$$i\hbar\frac{\partial\hat{\psi}_{\alpha}(t)}{\partial t} = \left[\hat{\psi}_{\alpha}(t),\hat{H}\right]_{-} = \epsilon_{\alpha}\hat{\psi}_{\alpha}(t) \; . \tag{6}$$

Since we have used the eigenstates of the annihilation operators to define the probability distribution $P[\phi^*,\phi;t]$, we also know that

$$\langle\hat{\psi}_{\alpha}(t)\rangle = \int d[\phi^*]d[\phi]\;\phi_{\alpha}P[\phi^*,\phi;t] \equiv \langle\phi_{\alpha}\rangle(t) \; , \tag{7}$$

where $\int d[\phi^*]d[\phi]$ denotes the (functional) integral over the complex functions $\phi(\mathbf{x})$. Combining the last two equations we thus find that

$$i\hbar\frac{\partial}{\partial t}\langle\phi_\alpha\rangle(t) = \epsilon_\alpha\langle\phi_\alpha\rangle(t)\ . \tag{8}$$

Moreover, by considering the Heisenberg equation for the creation operators $\hat{\psi}^\dagger_\alpha(t)$ we also obtain

$$i\hbar\frac{\partial}{\partial t}\langle\phi^*_\alpha\rangle(t) = -\epsilon_\alpha\langle\phi^*_\alpha\rangle(t)\ . \tag{9}$$

In this manner we have thus been able to derive the equation of motion for the first moments of the probability distribution. However, to arrive at a Fokker-Planck equation for $P[\phi^*,\phi;t]$ we also need to consider the higher moments [5], and in our case in particular $\langle|\phi_\alpha|^2\rangle(t)$. *A priori* we expect this expectation value to be related to the average occupation numbers $\langle\hat{N}_\alpha(t)\rangle$ and therefore that

$$i\hbar\frac{\partial}{\partial t}\langle|\phi_\alpha|^2\rangle(t) = 0\ . \tag{10}$$

Although the latter result is all that we need here, we need later on also the precise relation between $\langle|\phi_\alpha|^2\rangle(t)$ and the average occupation numbers, which can be shown to be given by

$$\langle|\phi_\alpha|^2\rangle(t) = \langle\hat{N}_\alpha(t)\rangle + \frac{1}{2}\ . \tag{11}$$

A derivation of this relation is complicated by the fact that the creation and annihilation operators do not commute at equal times. It can, however, be understood physically from the fact that the second moment of $P[\phi^*,\phi;t]$ should contain both classical as well as quantum fluctuations.

From Eqs. (8), (9) and (10) we conclude that the desired Fokker-Planck equation reads [6]

$$\begin{aligned} i\hbar\frac{\partial}{\partial t}P[\phi^*,\phi;t] = & -\sum_\alpha\frac{\partial}{\partial\phi_\alpha}(\epsilon_\alpha\phi_\alpha)\,P[\phi^*,\phi;t] \\ & +\sum_\alpha\frac{\partial}{\partial\phi^*_\alpha}(\epsilon_\alpha\phi^*_\alpha)\,P[\phi^*,\phi;t]\ . \end{aligned} \tag{12}$$

It thus contains in the right-hand side only 'streaming' terms and no 'diffusion' term. As a result there is no unique equilibrium and in fact any function of $|\phi_\alpha|^2$ is a stationary solution. Again, this makes sense physically, because without interactions there is no way in which the occupation numbers can relax to an equilibrium Bose-Einstein distribution. To include such relaxation into our discussion, we therefore now bring our ideal Bose gas into contact with a thermal reservoir.

3. Ideal Bose Gas in Contact With a Reservoir

As our reservoir we take an ideal Bose gas in a box with volume V. The reservoir is assumed to be sufficiently large so that it can be treated in the thermodynamic limit. Moreover, it is in equilibrium with a temperature T and a chemical potential μ. The states in this box are labeled by the momentum $\hbar\mathbf{k}$ and equal to $\chi_{\mathbf{k}}(\mathbf{x}) = e^{i\mathbf{k}\cdot\mathbf{x}}/\sqrt{V}$. They have an energy $\epsilon(\mathbf{k}) = \hbar^2\mathbf{k}^2/2m$. Finally, the reservoir is thought to be in contact with the trap discussed above, by means of the tunnel hamiltonian

$$\hat{H}^{\text{int}} = \frac{1}{\sqrt{V}} \sum_{\alpha} \sum_{\mathbf{k}} \left(t_\alpha(\mathbf{k}) \hat{\psi}_\alpha(t) \hat{\psi}^\dagger_{\mathbf{k}}(t) + t^*_\alpha(\mathbf{k}) \hat{\psi}_{\mathbf{k}}(t) \hat{\psi}^\dagger_\alpha(t) \right) . \tag{13}$$

Here $t_\alpha(\mathbf{k})$ are complex tunneling matrix elements that for simplicity are assumed to be almost constant for momenta $\hbar k$ smaller that a cutoff $\hbar k_c$ but to vanish rapidly for momenta larger than this ultraviolet cutoff. Moreover, we consider here only the low-temperature regime in which the thermal de Broglie wavelength $\Lambda = (2\pi\hbar^2/mk_BT)^{1/2}$ of the particles obeys $\Lambda \gg 1/k_c$, since this is the most appropriate limit for realistic atomic gases.

Due to this interaction the particles can tunnel back and forth from the trap to the reservoir, which results both in a shift in the energy as well as a finite lifetime of the state $\chi_\alpha(\mathbf{x})$. The energies of the states in the trap therefore become complex and equal to $\epsilon_\alpha + S_\alpha - iR_\alpha$, where the real and imaginary contributions to the shift can essentially be found from second-order perturbation theory. Denoting the Cauchy principle value part of an integral by $\mathcal{P}$, they obey

$$S_\alpha = \int \frac{d\mathbf{k}}{(2\pi)^3}\, t^*_\alpha(\mathbf{k}) \frac{\mathcal{P}}{\epsilon_\alpha + S_\alpha - \epsilon(\mathbf{k})} t_\alpha(\mathbf{k}) \tag{14}$$

and

$$R_\alpha = \pi \int \frac{d\mathbf{k}}{(2\pi)^3}\, \delta(\epsilon_\alpha + S_\alpha - \epsilon(\mathbf{k})) |t_\alpha(\mathbf{k})|^2 , \tag{15}$$

respectively.

Introducing the retarded and advanced self-energies $\hbar\Sigma^{(\pm)}_\alpha = S_\alpha \mp iR_\alpha$, we conclude from the above that Eqs. (8) and (9) now become

$$i\hbar \frac{\partial}{\partial t} \langle \phi_\alpha \rangle(t) = \left(\epsilon_\alpha + \hbar\Sigma^{(+)}_\alpha - \mu \right) \langle \phi_\alpha \rangle(t) \tag{16}$$

and

$$i\hbar \frac{\partial}{\partial t} \langle \phi^*_\alpha \rangle(t) = - \left(\epsilon_\alpha + \hbar\Sigma^{(-)}_\alpha - \mu \right) \langle \phi^*_\alpha \rangle(t) , \tag{17}$$

where we have also measured our energies relative to the chemical potential.

Next, we need to consider the fluctuations, i.e., the generalization of Eq. (10). This turns out to be given by

$$i\hbar\frac{\partial}{\partial t}\langle|\phi_\alpha|^2\rangle(t) = -2iR_\alpha\langle|\phi_\alpha|^2\rangle(t) - \frac{1}{2}\hbar\Sigma_\alpha^K , \tag{18}$$

with the so-called Keldysh self-energy equal to

$$\hbar\Sigma_\alpha^K = -2iR_\alpha\left(1 + 2N_\alpha^{\mathrm{eq}}\right) \tag{19}$$

and $N_\alpha^{\mathrm{eq}} = 1/(e^{(\epsilon_\alpha + S_\alpha - \mu)/k_BT} - 1)$ the equilibrium Bose-Einstein distribution function. How can this result be understood? The first term in the right-hand side of Eq. (18) follows simply from the fact that if we neglect correlations $\langle|\phi_\alpha|^2\rangle(t)$ is equal to $|\langle\phi_\alpha\rangle(t)|^2$. Furthermore, the second term in the right-hand side of Eq. (18) guarantees that if we make use of the relation between $\langle|\phi_\alpha|^2\rangle(t)$ and the average occupation numbers $N_\alpha(t) \equiv \langle\hat{N}_\alpha(t)\rangle$, we recover exactly the Boltzmann equation

$$\frac{\partial}{\partial t}N_\alpha(t) = -\Gamma_\alpha N_\alpha(t) + \Gamma_\alpha N_\alpha^{\mathrm{eq}} \tag{20}$$

with the correct transition rates $\Gamma_\alpha = 2R_\alpha/\hbar$ expected from Fermi's Golden Rule.

In a similar manner as for the isolated case, we now conclude from Eqs. (16), (17) and (18) that the Fokker-Planck equation for our trapped Bose gas becomes

$$\begin{aligned} i\hbar\frac{\partial}{\partial t}P[\phi^*,\phi;t] = & - \sum_\alpha \frac{\partial}{\partial\phi_\alpha}\left(\epsilon_\alpha + \hbar\Sigma_\alpha^{(+)} - \mu\right)\phi_\alpha P[\phi^*,\phi;t] \\ & + \sum_\alpha \frac{\partial}{\partial\phi_\alpha^*}\left(\epsilon_\alpha + \hbar\Sigma_\alpha^{(-)} - \mu\right)\phi_\alpha^* P[\phi^*,\phi;t] \\ & - \frac{1}{2}\sum_\alpha \frac{\partial^2}{\partial\phi_\alpha\partial\phi_\alpha^*}\hbar\Sigma_\alpha^K P[\phi^*,\phi;t] . \end{aligned} \tag{21}$$

It is interesting to note that this Fokker-Planck equation is equivalent to the Langevin equation

$$i\hbar\frac{\partial}{\partial t}\phi_\alpha(t) - \left(\epsilon_\alpha + \hbar\Sigma_\alpha^{(+)} - \mu\right)\phi_\alpha(t) = \eta_\alpha(t) , \tag{22}$$

where the Gaussian noise obeys

$$\langle\eta_\alpha^*(t)\eta_{\alpha'}(t')\rangle = \frac{i\hbar^2}{2}\hbar\Sigma_\alpha^K\delta_{\alpha,\alpha'}\delta(t-t') \tag{23}$$

and the fluctuation-dissipation theorem reads

$$\hbar\Sigma_\alpha^K = (1 + 2N_\alpha^{\text{eq}})\left(\hbar\Sigma_\alpha^{(+)} - \hbar\Sigma_\alpha^{(-)}\right) . \tag{24}$$

Clearly, due to the fluctuation-dissipation theorem the probability distribution $P[\phi^*, \phi; t]$ for the gas in the trap now relaxes to the correct equilibrium state

$$P[\phi^*, \phi; \infty] = \prod_\alpha \frac{1}{N_\alpha^{\text{eq}} + 1/2} \exp\left\{-\frac{1}{N_\alpha^{\text{eq}} + 1/2}|\phi_\alpha|^2\right\} . \tag{25}$$

More important for our purposes, however, is that the Fokker-Planck equation in Eq. (21) describes simultaneously both the incoherent (kinetic) as well as the coherent dynamics in the gas, since it incorporates the equations of motion for both $\langle \hat{N}_\alpha(t) \rangle$ and $\langle \hat{\psi}_\alpha(t) \rangle$, respectively. As mentioned previously, this is precisely what is needed for an accurate treatment of non-equilibrium phenomena in Bose condensed atomic gases.

4. Condensate Formation in an Interacting Bose Gas

We are now in a position to discuss an interacting Bose gas, because in an interacting Bose gas, the gas is, roughly speaking, its own thermal reservoir. The Fokker-Planck equation therefore turns out to be quite similar to the one presented in Eq. (21). To be more precise, however, we should mention that we aim in this section to describe the formation of a condensate in an interacting Bose gas under the conditions that have recently been realized in experiments with atomic ^{87}Rb [7], ^{7}Li [8] and ^{23}Na [9] gases. In these experiments the gas is cooled by means of evaporative cooling. Numerical solutions of the Boltzmann equation have shown that during evaporative cooling the energy distribution function is well described by an equilibrium distribution with time-dependent temperature $T(t)$ and chemical potential $\mu(t)$, that is truncated at high energies due to the evaporation of the highest energetic atoms from the trap [10]. Moreover, in the experiments of interest the densities just above the critical temperature are essentially always such that the gas is in the weak-coupling limit, which implies in this context that the average interaction energy per atom is always much less than the energy splitting of the one-particle states in the harmonic trapping potential.

Keeping the above remarks in mind, we find that near the critical temperature the non-equilibrium properties of the gas are, to an excellent approximation, described by a nonlinear Fokker-Planck equation with time-dependent selfenergies $\hbar\Sigma_\alpha^{(\pm),K}(t)$ and effective interaction matrix elements $V_{\alpha,\beta;\alpha',\beta'}^{(\pm),K}(t)$, for which, in the so-called ladder approximation, explicit expressions can be derived in terms of the average occupation numbers $N_\alpha(t)$

in the gas [3]. In full detail it reads

$$
\begin{aligned}
i\hbar\frac{\partial}{\partial t}P[\phi^*,\phi;t] = \\
&-\sum_\alpha \frac{\partial}{\partial\phi_\alpha}\left\{\left(\epsilon_\alpha + \hbar\Sigma_\alpha^{(+)} - \mu\right)\phi_\alpha + \sum_{\alpha',\beta,\beta'} V^{(+)}_{\alpha,\beta;\alpha',\beta'}\phi^*_\beta\phi_{\beta'}\phi_{\alpha'}\right\} P[\phi^*,\phi;t] \\
&+\sum_\alpha \frac{\partial}{\partial\phi^*_\alpha}\left\{\left(\epsilon_\alpha + \hbar\Sigma_\alpha^{(-)} - \mu\right)\phi^*_\alpha + \sum_{\alpha',\beta,\beta'} V^{(-)}_{\alpha',\beta';\alpha,\beta}\phi^*_{\alpha'}\phi^*_{\beta'}\phi_{\beta}\right\} P[\phi^*,\phi;t] \\
&-\frac{1}{2}\sum_{\alpha,\alpha'} \frac{\partial^2}{\partial\phi_\alpha\partial\phi^*_{\alpha'}}\left\{\hbar\Sigma^K_\alpha\delta_{\alpha,\alpha'} + \sum_{\beta,\beta'} V^K_{\alpha,\beta;\alpha',\beta'}\phi^*_\beta\phi_{\beta'}\right\} P[\phi^*,\phi;t] . \qquad (26)
\end{aligned}
$$

To understand how this equation describes the formation of the condensate, we make use of the fact that in the weak-coupling limit it is appropriate to solve the Fokker-Planck equation with the (Hartree-Fock) *ansatz*

$$P[\phi^*,\phi;t] = P_0[\phi^*_g,\phi_g;t]P_1[\phi'^*,\phi';t] , \qquad (27)$$

where the complex function $\phi'(\mathbf{x}) = \sum_{\alpha\neq g}\chi_\alpha(\mathbf{x})\phi_\alpha$ is associated with all the one-particle states except the groundstate $\chi_g(\mathbf{x})$. Substituting the above *ansatz* into the Fokker-Planck equation, we obtain the following results. First, the dynamics of the noncondensed cloud is determined by the quantum Boltzmann equation

$$\frac{\partial}{\partial t}N_\alpha(t) = -\Gamma^{\text{out}}_\alpha N_\alpha(t) + \Gamma^{\text{in}}_\alpha(1+N_\alpha) , \qquad (28)$$

with $\alpha \neq g$ and the scattering rates $\Gamma^{\text{out,in}}(t)$ of a similar form as in the Uehling-Uhlenbeck equation but with a cross-section which is proportional to $|V^{(+)}_{\alpha',\beta';\alpha,\beta}(t)|^2$. Second, this kinetic equation is coupled to a Fokker-Planck equation for the condensate, i.e., to

$$
\begin{aligned}
i\hbar\frac{\partial}{\partial t}P_0[\phi^*_g,\phi_g;t] = &- \frac{\partial}{\partial\phi_g}\left(\epsilon_g + \hbar\Sigma^{(+)}_g - \mu + V^{(+)}_{g,g;g,g}|\phi_g|^2\right)\phi_g P_0[\phi^*_g,\phi_g;t] \\
&+ \frac{\partial}{\partial\phi^*_g}\left(\epsilon_g + \hbar\Sigma^{(-)}_g - \mu + V^{(-)}_{g,g;g,g}|\phi_g|^2\right)\phi^*_g P_0[\phi^*_g,\phi_g;t] \\
&- \frac{1}{2}\frac{\partial^2}{\partial\phi_g\partial\phi^*_g}\hbar\Sigma^K_g P_0[\phi^*_g,\phi_g;t] . \qquad (29)
\end{aligned}
$$

Due to the fluctuation-dissipation theorem, the probability distribution for the condensate relaxes to the equilibrium solution

$$P_0[\phi^*_g,\phi_g;\infty] \propto \exp\left\{-\frac{1}{k_BT}\left(\epsilon_g + S_g - \mu + \frac{V^{(+)}_{g,g;g,g}}{2}|\phi_g|^2\right)|\phi_g|^2\right\} . \qquad (30)$$

The Landau free energy for the condensate order parameter thus equals

$$F_L[\phi_g^*, \phi_g] = (\epsilon_g + S_g - \mu)\,|\phi_g|^2 + \frac{V^{(+)}_{g,g;g,g}}{2}|\phi_g|^4 \tag{31}$$

and clearly shows a spontaneous breaking of symmetry if $\epsilon_g + S_g - \mu < 0$ and the effective interatomic interaction is repulsive, i.e., $V^{(+)}_{g,g;g,g} > 0$ [11]. In our formulation of the problem these quantities are a function of time, whose evolution is essentially determined by the quantum Boltzmann equation in Eq. (28). In particular, for Bose-Einstein condensation to occur, $\epsilon_g + S_g - \mu$ has to change sign during this evolution. As a first rough calculation of the condensate formation, we can assume that this change of sign takes place instantaneously. Introducing for convenience the dimensionless time $\tau = t(i\Sigma_g^K/8)(2V^{(+)}_{g,g;g,g}/k_BT)^{1/2}$ and the dimensionless condensate number $I = |\phi_g|^2(2V^{(+)}_{g,g;g,g}/k_BT)^{1/2}$, this assumption leads to a typical evolution for the probability distribution $P_0[I;\tau]$ that is shown in Fig. 1.

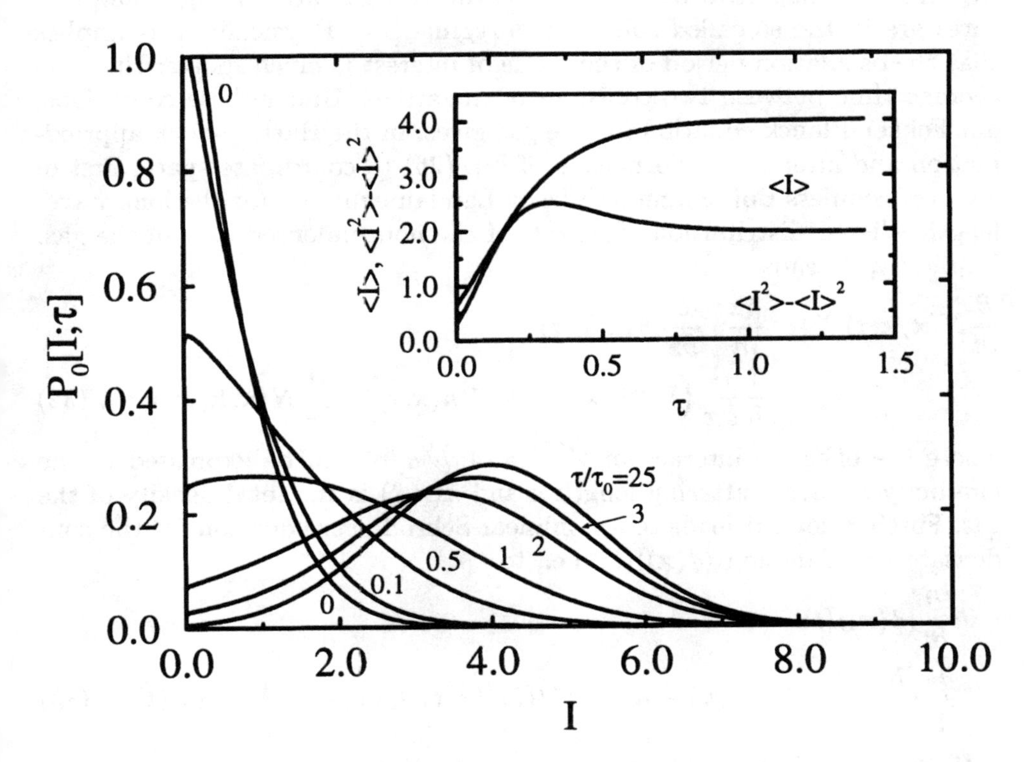

Figure 1. Typical evolution of the condensate probability distribution during Bose-Einstein condensation. The slowest relaxation rate is $1/\tau_0 \simeq 5.7$. The inset shows the evolution of the average condensate number and its fluctuations.

At this point we should mention that the solution of the condensate Fokker-Planck equation in Eq. (29) with a fixed and negative value of $\epsilon_g + S_g - \mu$, is equivalent to the theory recently put forward by Gardiner and coworkers using different methods [12]. It even turns out to agree qualitatively with experiment, although quantitatively some important discrepancies exist [13]. In our opinion these discrepancies are probably due to the fact that we should really solve the Fokker-Planck equation for the condensate together with the quantum Boltzmann equation for the thermal cloud. Moreover, an additional problem is that the experiments of interest here are not really in the weak-coupling limit, which substantially complicates the theory because more states are needed to describe the condensate.

5. Collective Modes

As another illustration of our general nonequilibrium approach we consider now the collective modes of a Bose condensed gas at such high temperatures that a substantial noncondensate fraction is present in the gas. The experiments that have been performed at these relatively high temperatures are in the so-called collisionless regime [14]. Physically, this implies that the oscillation period of the mode of interest is much shorter than the average time between two collisions of the atoms. Under these conditions, our Fokker-Planck equation for the gas gives, in the Hartree-Fock approximation and after a transformation of Eq. (26) to coordinate space, first of all a collisionless Boltzmann or Vlasov-Landau equation for the long wavelength Wigner distribution $N(\mathbf{x}, \mathbf{k}; t)$ of the noncondensed part of the gas. Explicitly, it reads

$$\begin{aligned}\frac{\partial}{\partial t}N(\mathbf{x},\mathbf{k};t) &+ \frac{\hbar\mathbf{k}}{m}\cdot\frac{\partial}{\partial\mathbf{x}}N(\mathbf{x},\mathbf{k};t) \\ &- \frac{1}{\hbar}\frac{\partial}{\partial\mathbf{x}}\left(V^{\text{trap}}(\mathbf{x}) + 2V^{(+)}n(\mathbf{x},t)\right)\cdot\frac{\partial}{\partial\mathbf{k}}N(\mathbf{x},\mathbf{k};t) = 0\ , \quad (32)\end{aligned}$$

where the effective interaction $V^{(+)} = 4\pi\hbar^2 a/m$ can be expressed in the two-body s-wave scattering length a and $n(\mathbf{x}, t)$ is the total density of the gas. Furthermore, it leads to a nonlinear Schrödinger equation for the condensate wavefunction $\langle\phi(\mathbf{x})\rangle(t)$, i.e., to

$$\begin{aligned}&i\hbar\frac{\partial}{\partial t}\langle\phi(\mathbf{x})\rangle(t) = \\ &\left\{-\frac{\hbar^2\nabla^2}{2m} + V^{\text{trap}}(\mathbf{x}) - \mu + V^{(+)}(2n'(\mathbf{x},t) + n_0(\mathbf{x},t))\right\}\langle\phi(\mathbf{x})\rangle(t)\ , \quad (33)\end{aligned}$$

with the condensate density $n_0(\mathbf{x}, t)) = |\langle\phi(\mathbf{x})\rangle(t)|^2$ and the noncondensate density $n'(\mathbf{x}, t) = \int d\mathbf{k}\ N(\mathbf{x}, \mathbf{k}; t)/(2\pi)^3$ of course adding up to the total density of the gas.

A numerical solution of the above coupled equations turns out to be surprisingly difficult. Therefore, to gain insight into the collective modes of the gas, we have recently put forward a variational approach [15], in which we apply a dynamical scaling *ansatz* on the ideal gas results for both the Wigner distribution $N(\mathbf{x}, \mathbf{k}; t)$ and the condensate wavefunction $\langle\phi(\mathbf{x})\rangle(t)$. The outcome of this calculation is presented in Fig. 2, where we also make a comparison with experiment which turns out to be quite reasonable in view of the simplicity of our scaling *ansatz*. The main discrepancies are the two measurements halfway between the $m = 0$ in and out-of-phase modes. This discrepancy is, however, presumably due to the fact that in the experiment both modes are excited simultaneously [16].

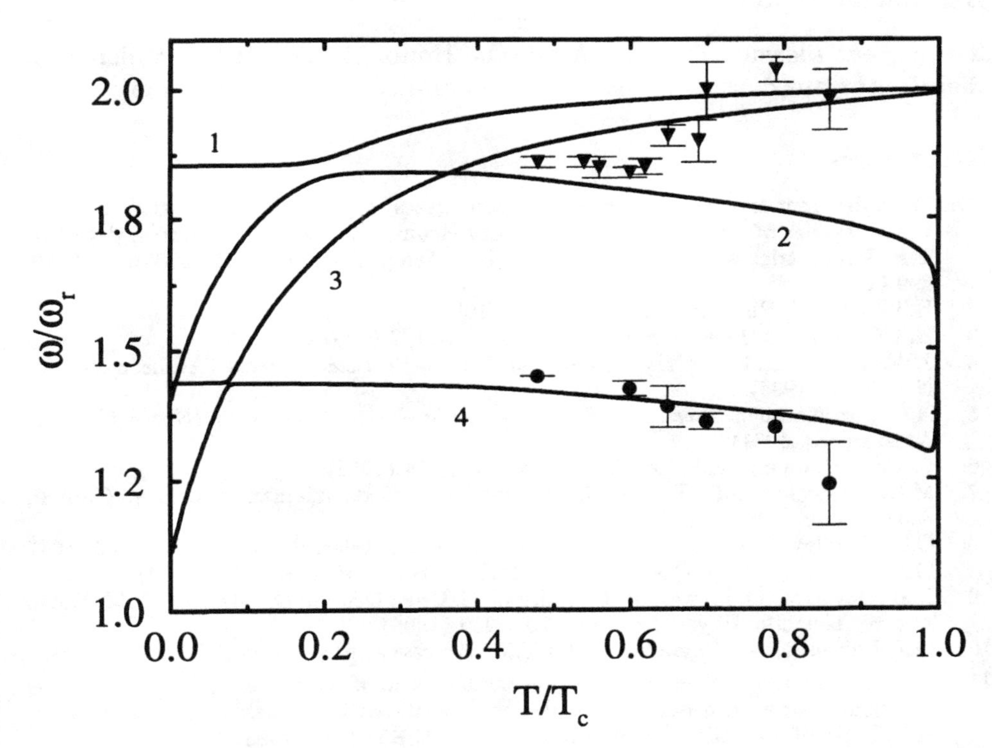

Figure 2. Collisionless modes in a Bose condensed ^{87}Rb gas. Curves 1 and 2 correspond to the $m = 0$ in and out-of-phase modes, respectively. Similarly, curves 3 and 4 give the frequencies of the $m = 2$ in and out-of-phase modes. The experimental data is also shown. Triangles are for an $m = 0$ mode and circles for an $m = 2$ mode.

6. Conclusions

In summary, we have presented a general framework in which we can discuss various non-equilibrium problems in Bose condensed atomic gases. As an example we have presented our first results on the formation of the condensate and the collisionless modes at nonzero temperatures. Other topics that are of interest are, for example, the condensate formation in a Bose gas with attractive interactions, the dynamics of vortices and spin domain walls, the collective modes in the hydrodynamic regime, the damping of collective modes, and last but not least the non-equilibrium dynamics of trapped atomic Fermi gases. In view of the many topics that remain to be explored, we hope that the present lecture may motivate some of the participants of this summerschool to enter into this, in our view, very exciting area of physics.

Acknowledgement

It is a great pleasure to thank Marianne Houbiers and Michiel Bijlsma for their important contributions to the topics discussed in this paper.

References

1. For a different treatment of trapped atomic gases see C.W. Gardiner and P. Zoller, Phys. Rev. A **55**, 2902 (1997). The homogeneous case was worked out in detail by T.R. Kirkpatrick and J.R. Dorfman, J. Low Temp. Phys. **58**, 301 (1985); *ibid.* **58**, 399 (1985).
2. H.T.C. Stoof, Phys. Rev. Lett. **78**, 768 (1997).
3. H.T.C. Stoof, J. Low. Temp. Phys. **114**, Nos. 1/2 (1999).
4. J.W. Negele and H. Orland, *Quantum Many-Particle Systems* (Addison-Wesley, New York, 1988).
5. N.G. van Kampen, *Stochastic Processes in Physics and Chemistry* (North-Holland, Amsterdam, 1981).
6. P. Carruthers and K.S. Dy, Phys. Rev. **147**, 214 (1966).
7. M.H. Anderson, J.R. Ensher, M.R. Matthews, C.E. Wieman, and E.A. Cornell, Science **269**, 198 (1995).
8. C.C. Bradley, C.A. Sackett, J.J. Tollett, and R.G. Hulet, Phys. Rev. Lett. **75**, 1687 (1995); C.C. Bradley, C.A. Sackett, and R.G. Hulet, *ibid.* **78**, 985 (1997).
9. K.B. Davis, M.-O. Mewes, M.R. Andrews, N.J. van Druten, D.S. Durfee, D.M. Kurn, and W. Ketterle, Phys. Rev. Lett. **75**, 3969 (1995).
10. O.J. Luiten, M.W. Reynolds, and J.T.M. Walraven, Phys. Rev. A **53**, 381 (1996).
11. For ^{7}Li the effective interatomic interaction is attractive and $V^{(+)}_{g,g;g,g} < 0$. This drastically alters the behavior of the condensate as is discussed in C.A. Sackett, H.T.C. Stoof, and R.G. Hulet, Phys. Rev. Lett. **80**, 2031 (1998).
12. C.W. Gardiner, P. Zoller, R.J. Ballagh, and M.J. Davis, Phys. Rev. Lett. **79**, 1793 (1997). A recent extension of this theory is presented in C.W. Gardiner, M.D. Lee, R.J. Ballagh, M.J. Davis, and P. Zoller, Phys. Rev. Lett. **81**, 5266 (1998).
13. H.-J. Miesner, D.M. Stamper-Kurn, M.R. Andrews, D.S. Durfee, S. Inouye, and W. Ketterle, Science **279**, 1005 (1998).
14. D.S. Jin, M.R. Matthews, J.R. Ensher, C.E. Wieman, and E.A. Cornell, Phys. Rev. Lett. **78**, 764 (1997).
15. M.J. Bijlsma and H.T.C. Stoof, Phys. Rev. A **60**, 3973 (1999).
16. E.A. Cornell, private communication.

PARTICIPANTS

Patrick Ahlrichs
Max Planck Inst. f. Polymer Res.
Ackermannweg 10
D-55128 Mainz
GERMANY

Reka Albert
Department of Physics
225 Nieuwland Science Hall
University of Notre Dame
Notre Dame, IN 46556
U.S.A

Ana Aparicio Munera
Avda. del Padre Piquer, 35 4 B
28024 Madrid
SPAIN

Toshihico Arimitsu
Institute of Physics
University of Tsukuba
Ibaraki 305-8571
JAPAN

Naoko Arimitsu
Div. of Elect. and Computer Eng.
Yokohama National University
79-5,Tokiwadai
Hodogaya-ku
Yokohama 240-8501
JAPAN

Birgir Orn Arnarson
Dept. of Th. and Appl. Mech.
Cornell University
212 Kimball Hall
Ithaca, NY 14853
U.S.A.

Ophir M. Auslaender
Dept. of Cond. Matter Physics
Weizmann Institute of Science
Rehovot 76100
ISRAEL

Peter Balint
Renyi Mathematical Institute of the
Hungarian Academy of Sciences
P.O.B. 127
H-1364, Budapest
HUNGARY

Eli Barkai
Dept. of Chem. and Center for Mat.
Sci. and Eng.
Massachusetts Inst. of Tech.
77 Massachusetts Ave.
Cambridge, MA 02139
U.S.A.

Henry van den Bedem
William E. Wecker Assoc., Inc.
505 San Marin Drive B200
Novato, CA 94945
U.S.A.

Henk van Beijeren
Institute for Th. Physics
Utrecht University
Princetonplein 5
3584 CC Utrecht
THE NETHERLANDS

Dietrich Belitz
Department of Physics
University of Oregon
Eugene, OR 97403-1274
U.S.A

Michiel J. Bijlsma
Institute for Th. Physics
Utrecht University
Princetonplein 5
3584 CC Utrecht
THE NETHERLANDS

Chris Bizon
Colorado Research Associates
3380 Mitchell Lane
Boulder, CO 80301
U.S.A.

Lydéric Bocquet
Laboratoire de Physique, ENS Lyon
46, Allee d'Italie
F-69364 Lyon Cedex 07
FRANCE

Arkadiusz Branka
Institute of Molecular Physics
Polish Academy of Sciences
Ul. Smoluchowskiego 17
60-179 Poznan
POLAND

J. Javier Brey
Fisica Teorica, Facultad de Fisica
Apdo Correos 1065 Sector Sur
E-41080 Sevilla
SPAIN

Manuel Osvaldos Caceres
Centro Atomico Bariloche
8400, Bariloche
ARGENTINA

Nikolai Chernov
Department of Mathematics
U. of Alabama/Birmingham
Campbell Hall 492A
Birmingham, AL 35294-1170
U.S.A

Bogdan Cichocki
Institute of Theoretical Physics
University of Warsaw
ul. Hoza 69
PL-00-681 Warsawa
POLAND

Piero Cipriani
C.S.S.
Via E.Nardi, 14-16
02047 Poggio Mirteto (RI)
ITALY

E.G.D. Cohen
Center for the Studies in Physics and Biology
The Rockefeller University
1230 York Ave.
New York, NY 10021
U.S.A

David Cubero Gomez
Física Teórica
Universidadde Sevilla
E-41080 Sevilla
SPAIN

Carl Philip Dettmann
The Rockefeller University
1230 York Ave.
New York, NY 10021
U.S.A.

Jan K.G. Dhont
Forschungszentrum Jülich
IFF
D-52425 Jülich
GERMANY

Anjani Kumar Didwania
Dept. of Mech. and Aerospace Eng.
U. of California, San Diego
La Jolla, CA 92093-0411
U.S.A.

J. Robert Dorfman
Department of Physics
University of Maryland
College Park, MD 20742
U.S.A

James W. Dufty
Department of Physics
University of Florida
215 Williamson Hall
Gainesville, FL 32611-2085
U.S.A

Matthieu H. Ernst
Instituut voor Th. Fysica
Utrecht University
Princetonplein 5
3584 CC Utrecht
THE NETHERLANDS

B.Ubbo Felderhof
Institut für Th. Physik
RWTH Aachen
Sommerfeldstrasse 26/28
D-52056 Aachen
GERMANY

Vladimir Filinov
FB Physik
Universität Rostock
Universitätplatz 3
D-18051 Rostock
GERMANY

Boris Fine
Department of Physics
University of Illinois
1110 W. Green St.
Urbana, IL 61801-3080
U.S.A.

Aldo Frezzotti
Dipartimento di Matematica
Politecnico di Milano
Piazza Leonardo da Vinci 32
20133 Milano
ITALY

Ramon Garcia Rojo
Facultad de Fisica
Depto. de Fisica, Atomica, Molecular y Nuclear - Area de Física Teórica
P.O. Box 1065
E-41080 Sevilla
SPAIN

Karl M. Garcia-Ruiz
Dept. de Fisica Y Quimica Teorica
Fac. De Quimica, UNAM
Cd. Universitaria 04510
Mexico D.F.
MEXICO

Pierre Gaspard
Departement de Physique
Universite Libre de Bruxelles
Boulevard du Triomphe
1050 Bruxelles
BELGIUM

Oleg I. Gerasimov
Dept. of General Physics
Odessa St. Hydrometeorological Inst.
Lvovskayastr. 15 Odessa 270016
UKRAINE

Thomas Gilbert
L.P.T.M.C.
Universite Paris 7 - Denis Diderot
case 7020
Pl. Jussieu
75251 Paris cedex 05
FRANCE

Maksim Yu. Gladkov
Institut für Theoretische Physik
Universität Regensburg
D-93040 Regensburg
GERMANY

Mathijs J.V. Goldschmidt
Universiteit Twente
Faculteit Chemische Technologie
Postbus 217
7500 AE Enschede
THE NETHERLANDS

Dmitri Grinev
Polymers and Colloids Group
Cavendish Laboratory
University of Cambridge
Madingley Road
Cambridge CB3 OHE
UNITED KINGDOM

Vadim L. Gurevich
Ioffe Institute
26 Polytekhnicheskaya
St. Petersburg 194021
RUSSIAN FEDERATION

Alex Hansen
Department of Th. Physics
University of Trondheim
N-7034 Trondheim
NORWAY

Jean-Pierre Hansen
Department of Chemistry
Cambridge University
Lensfield Road
Cambridge CB2 1EW
UNITED KINGDOM

Eivind Hauge
Department of Th. Physics
University of Trondheim
N-7034 Trondheim
NORWAY

Martin van Hecke
CATS/Niels Bohr Institute
Blegdamsvej 17
2100 DK Copenhagen
DENMARK

Heinz-Guenter Hermanns
Institut für Th. Physik A
RWTH Aachen
Templergraben 55
52056 Aachen
GERMANY

Jonathan M. Huntley
Department of Mech. Eng.
Loughborough University
LE11 3TU Leicestershire
UNITED KINGDOM

Martin Huthmann
Institut für Theoretische Physik
Universität Göttingen
Bunsenstrasse 9
D-37073 Göttingen
GERMANY

Marta Ibanes
Facultat de Fisica, Dept. E.C.M.
Universitat de Barcelona
Diagonal 647
08028 Barcelona
SPAIN

Alexander Iomin
Dept. of Physics
Technion
Haifa 32000
ISRAEL

Sudhir Jain
School of Eng. and Appl. Sciences
Aston University
Aston Triangle
Birmingham B4 7ET
UNITED KINGDOM

Richard J.M. Janssen
Particle Technology
Delft University of Technology
Julianalaan 136
2628 BL Delft
THE NETHERLANDS

James T. Jenkins
Dept. of Th. and Appl. Mech.
Cornell University
211 Kimball Hall
Ithaca, NY 14853
U.S.A

Robert B. Jones
Dept. of Physics
Queen Mary and Westfield College
Mile End Road
London E1 4NS
UNITED KINGDOM

John Karkheck
Department of Physics
Marquette University
P.O. Box 1881
Milwaukee, WI 53201-1881
U.S.A

Theodore Kirkpatrick
Department of Physics
University of Maryland
College Park, MD 20742
U.S.A

Kazuo Kitahara
Dept. of Applied Physics
Tokyo Institute of Technology
Ookayama 2-12-1, Meguro-ku
Tokyo 152-8550
JAPAN

Ulrich Krebs
Institut f. Theoretische Physik A
RWTH Aachen
D-52056 Aachen
GERMANY

Herman Kruis
Instituut Lorentz for Th. Physics
Universiteit Leiden
P.O. Box 9506
2300 RA Leiden
THE NETHERLANDS

V. Kumaran
Dept. of Chemical Engineering
Indian Institute of Science
Bangalore 560 012
INDIA

Anthony J.C. Ladd
Dept. of Chem. Eng., Rm 227
P.O. Box 116005
University of Florida
Gainesville, FL 32611-6005
U.S.A

Arnulf Latz
Institut für Physik
Johannes Gutenberg Universität
Staudinger Weg 7
55099 Mainz
GERMANY

Colin Marsh
The Technology Partnership
Melbourn Science Park
Melbourn, Royston
Herts. SG8 6EE
UNITED KINGDOM

Gerardo Martinez
Instituto de Fisica, UFRGS
Univ. Federal do Rio Grande do Sul
Campus do Vale, Caixa Postal 15051
CEP 91501-970, Porto Alegre - RS
BRAZIL

Sean McNamara
CECAM
ENS-Lyon
46 allee d'Italie
69364 Lyon Cedex 07
FRANCE

Ljubomir Milanovic
Institute for Experimental Physics
University of Vienna
Boltzmanngasse 5
A-1090 Vienna
AUSTRIA

Kunimasa Miyazaki
Department of Physical Chemistry
Nat. Inst. of Mat. and Chem. Res.
Tsukuba, Ibaraki 305-8565
JAPAN

Francisco Moreno Franco
Fisica Teorica
Facultad de Fisica
Aptado 1065
41080 Sevilla
SPAIN

Magamed Muradov
Politechnicheskaya 26
A.F. Ioffe Institute
194021 St. Petersburg
RUSSIA

Ihor Mryglod
Institute for Cond. Matter Physics
1 Svientsitskii St.
UA-79011 Lviv
UKRAINE

Peter Mueller
Institut für Theoretische Physik
Georg-August Universität
D-37073 Göttingen
GERMANY

Amador Muriel
Centre for Fl. Dyn. (World Lab)
U. of the Philippines at Los Banos
College, Laguna 4051
PHILIPPINES

Rajesh Narayanan
Department of Theoretical Physics
University of Oxford
Keble Road
Oxford 0X1 3NP
UNITED KINGDOM

Th.M. Nieuwenhuizen
Van der Waals-Zeeman Instituut
Valckenierstraat 65
1018 XE Amsterdam
THE NETHERLANDS

T.P.C. van Noije
Fluid Flow and Thermo., OGBE/6
Shell Global Solutions
Shell Res. and Tech. Ctr.
P.O. Box 38000
1030 BN Amsterdam
THE NETHERLANDS

Riza Ogul
Department of Physics
Selcuk University
42079 Konya
TURKEY

Seung-Hoon Oh
Dept. of Chemical Engineering
University of Florida
Gainesville, Fl 32611-6005
U.S.A

Tsuneyasu Okabe
U. of Waikato Language Institute
P.O. Box 1317
Waikato Mail Centre
Hamilton
NEW ZEALAND

Ignacio Pagonabarraga
Dept. of Physics and Astronomy
The University of Edinburgh
Edinburgh EH9 3JZ
UNITED KINGDOM

Debabrata Panja
Inst. for Physical Sci. and Tech.
University of Maryland
College Park, MD 20742
U.S.A

Ramon Peralta-Fabi
Departamento de Fisica
Facultad de Ciencias, UNAM
04510 Mexico DF
MEXICO

Jaroslaw Piasecki
Institute of Theoretical Physics
University of Warsaw
ul. Hoza 69
PL-00-681 Warsawa
POLAND

Thorsten Poeschel
Humboldt Universität zu Berlin
Institut für Physik
Invalidenstrasse 110
D-10115 Berlin
GERMANY

Harald Posch
Institut für Experimentalphysik
Universität Wien
Strudlhofgasse 4
A-1090 Wien
AUSTRIA

Rosa Ramirez Martinez
Centre Europeen de Calcul At. et Moleculaire (CECAM)
Ecole Normale Superieure
46, Allee d'Italie
69007 Lyon
FRANCE

Raul Rechtman
Centro de Investigacion en Energia
UNAM
Apdo. Postal 34
Temixco, Morelos 62580
MEXICO

Valeria Ricci
ENS-Cachan
CMLA
61, Avenue du President Wilson
94235 Cachan Cedex
FRANCE

Howard L. Richards
Department of Physics
University of Maryland
College Park, MD 20742-4111
U.S.A.

Marisol Ripoll Hernando
Depto. de Fisica Fundamental
Facultad de Ciencias UNED
C/ Senda del Rey 9
28040 Madrid
SPAIN

Maria J. Ruiz-Montero
Fisica Teorica
Facultad de Fisica
Apartado de Correos 1065
41080 Sevilla
SPAIN

Márton Sasvári
Dept. of Theoretical Physics
Institute of Physics
Budapest U. of Tech. and Economics
Budafoki ut 8.
H-1111
HUNGARY

Ignatz M. de Schepper
IRI, Delft University of Technology
Julianalaan 134 postbus 5
2600 AA Delft
THE NETHERLANDS

Andrew V. Sergeev
Department of Physics
Universität Regensburg
Universitätstr. 31
D-93035 Regensburg
GERMANY

Yakov Sinai
Department of Mathematics
Princeton University
708 Fine Hall
Princeton, NJ 08544
U.S.A

Vyacheslav M. Somsikov
Institute of Ionosphere
480020 Almaty
KAZAKSTAN

Dominique Spehner
Universidad Catolica
Facultad de Fisica
Vicuna Mackenna 4860, San Joaquim
Casilla 306, Santiago 22
CHILE

Vaclav Spicka
Institute of Physics
Acad. of Sci. of the Czech Repub.
Cukrovarnicka 10
162 53 Prague 6
CZECH REPUBLIC

Henk T.C. Stoof
Instituut voor Th. Fysica
Utrecht University
Princetonplein 5
3584 CC Utrecht
THE NETHERLANDS

Xin Sun
Physics Department
Fudan University
Shanghai 200433
CHINA

Harry L. Swinney
Department of Physics
University of Texas, Austin
Austin, TX 78712
U.S.A.

Piotr Szymczak
Institute of Theoretical Physics
Faculty of Physics, Warsaw U.
Hoza 69
00-681 Warsawa
POLAND

Shuichi Tasaki
Department of Applied Physics
School of Science and Engineering
Waseda University
3-4-1- Okubo, Tokyo 169-8555
JAPAN

Roumen Tsekov
Dept. of Physical Chemistry
University of Sofia
1126 Sofia
BULGARIA

Rolf Verberg
Dept. of Chemical Engineering
University of Florida
Gainesville, FL 32611
U.S.A.

Jun'ichi Wakou
Instituut voor Th. Fysica
Universiteit Utrecht
Postbus 80006
3508 TA Utrecht
THE NETHERLANDS

Ricky Wildman
Dept. of Mechanical Engineering
Loughborough University
Loughborough
Leicestershire LE11 3TU
UNITED KINGDOM

Ramses van Zon
Theoretical Chemical Physics
Department of Chemistry
Lash Miller Chemical Laboratories
80 St. George St.
University of Toronto
Toronto, Ontario, M5S 3H6
CANADA

INDEX

Printed in the USA
CPSIA information can be obtained
at www.ICGtesting.com
LVHW011558240923
759164LV00007B/1015